Botany: Science and Technology

Botany: Science and Technology

Edited by **Molly Ismay**

SYRAWOOD
PUBLISHING HOUSE
New York

Published by Syrawood Publishing House,
750 Third Avenue, 9th Floor,
New York, NY 10017, USA
www.syrawoodpublishinghouse.com

Botany: Science and Technology
Edited by Molly Ismay

International Standard Book Number: 978-1-68286-366-4 (Hardback)

Printed in China.

Contents

Preface

The discipline of botany enables us to understand the evolution of plants, their utility to humans and animals. It also encompasses techniques for comprehending life processes of plants. This book on botany is well equipped to educate the reader on various topics like plant breeding, genetics, diversity and development of new methodologies for the cultivation and engineering of plants. It also contains multiple topics which can be taken up for research by enthusiastic readers. The students of biology and botany will find in this book a perfect guide to their respective fields of study.

This book unites the global concepts and researches in an organized manner for a comprehensive understanding of the subject. It is a ripe text for all researchers, students, scientists or anyone else who is interested in acquiring a better knowledge of this dynamic field.

I extend my sincere thanks to the contributors for such eloquent research chapters. Finally, I thank my family for being a source of support and help.

Editor

Development and characterization of 96 microsatellite markers suitable for QTL mapping and accession control in an Arabidopsis core collection

Patrick Cosson[1,2], Véronique Decroocq[1,2] and Frédéric Revers[1,2*]

Abstract

Background: To identify plant genes involved in various key traits, QTL mapping is a powerful approach. This approach is based on the use of mapped molecular markers to identify genomic regions controlling quantitative traits followed by a fine mapping and eventually positional cloning of candidate genes. Mapping technologies using SNP markers are still rather expensive and not feasible in every laboratory. In contrast, microsatellite (also called SSR for Simple Sequence Repeat) markers are technologically less demanding and less costly for any laboratory interested in genetic mapping.

Results: In this study, we present the development and the characterization of a panel of 96 highly polymorphic SSR markers along the *Arabidopsis thaliana* genome allowing QTL mapping among accessions of the Versailles 24 core collection that covers a high percentage of the *A. thaliana* genetic diversity. These markers can be used for any QTL mapping analysis involving any of these accessions. We optimized the use of these markers in order to reveal polymorphism using standard PCR conditions and agarose gel electrophoresis. In addition, we showed that the use of only three of these markers allows differentiating all 24 accessions which makes this set of markers a powerful tool to control accession identity or any cross between any of these accessions.

Conclusion: The set of SSR markers developed in this study provides a simple and efficient tool for any laboratory focusing on QTL mapping in *A. thaliana* and a simple means to control seed stock or crosses between accessions.

Keywords: Microsatellite, *Arabidopsis thaliana*, QTL mapping, Accession control

Introduction

Identifying plant genes that control key traits is central to understanding the mechanism of their gene action and to target them in breeding programs. Few genes controlling quantitative traits have been identified in plants and most have been cloned by linkage mapping [1]. Mapping of Quantitative Trait Loci (QTL) is a standard procedure that requires both genotypic (i.e. based on molecular markers) and phenotypic data generated from a segregating population [2,3]. In a first step, mapped QTL intervals of several tens of centimorgans are identified for which a portion of the phenotypic variation of trait is explained. A fine mapping approach that aims to narrow down these QTL to candidate genes is then implemented.

This initial step necessitates having a set of molecular markers distributed along the plant genome and polymorphic between the two parents of the progeny. In *Arabidopsis thaliana*, major inroads have been made for gene mapping, through the identification of thousands of Single Nucleotide Polymorphism (SNP) markers [4-11] or through extensive genome re-sequencing of individual accessions [12-14]. However, mapping technologies using SNPs markers are still rather expensive and not applicable in every laboratory. On the other hand, due to their relative abundance in the genome and simple cost effective

* Correspondence: revers@bordeaux.inra.fr
[1]INRA, UMR 1332 de Biologie du fruit et Pathologie, Villenave d'Ornon F-33140, France
[2]Univ Bordeaux, UMR 1332 de Biologie du fruit et Pathologie, Villenave d'Ornon F-33140, France

detection microsatellite markers (also named simple sequence repeat or SSR markers) remain for many laboratories the markers of choice in a primary QTL mapping screen. Microsatellites are tandem repeat motifs of 1 to 6 bp that have frequent occurrence in all genomes analyzed to date [15,16]. The high variability of microsatellites is mainly due to different number of repeats in the region of the repeat motif but also to short insertion/deletion events [17]. Thus, SSR markers usually exhibit many alleles in comparison to SNP or other polymerase chain reaction (PCR)-based markers. Moreover, they have the advantage of being codominant. Consequently, only a limited number of markers are sufficient to detect polymorphism among parental lines. As the flanking regions of the repeat motif are in many cases highly conserved (particularly in the case of EST derived SSRs), microsatellite markers are easily and reproducibly amplified by PCR in most accessions and visualized by gel electrophoresis (agarose, polyacrylamide) or automated sequencing. In addition, their identification is relatively simple when plant genome sequences are available. In addition, in the case of *A. thaliana*, several studies showed that microsatellites are abundant and highly variable [18-20].

In this paper, we describe the development and the characterization of a panel of 96 SSR markers distributed along the *A. thaliana* genome suitable for QTL mapping in an Arabidopsis core collection.

Results and discussion

SSR development

The primary aim of this study was to produce molecular markers suitable for QTL mapping in *A. thaliana* without the use of expensive and specific technical platforms. Therefore, we decided to develop a panel of polymorphic SSR markers among accessions of the Versailles core collection 24 which maximizes genetic diversity in a limited number of individuals and has been used to generate F2 and Recombinant Inbred Lines (RIL) [21,22]. We showed previously in several studies that these accessions display a diversity in behavior toward plant viruses allowing development of mapping projects to identify plant genes involved in plant/virus interactions [23,24]. Our strategy was to identify markers distributed along the Arabidopsis genome, approximately every 4 Mb (approximately 20 cM), which are as polymorphic as possible among the Versailles core-24 accessions. Those criteria optimize their use for mapping analysis within progenies obtained from crosses between any of the 24 Versailles accessions [21]. Six other accessions (Col-0, C24, Ws-2, Nd-1, Bay-0 and Ler-2) were added to the study as they are frequently used as parents in our studies [23-27]. The criteria of 4-Mb intervals along each chromosome generated 8 regions (1.1 to 1.8) in chromosome 1 (chr1), 5 regions (2.1 to 2.5) in chr2, 6 regions (3.1 to 3.6) in chr3, 5 regions (4.1 to 4.5) in chr4 and 7 regions (5.1 to 5.7) in chr5 (Figure 1) in which we searched for tandem repeat motifs. Thirty-one SSR markers distributed throughout these regions were found in the Versailles MSAT database [28] or from The Arabidopsis Information Resource (TAIR) [29]. They displayed high genetic diversity (Additional file 1: Table S1) but they were not sufficient to cover the 31 genomic regions of interest (Figure 1). Moreover, they are not sufficiently polymorphic among our set of accessions (Additional file 1: Table S2). To complete this marker set, new loci containing tandem repeat motifs were identified within the Col-0 sequence as described in the Methods section. We preferentially selected perfect dinucleotide repeat motifs with a minimum of ten repeats except for a few regions where we had to select mono or trinucleotide repeat motifs or imperfect motifs (Additional file 1: Table S1). Primer sets were developed to amplify these repeat containing loci and tested among the 30 accessions. Our aim was to select SSRs in each region that displayed at least one polymorphism among all accessions and also for which polymorphism can be preferentially detected on agarose rather than polyacrylamide gels. Following this procedure, we generated 65 new SSRs, named BSATX.Y, X being the chromosome number and Y, the development rank number, with two to four markers generated in each selected chromosomal region. They are described in Additional file 1: Table S1 and their relative position on the five Arabidopsis chromosomes is presented in Figure 1. For some regions, especially those located around the centromere, we extended the size of the interval in order to identify suitable SSRs (Figure 1). Indeed, it has been previously shown that the occurrence of microsatellites is consistently lower in the centromeric regions than in the rest of the chromosomes [30]. To facilitate the use of these markers, primers were designed such that almost all SSRs can be amplified using the same PCR conditions. An example of SSR polymorphism displayed on an agarose gel is presented in Figure 2. The size of each SSR allele for each accession is presented in Additional file 1: Table S2. A few SSRs in some accessions could not be amplified at all, most likely due to polymorphisms or deletions in primer binding sites compared with the Col-0 sequence used for primer design. In other cases two amplified products were produced, most likely due to residual heterozygosity (Additional file 1: Table S2). From the size polymorphism of each SSR, several features were deduced. The average number of alleles for each marker is 14, ranging from 2 to 27 alleles per marker (Additional file 1: Table S1). The average polymorphism information content (PIC, which corresponds to the average allelic diversity of all the markers, see Methods) for the markers reaches 0.87 with

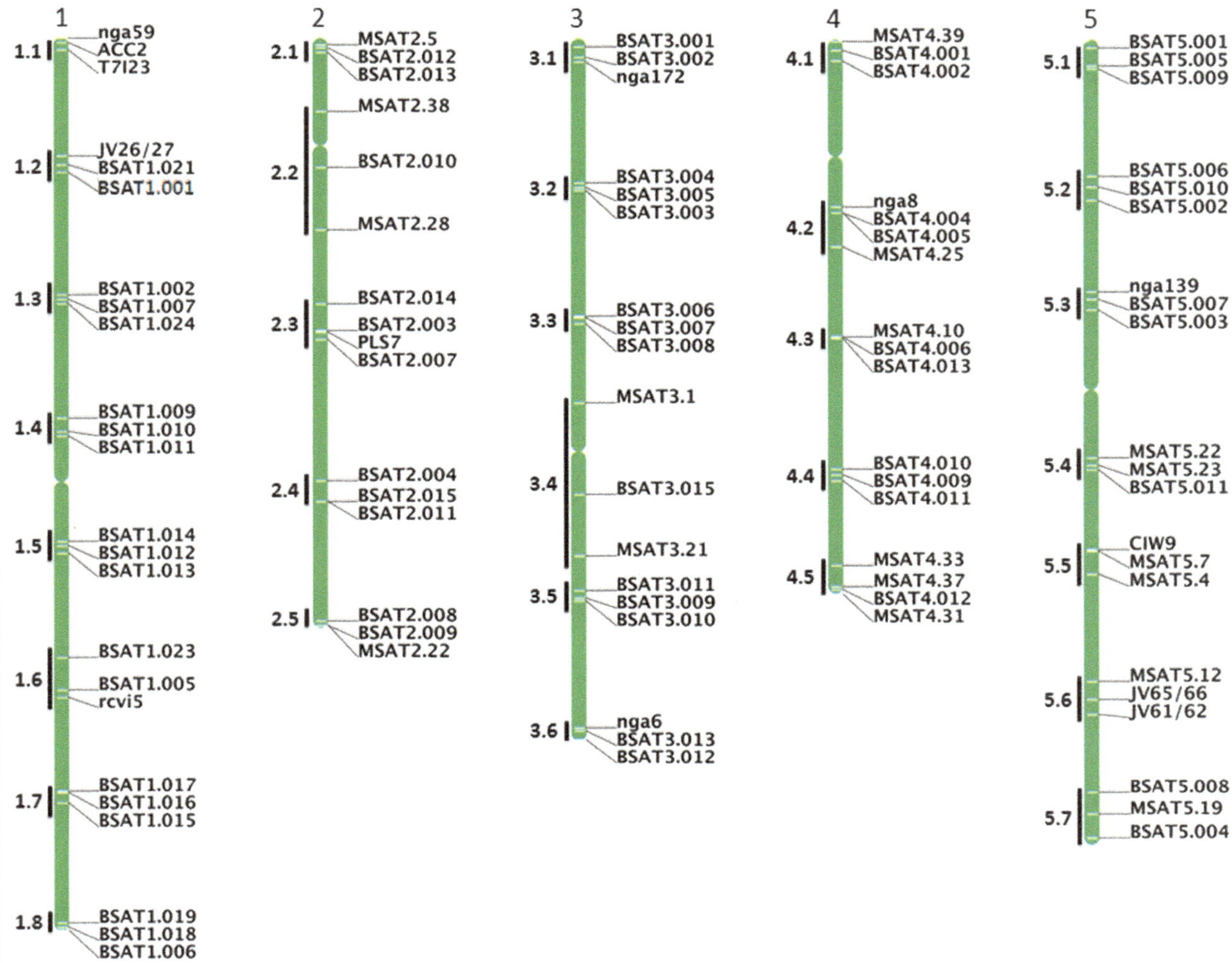

Figure 1 Genomic localization of the SSR markers and the selected genome regions. This map was made using the chromosome map tool from TAIR (http://www.arabidopsis.org/index.jsp). Number at the top of each chromosome is the number of each Arabidopsis chromosome. Markers are indicated on the right of each chromosome. The genome regions selected in this study are indicated on the left of the chromosome in front of the corresponding markers.

values ranging from 0.50 to 0.96 (Additional file 1: Table S1) and the percentage of polymorphic markers between each pair of Arabidopsis accessions ranges from 64% to 99% with an average value of 90% (Additional file 1: Table S3). Thus, except for only a very few chromosomal regions, the SSR set distinguishes all the accessions (Additional file 1: Table S2 and Additional file 1: Table S4).

With this panel of SSR markers, any primary QTL mapping can be easily achieved with progenies (F2 or RIL populations) issued from any cross between the 24 Versailles accessions or between these accessions and Col-0, C24, Ws-2, Nd-1, Bay-0 and Ler-2. As soon as two accessions diverge for a trait of interest, the genetic factors associated with this phenotype can be located on the five Arabidopsis chromosomes. In addition, to simply revealing SSR polymorphisms, these SSRs are amplified using similar PCR conditions and subsequently displayed on agarose gels.

The high level of genetic diversity of these SSRs is in agreement with the high genetic diversity of the Versailles core collection [22,31]. Within the collection these 96 markers display more than 1,300 different alleles. This high level of polymorphism also makes them markers of choice for other Arabidopsis segregating populations which do not belong to the Versailles collection as illustrated in our recent genetic mapping study [24]. In addition, as only PCR and electrophoresis are needed to reveal polymorphism, this set of SSR markers has significant advantages and is complementary to the use of SNP markers [31,32] or Simple Sequence Length Polymorphism (SSLP) markers which are based on indels [33].

Accession identification using a set of three highly polymorphic SSR markers

Due to the high level of genetic diversity displayed by the developed markers, we examined the possibility of

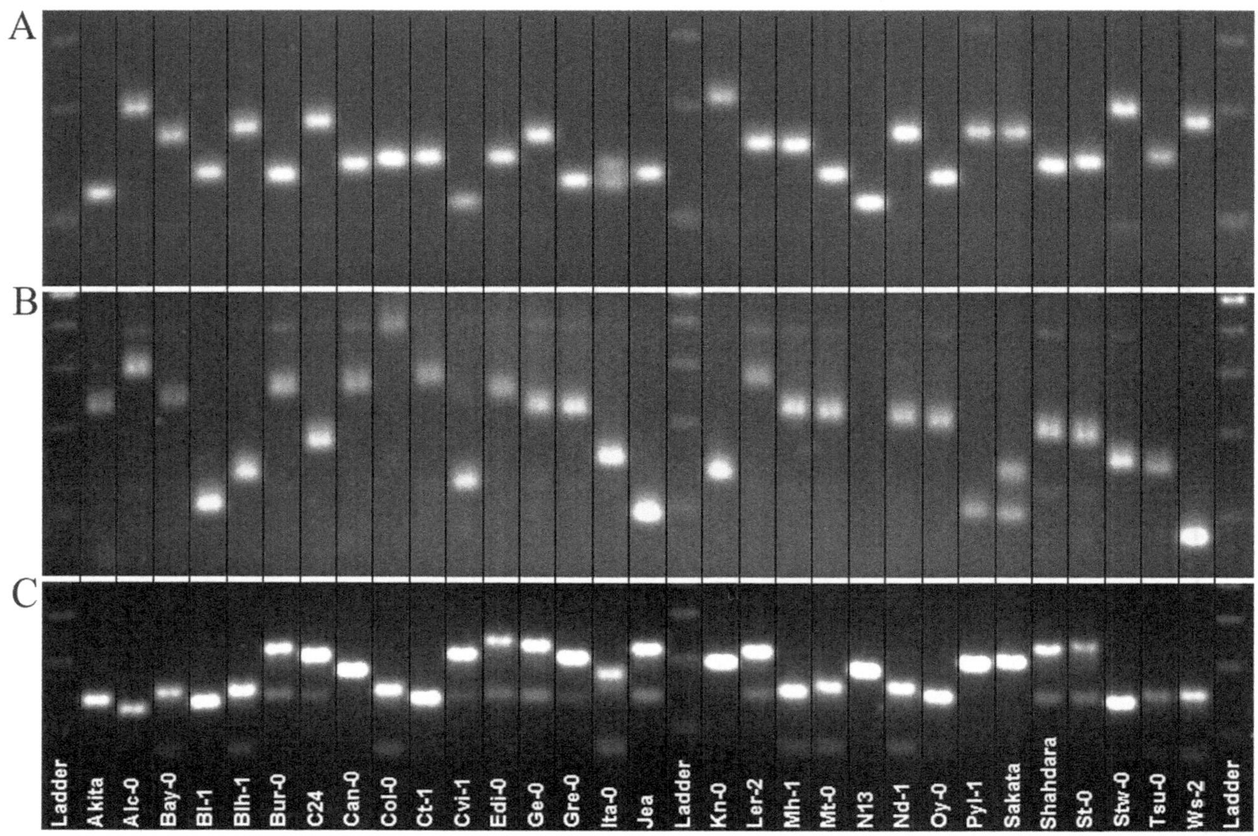

Figure 2 Polymorphism obtained with three SSR markers visualized using agarose gel. Agarose electrophoresis of the corresponding PCR products is shown for the thirty accessions (indicated at the bottom of the picture). **(A)** BSAT1.024; **(B)** MSAT3.1; **(C)** MSAT5.22.

differentiating each of the accessions with a minimum number of SSR markers. In this regard, we determined that using a select combination of only 3 SSR markers we could discriminate each accession from the 29 others. These markers are BSAT1.024, MSAT3.1 and MSAT 5.22 (Figure 2). Thus, these markers constitute a powerful tool to control seed stocks and crosses for the accessions studied in this work and likely for many other worldwide accessions as well.

Methods

Plant material

Accessions included in the Versailles core collection 24 were obtained from the INRA Versailles Centre [21]. Other accessions were obtained from the Nottingham Arabidopsis Stock Centre [34].

Mining of microsatellites

The Arabidopsis BAC clones file containing the GenBank versions (fasta format) of sequences that make up the TAIR AGI map [35] was subsequently searched for the presence of one to three perfect repeat motifs using the Sputnik software [36]. The minimum length for each type of SSR is set as follows: mononucleotide repeats ≥30 nucleotides; dinucleotide repeats ≥30 nucleotides and trinucleotide repeats ≥24 nucleotides.

Primer design and PCR conditions

Sequences of the SSRs that were identified in the genomic areas of interest were then subjected to similarity searches in the TAIR Whole Genome (BAC clones) dataset through the BLASTN algorithm. SSRs with several matches along the genome were eliminated. Oligonucleotide primers were designed for selected SSR loci using the PRIMER3 software [37] and were also subjected to BLASTN analysis. The parameters for primer design were as follows: amplicon size of 100–350 bp, primer size of 18–25 bp and primer melting temperature of 55–60°C with an optimum at 57°C.

Genomic DNA used for PCR was extracted from Arabidopsis young leaves using the NucleoSpin® Plant kit (Macherey-Nagel, Düren, Germany). PCRs were performed in a final volume of 15 µl consisting of 15 ng of DNA template, 1 U of Taq polymerase (Fermentas, Glen Burnie, MD, USA), 1x Taq reaction buffer, 0.2 mM dNTPs, 1.5 mM $MgCl2$, 200 nM of each primer. Conditions used for amplification were as follows: preincubation at 94°C for 2 min; followed by 35 cycles of denaturation at

94°C for 30 s; annealing at 57°C for the SSRs developed in this study, 50°C for the MSAT SSRs and 55°C for the other SSRs for 30 s; elongation at 72°C for 30s and a final extension step at 72°C for 10 min.

To reveal polymorphism, the PCR products were subsequently separated in 3% agarose gels. To determine the precise PCR product size, 6% acrylamide gels were used.

Genetic diversity

Polymorphic SSR markers were scored for the presence or absence of the corresponding bands among the tested accessions. Stutter and background bands were excluded. The scores '1' and '0' indicated presence and absence of bands, respectively. The polymorphism information content (PIC) of each marker was calculated according to [38] as follows: $PIC = 1-\Sigma Pi^2$, where Pi is the band frequency of the ith allele. PIC provides an estimate of the discriminatory power of a locus by taking into account, not only the number of alleles but also the relative frequencies of those alleles. PIC values vary from 0 (monomorphic) to 1 (very highly discriminative, with many alleles in equal frequencies).

Additional file

Additional file 1: Table S1. Description of the SSR markers. **Table S2:** Size of the SSR markers for all the 30 accessions. **Table S3:** Percentage of polymorphic markers between each pair of Arabidopsis accessions. **Table S4:** Non polymorphic regions identified among accessions.

Acknowledgements

The authors thank Thierry Mauduit, Marylin Roncoroni and Aurélie Bailly for the production and maintenance of the Arabidopsis plants and Albert G. Abbott for his critical reading of the manuscript.

References

1. Nordborg M, Weigel D: **Next-generation genetics in plants.** *Nature* 2008, **456:**720–723.
2. Borevitz JO, Chory J: **Genomics tools for QTL analysis and gene discovery.** *Curr Opin Plant Biol* 2004, **7:**132–136.
3. Mauricio R: **Mapping quantitative trait loci in plants: uses and caveats for evolutionary biology.** *Nat Rev Genet* 2001, **2:**370–381.
4. Clark RM, Schweikert G, Toomajian C, Ossowski S, Zeller G, Shinn P, Warthmann N, Hu TT, Fu G, Hinds DA, *et al*: **Common sequence polymorphisms shaping genetic diversity in arabidopsis thaliana.** *Science* 2007, **317:**338–342.
5. Bergelson J, Roux F: **Towards identifying genes underlying ecologically relevant traits in Arabidopsis thaliana.** *Nat Rev Genet* 2010, **11:**867–879.
6. Brachi B, Faure N, Horton M, Flahauw E, Vazquez A, Nordborg M, Bergelson J, Cuguen J, Roux F: **Linkage and association mapping of *Arabidopsis thaliana* flowering time in nature.** *PLoS Genet* 2010, **6:**e1000940.
7. El-Lithy ME, Bentsink L, Hanhart CJ, Ruys GJ, Rovito D, Broekhof JLM, van der Poel HJA, van Eijk MJT, Vreugdenhil D, Koornneef M: **New Arabidopsis recombinant inbred line populations genotyped using SNPWave and their use for mapping flowering-time quantitative trait loci.** *Genetics* 2006, **172:**1867–1876.
8. Filiault DL, Maloof JN: **A genome-wide association study identifies variants underlying the *Arabidopsis thaliana* shade avoidance response.** *PLoS Genet* 2012, **8:**e1002589.
9. Fournier-Level A, Korte A, Cooper MD, Nordborg M, Schmitt J, Wilczek AM: **A map of local adaptation in Arabidopsis thaliana.** *Science* 2011, **334:**86–89.
10. Seymour DK, Filiault DL, Henry IM, Monson-Miller J, Ravi M, Pang A, Comai L, Chan SWL, Maloof JN: **Rapid creation of Arabidopsis doubled haploid lines for quantitative trait locus mapping.** *Proc Natl Acad Sci* 2012, **109:**4227–4232.
11. Simon M, Loudet O, Durand S, Bérard A, Brunel D, Sennesal F-X, Durand-Tardif M, Pelletier G, Camilleri C: **Quantitative trait loci mapping in five new large recombinant inbred line populations of Arabidopsis thaliana genotyped with consensus single-nucleotide polymorphism markers.** *Genetics* 2008, **178:**2253–2264.
12. Cao J, Schneeberger K, Ossowski S, Gunther T, Bender S, Fitz J, Koenig D, Lanz C, Stegle O, Lippert C, *et al*: **Whole-genome sequencing of multiple Arabidopsis thaliana populations.** *Nat Genet* 2011, **43:**956–963.
13. Gan X, Stegle O, Behr J, Steffen JG, Drewe P, Hildebrand KL, Lyngsoe R, Schultheiss SJ, Osborne EJ, Sreedharan VT, *et al*: **Multiple reference genomes and transcriptomes for Arabidopsis thaliana.** *Nature* 2011, **477:**419–423.
14. **1001 Genomes.** *Adv Physiol Educ* [http://1001genomes.org/].
15. Jones N, Ougham H, Thomas H, Pašakinskienė I: **Markers and mapping revisited: finding your gene.** *New Phytol* 2009, **183:**935–966.
16. Kalia R, Rai M, Kalia S, Singh R, Dhawan AK: **Microsatellite markers: an overview of the recent progress in plants.** *Euphytica* 2011, **177:**309–334.
17. Decroocq V, Favé MG, Hagen L, Bordenave L, Decroocq S: **Development and transferability of apricot and grape EST microsatellite markers across taxa.** *Theor Appl Genet* 2003, **106:**912–922.
18. Casacuberta E, Puigdomènech P, Monfort A: **Distribution of microsatellites in relation to coding sequences within the Arabidopsis thaliana genome.** *Plant Sci* 2000, **157:**97–104.
19. Symonds VV, Lloyd AM: **An analysis of microsatellite loci in Arabidopsis thaliana: mutational dynamics and application.** *Genetics* 2003, **165:**1475–1488.
20. Zhang L, Yuan D, Yu S, Li Z, Cao Y, Miao Z, Qian H, Tang K: **Preference of simple sequence repeats in coding and non-coding regions of Arabidopsis thaliana.** *Bioinformatics* 2004, **20:**1081–1086.
21. **Versailles Arabidopsis stock center.** http://publiclines.versailles.inra.fr/.
22. McKhann HI, Camilleri C, Berard A, Bataillon T, David JL, Reboud X, Le Corre V, Caloustian C, Gut IG, Brunel D: **Nested core collections maximizing genetic diversity in Arabidopsis thaliana.** *Plant J* 2004, **38:**193–202.
23. Cosson P, Schurdi-Levraud V, Le QH, Sicard O, Caballero M, Roux F, Le Gall O, Candresse T, Revers F: **The RTM resistance to potyviruses in *Arabidopsis thaliana*: natural variation of the *RTM* genes and evidence for the implication of additional genes.** *PLoS One* 2012, **7:**e39169.
24. Pagny G, Paulstephenraj PS, Poque S, Sicard O, Cosson P, Eyquard J-P, Caballero M, Chague A, Gourdon G, Negrel L, *et al*: **Family-based linkage and association mapping reveals novel genes affecting Plum pox virus infection in Arabidopsis thaliana.** *New Phytol* 2012, **196:**873–886.
25. Decroocq V, Sicard O, Alamillo JM, Lansac M, Eyquard JP, Garcia JA, Candresse T, Le Gall O, Revers F: **Multiple resistance traits control Plum pox virus infection in Arabidopsis thaliana.** *Mol Plant Microbe Interact* 2006, **19:**541–549.
26. Revers F, Guiraud T, Houvenaghel M-C, Mauduit T, Le Gall O, Candresse T: **Multiple resistance phenotypes to lettuce mosaic virus among Arabidopsis thaliana accessions.** *Mol Plant Microbe Interact* 2003, **16:**608–616.
27. Sicard O, Loudet O, Keurentjes JJB, Candresse T, Le Gall O, Revers F, Decroocq V: **Identification of quantitative trait loci controlling symptom development during viral infection in Arabidopsis thaliana.** *Mol Plant Microbe Interact* 2008, **21:**198–207.
28. **The V.A.S.T lab.** http://www7.inra.fr/vast/msat.php.
29. **The Arabidopsis information resource.** http://www.arabidopsis.org/index.jsp.
30. Guo W-J, Ling J, Li P: **Consensus features of microsatellite distribution: microsatellite contents are universally correlated with recombination rates and are preferentially depressed by centromeres in multicellular eukaryotic genomes.** *Genomics* 2009, **93:**323–331.
31. Simon M, Simon A, Martins F, Botran L, Tisné S, Granier F, Loudet O, Camilleri C: **DNA fingerprinting and new tools for fine-scale discrimination of Arabidopsis thaliana accessions.** *Plant J* 2012, **69:**1094–1101.
32. Anastasio AE, Platt A, Horton M, Grotewold E, Scholl R, Borevitz JO, Nordborg M, Bergelson J: **Source verification of mis-identified Arabidopsis thaliana accessions.** *Plant J* 2011, **67:**554–566.
33. Păcurar DI, Păcurar ML, Street N, Bussell JD, Pop TI, Gutierrez L, Bellini C: **A collection of INDEL markers for map-based cloning in seven Arabidopsis accessions.** *J Exp Bot* 2012, **63:**2491–2501.

34. **The European Arabidopsis stock centre.** http://arabidopsis.info/.
35. **The Arabidopsis BAC clones file.** ftp://ftp.arabidopsis.org/home/tair/Sequences/clones/AGI_BAC.
36. Sputnik: **DNA microsatellite repeat search utility.** http://services.cbib.u-bordeaux2.fr/pise/sputnik.html.
37. **Primer3.** http://bioinfo.ut.ee/primer3-0.4.0/primer3/.
38. Smith JSC, Chin ECL, Shu H, Smith OS, Wall SJ, Senior ML, Mitchell SE, Kresovich S, Ziegle J: **An evaluation of the utility of SSR loci as molecular markers in maize (Zea mays L.): comparisons with data from RFLPS and pedigree.** *Theor Appl Genet* 1997, **95:**163–173.

2

ClearedLeavesDB: an online database of cleared plant leaf images

Abhiram Das[1], Alexander Bucksch[1,3], Charles A Price[4] and Joshua S Weitz[1,2*]

Abstract

Background: Leaf vein networks are critical to both the structure and function of leaves. A growing body of recent work has linked leaf vein network structure to the physiology, ecology and evolution of land plants. In the process, multiple institutions and individual researchers have assembled collections of cleared leaf specimens in which vascular bundles (veins) are rendered visible. In an effort to facilitate analysis and digitally preserve these specimens, high-resolution images are usually created, either of entire leaves or of magnified leaf subsections. In a few cases, collections of digital images of cleared leaves are available for use online. However, these collections do not share a common platform nor is there a means to digitally archive cleared leaf images held by individual researchers (in addition to those held by institutions). Hence, there is a growing need for a digital archive that enables online viewing, sharing and disseminating of cleared leaf image collections held by both institutions and individual researchers.

Description: The Cleared Leaf Image Database (ClearedLeavesDB), is an online web-based resource for a community of researchers to contribute, access and share cleared leaf images. ClearedLeavesDB leverages resources of large-scale, curated collections while enabling the aggregation of small-scale collections within the same online platform. ClearedLeavesDB is built on Drupal, an open source content management platform. It allows plant biologists to store leaf images online with corresponding meta-data, share image collections with a user community and discuss images and collections via a common forum. We provide tools to upload processed images and results to the database via a web services client application that can be downloaded from the database.

Conclusions: We developed ClearedLeavesDB, a database focusing on cleared leaf images that combines interactions between users and data via an intuitive web interface. The web interface allows storage of large collections and integrates with leaf image analysis applications via an open application programming interface (API). The open API allows uploading of processed images and other trait data to the database, further enabling distribution and documentation of analyzed data within the community. The initial database is seeded with nearly 19,000 cleared leaf images representing over 40 GB of image data. Extensible storage and growth of the database is ensured by using the data storage resources of the iPlant Discovery Environment. ClearedLeavesDB can be accessed at http://clearedleavesdb.org.

Keywords: Database, Images, Leaves, Digital archiving, Digital curation

Background

The leaf organ is the site of the majority of primary production within land plants. Leaves vary in shape, composition, color, as well as the structure of their venation networks [1]. Here, we focus on efforts to describe and analyze the last of these characteristics: the venation network. A leaf venation network represents the interconnected set of veins present in a leaf. Carbon, nutrients, and water are transported through xylem and phloem contained within the vascular bundles that constitute veins [2]. In addition, veins confer material strength to the leaf due to the higher stiffness of veins relative to other leaf tissue [3,4].

The analysis of leaf venation networks is of increasing interest due to multiple hypotheses linking venation networks to changes in leaf physiological rates [5,6] and to the diversification of early land plants [7-9]. For example, increases in the density of leaf veins are hypothesized to

* Correspondence: jsweitz@gatech.edu
[1]School of Biology, Georgia Institute of Technology, Atlanta, Georgia, USA
[2]School of Physics, Georgia Institute of Technology, Atlanta, Georgia, USA
Full list of author information is available at the end of the article

increase physiological rates, such as net photosynthesis, because of the decreasing distance that water molecules must travel between the vein network, terminal sites of photosynthesis and stomates [10-12]. Similarly, the early adaptive radiation of land plants coincided with the appearance of increasingly reticulate (i.e. loopy) venation networks, as inferred from fossil leaves [13,14]. Reticulate networks are known to confer added redundancy to leaves in the face of damage from herbivory or disease, providing a mechanism linking leaf venation structure to whole plant fitness [15].

In practice, the quantitative and qualitative analysis of venation networks requires the identification of veins as distinct from that of surrounding leaf tissue in areoles. Hence, a first step toward identifying a venation network is to "clear" a leaf [16,17]. Multi-decadal efforts to collect cleared leaves have been undertaken by individual researchers and institutions. For example, in the USA there are now multiple, independent cleared leaf collections with thousands of specimens, e.g., held at the Smithsonian Institution [18], Yale Peabody Museum [19], New York Botanical Garden [20], and UC-Berkeley [21]. In each case, digital images have been taken of the specimens and in many instances, institutional websites offer free access to all or a portion of these curated collections. However, these sites do not necessarily provide functionality for individual researchers unaffiliated with these institutions to store, manage, search, and analyze images online – indeed, such functionality is outside of the scope of each individual collection. There are also other generalized repositories of images of plants and plant organs such as CalPhoto [22], Morphbank [23], USDA Plant database [24], Encyclopedia of Life records [25] and Bisque [26]. Bisque is notable for its support for secure image storage, analysis and data management capabilities. Bisque is also hosted on the iPlant [27] environment giving users control over sharing of images and analysis results. While these repositories are freely accessible, they are quite general in scope and are not designed for the specific set of features and meta-data associated with cleared leaf images. Moreover, individual collection sites are not linked, they do not share a common platform to share data, and they do not enable the integration of cleared leaf images with processed images and analysis. Hence there is a need for a complementary approach to enhance the utility of pre-existing digital-physical collections: a trans-institutional online database by which individual research groups and institutions can store, view, and share their cleared leaf images with the community.

In order to address these issues, we have developed the Cleared Leaf Image Database (ClearedLeavesDB). ClearedLeavesDB enables researchers to contribute, access, and share cleared leaf images. The database houses existing curated collections, such as a major contribution from the Smithsonian Institution collection, and enables researchers to contribute small-scale curated collections for dissemination, viewing and re-use. The web interface of ClearedLeavesDB is built on open-source technologies, allowing users to: (i) create and manage leaf image collections; (ii) upload/download images and their metadata in single or batch mode; (iii) annotate images; (iv) search images; and (v) link analysis data with the original images. The database provides varying degree of access based on user privileges which enhances data quality and reliability. As we describe below, we have developed ClearedLeavesDB to accelerate basic science and public understanding of plant biology generally, and leaf veins specifically.

Construction and content

We describe a user-oriented view of the ClearedLeavesDB system, given that our focus in this paper is on the practical usage of the web application by and for plant biologists. Major sections of the application are described below, including how to create and manage image collections, search and browse images, manage and download marked lists of images, and user roles. For users interested in viewing and modifying the source code, we note that the application is developed using Drupal [28]; a content management system built upon Apache web server [29], MySQL [30] database server and PHP [31].

Create and manage collections

Image collections are sets of uploaded images that share some common attributes (e.g. geographical location, experimental treatment, species etc.). These collections are curated and linked to the owner of the collection (but stored centrally). While creating a collection, the owner has to provide a name and brief description of the collection. The owner also has the option to mark a collection as private or public and the license under which the data is released. Public collections can be accessed by all users and visitors of the database whereas private collections can only be accessed by the owner and associated members. A personalized access module allows the owner to manage a collection membership for all registered users (see the user roles section below for details). Once a collection is created, images are uploaded via a simple user interface. Users with required privileges are permitted to annotate images. The collection owner can change all initial preferences at any time via the "Collections" menu. At the time of submission, ClearedLeavesDB includes nearly 19,000 images (see Figure 1 for a sample of the images available) from over 144 countries (see Figure 2 for the current geographical distributions of source specimens). Additional, global information on the collections can be viewed via the "Charts" menu.

Figure 1 Representative images of cleared leaves in ClearedLeavesDB. 1. *Clerodendrum bungei Steud*. **2**. *Clerodendrum infortunatum L*. **3**. *Tilia miqueliana Maxim*. **4**. *Quercus laurifolia* **5**. *Pseudocalyx africanus S. Moore* **6**. *Dombeya burgessiae*

Search and browse images

A user searches for images or collections via a name or various metadata parameters. Current search fields include: class, order, family, genus, species, blade margin, blade shape, herbarium ID, identified by, specimen number, image resolution and whether or not the original image has processed images associated with it. A free text search is also available. When viewing a collection (Figure 3) a user sees thumbnails of all images. A mouse click on the thumbnail leads to a detailed image view with

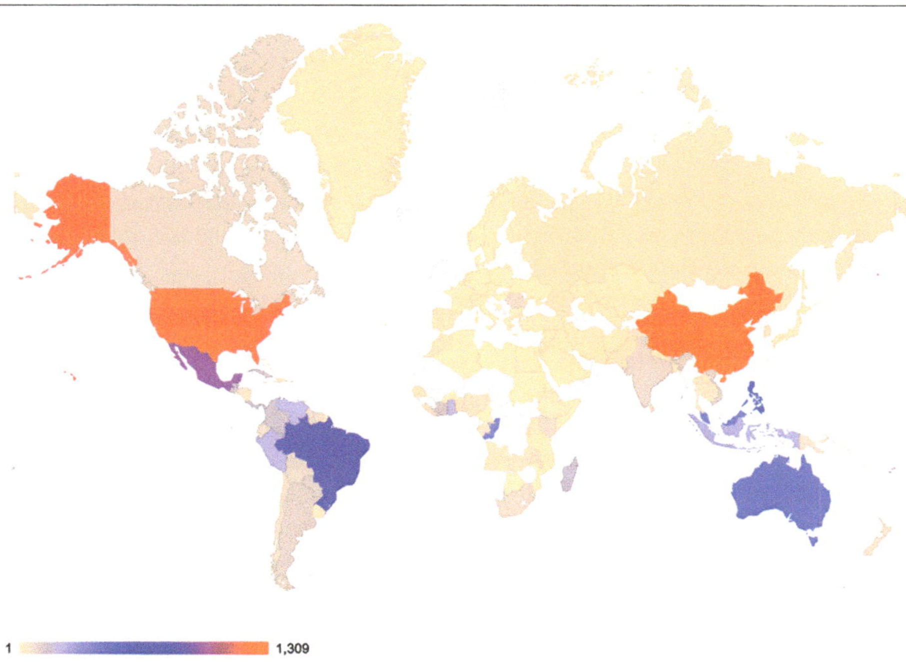

Figure 2 Geographic distribution of source specimens in the public collections of ClearedLeavesDB. Each country's color denotes the number of specimens for which images are available in the database as of submission. The world map projection is via Google geo chart: https://code.google.com/apis/ajax/playground/?type=visualization#geo_chart.

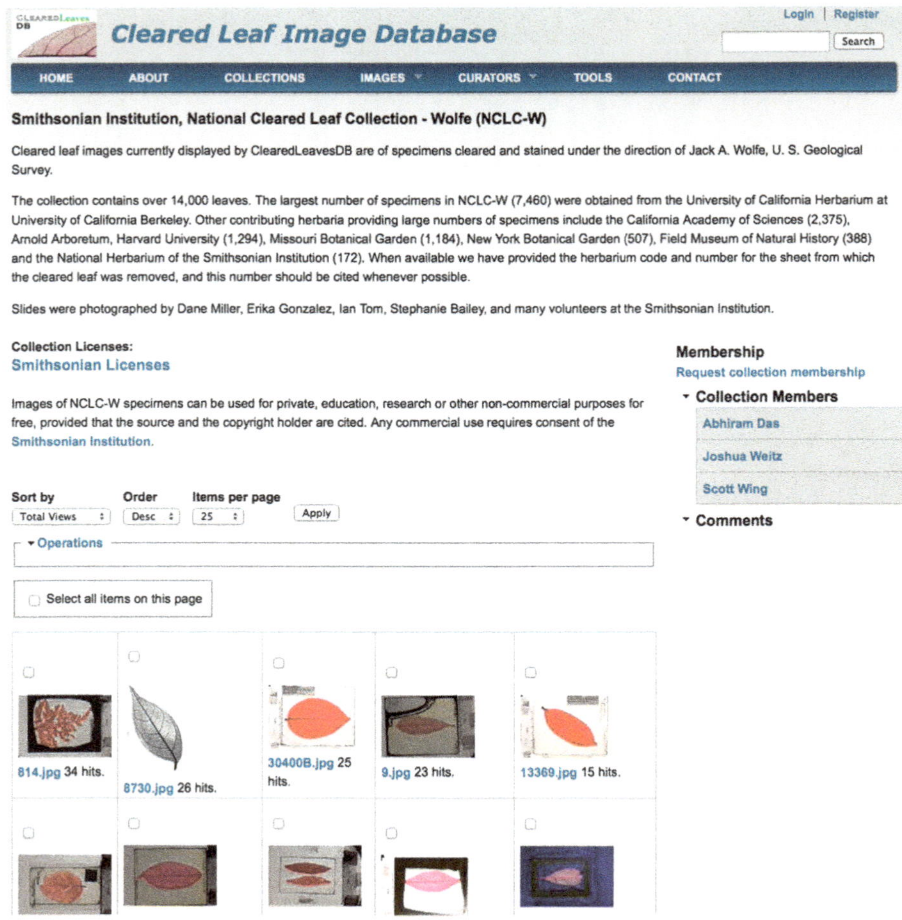

Figure 3 Collection view for curator user. Curator has the right to add image to the marked list, post comments on the collection or manage collection membership.

metadata. The search functionality is accessible via a form available in the "Images" menu.

Marked list

A marked list allows the user to maintain a list of images of interest as they search or browse. A user can add images to the marked list from the collection view module or via the search module. The marked list module also allows the user to download image groups (Figure 4) together with their associated metadata as a "zip" file. The downloaded zip file contains a comma-separated value (.csv) file containing the image name associated with the image metadata – such formats are compatible with all standard spreadsheet programs. Only registered users of the system have the privilege to maintain a marked list and batch download images. The metadata file contains a unique number i.e. the database identifier for each image downloaded. This number is required to associate processed images and their analysis data to the parent image in the database while uploading processed images through our desktop client interface and API. Users interested in batch uploading processed images to the database are advised to use the batch download functionality to obtain image metadata. To facilitate this, we introduce a java tool for automated external access to the database without using the web interface (described next).

Interface for linking processed data with original images

ClearedLeavesDB is capable of storing original and processed images along with their metadata. The types of processed images might be binary images resulting from manual or automated segmentation of leaf veins from the leaf background as in [5,6], skeletonized images of the approximate center-line of the leaf venation network [5], or aereole segmentation of the image [5]. Additional processed data might include the results of quantitative analysis of the structure of the cleared leaf. In order to link post-processed images and data to the original image, we provide tools to upload processed images and its analysis data to the database from a user's computer through a command line interface and application programming interface (API).

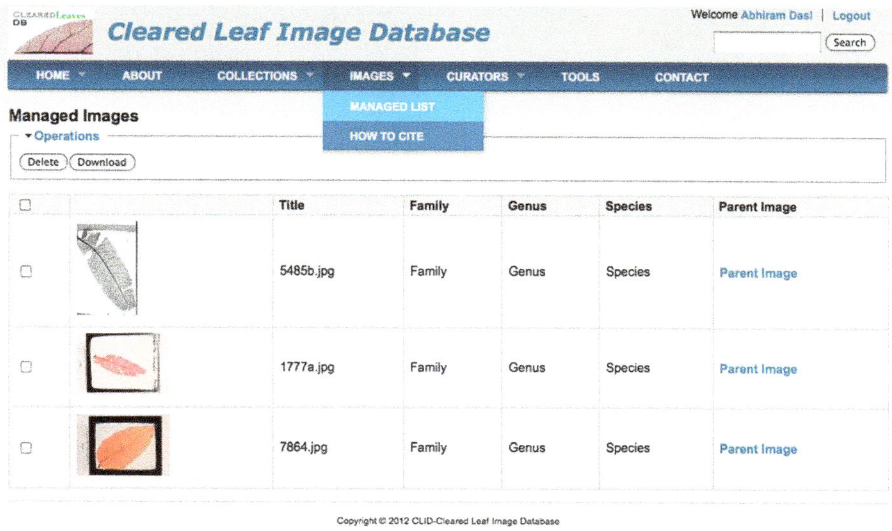

Figure 4 Marked list of images. User can bulk download images from the marked list with its metadata.

The command line interface is a Java client application that can be downloaded from the ClearedLeavesDB website (http://www.clearedleavesdb.org/sites/default/files/clid.jar). The downloaded file "clid.jar" is an executable Java Archive (jar) file that contains the client application and an open API. This application and API uses Representational State Transfer (REST) Web services to connect and upload processed data to the database. It requires Java Runtime Environment (JRE) version 1.6 or above to be installed on the user's desktop or laptop. Instructions on how to launch the command line application is described in detail on the web site and can be accessed by going to the "API" tab of the "TOOLS" menu. Database users can also upload processed images and its analysis data through the web interface. To do so, users should navigate to the image view page from search, collection view or marked list page and click on "Add new processed image" link. The API is provided through a Java class called "UploadProcessedImages". It exposes an interface called "ToCLID()" that requires three parameters:

1. CSV file name (String data type)
2. Database user name (String data type)
3. Database user password (String data type)

The following Java code snippet can be used to invoke the API:

```
import clid.rest.client.UploadProcessed
Images;
UploadProcessedImages.ToCLID(csvFile,
username, password);
```

Please note that the CSV file should be compatible to the format as mentioned on the website. A sample file can be downloaded from ClearedLeavesDB website (http://www.clearedleavesdb.org/sites/default/files/sample_processed_upload.csv).

Analyzing images with third party software such as LeafGUI [32], LAMINA [33] and LIMANI [34] is enabled by the integrated batch download functionality. After analysis, processed images and their corresponding analysis data can be uploaded. These uploaded data are associated with the original image, hence heterogeneous and non-standard processing types can be uploaded, though the processed analysis data cannot yet be searched, a trade-off given the design decision to allow greater flexibility in allowing users to associate processed data with the original images.

User roles

The database provides multiple levels of user access to cater to different needs of the community. The database provides the following three levels of users (see Table 1 for a list of user roles and associated functions): 1) guests, 2) registered users and 3) curators. A collection manager (i.e. a curator) can define custom roles and manage permissions for the collection irrespective of the global roles defined above. They can do so by clicking on the "Group" tab of the collection.

User scenarios

In this section we describe several examples to highlight the main use cases of our database: database browsing by a guest user; creating an image collection; managing collection membership, roles and permissions; and

Table 1 List of user roles and their associated permissions

Permissions	Guest	Registered User	Curator
Browse collections and images	√	√	√
View collections and its metadata	√	√	√
View images and its metadata	√	√	√
View high resolution original images	X	√	√
Create image collections	X	√	√
Upload images to the collection	X	√	√
Bulk upload images to the collection	X	X	√
Upload metadata to the collection	X	√	√
Comment on collections and images	X	√	√
Maintain a marked list of images	X	√	√
Download images and its metadata from the marked list	X	√	√
Image upload quota	0 KB	1 GB	20 GB
Add processed images and results	X	√	√

uploading processed images through the REST Web Services client.

A guest browsing the database

A user accessing the database will first arrive on the home page displaying the welcome message and a list of public collections available on the database. The user can either select any one of the collections shown or click on the "Collections" menu to search for a public collection and then click on it to browse the collection. The collection view (Figure 3) displays thumbnail images in the collection in a grid format along with members of the collection and comments posted by different users. The guest user does not have rights to post comments, add images or request membership to the group. Guests can select an image and view its metadata but cannot view full size images, post comments on an image or edit metadata. Guests can click on the "Images" menu to search for images with certain filters. From the image search page user can also navigate to the image details view page as described earlier. Guests do not have access to the marked list functionality i.e. adding images to the marked list from collection or search page.

Registered user and curator creating image collection

Here we describe the difference in creating an image collection by an authenticated user and a curator. Both these user roles have the rights to create image collections but registered users do not have rights to bulk upload images to the collection. We demonstrate this functionality using a registered user and a curator. Both these users have access to the "Create Collection" sub-menu in "Collections" menu on the database but the registered user does not have access to "Bulk Upload Images" sub-menu. After creating a collection, the registered user is redirected to the single image upload page whereas the curator is redirected to the bulk upload page.

Managing collection membership, roles and permissions

The founding curator (i.e., "author") of a collection has all the rights to manage its members, roles and permissions. The author can visit the "Group" tab and click on the "Add People" link to add people to the collection and click on the "People" link to manage membership. Similarly "Roles" and "Permissions" link can be followed to manage roles and permissions of the collection.

Uploading processed images through REST web services client

Our client application can be used to upload processed images and its analysis data to the database. A registered user or a curator can add images to its marked list (Figure 4) and batch download the images with its metadata using the download feature. After the images are downloaded a user can use any image processing software to analyze image traits and produce processed images and analysis data files. These processed images and analysis data files can be uploaded to the database using the Web services client. The client application reads a csv file containing the parent image id and a corresponding processed image and data file. A sample csv file is available for download at the website in the FAQ section (i.e. CSV file format) of the 'About' page. The image name and ID in the csv file are obtained from the metadata file downloaded along with the images. In this case study, we downloaded three images from a user's marked list (Figure 4), processed them using Leaf GUI [32] and uploaded the analysis data using the client application. A user can view the processed images and their analysis data by navigating to the parent image's view page (Figure 5).

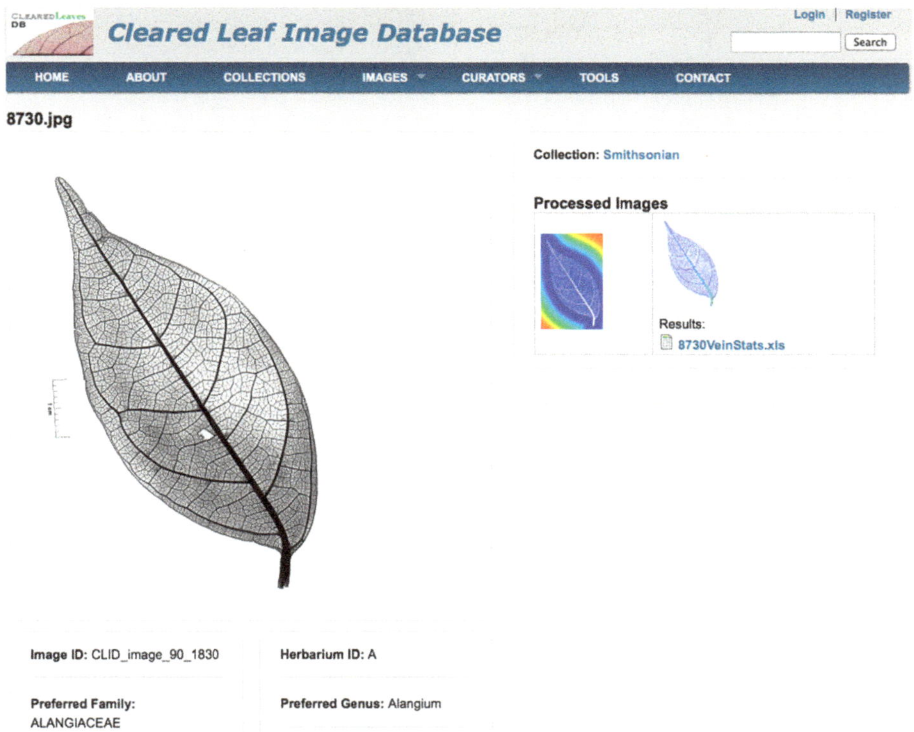

Figure 5 View of the image showing associated processed image and its analysis data.

Datasets

ClearedLeavesDB is seeded with five distinct collections. The largest is from the National Cleared Leaf Collection housed at the Museum of Natural History, Smithsonian Institution, including specimens cleared and stained under the direction of Jack A. Wolfe, U.S. Geological Survey. The other collections are of (i) *Arabidopsis thaliana* specimens from a study of variation in venation network traits for which data include ecotypes, RILs, NILs, and vascular patterning mutants (courtesy of Benjamin Blonder, unpublished); (ii) specimens from multiple *Populus tremuloides* clones from the Colorado Rocky Mountains [35]; (iii) specimens taken from the University of Arizona arboretum [36]; (iv) specimens taken from oak trees of different species on the campus of the Georgia Institute of Technology (courtesy of Charles Price, unpublished).

Discussions and conclusions

ClearedLeavesDB provides a range of practical tools to store, manage and access cleared leaf images. The web interface for the database is built on open source technologies and is freely accessible online. At present, the database is seeded with over 40 GB of primary images of cleared leaves representing a total of 19,000 images. In doing so, ClearedLeavesDB provides a means to connect researchers, institutional repositories, and the public in accessing, sharing, and analyzing the biology of plant leaves. In this sense, ClearedLeavesDB is complementary to pre-existing websites that enable access to images of plants and plant organs, and specialized repositories of cleared leaf images. ClearedLeavesDB aims to bridge the gap between these two types of platforms by offering individuals and institutions a common platform, built on open source technology, to store, manage, share, view and analyze cleared leaf images. At present, ClearedLeavesDB leverages the infrastructure of iPlant [27] for flexible storage and access of third-party developed tools. In moving forward, we plan to extend the current system to enable further, integrated analysis of cleared leaf images and associated trait data. First, we plan to enable the association of images stored on ClearedLeavesDB with trait data pertaining to the original leaf and plant specimens, e.g., as stored on TRY-db.org [37], a global database for plant traits. Second, we plan to extend the current system to integrate the database with iPlant's Data Store [27] built using an integrated rule-oriented data-management system (iRODS) [38]. Hence, future efforts to analyze large-scale datasets of cleared leaves may benefit from bringing the software analysis tools to the data, rather than the other way around, as well as leveraging many of the other benefits of a scalable infrastructure.

Availability and requirements

The database is named as "ClearedLeavesDB" and is accessible at http://clearedleavesdb.org. The database can be accessed through any web browser; however it has been tested on Firefox (Version 27.0.1), Chrome (Version 33.0.1750.117), Safari (Version 6.1.1 and Version 7.0.1) and Internet Explorer (Version 11.0.9.x). A quick start guide is available on the website and is included here as Supplementary File 1. The public datasets present in the database are freely accessible but carry copyright information defined by the collection author.

Availability of supporting data

The data sets supporting the results of this article are available in the ClearedLeavesDB repository, http://clearedleavesdb.org.

Abbreviations

REST: Representational State Transfer; JVM: Java Virtual Machine.

Competing interests

The authors declare that they have no competing interests.

Authors' contributions

AD designed the database, developed the database and web interface, and co-wrote the manuscript. AB designed the database. CAP conceived of the project plan and contributed to writing the manuscript. JSW conceived of the project plan, contributed to web interface design, and co-wrote the manuscript. All authors read and approved the final manuscript.

Acknowledgements

The authors thank Scott Wing and Benjamin Blonder for their contributions of datasets to the initial release of Cleared Leaves DB and also thank Troy Hilley for technical contributions to database testing. This project was supported, in part, by a seed grant from the Center for Data Analytics at the Georgia Institute of Technology (to AB and JSW). Joshua S Weitz, PhD, holds a Career Award at the Scientific Interface from the Burroughs Wellcome Fund. CAP is supported by a Discovery Early Career Research Award from the Australian Research Council.

Author details

[1]School of Biology, Georgia Institute of Technology, Atlanta, Georgia, USA. [2]School of Physics, Georgia Institute of Technology, Atlanta, Georgia, USA. [3]School of Interactive Computing, Georgia Institute of Technology, Atlanta, Georgia, USA. [4]School of Plant Biology, University of Western Australia, Perth, Australia.

References

1. Ellis B, DCD, Hickey LJ, Johnson KR, Mitchel JD, Wilf P, Wing SL: *Manual of Leaf Architecture*. Ithaca, NY: Cornell University Press; 2009. ISBN 978-0801475184.
2. Sack L, Holbrook NM: **Leaf hydraulics.** *Annu Rev Plant Biol* 2006, **57**:361–381.
3. Niinemets U, Portsmuth A, Tobias M: **Leaf size modifies support biomass distribution among stems, petioles and mid-ribs in temperate plants.** *New Phytol* 2006, **171**:91–104.
4. Niinemets U, Portsmuth A, Tobias M: **Leaf shape and venation pattern alter the support investments within leaf lamina in temperate species: a neglected source of leaf physiological differentiation?** *Funct Ecol* 2007, **21**:28–40.
5. Brodribb TJ, Feild TS: **Stem hydraulic supply is linked to leaf photosynthetic capacity: evidence from New Caledonian and Tasmanian rainforests.** *Plant Cell and Environment* 2000, **23**:1381–1388.
6. Brodribb TJ, Feild TS, Jordan GJ: **Leaf maximum photosynthetic rate and venation are linked by hydraulics.** *Plant Physiol* 2007, **144**:1890–1898.
7. Boyce CK, Brodribb TJ, Feild TS, Zwieniecki MA: **Angiosperm leaf vein evolution was physiologically and environmentally transformative.** *Proceedings of the Royal Society B-Biological Sciences* 2009, **276**:1771–1776.
8. Brodribb TJ, Feild TS: **Leaf hydraulic evolution led a surge in leaf photosynthetic capacity during early angiosperm diversification.** *Ecol Lett* 2010, **13**:175–183.
9. Feild TS, Brodribb TJ, Iglesias A, Chatelet DS, Baresch A, Upchurch GR Jr, Gomez B, Mohr BA, Coiffard C, Kvacek J, Jaramillo C: **Fossil evidence for Cretaceous escalation in angiosperm leaf vein evolution.** *Proc Natl Acad Sci U S A* 2011, **108**:8363–8366.
10. Price CA, Wing S, Weitz JS: **Scaling and structure of dicotyleconous leaf venation networks.** *Ecol Lett* 2012, **15**:87–95.
11. Raven JA: **The evolution of vascular plants in relation to quantitative functioning of dead water-conducting cells and stomata.** *Biol Rev Camb Phil Soc* 1993, **68**:337–363.
12. Sack L, Tyree MT, Holbrook NM: **Leaf hydraulic architecture correlates with regeneration irradiance in tropical rainforest trees.** *New Phytol* 2005, **167**:403–413.
13. Kull U, Herbig A: **Leaf venation of angiosperms - form and evolution.** *Naturwissenschaften* 1995, **82**:441–451.
14. Sack L, Scoffoni C, McKown AD, Frole K, Rawls M, *et al*: **Developmentally based scaling of leaf venation architecture explains global ecological patterns.** *Nat Commun* 2012, **3**. doi:10.1038/ncomms1835.
15. Sack L, Dietrich EM, Streeter CM, Sanchez-Gomez D, Holbrook NM: **Leaf palmate venation and vascular redundancy confer tolerance of hydraulic disruption.** *Proc Natl Acad Sci U S A* 2008, **105**:1567–1572.
16. Bates JC: **A method for clearing leaves.** *Am Nat* 1931, **65**:288–288.
17. Gardner RO: **Overview of botanical clearing technique.** *Stain Tech* 1975, **50**:99–105.
18. *Smithsonian Institution*. 2014. http://www.si.edu/.
19. *Yale Peabody Museum*. 2014. http://peabody.yale.edu/.
20. *New York Botanical Garden*. 2014. http://www.nybg.org.
21. *UC Berkeley Cleared Leaf Collections*. 2014. http://www.ucmp.berkeley.edu/collections/plants/clearedleaf.html.
22. CalPhotos: *A database of photos of plants, animals, habitats and other natural history subjects*. Berkeley: BSCIT, University of California; 2014. http://calphotos.berkeley.edu/.
23. Morphbank: *Biological Imaging*. USA: Florida State University, Department of Scientific Computing, Tallahassee, FL; 2014:32306–4026. http://www.morphbank.net.
24. The PLANTS Database: *USDA, NRCS*. NC 27401-4901USA: National Plant Data Team, Greensboro; 2014:27401–4901. http://plants.usda.gov.
25. *Encyclopedia of Life*. 2014. http://www.eol.org.
26. Kvilekval K, Fedorov D, Obara B, Singh A, Manjunath BS: **Bisque: a platform for bioimage analysis and management.** *Bioinformatics* 2010, **26**:544–552.
27. Jorgensen RA, Stein L, Rain S, Andrews G, Chandler V: **The iPlant collaborative: a cyberinfrastructure-centered community for a new plant biology.** *In Vitro Cellular & Developmental Biology-Animal* 2008, **44**:S26–S26.
28. *Drupal*. http://drupal.org/.
29. *The Apache HTTP Server*. http://httpd.apache.org/.
30. *MySQL Database Server*. http://www.mysql.com/.
31. *PHP*. http://php.net/.
32. Price CA, Symonova O, Mileyko Y, Hilley T, Weitz JS: **Leaf extraction and analysis framework graphical user interface: segmenting and analyzing the structure of leaf veins and areoles.** *Plant Physiol* 2011, **155**:236–245.
33. Bylesjo M, Segura V, Soolanayakanahally RY, Rae AM, Trygg J, Gustafsson P, Jansson S, Street NR: **LAMINA: a tool for rapid quantification of leaf size and shape parameters.** *Bmc Plant Biology* 2008, **8**:82. doi:10.1186/1471-2229-8-82.
34. Dhondt S, Van Haerenborgh D, Van Cauwenbergh C, Merks RMH, Philips W, Beemster GT, Inzé D: **Quantitative analysis of venation patterns of Arabidopsis leaves by supervised image analysis.** *Plant J* 2012, **69**:553–563.
35. Blonder B, Violle C, Enquist BJ: **Assessing the causes and scales of the leaf economics spectrum using venation networks in Populus tremuloides.** *J Ecol* 2013, **101**:981–989.
36. Blonder B, Violle C, Bentley LP, Enquist BJ: **Venation networks and the origin of the leaf economics spectrum.** *Ecol Lett* 2011, **14**:91–100.
37. Kattge J, Díaz S, Lavorel S, Prentice IC, Leadley P, Bönisch G: **TRY – a global database of plant traits.** *Glob Chang Biol* 2011, **17**:2905–2935.
38. *iRODS (Integrated Rule-Oriented Data System*. http://www.irods.org.

3

FluxTransgenics: a flexible LIMS-based tool for management of plant transformation experimental data

Lucas AF Hanke[1], Cristiano S Botelho[1], Fernando AF Braz[1], Paulo HS Batista[1], Aurea V Folgueras-Flatschart[3], Roberto W Noda[2], Andrea A Carneiro[2], Alessandra C Faria-Campos[1,3*] and Sérgio VA Campos[1]

Abstract

Background: The production and commercial release of Genetically Modified Organisms (GMOs) are currently the focus of important discussions. In order to guarantee the quality and reliability of their trials, companies and institutions working on this subject must adopt new approaches on management, organization and recording of laboratory conditions where field studies are performed. Computational systems for management and storage of laboratory data known as Laboratory Information Management Systems (LIMS) are essential tools to achieve this.

Results: In this work, we have used the SIGLa system – a workflow based LIMS as a framework to develop the FluxTransgenics system for a GMOs laboratory of Empresa Brasileira de Pesquisa Agropecuária (EMBRAPA) Maize and Sorghum (Sete Lagoas, MG - Brazil). A workflow representing all stages of the transgenic maize plants generation has been developed and uploaded in FluxTransgenics. This workflow models the activities involved in maize and sorghum transformation using the *Agrobacterium tumefaciens* method. By uploading this workflow in the SIGLa system we have created *Fluxtransgenics*, a complete LIMS for managing plant transformation data.

Conclusions: FluxTransgenics presents a solution for the management of the data produced by a laboratory of genetically modified plants that is efficient and supports different kinds of information. Its adoption will contribute to guarantee the quality of activities and products in the process of transgenic production and enforce the use of Good Laboratory Practices (GLP).
The adoption of the transformation protocol associated to the use of FluxTransgenics has made it possible to increase productivity by at least 300%, increasing the efficiency of the experiments from between 0.5 and 1 percent to about 3%. This has been achieved by an increase in the number of experiments performed and a more accurate choice of parameters, all of which have been made possible because it became easier to identify which were the most promising next steps of the experiments. The FluxTransgenics system is available for use by other laboratories, and the workflows that have been developed can be adapted to other contexts.

Keywords: Laboratory information management system, Workflow, Plant transformation, Maize, Sorghum

*Correspondence: alessa@dcc.ufmg.br
[1] Department of Computer Science, Universidade Federal de Minas Gerais, Antônio Carlos Avenue, 6627, Pampulha, 30123-970 Belo Horizonte MG, Brazil
[3] INMETRO, Nsa.Sra.das Graças Avenue, 50, 25250-020 Xérem-Duque de Caxias RJ, Brazil
Full list of author information is available at the end of the article

Background

The amount of data generated by experimental processes has been growing significantly. As a consequence, the management of data and processes from laboratories has been the subject of debate. Stafford [1] has discussed the need of computational tools designed to manage all laboratory information — including, but not limited to, the data produced by these laboratories — highlighting the relevance of such systems. These applications are referred to as Laboratory Information Management Systems (LIMS) and are characterized by data storage and tracking functionalities, management of the laboratory processes, quality assurance and integration with other systems and equipments. Several LIMS are currently available as academic, proprietary and open source solutions. Some examples of these include SQL LIMS [2], LabSoft [3], LabWare [4], FreeLIMS [5] and the systems developed by Hendrick [6], Quo [7], Tharayil [8] and Sanchez [9].

The available LIMS are usually restricted to the needs of the laboratories involved in their development and are specific to the context of these laboratories. SQL LIMS, for example, has distinct solutions for pharmaceutical, chemical, nourishing, forensic and water analysis laboratories. Other systems present functionalities varying from data management from an academic micro chip fabrication facility [6] or a cancer research laboratory [7], to a maize mapping project [9] or a service based LIMS focused on the integration of the data stored in biological databases [8]. These systems have been designed to manage specific data for one kind of laboratory and therefore present some limitations when customization is necessary.

Although a significant number of LIMS is available, its use is still limited since most of them are designed especially to the needs of the supported laboratory. Laboratories of transgenic production and analysis are among the laboratories for which only a small number of LIMS have been developed. One example of such system is Villager (Villager Transgenic Animal Management System), a proprietary solution for the management of information related to the development of genetically modified animals [10]. Currently there are only a few similar systems for laboratories working on transgenic plants, although most of them can not be characterized as LIMS. These systems include Phytotracker [11] and QTreds [12], which we compare with the SIGLa system in Table 1. Flexible LIMS can be used to manage these kind of data. Melo [13] has proposed the **SIGLa** system, a flexible workflow based LIMS designed to allow its adaptation to activities and processes of different types of laboratories. A workflow can be defined as the tasks executed according to a set of rules and procedures in order to conclude a process. A workflow consists of a sequence of activities, or a complex set of processes occurring concurrently and eventually impacting in others [14]. Other LIMS incorporate this concept, such as the system developed by Wilkins [15], which manage data from proteomics laboratories.

Furthermore, the SIGLa system shares some key features with Electronic Laboratory Notebooks (ELN), such as record keeping, online data sharing and retrieval, support to different types of data formats, collaboration, permissions system, inventory and file management and audit trail. The SIGLa system is web based but can also operate offline by simply installing the complete client-server architecture in a stand alone machine. There are other ELNs available, such as eJournal [16,17], eCAT [18] and E-WorkBook Suite (EWBS) [19], and even the web application Evernote [20] that can be used to manage data. However, these systems lack the flexibility and workflow perspective of LIMS-based systems such as SIGLa.

In this work, we propose the construction of workflows to be used in the SIGLa system for data management of a plant transformation facility and modifications in SIGLa to comply with the needs of these complex processes. By uploading the workflows to the SIGLa system the LIMS system is created. These workflows can serve as a model for similar laboratories. The system proposed is called *FluxTransgenics* and has been designed to manage data of the Cellular Biology and Plant Tissue Culture Laboratory at EMBRAPA Maize and Sorghum — MG — Brazil. This facility works on the analysis and production of transgenic maize and sorghum plants. A complex workflow has been constructed for data management considering one of the main methods used for maize transformation: the use of the bacteria *Agrobacterium tumefaciens*. Transformation through *A. tumefaciens* uses the bacteria natural system of gene transferring [21]. This process has a significant number of activities with specific attributes that are defined with great detail during workflow construction.

FluxTransgenics has been developed in collaboration with EMBRAPA Maize and Sorghum — MG — Brazil, an institution that works, among other things, in the production of transgenic plants. EMBRAPA has initiated the use of the Good Laboratory Practices (GLP) to fulfill the demand for quality at the production and commercialization of genetically modified organisms. GLP are a set of management principles for research laboratories activities that provide a framework within which studies are planned and performed in order to ensure results consistency and reliability [22]. In order to enforce the GLP adoption it is important to demonstrate that information produced and processes applied to obtain it are correct and traceable. The use of automated systems as **FluxTransgenics** to follow each step of the research in a laboratory is essential to ensure the quality of the production.

The *FluxTransgenics* system has been used to manage data from experiments starting on September 2011. Up

Table 1 Table containing a comparison between *FluxTransgenics* and similar systems

Feature	*FluxTransgenics*	*Phytotracker*	*QTreds*
Operational system	Web Application, can be accessed from any OS.	Mac OS or Microsoft Windows.	Web Application, can be accessed from any OS.
Data storage	MySQL Relational Database.	FileMaker Pro.	MySQL Relational Database.
Process support	Most laboratory processes. Adaptable to any workflow.	Pre-defined workflow.	Adaptable to any workflow.
Access control	Permissions are granted to user or roles. Roles, or groups, can be created by administrator user.	Permissions are granted only to users. The concept of roles is non existent.	Permissions are granted to user or roles. But the roles are fixed (cannot be created).

to September 2012, it has stored data of 67 experiments, including complete and ongoing experiments. From these experiments, 20 transgenic plants have been successfully generated. These 67 experiments have been performed during the initial setup phase of the transformation laboratory. The small number of plants generated is a consequence of the problems encountered during the protocol standardization. Currently the laboratory produces between 6 to 15 plants per experiment, which generates an amount of data that requires more efficient management tools.

Our next step in the research is to perform a data mining analysis of this data, searching for patterns that could identify the ideal conditions for a successful experiment. It is important to note that this analysis will be possible only because the data is stored in a structured representation, where not only each experiment is stored, but also which experiment is derived from which one, making it possible to track and relate all experimental data.

Access to the *FluxTransgenics* system is available at the address http://www.luar.dcc.ufmg.br/FluxTransgenics (login/password: guest/guest). The definition of the transformation process is completely included in the workflow, modifications for other protocols require only a change in the workflow definition. So, adaptations for other laboratories do not require changes in the source code of the system, only in its description. The workflow definition is available for download at the address above. In addition to this, the source code of the SIGLa system is also available for code developers as well.

Methods

The SIGLa system

SIGLa has been constructed using *Java* technology and uses MySQL as a database server and Apache Tomcat as a Web server. The system has a web interface which is platform-independent and can be accessed using the main browsers (Internet Explorer, Google Chrome, Mozilla Firefox and Safari). Workflow files are uploaded in the system through its web interface. The activities created in the workflow construction are interpreted by the system as links. The workflow of **FluxTransgenics** has been constructed using the Together Workflow Editor (TWE) [23] in the XML Process Definition Language (XPDL) file format. The SIGLa system can represent different types of workflows which have been modeled in TWE. In this application, the protocols are defined as processes that are composed of steps, referred to as activities. An activity represents events of a process and, as such, has transitions, actors and rules. This information is represented as attributes in the workflow definition. Therefore, the user can define the characteristics of the attributes of each activity, such as its types, the range of values that each attribute can assume, its formats or even define auto-calculated attributes derived from other attributes. In TWE, inputs and outputs of each activity are also defined, including the relation of these with the experiments. During workflow definition it is also possible to assign to each activity a documentation that contains standard protocols, instruments calibration, procedures and records associated with the activities.

All this information is stored in a entity-relationship database. However, that it is not a traditional entity-relationship diagram, because the SIGLa system is completely dynamic, and a hard-coded ER model would not allow different activities with different attributes to be stored without modifications to the source code. Instead, an activities table store all the data, using metadata to indicate, whenever necessary, to which experiment or activity each data belongs to. A simplified entity-relationship diagram is show in Figure 1.

The construction of the FluxTransgenics LIMS

The construction of the FluxTransgenics LIMS consisted of the development of the plant transformation workflow and its upload to the SIGLa system. In this workflow each activity corresponds to an experimental step, and its attributes identify the types of information that are generated in this experimental step. The sequence of activities then contain all the steps of the transformation protocol. The attributes in each activity store the specific information of that activity. As a consequence, the workflow has

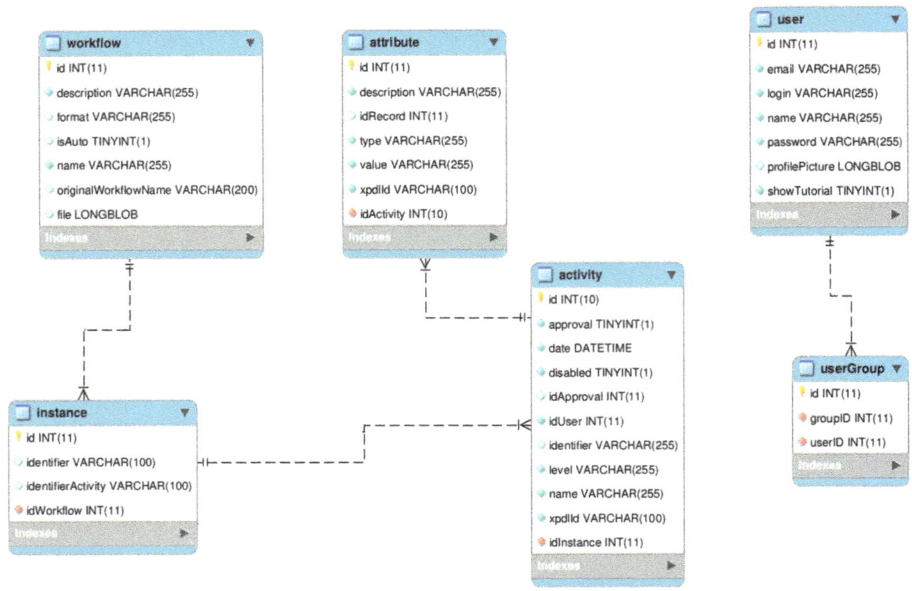

Figure 1 ER Diagram. Simplified Entity-relationship Diagram with the main entities fo the system.

the complete description of the data being managed, all the steps and for each step the required data.

Because of this, the FluxTransgenics system has been fully developed by specifying the workflow, without changing the code of the SIGLa system. Other workflows, or adaptations to this workflow can be developed by changing the workflow files in the editor, without changing the SIGLa system. The information managed by the FluxTransgenics system is completely contained in its workflow, the SIGLa system is the engine that drives the workflow, in the same way as experimental data can be stored in an Excel spreadsheet file. This file contains the data and all of the formulas that are required to understand it, the Excel program is merely an engine to understand this data.

There are several different types of attributes, each fulfilling a specific need in storing and managing laboratory data. For example, to specify which reagents are used in an experimental step, one attribute "reagent" can be specified. This information is entered by the user when registering the experiment. There are several ways in which a reagent can be specified. One way is to let the user type a string with the reagent name. Another is to let the name be chosen from a list of specified names. Yet another is to let the user determine which are the valid reagent names in another menu. Which option will be used is specified in the workflow.

Other features of the SIGLa system can be defined in the workflow, such as the protocol used, or the callus identifiers. Protocols are usually specified in a type of attribute, called *register*, which lets the user choose from a list of pre-specified names. Each name, when defined, can allow a text file (in plain text, Word or any other format) to be stored with the description of the protocol. In this way, a list of allowed protocols is stored, together with the full information about the protocol.

Callus identifiers use a different feature, called *auto increment*. When an attribute has this type, its value is not entered by the user, but defined when the activity is registered, as a unique identifier generated by the SIGLa engine. In this way the system guarantees its uniqueness, and can even generate bar codes to be attached to the samples, if needed.

This information is stored by the SIGLa system in dynamically defined tables in the database. The entities in this database are activities, which are related to one another by the relation *is derived from*, with a cardinality of 1:n, i.e., an activity can generate several "daughter" activities, but each activity has exactly one "parent" activity, except for the first activity. For example, the second activity in an experiment *is derived* from the first activity, but this first activity is not derived from any other.

Tables have a dynamic content, because each activity is defined by different attributes. So each item stored in the attributes table is a line containing, basically, the tuple (*activity*, *attribute_name*, *value*), where *activity* defines which activity in the workflow it refers to, *attribute_name* defines which attribute, and *value* defines the value of the attribute entered by the user. The SIGLa system is responsible for reconstructing, from the activities table and an experiment identifier, the data pertaining to that experiment.

It is important to emphasize that the SIGLa system is used in this work as an *engine* that drives the plant transformation workflow creating the FluxTransgenics LIMS. All of the information relating to plant transformation and experimental data is stored in the workflow.

Workflow construction and transformation protocol

The workflow has been constructed considering all steps of a transformation method, as shown in Figure 2. The protocol used for *Agrobacterium*-mediated transformation of immature zygotic embryos is based on the work of Frame *et al.* [24] and Vega *et al.* [25]. For details on the protocol used, please see the supplementary material available at the address `http://www.luar.dcc.ufmg.br/FluxTransgenics`.

FluxTransgenics contains a workflow constructed to represent all the steps of the transformation process performed at the Cellular Biology and Plant Tissue Culture Laboratory at Embrapa Maize and Sorghum using *A. tumefaciens*. This workflow allows recording and managing reagents and protocols used in the transformation process, including the clones and modified organisms produced. The initial activity in the workflow is *Planning* (Figure 3), which is used to record general information on the plant transformation experiment, such as genotype, experiment number, username, genetic construction name, etc. In this activity, an experiment identifier (id) is assigned to each experiment (using the *auto increment* feature), which is used to link all data related to the experiment and allow experiment tracking. In the transformation using *A. tumefaciens* (TAT), the activity *Planning* includes the input of information regarding the Genetic Construction that is selected from a list of constructions names previously stored in a table of the system's database (using the *register* feature).

The next activities are *Agrobacterium subculture* and *Agrobacterium inoculation* in which the system stores the procedure date and the protocols used. In *Embryo harvest*, the main information is the collection date. In *Embryo Co-cultivation*, the collected embryos from the previous activity are genetically modified and all information on those is stored, such as length and amount of embryos, culture medium and confirmatory test (Figure 4). After the execution of the *Embryo Co-cultivation* activity, there are two different paths. In some cases, a *GUS Test* is performed to check the quality of the infectious process and, only after this test, the next activity is executed, while in other cases, the *GUS Test* is not performed and the next activity is *Transfer to the Resting Medium*. In the maize and sorghum transformation experiments, the embryos remain for some time in a culture medium and can be transferred from up to five different media to perform the selection. The transfer to each medium constitutes a new activity in the workflow with its attributes.

In each transfer to a different medium, a control of the generated calluses is performed through the recording of inputs and outputs. This information contains the callus' *ID* and the number of the copies of each callus (Figure 5).

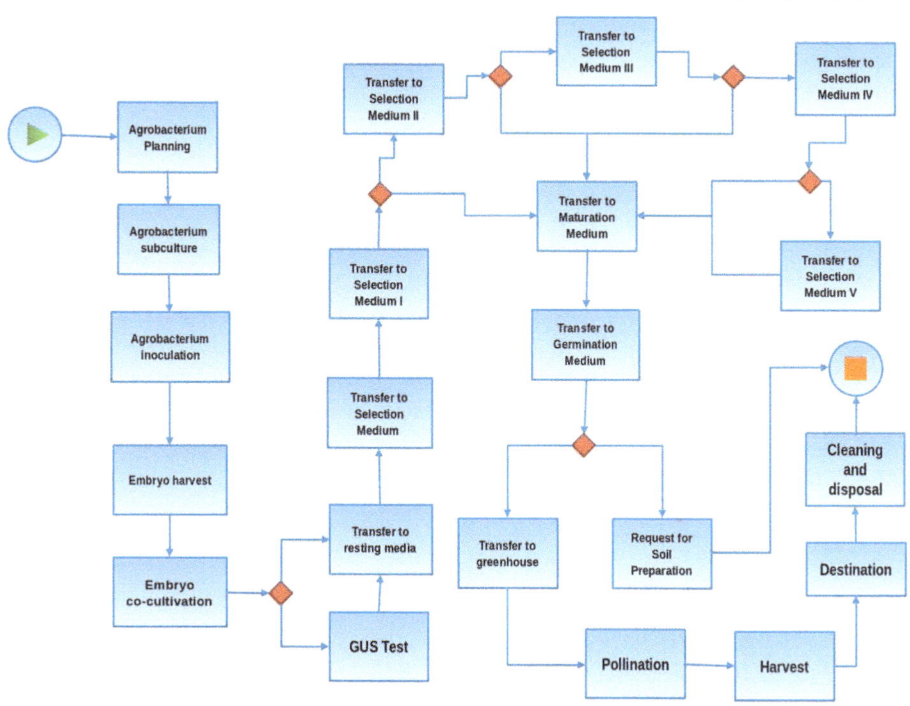

Figure 2 ***Agrobacterium tumefaciens's* Transformation *Workflow*.** *Workflow* of the maize and sorghum transformation process by *Agrobacterium tumefaciens*. Each box represents an activity in the process and the arrow from an activity to the other represents a transition.

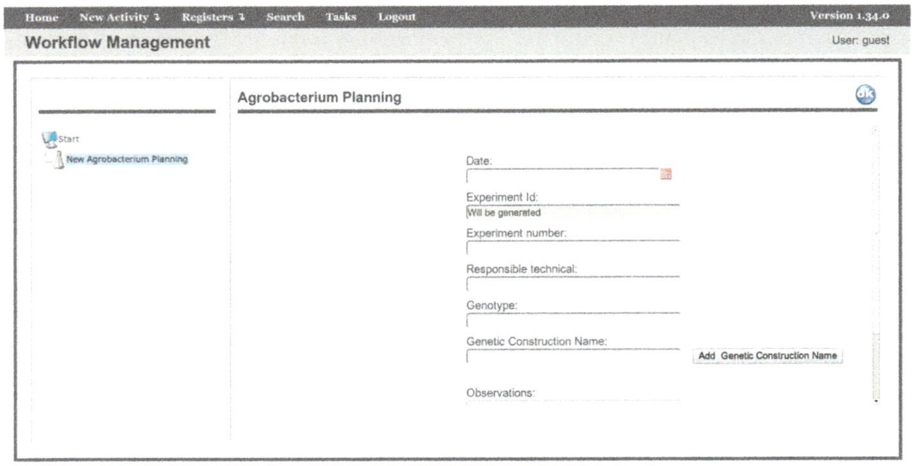

Figure 3 Planning Activity. Interface of the FluxTransgenics system showing the execution of the first activity of the workflow — *Planning*.

In the *Transfer to A Medium* activity, the stored calluses are identified and used (all or part of it) as input to the next activity. This behavior is repeated until the activity *Transfer to Greenhouse.* The following activity is the *Pollination,* where the pollination type used (self, cross or sibling) can be chosen from a drop-down menu. *Pollination* outputs are the id of the maize ear and the number of seeds. After the *Pollination,* the next activity is *Harvesting.* In this activity, the amount of picked seeds is stored as an output, including the date and the protocol used. The final activity is *Destination,* which stores the date and the protocol of the execution, allowing the user to record the destination of the seeds. This *input/output* feature is an important issue in FluxTransgenics. Through this procedure, the user can keep track of which calluses will generate the final genetically modified plants.

During the laboratory routine, the user can access the system and verify, through the the tasks' screen, which activities have been executed, which are available e which are scheduled to be executed (Figure 6).

The FluxTransgenics workflow — developed to represent the transformation process — presents a comprehensive set of attributes designed to allow recording and managing of all activities performed during the transformation process and the definition of the details regarding those activities for an easy retrieval by the user for reports production.

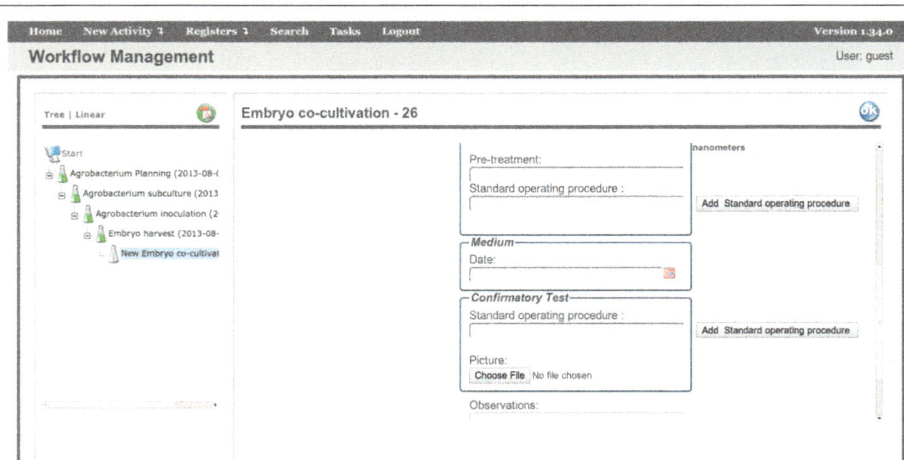

Figure 4 Embryo Co-cultivation Activity. Interface of the FluxTransgenics system showing the Embryo co-cultivation activity. After it has been executed, information about absorbance, amount and size of embryos, confirmatory test, protocols, etc is stored. In the user interface, activities that have already been executed are represented as green (filled) tubes, while activities that are available to be executed are shown as white (empty) tubes.

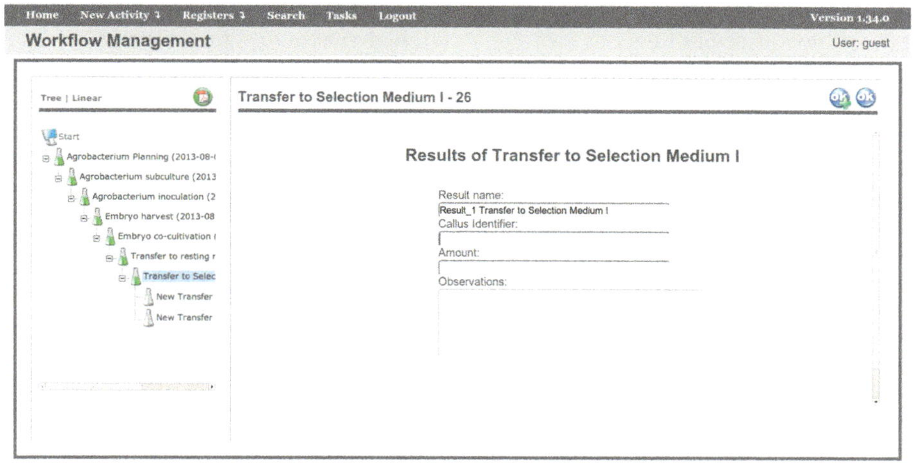

Figure 5 Interface for registering calluses. Interface of the FluxTransgenics system showing the forms for filling the outputs of the Transfer to Selection Medium I.

Discussion

The *FluxTransgenics* system has been successfully tested by the researchers at EMBRAPA Maize and Sorghum as a tool to track the plant transformation process.

The process of producing GMO plants is complex involving a large number of steps and it may take a long time. For maize, for example, it may take up to six months to complete an experiment. During this time, researchers are looking for what is called an elite event that is, a result that has the ideal characteristics to produce the desired outcome. An elite event must contain a small number of copies of the implanted gene, express the heterologous protein at high levels, and this protein must be the only difference between this new lineage and the previous one. In our lab, about twenty events are produced for each gene, in search of an elite event.

However, due to the long duration of the experiments, it is not always possible to wait until it is known that an event will be a successful one before initiating a new one and therefore several experiments are performed simultaneously. An accurate management of the samples and experimental steps is necessary to be able to identify the more promising results. However, with so many different ongoing experiments being performed by different laboratory analysts, the link between an elite event and its related experiments can be easily lost. The use of standard protocols and computational aids are essential to follow results efficiently and accurately.

By using the protocol developed by [24-26], and implementing it using FluxTransgenics, EMBRAPA'S lab has been able to increase the number of experiments performed at it from one to two a week to up to 16 (4 transformations with 4 different genic constructions) after protocol standardization. It has become much more efficient to identify the best conditions to generate a transgenic event and to coordinate the different steps of

Figure 6 Tasks's Screen. Interface of the Tasks' screen showing the executed activities.

the various experiments to be performed daily. Moreover, modifications of the protocols can not only be easily implemented, but also can be easily related to which experiments use it instead of the original ones.

Previously, the efficiency of the transformation process at EMBRAPA'S lab was very low, only between 0.5 and 1 percent of the experiments were successful. As a consequence of the work described, the efficiency has increased to about 3 percent, an increase of 300 to 600 percent.

Sixty-seven experiments have been performed and twenty transgenic plants have been produced using *FluxTransgenics* to manage the steps of the process. These experiments represent one year of experimental data that have been successfully stored in the system.

In spite of the low productivity of the initial experiments, the use of an automated management system was very important. Because so few experiments were successful, identifying those and the reasons for their success was crucial for the establishment of a productive laboratory. Current productivity in the laboratory has since increased to performing on average one experiment a day, and each experiment generating from 6 to 15 plants.

A new version of the system is under development with additional features, including a workflow for management of data from transformation performed using microparticles bombardment and the design of new workflows to manage processes that are indirectly related to the processes of plant transformation.

FluxTransgenics is a LIMS that has been customized to attend the needs of the Cellular Biology and Plant Tissue Culture Laboratory at EMBRAPA Maize and Sorghum. It is currently being used in its production. It can be accessed at the URL: `http://www.luar.dcc.ufmg.br/FluxTransgenics` (login/password: guest/guest). At this address the system can be used to store a limited amount of experimental data, and to access the system source code.

The system can be easily adapted to similar laboratories by modifying the workflow description. A description of how to adapt the workflow to other laboratories is outside the scope of this paper, but is available at the address given. Potential users are welcome to use the system and contact the authors for modifications for other processes and different laboratories.

Conclusion

The production and commercial release of Genetically Modified Organisms (GMOs) are currently the focus of important discussions. Thus, efficient management methods to assess the conditions under which GMO laboratories experiments are carried out are needed. As a developer of GMOs, EMBRAPA Maize and Sorghum (Brazil) has enforced the use of the Good Laboratory Practices (GLP) to ensure quality on the processes of producing modified genetically plants and needs a computational system to support the use of GLP.

In this work, we present *FluxTransgenics*, a system for management of data on the process of plant transformation by *A. tumefaciens* to be used as a computational aid. The system is able to manage the information and the correct sequence of activities of the transformation performed at the Cellular Biology and Plant Tissue Culture Laboratory at EMBRAPA Maize and Sorghum. It is possible by using *FluxTransgenics* to track the stored data during the execution of each activity of the plant transformation, ensuring the quality required by GLP.

Competing interests

The authors declare that they have no competing interests.

Authors' contributions

RWN and AAC and AVFF were involved in modeling, validation of the workflows and tests of the LIMS. LAFH, ACFC and CSB were directly involved in workflow construction, contributed in SIGLa System alterations and participated of all other previously cited work phases. PHSB, FAFB and CSB have modified SIGLa code to develop new functions for the system, according to the transgenic laboratory needs. SVAC is the principal investigator and supervised all the work as well as manuscript preparation. All authors read and approved the final manuscript.

Acknowledgements

The authors wish to thank the Brazilian agencies CAPES (Coordenação de Aperfeiçoamento de Pessoal de Nível Superior), FAPEMIG (Fundação de Amparo à Pesquisa do Estado de Minas Gerais) and CNPq (Conselho Nacional de Desenvolvimento Científico e Tecnológico) for financial support.

Author details

[1]Department of Computer Science, Universidade Federal de Minas Gerais, Antônio Carlos Avenue, 6627, Pampulha, 30123-970 Belo Horizonte MG, Brazil. [2]Embrapa Maize and Sorghum, MG 424 Highway KM 45, 35701-970 Sete Lagoas MG, Brazil. [3]INMETRO, Nsa.Sra.das Graças Avenue, 50, 25250-020 Xérem-Duque de Caxias RJ, Brazil.

References

1. Stafford JEH: **Lims: an automating or informating technology?** *Lab Automation Inf Manage* 1998, **33**(3):163–168.
2. **SQL*LIMS** [http://www.sqllims.com]
3. **Computing Solutions I** [http://www.labsoftlims.com]
4. **LabWare LIMS Solutions** [http://www.labware.com]
5. **Geeknet, Inc - FreeLIMS** [http://sourceforge.net/projects/freelims/]
6. Hendricks RW, Learn MR: **A laboratory information management system (lims) for an academic microchip fabrication facility.** In *University/Government/Industry Microelectronics Symposium*. 2003:88–93.
7. Quo B, Wu C: **Development of a laboratory information system for cancer collaboration projects.** *27th Annual Int Conf Eng Med Biol Soc* 2005, **3:**2859–2862.
8. Tharayil SM, Kalbfleisch T, Kumar A: **Service-oriented laboratory information management system for life sciences research.** In *IEEE International Conference on Services Computing*. 2007:621–627.
9. Sanchez-Villeda H, Schroeder S, Polacco M, McMullen M, Havermann S, Davis G, Vroh-Bi I, Cone K, Sharopova N, Yim Y, Schultz L, Duru N, Musket T, Houchins K, Fang Z, Gardiner J, Coe E: **Development of an integrated laboratory information management system for the maize mapping project.** *Bioinformatics* 2003, **19**(16):2022–2030.
10. Transgenic Software Inc.: **Villager transgenic animal management system.** [http://www.trademarkia.com/villager-transgenic-animal-management-system-78363691.html]

11. Nieuwland J, Sornay E, Marchbank A, de Graaf BH, Murray JA: **Phytotracker, an information management system for easy recording and tracking of plants, seeds and plasmids.** *Plant Methods* 2012, **8**(1):43.
12. Palla P, Frau G, Vargiu L, Rodriguez-Tomé P: **QTreds: a flexible LIMS for omics laboratories.** *EMBnet.journal* 2012, **18**(B):41.
13. Melo A, Faria-Campos A, DeLaat D, Keller R, Abreu V, Campos S: **Sigla: an adaptable lims for multiple laboratories.** *BMC Genomics* 2010, **11**(Suppl 5):8.
14. **Workflow Management Coalition** [http://www.wfmc.org/]
15. Wilkins MR, Arthur JW, Junius FK, Harrison MJ, Traini MD, Bizannes P, Doggett PE, McDonald W: **Information management for proteomics: a perspective.** *Expert Rev Proteomics* 2008, **5**(5):663–678.
16. Voegele C, Bouchereau B, Robinot N, McKay J, Damiecki P, Alteyrac L: **A universal open-source electronic laboratory notebook.** *Bioinformatics* 2013, **29**(13):1710–1712.
17. Fakas GJ, Nguyen AV, Gillet D: **The electronic laboratory journal: a collaborative and cooperative learning environment for web-based experimentation.** *Comput Supported Cooperative Work (CSCW)* 2005, **14**(3):189–216.
18. Goddard NH, Macneil R, Ritchie J: **eCAT: online electronic lab notebook for scientific research.** *Autom Exp* 2009, **1**(1):4.
19. Beato B, Pisek A, White J, Grever T, Engel B, Pugh M, Schneider M, Carel B, Branstrator L, Shoup R: **Going paperless: implementing an electronic laboratory notebook in a bioanalytical laboratory.** *Bioanalysis* 2011, **3**(13):1457–1470.
20. Walsh E, Cho I: **Using Evernote as an electronic lab notebook in a translational science laboratory.** *J Lab Autom* 2013, **18**(3):229–234.
21. Gelvin SB: **Agrobacterium-mediated plant transformation: the biology behind the "gene-jockeying" tool.** *Microbiol Mol Biol Rev* 2003, **67**(1):16–37.
22. Santos PE: **Boas Práticas de Laboratório (BPL), – Uma questão de qualidade.** *Revista Intertox de Toxicologia, Risco Ambiental e Sociedade* 2010, **3**(2):37–39.
23. **Together Workflow Editor** [http://www.together.at/prod/workflow/twe]
24. Frame BR, Shou H, Chikwamba RK, Zhang Z, Xiang C, Fonger TM, Pegg SE, Li B, Nettleton DS, Pei D, Wang K: **Agrobacterium tumefaciens-mediated transformation of maize embryos using a standard binary vector system.** *Plant Physiol* 2002, **129**(1):13–22.
25. Vega JM, Yu W, Kennon AR, Chen X, Zhang ZJ: **Improvement of Agrobacterium-mediated transformation in Hi-II maize (Zea mays) using standard binary vectors.** *Plant Cell Rep* 2008, **27**(2):297–305.
26. Petrillo CP, Carneiro NP, Purcino AAC, Carvalho CHS, Alves JD, Carneiro AA: **Optimization of particle bombardment parameters for the genetic transformation of Brazilian maize inbred lines.** *Pesquisa Agropecuária Brasileira* 2008, **43**(3):371–378.

4

Rapid bioassay to measure early reactive oxygen species production in *Arabidopsis* leave tissue in response to living *Pseudomonas syringae*

John M Smith[1,2,3] and Antje Heese[1,3*]

Abstract

Background: *Arabidopsis thaliana* and *Pseudomonas syringae* pathovar *tomato* (*Pto*) provide an excellent plant-bacteria model system to study innate immunity. During pattern-triggered immunity (PTI), cognate host receptors perceive pathogen-associated molecular patterns (PAMPs) as non-self molecules. *Pto* harbors many PAMPs; thus for experimental ease, many studies utilize single synthesized PAMPs such as flg22, a short protein peptide derived from *Pseudomonas* flagellin. Flg22 recognition by *Arabidopsis* Flagellin Sensing 2 (FLS2) initiates a plethora of signaling responses including rapid production of apoplastic reactive oxygen species (ROS). Assessing flg22-ROS has been instrumental in identifying novel PAMP-signaling components; but comparably little is known whether in *Arabidopsis*, ROS is produced in response to intact live *Pto* and whether this response can be used to dissect genetic requirements of the plant host and live bacterial pathogens *in planta*.

Results: Here, we report of a fast and robust bioassay to quantitatively assess early ROS in *Arabidopsis* leaves, a tissue commonly used for pathogen infection assays, in response to living bacterial *Pto* strains. We establish that live *Pto* elicits a transient and dose-dependent ROS that differed in timing of initiation, amplitude and duration compared to flg22-induced ROS. Our control experiments confirmed that the detected ROS was dependent on the presence of the bacterial cells. Utilizing *Arabidopsis* mutants previously shown to be defective in flg22-induced ROS, we demonstrate that ROS elicited by live *Pto* was fully or in part dependent on *RbohD* and *BAK1*, respectively. Because *fls2* mutants did not produce any ROS, flagellin perception by FLS2 is the predominant recognition event in live *Pto*-elicited ROS in *Arabidopsis* leaves. Furthermore using different *Pto* strains, our *in planta* results indicate that early ROS production appeared to be independent of the Type III Secretion System.

Conclusions: We provide evidence and necessary control experiments demonstrating that *in planta*, this ROS bioassay can be utilized to rapidly screen different *Arabidopsis* mutant lines and ecotypes in combination with different bacterial strains to investigate the genetic requirements of a plant host and its pathogen. For future experiments, this robust bioassay can be easily extended beyond *Arabidopsis-Pto* to diverse plant-pathosystems including crop species and their respective microbial pathogens.

Keywords: ROS, Reactive oxygen, PAMP, FLS2, RbohD, BAK1, flg22, *Pseudomonas syringae*, DC3000, *hrcC*

* Correspondence: heesea@missouri.edu
[1]Division of Biochemistry, University of Missouri-Columbia, 117 Schweitzer Hall, Columbia, MO 65211, USA
[3]Interdisciplinary Plant Group (IPG), University of Missouri-Columbia, Columbia, MO 65211, USA
Full list of author information is available at the end of the article

Background

Eukaryotes have developed highly effective immune mechanisms for protection against microbial pathogens using pattern-triggered immunity (PTI) as the first line of defense. Pathogen-associated molecular patterns (PAMPs), also referred to as microbe-associated molecular pattern (MAMPs), are highly conserved and essential molecules common to entire classes of microbes but are absent from the host. Host cells utilize pattern recognition receptors (PRRs) to recognize PAMPs as non-self to initiate a large number of signaling responses that contribute to growth restriction of microbial pathogens [1-3]. To evade these host immune responses, pathogenic microbes express and deliver effector molecules into host cells to interfere with PTI [4]. For example, the virulent model bacterium *Pseudomonas syringae* pathovar *tomato* (*Pto*) DC3000 translocates 28 or more effector proteins into plant cells via the type III secretion system (T3SS), some of which are known to suppress PTI [5-7]. Some effector proteins, however, betray the pathogen due to their direct or indirect recognition by cytosolic host resistant proteins resulting in Effector-Triggered Immunity (ETI) [1,5-8]. Non-pathogenic strains lacking functional T3SS such as *Pto* DC3000 *hrcC*$^-$ (*Pto hrcC*$^-$) do not suppress PTI because of their inability to deliver effectors into host cells [6,8].

In plants, only few PAMP/PRR pairs involved in PTI have been identified and characterized [1,2,9,10]. In the model plant *Arabidopsis thaliana*, the best studied PTI-system is perception of bacterial flagellin by Flagellin Sensing 2 (FLS2), a plasma membrane localized PRR [2,11]. Flagellin is the main proteinaceous component of extracellular flagellum filaments essential for the mobility and ability of bacteria such as *Pto* to infect host plants [12,13]. Binding of flagellin or flg22, a 22-amino acid peptide derived from the consensus sequence for the most highly conserved region in the N-terminus of eubacterial flagellin [14], to the extracellular domain of FLS2 occurs within the plant apoplast and leads to a variety of early and late signaling responses [11,15,16], all of which are dependent on BRI1-Associated Receptor Kinase 1 (BAK1) [17,18]. One of the best characterized and robust early PAMP-signaling events is the rapid and transient accumulation of apoplastic reactive oxygen species (ROS). Assessing PAMP-elicited ROS has proven to be a valuable tool in identifying and characterizing novel PAMP-signaling components and specific amino acids necessary for their function [18-23]. Production of rapid apoplastic ROS in response to PAMP peptides is solely dependent on the plasma membrane localized NADPH respiratory burst oxidase homolog D (RbohD) [22,24,25]. Although the exact role of ROS production in innate immunity remains unclear, ROS initiates a plethora of downstream signaling events, some of which are essential in establishing defense mechanisms to prevent the spread of bacterial pathogens [26,27].

Over the past decade, studies utilizing commercially synthesized flg22 peptide have greatly aided in increasing our insight into early and late signaling events and in identifying required signaling components and their contribution to PTI [2,28]. Other efforts exploited boiled bacterial extract to investigate various PAMP-induced responses in leaves or cultured plant cells [14,29-31]. Boiling bacterial cells results in the release of both extra- and intracellular PAMPs, thus making it difficult to determine the biologically relevant order of PAMP recognition by specific host PRRs. The disadvantage of utilizing cultured cells as opposed to leaf tissue is that because of the unavailability of *Arabidopsis* mutant cell culture lines, cell culture limits the ability to assess the genetic requirement of the plant host for the response of interest. In contrast, the genetic plant host-pathogen interplay between *Arabidopsis* plants and *Pseudomonas* can be interrogated *in planta* due to the availability of large number of *Arabidopsis* ecotypes and mutant lines. When investigating responses induced by living bacteria in plant tissue (*in planta*), most efforts have focused on later responses such as accumulation of the defense hormone salicylic acid (SA) [12-24 hours post infection (hpi)], transcriptional changes of the late gene marker *PR1* (24 hpi) or changes in resistance to bacterial infection measured 3 days post-infection (dpi). Only more recently, attention has been given to identifying early signaling events and their genetic requirements induced by living bacterial pathogens on plant host leaves [32-35], the tissue that serves as the primary source for bacterial pathogen infection.

Here, we describe advancement of a fast and convenient *in planta* bioassay that allows for quantitative assessment of early ROS production in *Arabidopsis* leaf tissue, the primary site of *Pto* infection, induced by living *Pto* strains. Importantly, we provide necessary control experiments showing that *in planta*, early ROS production was dependent on the presence of *Pto* cells. By utilizing *Arabidopsis* mutants previously shown to be affected in flg22-induced ROS production, we demonstrate that early ROS produced in response to live *Pto* strains was fully dependent on *RbohD* and *FLS2* and partially dependent on *BAK1*. No statistical differences were observed between ROS induced by *Pto* DC3000 and *hrcC*$^-$ cells, thus the virulence-promoting T3SS does not appear to have an influence on early ROS production. Because of the ease in setting up ROS assays in a 96-well plate assay, this quantitative analysis is highly suitable to screen within a relatively short period of time large populations of *Arabidopsis* accession lines in combination with diverse *Pto* mutant strains to define the genetic requirements of host and bacterial pathogen.

Results and discussion

Our goal was to establish a rapid and robust *in planta* method to quantitatively measure early ROS production in response to live *Pto* strains in *Arabidopsis* leaves, a tissue commonly used for bacterial infection assays. To this end, we adapted the luminol-based ROS assay used for PAMP-elicitation [18,21,36,37] in that we substituted the synthetic PAMP peptide with bacterial cells of *Pto* strains that were freshly grown for 36-48 hours at 22°C on King's B media (KBM) plates containing appropriate antibiotics [Figure 1; see Methods for details]. In brief, *Arabidopsis* leaf discs obtained from 5 week old plants were cut in half with a sharp razor blade to increase surface area, and each half was placed into a distilled water (dH_2O)-containing well of a 96-well titer plate for at least 20 hours to reduce wounding effect [Figure 1A]. Immediately prior to elicitation, bacteria were harvested directly from plates [38] and washed twice in sterile dH_2O. After the second wash step, bacterial cells were resuspended in sterile dH_2O [38], and their cell densities (OD_{600}) were adjusted between OD_{600} = 0.001 to OD_{600} = 0.1 [Figure 1B]. The OD_{600} can be used as a rapid means to provide an estimation of bacterial cell density (as colony-forming units (cfu)/mL) [38,39]. After Luminol and Horseradish Peroxidase (HRP) was given to the *Pto*-containing solution, this Elicitation Solution was added to *Arabidopsis* leaf halves to measure luminol-based ROS production between 0 and 80 minutes using a luminometer [Figure 1C].

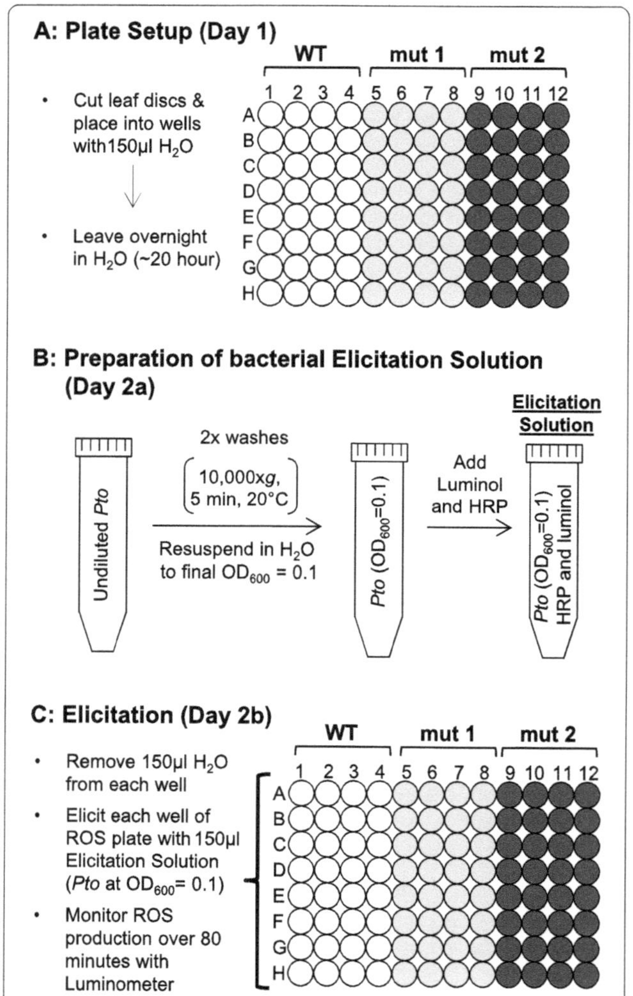

Figure 1 Outline for plate-setup and preparation of bacterial elicitation solution to measure luminol-based ROS production in leaf disc halves of *Arabidopsis* wildtype (Col-0) and mutant lines in response to live bacterial *Pto* cells. A: Preparation of a 96-well assay plate containing leaf disc halves of wildtype (WT, white) and mutant *Arabidopsis* lines (mut 1, gray and mut 2, black). **B**: Preparation of Elicitation Solution containing live *Pto* cells. **C**: Elicitation of wild-type and mutant tissue with Elicitation Solution to measure ROS response. WT, wild-type; mut, mutant; OD, optical density.

First, we established that virulent live *Pto* DC3000 cells induced ROS-production in wild-type *Arabidopsis* Col-0 leaves. As shown in Figure 2A, ROS production after elicitation with *Pto* DC3000 (OD_{600} = 0.1) was transient, in that increases in ROS was first detected around 20 minutes, peaking around 35-40 minutes and returning to near-basal levels around 70 minutes post-elicitation. Thus, it was delayed compared to that of PAMP peptide flg22-induced ROS which can be detected within 2-4 minutes, peaks at 10-14 minutes and returns to basal levels around 30-35 minutes [Additional file 1]. An explanation for the delayed ROS response to *Pto* DC3000 cells may be that in contrast to pure and short PAMP peptide(s), the PAMP(s) derived from living bacterial cells may not be readily accessible but may need to be released and/or processed for perception by the plant PRR(s) as suggested for flagellin-FLS2 system [40]. The reduced amplitude in response to *Pto* DC3000 cells may be an indication that living bacteria possibly release lower concentrations of PAMP(s) into the host apoplast compared to the concentration of applied synthesized flg22 peptide.

It is also possible that the reduced *Pto* DC3000-induced ROS amplitude may be due to suppression of ROS by bacterial effectors delivered into host cells. To address this hypothesis, we measured ROS production in response to non-pathogenic *Pto hrcC⁻* cells (OD_{600} = 0.1) that lacked functional T3SS and are defective in effector delivery [Figure 2B]. Importantly, ROS induced by *Pto hrcC⁻* was similar in the time of initiation, amplitude and duration compared to that by virulent *Pto DC3000* [Figure 2B; also see Figure 2C]. ROS production at their peaks (at 35 minutes) and total ROS productions over the 80 minute time-course did not show any statistically significant difference between these two *Pto* strains [Figure 2C], indicating that this *Pto*-elicited ROS production is independent of a functional T3SS. These results also suggest that compared to PAMP-induced ROS, the lower level ROS amplitude in response to *Pto* cells is unlikely due to interference of ROS by bacterial effectors. Similar ROS results were obtained over 75 min time-course after treatment with *Pto avrRpm1*

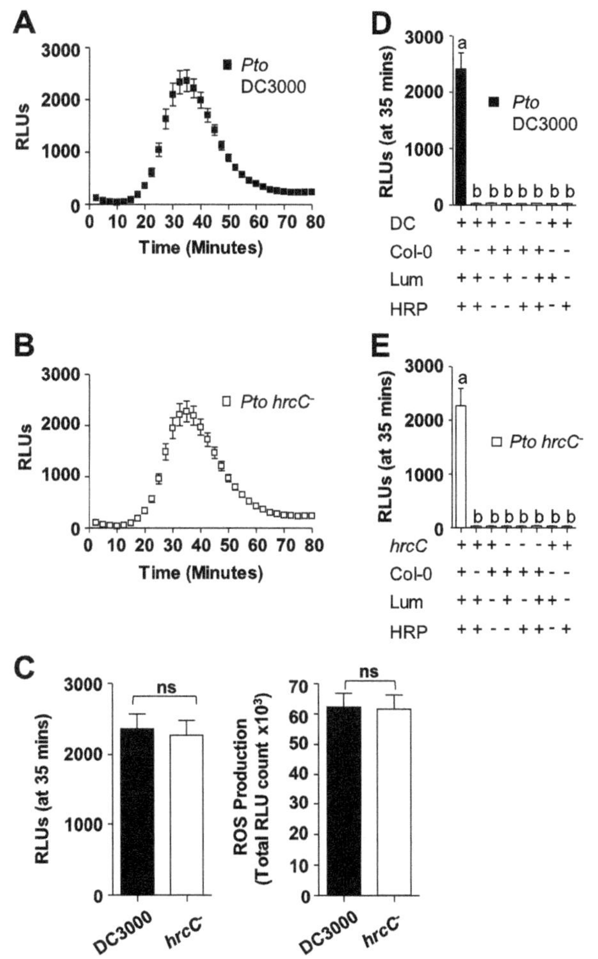

Figure 2 Transient ROS production in response to live *Pto* DC3000 and *Pto hrcC*⁻ in wildtype *Arabidopsis* Col-0. A: Time-course of ROS production in response to *Pto* DC3000 (n = 48/treatment). **B**: Time-course of ROS production in response to *Pto hrcC*⁻ (n = 48/treatment). **C**: ROS production induced by *Pto* DC3000 (black bar) and *Pto hrcC*⁻(white bar) shown at peak ROS production (35 minutes post-elicitation) and as total Relative Light Units (RLUs) over 80 minutes following elicitation with indicated bacteria (n = 48/treatment). **D**: Specificity of ROS production based on presence of *Arabidopsis* tissue and *Pto* DC3000 cells (n = 12/treatment). **E**: Specificity of ROS production based on presence of *Arabidopsis* tissue and *Pto hrcC*⁻ cells (n = 12/treatment). To allow direct comparisons, ROS experiments in **(A, B and C)**, in **D** or in **E** were performed in the same 96-well plate at the same time. Values are mean ± SE, means with different letters denote a significance difference while similar letters denote no significance (Two tailed student's *t*-test, P < 0.0001). For all experiments, bacterial solutions were used at OD_{600} = 0.1. Experiments were repeated more than 3 times with similar results. RLU, Relative Light Units; ns, no significance; DC, *Pto* DC3000; hrcC, *Pto hrcC*⁻; Lum, Luminol; HRP, Horseradish Peroxidase.

and *Pto avrRps4*, two avirulent bacterial strains known to inject the avirulence proteins AvrRpm1 and AvrRps4, respectively, into host cells resulting in ETI-dependent responses [5-8]. Comparing *Pto avrRpm1* and *Pto avrRps4* to *Pto* DC3000, no difference in ROS initiation, amplitude and attenuation as well as total ROS production was observed over the 80 min time-course [Additional file 2]. Taken together, our results using virulent, non-pathogenic and avirulent *Pto* strains are in agreement that the observed ROS response was due to PTI-dependent events and are consistent with *Pto* effector delivery into host cells occurring at significantly later times post-infection (> 3 hours) [41] than the ROS response investigated in our study.

In control experiments, ROS production was measured using Elicitation Solutions that contained or lacked *Pto* strain (*Pto*), luminol, HRP and Col-0 leaf discs (Col-0) in different combinations. As evident in Figure 2D and E, ROS was produced only when the *Pto* strain, luminol, HRP and the leaf discs were present. Lack of any of these components did not result in any significant ROS production. No difference in the ROS production was observed in *Pto* strains resuspended in dH_2O or 10 mM $MgCl_2$ [Additional file 3].

Next, we showed that *Pto*-dependent ROS production was dose-dependent by eliciting Col-0 leaf tissue with different bacterial cell densities between OD_{600} = 0.001 and 0.1 [Figure 3]. Either strain, *Pto* DC3000 or *Pto hrcC*⁻, elicited a ROS response in a dose-dependent manner [Figure 3A or B, respectively]. Consistent with Figure 2C, we did not observe any statistically significant difference when comparing ROS production elicited by *Pto* DC3000 and *Pto hrcC*⁻ at comparable optical densities [Figure 3C]. Furthermore, *Pto* optical densities of OD = 0.001 or below (data not shown) did not elicit any significant detectable ROS. *Pto* cells with an OD_{600} = 0.1 yielded the most consistent and robust ROS response, thus we used this bacterial cell density for all subsequent experiments.

To investigate the genetic requirement of the plant host for *Pto*-induced ROS, we made use of previously characterized *Arabidopsis* mutants functioning in or upstream of apoplastic ROS production. A *rbohD* null mutant plant line in Col-0 background [22,25] was used to investigate dependency on *RbohD*, the plasma membrane localized protein previously shown to be the sole NADPH oxidase required for rapid induction of apoplastic ROS after elicitation with PAMP peptides [22,24,26,33]. In response to either live *Pto* DC3000 or *hrcC*⁻ cells, no ROS production was detected in *rbohD* compared to wild-type Col-0 for the 80 min time-course [Figure 4A or B, respectively; see also Figure 4C for peak comparison at 35 min post-elicitation]. We conclude that similar to the synthetic flg22 peptide, *Pto* DC3000 or *hrcC*⁻ elicited ROS production was fully dependent on *RbohD*. These result also suggested that the ROS measured in our assays was unlikely due to cell death-associated ROS.

Next, we determined whether *Pto*-elicited ROS was dependent on *BAK1*, the receptor-like kinase known to be required very early in initiating signaling after flg22-

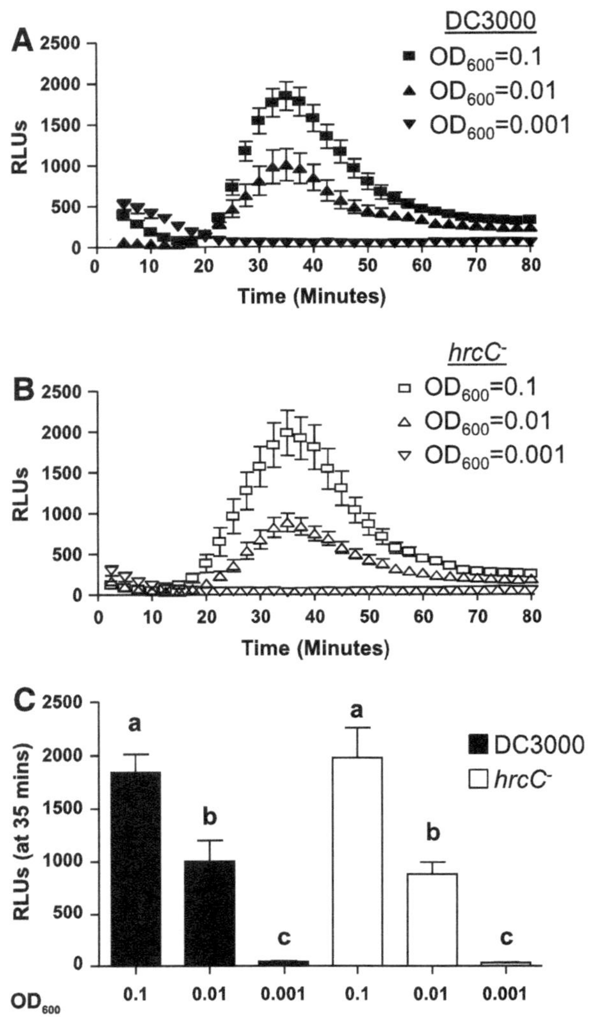

Figure 3 Dose-dependent ROS production in response to live *Pto* DC3000 and *Pto hrcC*⁻ in wildtype Col-0. A: Time-course of ROS production in response to *Pto* DC3000 cells with an OD_{600} of 0.001, 0.01 and 0.1. **B**: Time-course of ROS production in response to *Pto hrcC*⁻ cell with an OD_{600} of 0.001, 0.01 and 0.1. **C**: Comparison of peak ROS production for each concentration of *Pto* DC3000 (black bar) or *Pto hrcC*⁻ (white bar) shown at the point of maximum ROS production 35 minutes post-elicitation). For ROS production, Col-0 leaf disk halves were elicited with DC3000 (filled shape) or *hrcC*⁻ (open shapes) with the indicated bacterial cell density at 0 min (n = 16/treatment). To allow direct comparisons, all ROS experiments **(A-C)** were performed in the same 96-well plate at the same time. Values are mean ± SE, means with different letters denote a significance difference while similar letters denote no significance (Two tailed student's *t*-test, $P < 0.005$). Experiments were repeated more than 3 times with similar results. RLU, Relative Light Units; OD, Optical density.

elicitation [18,20,42]. Underlining its crucial role in PAMP-signaling and PTI, BAK1 forms a ligand-induced receptor complex with FLS2 within seconds [18,20,42,43]. In response to *Pto* DC3000 or *hrcC*⁻, we observed an increase in ROS production over time in *bak1-4* null mutant leaf discs; but importantly, the ROS amplitude was significantly reduced compared to Col-0 [Figure 4D or E, respectively]. No statistical significant difference in ROS production was observed in *bak1-4* mutant tissue in response to *Pto* DC3000 or *Pto hrcC*⁻ [Figure 4F]. Taken together, these results suggest that *Pto*-elicited ROS was only in part dependent on *BAK1*. Our studies are consistent with previous reports showing that full signaling responses to the bacterial PAMP peptides flg22 require other proteins in addition to BAK1 [18,20,42].

Previous studies have shown that *Arabidopsis* mutants lacking functional FLS2 receptor are unable to perceive flg22 which results in a complete loss of any flg22/FLS2-dependent signaling event including lack of apoplastic ROS production [16,31,36]. In addition, FLS2-dependent responses contribute to restriction of *Pto* DC3000 growth [31]. However when treated with crude boiled *Pto* DC3000 extracts, *fls2* null mutants are still able to elicit signaling responses including ROS production [29,31]. These results are consistent with the idea that in addition to flagellin (flg22), *Pto* DC3000 contains multiple extra- and intracellular PAMPs that are exposed upon boiling of bacterial strains. To determine whether FLS2 contributed to ROS production induced by intact living *Pto* DC3000 or *hrcC*⁻, we utilized previously characterized *fls2* null mutants [18,31,36]. Interestingly, *fls2* null mutants in a Col-0 background did not elicit any ROS in response to either *Pto* strain [Figure 5A-C]. To confirm this observation, we challenged Wassilewskija (Ws-0), an *Arabidopsis* ecotype that is flagellin-insensitive and considered a natural *fls2* mutant due to an early stop mutation in the *FLS2* gene [31], with *Pto* DC3000 or *Pto hrcC*⁻. Similar to the *fls2* null mutant in Col-0, Ws-0 displayed a complete lack of ROS production in response to either *Pto* strains [Additional file 4]. Based on our studies utilizing two different *Arabidopsis* ecotypes lacking functional FLS2, the flagellin/flg22 receptor FLS2 appeared to be the predominant host receptor responsible for initiating early ROS production in response to intact, living *Pto* DC3000 and *hrcC*⁻. Our results are consistent with observations that in response to intact *Pto* DC3000 and DC3118 at 1 to 2 hours post-treatment, stomatal closure of epidermal leaf peels is entirely dependent on the presence of functional FLS2 [33,34]. Thus, these and our present studies suggest that not all potential *Pto* PAMPs appear to be exposed simultaneously during the initial bacterial infection. Our results further underline the advantages of utilizing living intact *Pto* cells in contrast to crude boiled bacterial extract [14,29,31] when studying early signaling responses induced by bacteria *in planta*. Boiling of bacteria likely releases both extra- and intracellular localized PAMPs simultaneously. In contrast, utilizing intact live bacterial cells enables researchers to determine the biologically relevant contribution and/or order of PAMP recognition by specific host PRRs.

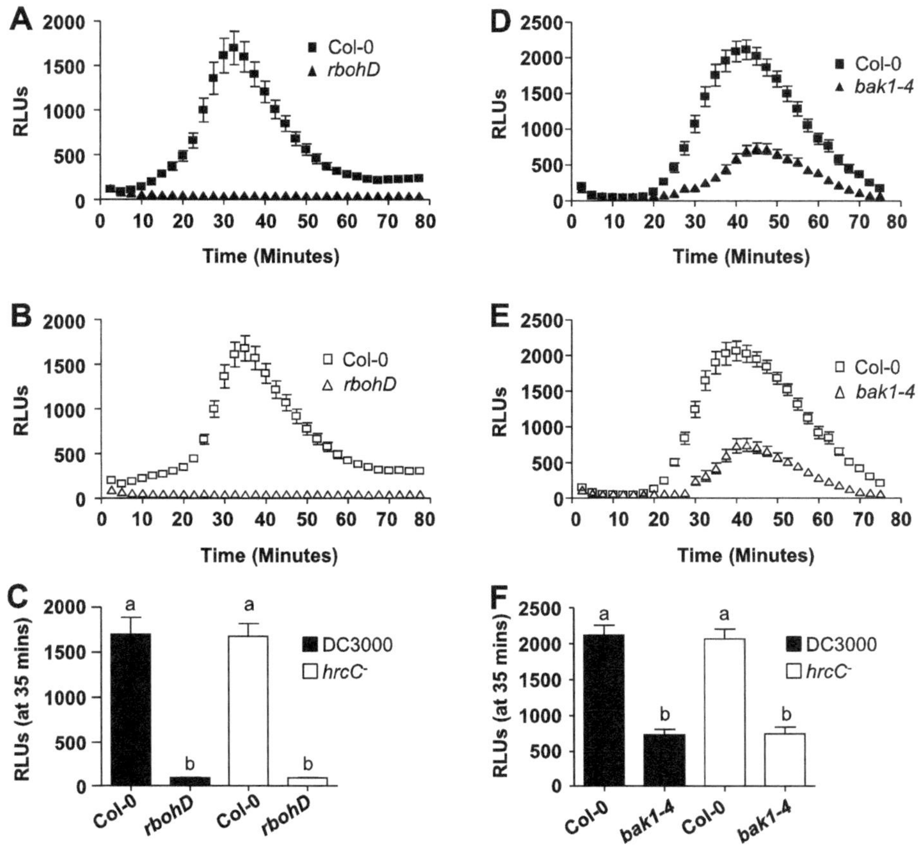

Figure 4 ***RbohD*** **and** ***BAK1*****-dependency of ROS production in response to live** ***Pto*** **DC3000 and** ***Pto hrcC$^-$*** **cells. A and B**: Time-course of ROS production in response to *Pto* DC3000 (**A**; closed symbols) or *Pto hrcC$^-$* (**B**; open sympols) in Col-0 (square) and *rbohD* null mutant (triangle) leaf disc halves (n = 24/treatment). **C**: Comparison of peak ROS production between *Pto* DC3000 or *Pto hrcC$^-$* in Col-0 and *rbohD* shown at the point of maximum ROS production (35 minutes post-elicitation from **A and B**). **D and E**: Time-course of ROS production in response to *Pto* DC3000 (**D**; closed symbols) or *Pto hrcC$^-$* (**E**; open sympols) in Col-0 (square) and *bak1-4* null mutant (triangle) leaf disc halves (n = 24/treatment). **F**: Comparison of peak ROS production between *Pto* DC3000 or *Pto hrcC$^-$* in Col-0 and *bak1-4* shown at the point of maximum ROS production (35 minutes post-elicitation from **D and E**). To allow for direct comparisons, all ROS experiments shown in **(A, B and C)** and **(D, E and F)** were performed in the same 96-well plate at the same time. Values are mean ± SE, means with different letters denote a significance difference while similar letters denote no significance (Two tailed student's *t*-test, $P < 0.0001$). For all experiments, bacterial solutions were used at $OD_{600} = 0.1$. Experiments were repeated more than 3 times with similar results. RLU, Relative Light Units.

Conclusions

In this study, we report the advancement of a rapid and convenient bioassay allowing the quantitative assessment of ROS production between *Arabidopsis* leaf tissue (the primary site of *Pto* infection) and living *Pto* bacterial strains. Because of the ease in setting up ROS assays in a 96-well plate assay, this quantitative analysis is highly suitable to screen large populations of *Arabidopsis* accession lines in combination with diverse *Pto* mutant strains to define the genetic requirements of host and bacterial pathogen. In future experiments, this ROS bioassay may also allow addressing which PAMP/PRR pair quantitatively contributes to early signaling in response to bacterial pathogens in leave tissue. Furthermore, the utility of this bioassay can be easily extended beyond *Arabidopsis* and *Pto* to diverse model or crop species and their cognate microbial pathogen to define components required for early ROS responses.

Methods

Plant material and growth conditions

Arabidopsis seeds were sterilized with 10% bleach + 1% Triton X-100 for 1 hour, rinsed with water and plated aseptically on 0.5% agar containing 2.14 g L^{-1} Murashige and Skoog (MS) salts (Sigma Chemical Company, St. Louis, MO, USA, http://www.sigmaaldrich.com/) + 1% sucrose, pH 5.7. Following stratification for 2 days at 4°C, seedlings were germinated in Percival CU-36 L4 growth chambers (Percival, Perry, IA) under continuous light at 22°C [18]. Seven day-old seedlings were transplanted in soil and grown in an 8-h light/16-h dark photoperiod at 82 μmol m^{-2} s^{-1}. Fully expanded rosette leaves were

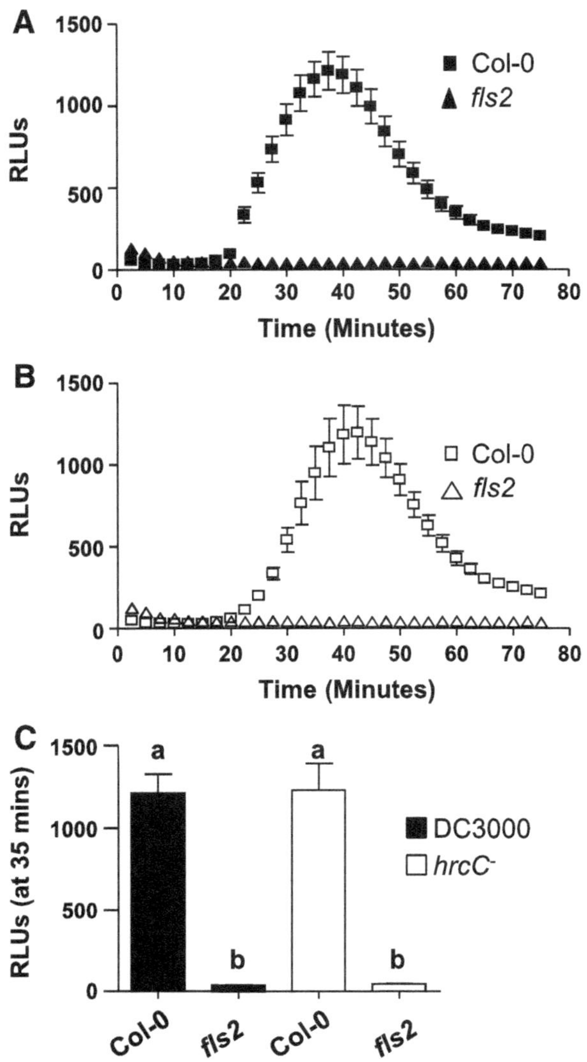

Figure 5 ***FLS2*-dependency of ROS production in response to live *Pto* DC3000 and *Pto hrcC*⁻ cells.** Time-course of ROS production in response to *Pto* DC3000 (**A**; closed symbols) or *Pto hrcC*⁻ (**B**; open sympols) in Col-0 (square) and *fls2* null mutant (triangle) leaf disc halves (n = 24/treatment). **C**: Comparison of peak ROS production between *Pto* DC3000 or *Pto hrcC*⁻ in Col-0 and *fls2* shown at the point of maximum ROS production (35 minutes post-elicitation from **A and B**). To allow for direct comparisons, all ROS experiments shown in **(A, B and C)** were performed in the same 96-well plate at the same time. Values are mean ± SE, means with different letters denote a significance difference while similar letters denote no significance (Two tailed student's *t*-test, P < 0.0001). For all experiments, bacterial solutions were used at $OD_{600} = 0.1$. Experiments were repeated more than 3 times with similar results. RLU, Relative Light Units.

used from 4-5 week old plants for all ROS experiments. The Ws-0, *fls2Δ* (Col-0), *bak1-4* (Col-0), and *rbohD* (Col-0) mutants have been previously described [18,22,25].

Chemicals

Synthetic flg22 peptide [21] was made by GenScript (Scotch Plains, NJ) and used at indicated concentrations. Horseradish Peroxidase (HRP; Sigma, catalog # P6782) was prepared as a 500x HRP stock solution by dissolving 10 mg/mL in sterile H_2O. Aliquots (10-30 μL) of the 500x HRP stock solution were stored at -20°C and used at a final concentration of 20 μg/mL. For the 500x Luminol stock solution, 17 mg Luminol (≥ 97% purity-HPLC; Sigma; catalog # A8511) was completely dissolved in 1 ml of 200 mM KOH and used at a final concentration of 0.2 μM. For example, 10 μl of each of the 500x stocks of luminol and HRP were added to 5 ml of resuspended bacterial solution. Because Luminol is light-sensitive, all solutions containing Luminol must be protected from light by wrapping tubes in aluminum foil. The 500x Luminol stock solution is made fresh prior to use and discarded daily.

Bacterial preparation

Two days prior to ROS experiments, *Pseudomonas syringae* pv. *tomato* strain DC3000 (*Pto* DC3000) or *Pto hrcC*⁻ was streaked from glycerol stocks (stored at −80°C) onto King's B medium (KBM) agar plates containing 50 μg ml^{-1} kanamycin and 30 μg ml^{-1}rifampicin (*Pto* DC3000) or 30 μg ml^{-1}rifampicin (*Pto hrcC*⁻) [38]. KMB plates containing bacteria were incubated for 36 to 48 hours at room temperature. Prior to elicitation, bacteria were scraped from plates and washed twice in sterile dH_2O by repeated centrifugation at 10,000×*g* for 5 minutes. After the second wash step, a 1:10 dilution of the bacterial solution was made and its optical density (OD_{600}) was measured using a spectrophotometer as a means to provide an approximate quantification of bacterial cell density [38,39]. The final bacterial elicitation solution was adjusted to an OD_{600} between 0.001 and 0.1, which under our conditions equated to 1×10^6 to 10^8 colony forming units (cfu)/mL based on serial dilution plating. The OD_{600} of the final bacterial elicitation solution was measured again to ensure accuracy of the dilution. To accurately compare the dose-dependency of the ROS response [Figure 3], bacteria were serially diluted from a stock solution.

Measurement of apoplastic ROS production

One day before the ROS assay, leaf disks (1.1 cm^2) from 4-5 week old plants were cut into two equal halves with a sharp razor blade to increase the cellular surface area exposed to elicitation solution, an important step for obtaining reproducible responses with less variability within and between experiment. Each leaf disc half was floated adaxial side up in an individual well of a 96-well microtiter plate (Costar; Fisher Scientific, catalog # 3912) containing 150 μl dH_2O and then incubated overnight at 22°C in continuous light for 20 to 24 hours to reduce the wounding response. Prior to elicitation, the Elicitation Solution was prepared containing bacteria, Luminol and

HRP. For a 10 ml Elicitation Solution, 20 μl of 500x HRP stock solution and 20 μl of the 500x Luminol stock solution is added to 10 ml of bacterial cells that have been already diluted to the desired concentration. For flg22-induced ROS production, flg22 peptide was used instead of bacteria in the Elicitation Solution. All Elicitation Solutions were kept at room temperature. Immediately prior to the elicitation, the incubating dH_2O solution was carefully removed from each well avoiding any tissue damage or desiccation. Then using a multichannel pipetman, 100 μl of the Elicitation Solution was quickly added to each well containing leaf disc half. For Luminol-based ROS production, the plate was placed without delay into a GloMax® 96-well microplate luminometer (Promega, Madison, USA) to measure *Pto*-induced ROS production between 0 and 80 minutes.

Statistical analysis

Each experiment was done at least 3 independent times with similar results. Statistical significances based on unpaired two sample *t*-test were determined with Graph Pad Prism4 software (La Jolla, CA).

Additional files

Additional file 1: Dose-dependent ROS production in response to synthetic flg22 peptide. A: Time-course of ROS production in response to 1, 10 and 100 nM flg22 in wild-type Col-0 leaf disc halves (n = 16/treatment). B: Bar graph representation of peak ROS production for each flg22 concentration from experiment shown in A. To allow direct comparisons, all ROS experiments were performed in the same 96-well plate at the same time. Values are mean ± SE, means with different letters denote a significance difference (Two tailed student's *t*-test, P < 0.0001). Experiment was repeated more than 3 times with similar results. RLU, Relative Light Units.

Additional file 2: Transient ROS production in response to live *Pto* DC3000, *Pto* DC3000 *avrRpm1* and *Pto* DC3000 *avrRps4* in Col-0. A: Time-course of ROS production between *Pto* DC3000 and *Pto* DC3000 *avrRpm1* (n = 24/treatment). B: Timecourse of ROS production between *Pto* DC3000 and *Pto* DC3000 *avrRps4* (n = 24/treatment). C: ROS production induced by *Pto* DC3000 (black bar), *Pto* DC3000 *avrRpm1* (white bar) and *Pto* DC3000 *avrRps4* (gray bar) shown at peak ROS production (35 minutes post-elicitation) and as total Relative Light Units (RLUs) over 75 minutes following elicitation with indicated bacteria (n = 24/treatment). To allow direct comparisons, ROS experiments in (A, B and C) were performed in the same 96-well plate at the same time. Values are mean ± SE, means with different letters denote a significance difference while similar letters denote no significance (Two tailed student's *t*-test, P ≥ 0.4084). For all experiments, bacterial solutions were used at OD600 = 0.1. Experiments were repeated more than 3 times with similar results. RLU, Relative Light Units.

Additional file 3: ROS production in response to live *Pto* DC3000 cells resuspended in dH_2O or 10 mM $MgCl_2$. A: Time-course of ROS production in response to *Pto* DC3000 resuspended in sterile dH_2O (closed symbols) or 10 mM $MgCl_2$ (open sympols) in Col-0 leaf disc halves (n = 24/treatment). B: Comparison of peak ROS production between *Pto* DC3000 resuspended in sterile dH_2O or 10 mM $MgCl_2$ at the point of maximum ROS production (35 minutes post-elicitation from A). To allow for direct comparisons, all ROS experiments shown in (A and B) were performed in the same 96-well plate at the same time. Values are mean ± SE, means with different letters denote a significance difference while similar letters denote no significance (Two tailed student's *t*-test, P = 0.59). For all experiments, bacterial solutions were used at OD600 = 0.1. Experiment was repeated more than 3 times with similar results. RLU, Relative Light Units.

Additional file 4: Lack of ROS production in *fls2* mutant of different *Arabidopsis* ecotypes in response to live *Pto* DC3000 and *Pto hrcC⁻* cells. A and B: Time-course of ROS production in response to *Pto* DC3000 (A; closed symbols) or *Pto hrcC⁻* (B; open sympols) in leaf disc halves of Col-0 (square), *fls2* (Col-0) null mutant (triangle) and Ws-0 (upside down triangle) (n = 16/treatment). C: Comparison of peak ROS production between *Pto* DC3000 or *Pto hrcC⁻* in Col-0, *fls2* (Col-0) and Ws-0 shown at the point of maximum ROS production (35 minutes post-elicitation from A and B). To allow for direct comparisons, all ROS experiments shown in (A, B and C) were performed in the same 96-well plate at the same time. Values are mean ± SE, means with different letters denote a significance difference while similar letters denote no significance (Two tailed student's *t*-test, P < 0.0001). For all experiments, bacterial solutions were used at OD600 = 0.1. Experiments were repeated more than 3 times with similar results. RLU, Relative Light Units.

Competing interests

The authors declare that they do not have any financial or other competing interests.

Authors' contributions

JMS performed experiments and made figures. JMS and AH designed experiments, analyzed and interpreted data, and wrote the manuscript. Both authors read and approved the final manuscript.

Acknowledgements

The authors thank current and former Heese lab members and Drs. Jeff Anderson and Walter Gassmann (University of Missouri, MU) for discussion. Funding was provided by MU start-up funds (AH) and the Millikan Graduate Fellowship, MU-Division of Plant Sciences (JMS).

Author details

[1]Division of Biochemistry, University of Missouri-Columbia, 117 Schweitzer Hall, Columbia, MO 65211, USA. [2]Division of Plant Sciences, University of Missouri-Columbia, Columbia, MO 65211, USA. [3]Interdisciplinary Plant Group (IPG), University of Missouri-Columbia, Columbia, MO 65211, USA.

References

1. Dodds PN, Rathjen JP: **Plant immunity: towards an integrated view of plant–pathogen interactions.** *Nat Rev Genet* 2010, **11**:539–548.
2. Boller T, Felix G: **A renaissance of elicitors: perception of microbe-associated molecular patterns and danger signals by pattern-recognition receptors.** *Annu Rev Plant Biol* 2009, **60**:379–406.
3. Monaghan J, Zipfel C: **Plant pattern recognition receptor complexes at the plasma membrane.** *Curr Opin Plant Biol* 2012, **15**:349–357.
4. Boller T, He SY: **Innate immunity in plants: an arms race between pattern recognition receptors in plants and effectors in microbial pathogens.** *Science* 2009, **324**:742–744.
5. Lindeberg M, Cunnac S, Collmer A: **Pseudomonas syringae type III effector repertoires: last words in endless arguments.** *Trends Microbiol* 2012, **20**:199–208.
6. Xin X-F, He SY: **Pseudomonas syringae pv. Tomato DC3000: a model pathogen for probing disease susceptibility and hormone signaling in plants.** *Annu Rev Phytopathol* 2013, **51**:473–498.
7. Munkvold KR, Martin GB: **Advances in experimental methods for the elucidation of Pseudomonas syringae effector function with a focus on AvrPtoB.** *Mol Plant Pathol* 2009, **10**:777–793.
8. Mansfield JW: **From bacterial avirulence genes to effector functions via the hrp delivery system: an overview of 25 years of progress in our understanding of plant innate immunity.** *Mol Plant Pathol* 2009, **10**:721–734.
9. Newman M-A, Sundelin T, Nielsen JT, Erbs G: **MAMP (Microbe-Associated Molecular Pattern) triggered immunity in plants.** *Front Plant Sci* 2013, **4**:139.
10. Gust AA, Willmann R, Desaki Y, Grabherr HM, Nürnberger T: **Plant LysM proteins: modules mediating symbiosis and immunity.** *Trends Plant Sci* 2012, **17**:495–502.

11. Robatzek S, Wirthmueller L: **Mapping FLS2 function to structure: LRRs, kinase and its working bits.** *Protoplasma* 2012, **250:**671–681.
12. Bardy SL, Ng SYM, Jarrell KF: **Prokaryotic motility structures.** *Microbiology* 2003, **149:**295–304.
13. Clarke CR, Chinchilla D, Hind SR, Taguchi F, Miki R, Ichinose Y, Martin GB, Leman S, Felix G, Vinatzer BA: **Allelic variation in two distinct Pseudomonas syringae flagellin epitopes modulates the strength of plant immune responses but not bacterial motility.** *New Phytol* 2013, **200:**847–860.
14. Felix G, Duran JD, Volko S, Boller T: **Plants have a sensitive perception system for the most conserved domain of bacterial flagellin.** *Plant J* 1999, **18:**265–276.
15. Sun W, Cao Y, Jansen Labby K, Bittel P, Boller T, Bent AF: **Probing the Arabidopsis flagellin receptor: FLS2-FLS2 association and the contributions of specific domains to signaling function.** *Plant Cell Online* 2012, **24:**1096–1113.
16. Gómez-Gómez L, Boller T: **FLS2: an LRR receptor-like kinase involved in the perception of the bacterial elicitor flagellin in arabidopsis.** *Mol Cell* 2000, **5:**1003–1011.
17. Chinchilla D, Bauer Z, Regenass M, Boller T, Felix G: **The Arabidopsis receptor kinase FLS2 binds flg22 and determines the specificity of flagellin perception.** *Plant Cell* 2006, **18:**465–476.
18. Heese A, Hann DR, Gimenez-Ibanez S, Jones AM, He K, Li J, Schroeder JI, Peck SC, Rathjen JP: **The receptor-like kinase SERK3/BAK1 is a central regulator of innate immunity in plants.** *Proc Natl Acad Sci U S A* 2007, **104:**12217–12222.
19. Boutrot F, Segonzac C, Chang KN, Qiao H, Ecker JR, Zipfel C, Rathjen JP: **Direct transcriptional control of the Arabidopsis immune receptor FLS2 by the ethylene-dependent transcription factors EIN3 and EIL1.** *Proc Natl Acad Sci* 2010, **107:**14502–14507.
20. Chinchilla D, Zipfel C, Robatzek S, Kemmerling B, Nurnberger T, Jones JD, Felix G, Boller T: **A flagellin-induced complex of the receptor FLS2 and BAK1 initiates plant defence.** *Nature* 2007, **448:**497–500.
21. Korasick DA, McMichael C, Walker KA, Anderson JC, Bednarek SY, Heese A: **Novel functions of Stomatal Cytokinesis-Defective 1 (SCD1) in innate immune responses against bacteria.** *J Biol Chem* 2010, **285:**23342–23350.
22. Nühse TS, Bottrill AR, Jones AME, Peck SC: **Quantitative phosphoproteomic analysis of plasma membrane proteins reveals regulatory mechanisms of plant innate immune responses.** *Plant J* 2007, **51:**931–940.
23. Ranf S, Eschen-Lippold L, Pecher P, Lee J, Scheel D: **Interplay between calcium signalling and early signalling elements during defence responses to microbe- or damage-associated molecular patterns.** *Plant J* 2011, **68:**100–113.
24. Zhang J, Shao F, Li Y, Cui H, Chen L, Li H, Zou Y, Long C, Lan L, Chai J, Zhang J, Shao F, Li Y, Cui H, Chen L, Li H, Zou Y, Long C, Lan L, Chai J, Chen S, Tang X, Zhou JM: **A pseudomonas syringae effector inactivates MAPKs to suppress PAMP-induced immunity in plants.** *Cell Host Microbe* 2007, **1:**175–185.
25. Torres MA, Dangl JL, Jones JDG: **Arabidopsis gp91phox homologues AtrbohD and AtrbohF are required for accumulation of reactive oxygen intermediates in the plant defense response.** *Proc Natl Acad Sci U S A* 2002, **99:**517–522.
26. Torres MA: **ROS in biotic interactions.** *Physiol Plant* 2010, **138:**414–429.
27. Lamb C, Dixon RA: **The oxidative burst in plant disease resistance.** *Annu Rev Plant Physiol Plant Mol Biol* 1997, **48:**251–275.
28. Nicaise V, Roux M, Zipfel C: **Recent advances in PAMP-triggered immunity against bacteria: pattern recognition receptors watch over and raise the alarm.** *Plant Physiol* 2009, **150:**1638–1647.
29. Gimenez-Ibanez S, Ntoukakis V, Rathjen JP: **The LysM receptor kinase CERK1 mediates bacterial perception in Arabidopsis.** *Plant Signal Behav* 2009, **4:**539–541.
30. Kunze G, Zipfel C, Robatzek S, Niehaus K, Boller T, Felix G: **The N terminus of bacterial elongation factor Tu elicits innate immunity in Arabidopsis plants.** *Plant Cell* 2004, **16:**3496–3507.
31. Zipfel C, Robatzek S, Navarro L, Oakeley EJ, Jones JDG, Felix G, Boller T: **Bacterial disease resistance in Arabidopsis through flagellin perception.** *Nature* 2004, **428:**764–767.
32. de la Torre F, Gutiérrez-Beltrán E, Pareja-Jaime Y, Chakravarthy S, Martin GB, del Pozo O: **The tomato calcium sensor Cbl10 and its interacting protein kinase Cipk6 define a signaling pathway in plant immunity.** *Plant Cell Online* 2013, **25:**2748–2764.
33. Mersmann S, Bourdais G, Rietz S, Robatzek S: **Ethylene signaling regulates accumulation of the FLS2 receptor and is required for the oxidative burst contributing to plant immunity.** *Plant Physiol* 2010, **154:**391–400.
34. Zeng W, He SY: **A prominent role of the flagellin receptor FLAGELLIN-SENSING2 in mediating stomatal response to pseudomonas syringae pv tomato DC3000 in Arabidopsis.** *Plant Physiol* 2010, **153:**1188–1198.
35. Zhang J, Li W, Xiang T, Liu Z, Laluk K, Ding X, Zou Y, Gao M, Zhang X, Chen S, Mengiste T, Zhang Y, Zhou JM: **Receptor-like cytoplasmic kinases integrate signaling from multiple plant immune receptors and Are targeted by a Pseudomonas syringae effector.** *Cell Host Microbe* 2010, **7:**290–301.
36. Gómez-Gómez L, Felix G, Boller T: **A single locus determines sensitivity to bacterial flagellin in Arabidopsis thaliana.** *Plant J* 1999, **18:**277–284.
37. Smith JM, Salamango DJ, Leslie ME, Collins CA, Heese A: **Sensitivity to Flg22 is modulated by ligand-induced degradation and de novo synthesis of the endogenous flagellin-receptor FLAGELLIN-SENSING2.** *Plant Physiol* 2014, **164:**440–454.
38. Katagiri F, Thilmony R, He SY: **The Arabidopsis thaliana-Pseudomonas syringae interaction.** In *The Arabidopsis Book.* doi/10.1199/tab.0039.
39. Ishiga Y, Ishiga T, Uppalapati S, Mysore K: **Arabidopsis seedling flood-inoculation technique: a rapid and reliable assay for studying plant-bacterial interactions.** *Plant Methods* 2011, **7:**32.
40. Albert M, Jehle AK, Mueller K, Eisele C, Lipschis M, Felix G: **Arabidopsis thaliana pattern recognition receptors for bacterial elongation factor Tu and flagellin can be combined to form functional chimeric receptors.** *J Biol Chem* 2010, **285:**19035–19042.
41. Schechter LM, Roberts KA, Jamir Y, Alfano JR, Collmer A: **Pseudomonas syringae type III secretion system targeting signals and novel effectors studied with a Cya translocation reporter.** *J Bacteriol* 2004, **186:**543–555.
42. Roux M, Schwessinger B, Albrecht C, Chinchilla D, Jones A, Holton N, Malinovsky FG, Tör M, de Vries S, Zipfel C: **The arabidopsis leucine-rich repeat receptor–like kinases BAK1/SERK3 and BKK1/SERK4 Are required for innate immunity to hemibiotrophic and biotrophic pathogens.** *Plant Cell Online* 2011, **23:**2440–2455.
43. Schulze B, Mentzel T, Jehle A, Mueller K, Beeler S, Boller T, Felix G, Chinchilla D: **Rapid heteromerization and phosphorylation of ligand-activated plant transmembrane receptors and their associated kinase BAK1.** *J Biol Chem* 2010, **285:**9444–9451.

Newly developed quantitative transactivation system shows difference in activation by *Vitis* CBF transcription factors on DRE/CRT elements

Annette Nassuth*, Mahbuba Siddiqua, Huogen Xiao, Michelle A Moody and Chevonne E Carlow

Abstract

Background: Agroinfiltration-based transactivation systems can determine if a protein functions as a transcription factor, and via which promoter element. However, this activation is not always a yes or no proposition. Normalization for variation in plasmid delivery into plant cells, sample collection and protein extraction is desired to allow for a quantitative comparison between transcription factors or promoter elements.

Results: We developed new effector and reporter plasmids which carry additional reporter genes, as well as a procedure to assay all three reporter enzymes from a single extract. The applicability of these plasmids was demonstrated with the analysis of CBF transcription factors and their target promoter sequence, DRE/CRT. Changes in the core DRE/CRT sequence abolished activation by *Vitis* CBF1 or *Vitis* CBF4, whereas changes in the surrounding sequence lowered activation by *Vitis* CBF1 but much less so for *Vitis* CBF4. The system also detected a reduction in activation due to one amino acid change in *Vitis* CBF1.

Conclusions: The newly developed effector and reporter plasmids improve the ability to quantitatively compare the activation on two different promoter elements by the same transcription factor, or between two different transcription factors on the same promoter element. The quantitative difference in activation by VrCBF1 and VrCBF4 on various DRE/CRT elements support the hypothesis that these transcription factors have unique roles in the cold acclimation process.

Keywords: CBF, DRE/CRT, Agroinfiltration normalization, Transactivation effector and reporter plasmids

Background

Transient transactivation systems have been developed to evaluate the activation of different promoters by transcription factors. They have been used successfully to analyze relative promoter strengths [1,2], which are reportedly similar to those in transgenic systems [3]. Transient expression systems are preferred for the analysis of the sequence targeted by a transcription factor because genes that are directly activated by the transcription factor will produce transcripts within the time period between infiltration and harvest, whereas genes that are indirectly activated will take a longer time. Synthetic promoters containing (multiple copies of) defined regulatory elements are often used to avoid complications due to a combinatorial effect of various elements in a natural promoter, and have allowed the confirmation of cis-regulatory elements important for promoter activation by pathogens [4,5].

DNA from promoter- and transcription factor-plasmids of interest have been introduced into plant cells by a wide variety of means but the most successful methods involve electroporation or PEG treatment for introduction into protoplasts [6-8], and particle bombardment [9,10] or agroinfiltration for introduction into plant tissues [11,12]. Studies on grape genes have employed particle bombardment of grapevine callus to investigate transactivation by transcription factors [13] and, with varying success, vacuum infiltration of grapevine leaves from *in vitro* grown plants to investigate gene function in fungal defense [14,15] or agroinfiltration to study subcellular localization [16] or silencing constructs [17].

The CBF pathway in plants ultimately results in the expression of cold regulated (*COR*) genes which encode proteins that are thought to help the plant survive frost

* Correspondence: anassuth@uoguelph.ca
Department of Molecular and Cellular Biology, University of Guelph, Guelph, Ontario N1G 2 W1, Canada

[18,19]. The name of the pathway derives from CBFs (CRT binding factors), the transcription factors initially discovered in Arabidopsis to be directly responsible for the activation of many *COR* genes at low temperatures by binding to CRT (defined as GCCGAC) elements in their promoters [20,21]. The same proteins were also discovered as DRE-binding transcription factors 1 (DREB1s), reported to bind to drought responsive elements (DRE; defined as TACCGACAT) [22,23]. As a result reference is often made to CBF/DREB1 factors (AtCBF1/DREB1B, AtCBF2/DREB1A, AtCBF3/DREB1C) that bind to CRT/DRE elements with the sequence A/GCCGAC [18]. CBF/DREB1 proteins have now been reported for a wide variety of plants [19,24] and these appear to also bind and activate via the CRT/DRE sequence. However, not all CBF proteins have the same affinity and specificity for a certain CRT sequence. For example, the *Brassica napus* BNCBF17 has a lower sequence binding specificity than BNCBF5 [25] whereas the barley HvCBF1 has a binding preference for an element, TTGCCGACAT, containing the GCCGAC (CRT) core sequence over a sequence with the ACCGAC (DRE) core [26]. The results with Chrysanthemum *DREB1A*- or *DREB1B*-overexpressing *Arabidopsis* showed that these CBFs activate different, overlapping regulons, which is in agreement with preferences of these CBF-like proteins for different promoter elements [27]. Also analysis of the promoters from genes that were induced in AtCBF-overexpressing Arabidopsis revealed that variations in the sequence surrounding the CRT element might affect activation by various CBFs [28,29]. Together these results suggest that different CBF paralogs in a plant, and possibly orthologs from different species, have unique preferences for CRT-like sequences but more research is needed to investigate this further. Our lab successfully applied agroinfiltration of tobacco leaves to show that CRT promoter elements are required for regulation of gene expression by grape CBF transcription factors [30,31]. The results also suggested that CBF4 activates better than CBF1 however our analyses did not consider differences in infiltration and extraction that might occur between separate events.

The goal of the present study was to introduce an optimized dual luciferase reporter assay system that allows a better quantitative comparison of gene expression between different combinations of transcription factors (TFs) and promoter elements. The resulting system was used to analyze the activation by grape CBF1 and CBF4 on artificial promoters containing variations of the CRT sequence, and to compare the activation by CBF1 of the wild grape *Vitis riparia* (VrCBF1) and a VrCBF1 with one amino acid mutated into the amino acid present in the CBF1 of the more freezing sensitive winegrape *V. vinifera*.

Results

Development of a quantitative dual luciferase transactivation system

Effector plasmids were prepared starting from pCAMBIA 1305.1 (Figure 1A). For the VrCBF4 effector plasmid (Figure 1B), this involved adding a 35S::VrCBF4-nos terminator cassette into the multiple cloning site (MCS). The VrCBF4 open reading frame (ORF) was replaced by the VrCBF1 ORF for the VrCBF1 effector plasmid. To prepare the reporter plasmid, the VrCBF4 effector plasmid was altered in several aspects. The GUSPlus ORF was replaced by the firefly luciferase (FiLUC) ORF, the VrCBF4 ORF was replaced by the renilla luciferase (RiLUC) ORF, and its 35S promoter was replaced by a 4XCRTmin35S promoter (Figure 1C). As a result, the GUS activity can be taken as an indicator of the amount of effector plasmid, and the FiLUC activity as an indicator of the amount of reporter plasmid.

Various assay conditions were tested for compatibility with an analysis of the activities of the beta-glucuronidase, renilla luciferase and firefly luciferase reporter enzymes in a single extract. It was determined that extracts prepared in CCLR and diluted in PLB buffer (both from Promega) could be used for either glucuronidase or luciferase assays. Dilutions between 75 and 100x gave values below the maximum value for the fluorescence reader and were in a linear range. This indicates that all reporter enzyme activity values could be determined from the same extract.

The GUS/protein and FiLUC/protein values were expected to be the same for each effector/reporter infiltration treatment if the amount of DNA taken up and expressed in leaves infiltrated with VrCBF4 effector and reporter plasmid-containing agrobacteria (reporters containing the M1, M2, M4 and M5 variants of the DRE/CRT element, for details on these variants see a later section), was similar. The results show that both GUS/protein and FiLUC/protein values vary (Figure 2A), indicating that the amount of DNA transferred and expressed in the plant cells after the different infiltrations varied for both effector (GUS) and reporter (FiLUC) plasmid. Also the FiLUC/GUS values vary (Figure 2B), indicating that the ratio between effector DNA and reporter DNA uptake and expression varies. This means that the amounts of CBF protein and reporter gene in the cells might vary independently from each other, and this can cause variations in activation between separate infiltrations. The RiLUC/FiLUC ratio can therefore not be taken as a true measure of transactivation (Figure 2C). Instead it is better to normalize for the amount and expression of effector plasmid (represented by GUS) as well and therefore the RiLUC/FiLUC/GUS was taken as a measure of transactivation (Figure 2D).

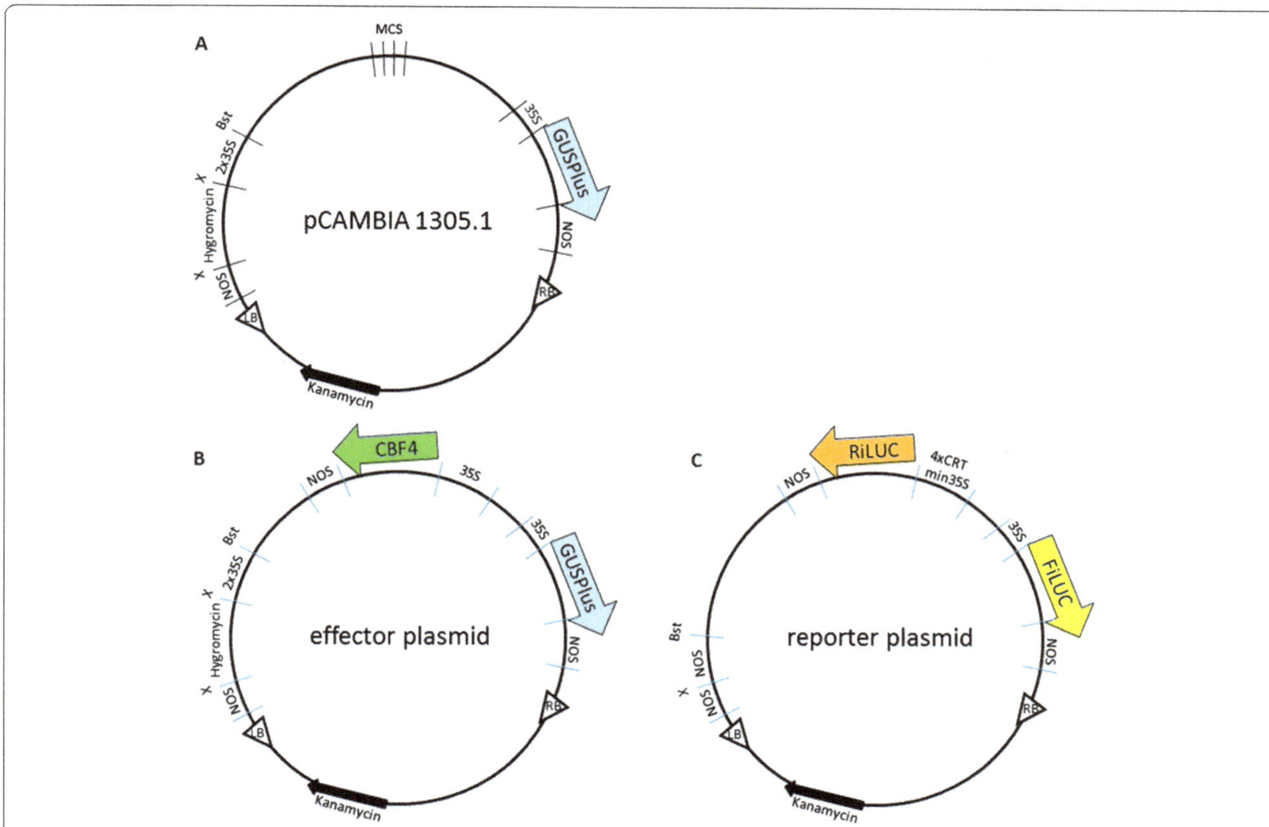

Figure 1 Schematic representation of [A] pCAMBIA1305.1, [B] GUS effector and [C] dual luciferase reporter plasmids. MCS = multiple cloning site. See Materials and Methods text for further details.

VrCBF1 and VrCBF4 require the conserved DRE/CRT core sequence to transactivate

To determine whether the transactivation by VrCBF1 and VrCBF4 has strict requirements regarding the TACCGACAT sequence present in the RiLUC reporter gene promoter, various mutations were made in this sequence and tested for transactivation. Examination of the results with the regular CRT sequence (GCCGAC = M2) showed that both VrCBF1 and VrCBF4 gave higher RiLUC/FiLUC/GUS values, respectively about 6 and 14 times higher, than the values obtained in the absence of a CBF (Figure 3). Mutations in the CCGAC core DRE/CRT sequence significantly reduced the activation by VrCBF1 and VrCBF4 to values that were not statistically different from those obtained without CBF.

The DRE sequence is ACCGAC, whereas the CRT sequence has a G instead of an A to give GCCGAC [20-23]. The question that was posed is whether the different *Vitis* CBFs have different affinities for these 2 sequences, and whether another change of the initial nucleotide has an effect. Figure 4 shows that VrCBF1 and VrCBF4 activated reporters with either TACCGACAT (M1) or TGCCGACAT (M2) elements to similar levels, but when the first nucleotide (A or G) is mutated to C (M10:CCCGAC) or T (M11:TCCGAC) the RiLUC/FiLUC/GUS values drop to control levels. This result, and also the results from Figures 3 and 5 support our previous suggestion that CBF4 activates better than CBF1 [31]. Five independent replicates of this experiment showed a higher activation by VrCBF1 on CRT (M2) over that on DRE (M1), and this difference was significant in three experiments (see also Figure 5).

Nucleotides flanking DRE sequence also affect activation levels

Nucleotides around the DRE sequence were changed to determine if such changes affect transactivation levels. These sequence variants were made to reflect CRT/DRE elements in Arabidopsis genes that had been reported targets of DREBs. These included elements found in genes as reported by Seki and colleagues [28], namely *RD29A/COR78*, (M1: TACCGACAT), *RD17/COR47* (M3: GACCGACAT) and *RD17/COR47* (M4: TACCGACTT). Two other chosen variants were based on the frequency logo determined for cold responsive genes, as reported by Wang and colleagues [32] (M5: GACCGACAA) or drought responsive genes (M6: GACCGACTC). In the

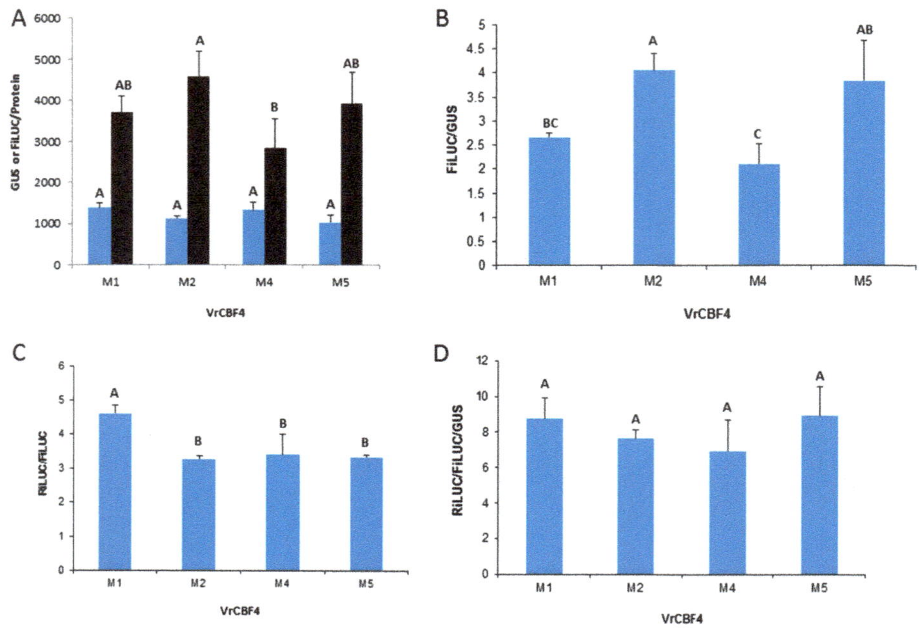

Figure 2 Analysis of reporter gene expression after agroinfiltration of tobacco leaves with VrCBF4 effector in combination with four different 4XCRTmin35S reporter constructs. [A] GUS/protein (blue bars) and FiLUC/protein (black bars), **[B]** FiLUC/GUS, **[C]** RiLUC/FiLUC, **[D]** RiLUC/FiLUC/GUS. Shown are the averages of three technical replicates and their standard deviation. Different letters indicate statistically significant differences (ANOVA $p < 0.05$).

absence of CBF effector, M1 to M6 gave variable activation values which were higher than background activation values with M7, the negative control CRT variant (Figure 5). Even higher activation values were obtained with VrCBF1 or VrCBF4 on all CRT variants except for the negative control M7. Both VrCBF1 and VrCBF4 activation on sequence variant M2 was among the highest whereas activation on sequence variant M4 was the lowest in all experiments and this was significant in two out of three independent experiments (Figure 5). However, activation by VrCBF1 was more affected by nucleotide changes around the DRE/CRT sequence than activation by VrCBF4 which resulted, generally speaking, in higher induction of transcription by VrCBF4 compared to VrCBF1 irrespective of the DRE/CRT variant present in the reporter plasmid.

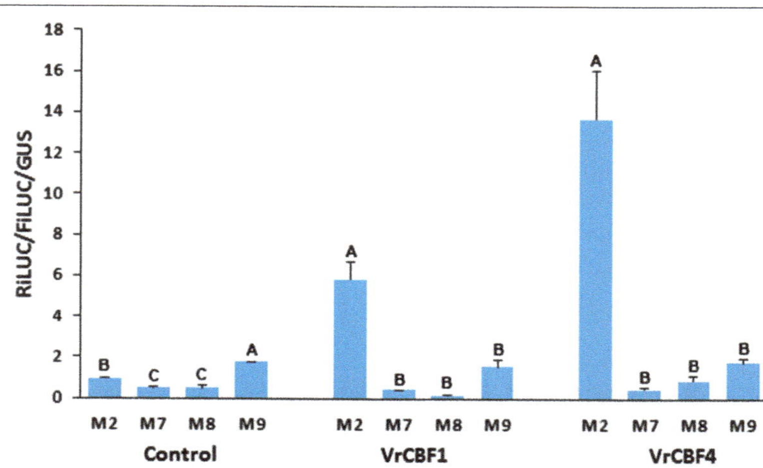

Figure 3 Activation by VrCBF1 or VrCBF4 on CRT (M2) and mutated core CRT sequence. M2: TGC*CGAC*AT, M7: TGAAGACAT, M8: TGCCGCCAT, M9: TGCCGAAAT. Error bars represent the standard deviation. Infiltrations without CBF effector were included as control. Statistical analysis was performed on the set of no CBF, VrCBF1 and VrCBF4 data separately and significantly different activation values (ANOVA $p < 0.05$) are indicated by different letters. Similar results were obtained for two other independent experiments.

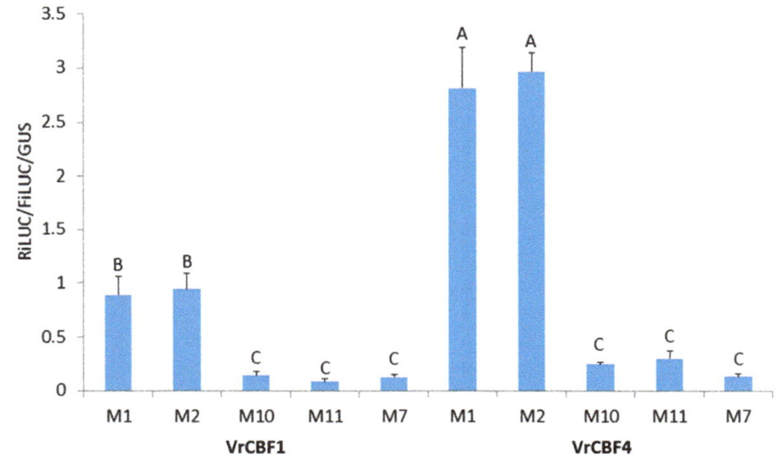

Figure 4 Activation by VrCBF1 or VrCBF4 on a reporter with various initial nucleotides in the DRE/CRT sequence. M1 = DRE: TA*CCGAC*AT, M2 = CRT: TG*CCGAC*AT, M10: TC*CCGAC*AT, M11: TT*CCGAC*AT (core CRT/DE sequence in italics), and the negative control CRT variant M7: TGAAGACAT Statistical analysis was performed on the set of no CBF, VrCBF1 and VrCBF4 data separately and significantly different activation values (ANOVA $p < 0.05$) are indicated by different letters. Error bars represent the standard deviation. Similar results were obtained in three other independent experiments.

Change in CBF amino acid sequence affects activation levels

Based on the higher transactivation by CBF1 from *V. riparia* compared to activation by CBF1 from *V. vinifera* in our previous transactivation system [30], we predicted that a change of the glutamic acid (E) in the AP2 DNA binding domain of VrCBF1 at position 85 to a lysine (K), as is present in VvCBF1, would decrease the transactivation. Therefore, transactivation by VrCBF1 was compared to that by the mutant VrCBF1-E85K in the newly developed dual luciferase transactivation system, on the two DRE/CRT variants that had given the highest values (M2 and M5, see Figure 5). An empty control effector plasmid was included to confirm that the activations observed with either CBF1 or CBF4 are higher than those observed in the presence of endogenous tobacco transcription factor only (control). The results showed that the mutant VrCBF1. E85K indeed had a lower activation of the reporter with either the M2 or M5 variants (4.0x or 1.1x), as compared to the wild type VrCBF1 (5.0x or 2.2x) (Figure 6). This difference in activation was observed in three independent experiments and was statistically significant in two of these.

Discussion

New effector and reporter plasmids were developed for transactivation analyses in plant tissues. The advantages of this new transactivation system are: (1) Two reporter genes, GUSPlus and FiLUC, were included in the vector

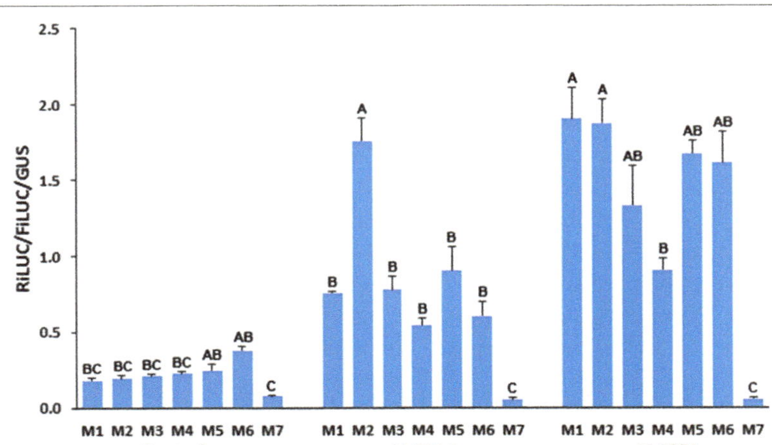

Figure 5 Activation by VrCBF1 and VrCBF4 on promoters with nucleotide variations around the DRE sequence (in italics). M1 = DRE: TA*CCGAC*AT, M2 = CRT: TG*CCGAC*AT, M3: GA*CCGAC*AT, M4: TA*CCGAC*TT, M5: GA*CCGAC*AA, M6: GA*CCGAC*TC. The M7 CRT variant (TGAAGACAT) was included as a negative reporter control, whereas mixtures without a CBF were included as a negative effector control. Error bars represent the standard deviation. Statistical analysis was performed on the set of no CBF, VrCBF1 and VrCBF4 data separately and significantly different activation values (ANOVA $p < 0.05$) are indicated by different letters. Similar results were obtained in two other independent experiments.

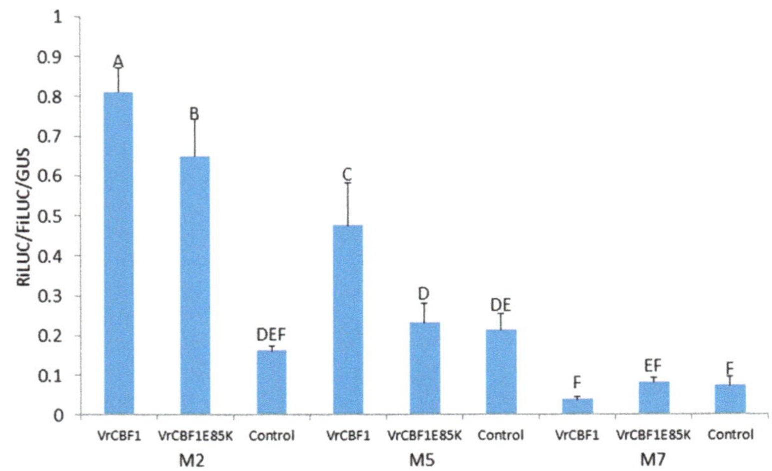

Figure 6 Comparison of activation by VrCBF1and VrCBF1E85K on the M2 and M5 CRT/DRE variants. M2 = CRT: *TGCCGACAT*, M5: *GACCGACAA*. The CRT variant M7 (TGAAGACAT) was included as a negative reporter control, whereas an empty effector was included as a negative effector control. Error bars represent the standard deviation. Statistical analysis was performed on all data together and significantly different activation values (ANOVA $p < 0.05$) are indicated by different letters. Similar results were obtained in two other independent experiments.

plasmids to be able to normalize for the amount of these proteins and be used as an indicator for the amount of respectively effector and reporter DNA transferred and expressed in the leaf cells. Other researchers examining either plant or animal systems have included a constitutively expressed reporter gene (35S::LUC cassette) on a separate plasmid [33,13,34] to normalize for differences in infiltration between samples. This was under the assumption that both plasmids are delivered in similar quantities into the cells. Our results show that this assumption is not true (Figure 2). (2) The chosen reporter genes have an intron, which cannot be spliced out by bacteria [35], therefore no enzyme is translated from erroneous transcripts that might be produced in the numerous Agrobacteria present in the infiltrated leaf tissue [36]. Commonly used reporter genes for plant tissues are β-glucuronidase (GUS) [37,38], green fluorescent protein (GFP) [13,39,40], firefly (*Photinus pyralis*) luciferase (FLUC) [41], and sea pansy (*Renilla reniformis*) luciferase (RLUC) [42]. The understanding that the inclusion of an intron is important led to the development of intron-containing reporters such as GUSi [34], GUSINT [36], GUSPlus (CAMBIA, Canberra, Australia and [30]), FiLUC [43] and RiLUC [44]. (3) Beta glucuronidase (GUSPlus), renilla luciferase (RiLUC) and firefly luciferase (FiLUC) were chosen as reporters as they all can be measured by a similar procedure. We avoided green fluorescent protein (GFP) as a reporter, since this can diffuse out of the cell [9]. Previously, Renilla luciferase has been used as the normalizer [2], but studies have shown that this luciferase has a 100 fold higher signal when compared to firefly luciferase [45] which gives it a wider range. Therefore, RiLUC was chosen to quantify differences in activation for our system. (4) The quantification procedure uses the same extract for the analysis of all reporter activities. This means that the activation value, RiLUC/FiLUC/GUS, is normalized for variation that may occur through infiltration, DNA uptake or protein extraction. The fact that new substrate solution for all enzymes has to be prepared fresh for every experiment means that some variation will exist between experiments and this can affect activation values. We therefore suggest that only effector and reporter combinations that have been analyzed in the same experiment be compared to each other.

The reporter construct was designed to contain 4 DRE/CRT sequence repeats combined with a minimal 35S CaMV promoter. The 46 nt minimal 35S promoter is one of the best characterized plant core regulatory promoter domains [46,47] with a reportedly very low basal transcription level in the absence of additional upstream regulatory elements [48], and has already been used successfully in our previous experiments [30,31]. Although activation by transcription factors of promoters containing only 1 binding domain was shown to be detectable [49], adding additional (4) repeats of a DRE/CRT element was considered appropriate since it would give a stronger activation and therefore higher reporter enzyme activity which would make differences in activation efficiency easier to detect [4]. The low level of activation on mutant DRE/CRT reporter constructs (Figures 3 and 4) showed that there is no significant contribution from any other potential enhancer elements on the reporter vector (for example, in the 35S promoter) to the RiLUC reporter activities caused by the *Vitis* CBFs. The low RiLUC/FiLUC/GUS values on various DRE/CRT variants suggests that there is some background activation by tobacco transcription factors especially when compared to the values on the M7 mutant CRT/DRE sequence, when no grape CBF-producing plasmid was included (see especially M6 and M9, Figures 3 and 5).

An advantage of a transient transactivation system is that one can detect an increase in transcripts from genes that are directly activated by the transcription factor under study, even if this activation is temporary and therefore not detectable in transgenic plants. Another advantage is that transcripts of indirectly activated genes, which can be detected in transgenic plants, are likely absent in the transient system. The transient expression system can therefore assist to interpret results from transgenic plants. For example, we previously reported that compared to wild type Arabidopsis, VrCBF4- but not VrCBF1-overexpressing plants have increased *AtRGL3* expression [50]. One might speculate that this is due to a preference by VrCBF4 and not VrCBF1 for the DRE/CRT-like sequence CCGCC in the *AtRGL3* promoter. However, the presented transient expression results showed that the M8 reporter construct (containing CCGCC) is not activated much by either VrCBF1 or VrCBF4. Similarly, because VrCBF1- but not VrCBF4-overexpressing plants had an increased *RD29A* (*COR78*) expression, it could be argued that this was due to a preference by VrCBF1 and not VrCBF4 for the CCGAC sequence present twice in the promoter of this gene [50]. However, the transient expression system results shows that all reporter constructs containing CCGAC (M1 to M6) are activated better by VrCBF4 than by VrCBF1 (Figures 4 and 5). This suggests that the induction of *AtRGL3* or *RD29A* in the transgenic Arabidopsis is due to an indirect effect, although it is also possible that the CBFs activate these genes via a DRE/CRT element outside of the "promoter" region that was examined for sequence elements, about 1 kb upstream of the ATG start codon [50].

All presented agroinfiltrations were performed in tobacco leaves, by a relatively easy procedure. We were not successful in infiltrating leaves from grapes grown under growth chamber conditions, despite trying various methods including vacuum infiltration. Indeed, also other researchers reported their failure to do so and were only successful if in plants were grown *in vitro* [14,15]. This has not been pursued further at this time since *in vitro* culture is labour-intensive and transactivation in grape leaves would only be necessary if one wanted to analyze the transactivation of endogenous grape genes.

Changes in the core DRE/CRT sequence greatly reduced the transactivation values (Figures 3 and 4), confirming that the complete core DRE/CRT sequence is required for binding by *Vitis* CBF1 or 4. This is in line with the report that Arabidopsis DREB1A (AtCBF3) and DREB2A bind weakly or not at all when the core CRT sequence is altered [51]. Binding by the Arabidopsis DREB proteins was not affected by changes in the surrounding sequence [51] however our transactivation results show that this is different for the *Vitis* CBFs, especially for *Vitis* CBF1 (Figure 5). Also of note, is that CRT variant M8 contains the core sequence of the GCC box (GCCGCC) which is known to interact with ERF transcription factors of the ethylene signalling pathway [52] but not with Arabidopsis DREB proteins [51] and, as shown here, also not with *Vitis* CBF1 and CBF4 (Figure 3).

Inclusion of the VrCBF4 effector plasmid generally resulted in higher activation of the DRE/CRT variants than inclusion of the VrCBF1 effector plasmid (Figures 4 and 5). This supports our suggestion that VrCBF4 is a better activator than VrCBF1 on CRT variant M1 based on previous experiments using a different transactivation system [31]. In principle, there are several possible explanations for this phenomenon besides an inherent better activation capability for the VrCBF4 protein. It is possible that different amounts of protein are produced for each because, even though we used the same 5′UTR and 3′UTR sequences for each construct, the coding sequence can also affect translation efficiency [53]. Other possible explanations include a difference in the half-life of the RNA or protein. Translation efficiency and RNA or protein stability likely differ between different tissues and conditions (e.g. ambient vs cold treatment), and the situation found here in tobacco leaves might therefore not reflect the conditions that exist when VrCBF1 and VrCBF4 are expressed in grape tissues. Quantification of CBF protein levels would be possible by Western blot analysis with antibodies specific for each CBF or to a tag added to each CBF. However, the results of the experiment shown in Figure 5 suggest that possible differences in protein quantity are not the main reason for the observed differences in activation by VrCBF1 and VrCBF4. In this experiment the same experimental parameters (bacterial cultures, tobacco plants, length of incubation etc.) were used for all reporters but not all reporters show a lower activation with VrCBF1 than with VrCBF4. A more likely explanation for these results is that VrCBF1 has a preference for the M2 sequence whereas VrCBF4 is more promiscuous. The higher activation by VrCBF1 vs. VrCBF1E85K supports the hypothesis that this amino acid difference contributes to the difference in freezing tolerance between *V. riparia* and *V. vinifera*. The ability to detect this difference shows the sensitivity of the transactivation system to detect changes in activation due to single amino acid sequence differences.

Conclusions

Here we describe the development of a novel set of effector and reporter plasmids for transient expression studies using agroinfiltration. The use of intron-containing reporter genes allow for normalization of transactivation values for variation in plasmid entry into the plant cells, sample collection and extract preparation. The ability to distinguish between activation by plasmids with minor sequence variations in DRE/CRT promoter elements or a CBF transcription factor suggests that this system could be valuable in examining a variety of transcription factors and their putative target promoter sequences. The results with VrCBF1 and VrCBF4 activation on DRE/CRT variants suggests that these two transcription factors likely activate different overlapping sets of genes, and therefore have unique roles in cold acclimation.

Materials and methods

Preparation of effector and reporter plasmids

The pCAMBIA 1305.1 binary vector containing a multiple cloning site (MCS) and a 35S::GUSPlus reporter gene (http://www.cambia.org/daisy/cambia/585.html), a gene with a catalase intron, was taken as the starting point for the creation of effector constructs first (Figure 1A). This is a multicopy plasmid, in contrast to the previously used pBI121 [30,31], and thus easier to use. The original pCAMBIA plasmid, which does not encode any CBF, was used as a negative control effector. A *Hind*III/*Eco*RI cassette containing a 35S promoter, VrCBF4 coding region with 5′ ribosome binding site (rbs), and Nos terminator sequence, obtained from a previously prepared pBI121-based effector [31], was inserted into the MCS of pCAMBIA, yielding the 35S::VrCBF4 pCAMBIA effector (Figure 1B). Preparation of 35S::VrCBF1 pCAMBIA effector plasmid involved simply replacing the BamHI/SacI fragment containing the 5′ ribosome binding site and VrCBF4 coding sequence [31] with a similar fragment containing VrCBF1 coding sequence [30].

The reporter construct was prepared from these pCAMBIA effector constructs in several steps. First the GUSPlus reporter sequence was replaced with a FiLUC (Firefly luciferase coding sequence including the PIV intron from GUS^{INT}) reporter sequence. To this end, NcoI/PmlI digested pCAMBIA effector, i.e. without the GUSPlus sequence, was ligated to an NcoI/PmlI fragment containing the FiLUC sequence which had been amplified from pLUC07 [42] using primers that introduce these restriction sites (FiLUC-H-2 + NcoI: 5′AGGTAAGCCATGGAAGA CGCCAA 3′ and FiLUC-C1842 + PmlI: 5′TACACGTGT TACAATTTGGACTTTCCGC 3′). Second, the VrCBF4 coding region was replaced with the RiLUC reporter (*Renilla reniformis* luciferase coding sequence including a modified intron from the castor bean catalase gene). This was accomplished by ligating a BamHI/SacI RiLUC fragment amplified from RiLUC plasmid [44] using primers that introduce these restriction sites (RiLUC-H1 + BamHI: 5′ATGGATCCAAGGAGATATAACAAT GACTTCGAAAGTTTATGATCC 3′ and RiLUC-C936 + SacI: 5′CGTTGACGAGCTCTTATTGTTCATTTTTGA GAACTCG 3′) to *Bam*HI and *Sac*I digested, FiLUC containing pCAMBIA, similar to the CBF fragments earlier. Third, vectors with the RiLUC reporter driven by a 4xCRTmin35S promoter, consisting of 4x TACCGACAT [30] plus 46 nucleotides of the 35S promoter [46,47], were constructed. To this end the HindIII-BamHI 35S promoter-containing fragment was replaced with a HindIII-BamHI fragment containing 4xCRTmin35S promoter (obtained from plasmids described in [30]). Finally, the 2x35S::hygromycin fragment was deleted by digestion with *Xho*I and *Bst*XI and replaced by a nos promoter fragment amplified from pBI121 using primers that introduce these restriction sites (NosproH1+ BstXI: 5′ACCACCATGTTGGGATCATGAGCGGAGAA TTAAG 3′ and NosproC307 + XhoI: 5′GCAGGCTCGA GAGATCCGGTGCAGATTATTT 3′), and the completed reporter plasmid was ready for use (Figure 1C). Effector and reporter plasmids were introduced into *A. tumefaciens* strain EHA105 according to the freeze-thaw method described by Höfgen and Willmitzer [54].

Reporter plasmids with altered CRT sequences were prepared by a quick change protocol essentially according to the procedure described by Stratagene on a smaller "35S cloning" plasmid [30] and subsequent subcloning of the new promoter fragment into the reporter plasmid. All primers that were used to prepare the various reporter plasmids are listed in Additional file 1: Table S1.

Agroinfiltration of tobacco leaves

Agroinfiltration was performed based on the protocols developed by Bendahmane and colleagues [55] and Vaucheret [56], essentially as described previously [30,31]. *Nicotiana benthamiana* plants were grown at 22°C for 16 hours of light and 20°C for 8 hours of dark until they reached a six leaf stage (approximately 4 weeks). Equal volumes of *A. tumefaciens* with reporter construct and of *A. tumefaciens* with effector construct were mixed to produce a final OD600 of 0.5 for each construct and the youngest two fully expanded leaves (leaves 3 and 4) of three different plants were infiltrated from the abaxial (lower) side with this mixture using a syringe. After 40-hour co-cultivation, one disc was taken from each infiltrated leaf, for a total of 6 discs per condition, frozen in liquid nitrogen and stored at −80°C. This was repeated 3–4 times to obtain 3–4 biological replicates for each mixture. Each experiment was repeated at least 2 times.

Preparation of extracts

Several protocols were tested for the extraction of leaf tissue to identify a procedure that is compatible with both glucuronidase and luciferase activity measurements, so that all reporter enzymes could be analyzed for the same extract. The harvested, frozen leaf discs were ground to powder with liquid N_2 and then 9 µl extraction buffer was added per mg tissue (300 µl buffer for six ~6 mm leaf discs). As extraction buffer we tried either GUS extraction buffer (25 mM potassium phosphate pH7.8, 1 mM EDTA, 7 mM 2-mercaptoethanol, 1% Triton X-100, and 10% glycerol) or 1X Cell Culture Lysis Reagent (CCLR; Luciferase assay systems, Promega). The glucuronidase and luciferase activities determined for the extracts were more consistent and higher when CCLR had been used for the preparation and PLB (Promega) for the dilution of the extracts, and therefore CCLR and PLB were used for all further experiments. Extracts were incubated on ice for 1 hour and cell debris was pelleted by centrifugation for 10 min at 13000 × g (SpeedFuge ® SFR13K, Savant). The supernatant was then diluted 75x with PLB and used for protein, glucuronidase and luciferase assays.

Analysis of protein quantities

Protein quantities were determined using the Bio-Rad dye-binding assay essentially according to the procedure described by the manufacturer (Bio-Rad), based on the method of Bradford [57], with BSA as standard.

Analysis of GUS, RiLUC and FiLUC activities

GUS activity was determined after incubating a mixture of 10 µl of the 75x diluted sample and 90 µl of Assay buffer (50 mM pH 7 Sodium phosphate buffer, 10 mM EDTA, 0.1 Triton X-100, 0.1% N-Lauroylsarcosine Sodium Salt, 10 mM 2-mercaptoethanol, 40 mM 4-MUG) for 30 minutes at 37°C. Each reaction was stopped by adding 900 µl of 0.2 M Na_2CO_3 and fluorescence caused by the conversion of 4-MUG to MUG was measured in a polystyrene flatbottom 96-well plate (Sarstedt) using a POLARStar Omega (BMG Labtech) microplate reader with the excitation filter set at 360 nm and the emission filter at 460–10, and orbital shaking at 300 rpm. Each extract was analyzed in triplicate and the average measurement was taken as the value for one replicate.

The dual luciferase protocol based on the Dual Luciferase reporter assay system from Promega, as reported by Cazzonelli and colleagues [43], was used essentially unchanged to quantify the amount of RiLUC and FiLUC expression. 75x diluted sample (10 µl) was added to Luciferase reagent II (50 µl, LARII) in a polystyrene flatbottom 96-well plate (Sarstedt) and mixed by pipetting. FiLUC fluorescence, indicative of *FiLUC* gene expression, was measured immediately in a POLARStar Omega microplate reader with the setting on Luminescence (end point). To measure RiLUC fluorescence, indicative of *RiLUC* gene expression, 50 µl of Stop and Glo reagent (20 µl substrate into 1 ml S&G buffer and mixed well, freshly prepared according to the instructions from Promega) was added to each sample, mixed by pipetting and returned to the luminometer for a second measurement. Both measurements were set on 0.2 sec delay with 10 flashes per well. Activities for all replicates of each extract were measured consecutively before a second set of replicates was prepared and measured.

Transactivation was expressed as RiLUC/FiLUC/GUS x 100000. The resulting data were analyzed by a one-way analysis of variance (ANOVA). Statistical differences amongst the means for each reporter/effector combination within an experiment were determined by the Tukey–Kramer HSD tests ($P < 0.05$) using JMP (version 11.1.1; SAS Institute) statistical software.

Additional file

Additional file 1: Table S1. List of primers used to prepare effector and reporter constructs.

Competing interests

The authors declare that they have no competing interests.

Authors' contributions

MAM prepared and tested the original effector and reporter constructs, and participated in the development of the assays, MS participated in the development of the constructs and assays and performed the statistical analyses, HX participated in the preparation of some constructs and subsequent assays, CC participated in the assays and statistical analyses, AN conceived the study, participated in its design and coordination, and wrote the initial manuscript. All authors read and approved the final manuscript.

Acknowledgements

The authors thank Jeff Velten and Luc Mankin for a vector with the RiLUC sequence, and FiLUC sequence respectively. Zamir Jetha provided technical support developing extraction and dilution protocols optimal for reporter enzyme assays. M. Atikur Rahman assisted with the construction of the M10 and M11 reporter constructs. This work was funded by grants from ORF and NSERC to AN. CC was partially supported by an Ontario Graduate Fellowship.

References

1. Chiera JM, Bouchard RA, Dorsey SL, Park E, Buenrostro-Nava MT, Ling PP, Finer JJ: **Isolation of two highly active soybean (Glycine max (L.) Merr.) promoters and their characterization using a new automated image collection and analysis system.** *Plant Cell Rep* 2007, **26**:1501–1509.
2. Hellens RP, Allan AC, Friel EN, Bolitho K, Grafton K, Templeton M, Karunairetnam S, Gleave AP, Laing WA: **Transient expression vectors for functional genomics, quantification of promoter activity and RNA silencing in plants.** *Plant Methods* 2005, **1**:13.
3. Hernandez-Garcia CM, Martinelli AP, Bouchard RA, Finer JJ: **A soybean (Glycine max) polyubiquitin promoter gives strong constitutive expression in transgenic soybean.** *Plant Cell Rep* 2009, **28**:837–849.
4. Rushton PJ, Reinstaedler A, Lipka V, Lippok B, Somssich IE: **Synthetic plant promoters containing defined regulatory elements provide novel insights into pathogen- and wound-induced signalling.** *Plant Cell* 2002, **14**:749–762.
5. Liu W, Mazarei M, Rudis MR, Fethe MH, Stewart CN: **Rapid *in vivo* analysis of synthetic promoters for plant pathogen phytosensing.** *BMC Biotech* 2011, **11**:108.
6. Christensen AH, Sharrock RA, Quail PH: **Maize polyubiquitin genes: structure, thermal perturbation of expression and transcript splicing, and promoter activity following transfer to protoplasts by electroporation.** *Plant Mol Biol* 1992, **18**:675–689.
7. Hartmann U, Valentine WJ, Christie JM, Hays J, Jenkins GI, Weisshaar B: **Identification of UV/blue light-response elements in the Arabidopsis**

thaliana chalcone synthase promoter using a homologous protoplast transient expression system. *Plant Mol Biol* 1998, **36**:741–754.

8. Pitzschke A, Persak H: **Poinsettia protoplasts - a simple, robust and efficient system for transient gene expression studies.** *Plant Methods* 2012, **8**:14.
9. Hernandez-Garcia CM, Bouchard RA, Rushton PJ, Jones ML, Chen X, Timko MP, Finer JJ: **High level transgenic expression of soybean (Glycine max) GmERF and Gmubi gene promoters isolated by a novel promoter analysis pipeline.** *BMC Plant Biol* 2010, **10**:237.
10. Rolfe SA, Tobin EM: **Deletion analysis of a phytochrome-regulated monocot rbcS promoter in a transient assay system.** *Proc Natl Acad Sci U S A* 1991, **88**(2):683–2686.
11. Cazzoneli CI, Velten J: **In vivo characterization of plant promoter element interaction using synthetic promoters.** *Transgenic Res* 2008, **17**:437–457.
12. Van Moerkercke A, Haring MA, Schuurink RC: **The transcription factor EMISSION OF BENZENOIDS II activates the MYB ODORANT1 promoter at a MYB binding site specific for fragrant petunias.** *Plant J* 2011, **67**:917–928.
13. Bogs J, Jaffe FW, Takos AM, Walker AR, Robinson SP: **The grapevine transcription factor VvMYBPA1 regulates proanthocyanidin synthesis during fruit development.** *Plant Physiol* 2007, **143**:1347–1361.
14. Santos-Rosa M, Poutaraud A, Merdinoglu D, Mestre P: **Development of a transient expression system in grapevine via agro-infiltration.** *Plant Cell Rep* 2008, **27**:1053–1063.
15. Guan X, Zhao H, Xu Y, Wang Y: **Transient expression of glyoxal oxidase from the Chinese wild grape *Vitis pseudoreticulata* can suppress powdery mildew in a susceptible genotype.** *Protoplasma* 2011, **248**:415–423.
16. Zottini M, Barizza E, Costa A, Formentin E, Ruberti C, Carimi F, Lo Schiavo F: **Agroinfiltration of grapevine leaves for fast transient assays of gene expression and for long-term production of stable transformed cells.** *Plant Cell Rep* 2008, **27**:845–853.
17. Bertazzon N, Raiola A, Castiglioni C, Gardiman M, Angelini E, Borgo M, Ferrari S: **Transient silencing of the grapevine gene VvPGIP1 by agroinfiltration with a construct for RNA interference.** *Plant Cell Rep* 2012, **31**:133–143.
18. Mizoi J, Shinozaki K, Yamaguchi-Shinozaki K: **AP2/ERF family transcription factors in plant abiotic stress responses.** *Biochim Biophys Acta* 1819, **2012**:86–96.
19. Wisniewski M, Nassuth A, Teulieres C, Marque C, Rowland J, Cao P-B, Brown A: **Genomics of cold hardiness in woody plants.** *Crit Rev Plant Sci* 2013, **33**:92–124.
20. Stockinger EJ, Gilmour SJ, Thomashow MF: **Arabidopsis thaliana CBF1 encodes an AP2 domain-containing transcriptional activator that binds to the Crepeat/DRE, a cis-acting DNA regulatory element that stimulates transcription in response to low temperature and water deficit.** *Proc Natl Acad Sci U S A* 1997, **94**:1035–1040.
21. Gilmour SJ, Zarka DG, Stockinger EJ, Salazar MP, Houghton JM, Thomashow MF: **Low temperature regulation of the Arabidopsis CBF family of AP2 transcriptional activators as an early step in cold-induced COR gene expression.** *Plant J* 1998, **16**:433–442.
22. Liu Q, Kasuga M, Sakuma Y, Abe H, Miura S, Yamaguchi-Shinozaki K, Shinozaki K: **Two transcription factors, DREB1 and DREB2, with an EREBP/AP2 DNA binding domain separate two cellular signal transduction pathways in drought- and low temperature- responsive gene expression, respectively, in Arabidopsis.** *Plant Cell* 1998, **10**:1391–1406.
23. Kasuga M, Liu Q, Miura S, Yamaguchi-Shinozaki K, Shinozaki K: **Improving plant drought, salt, and freezing tolerance by gene transfer of a single stress-inducible transcription factor.** *Nature Biotech* 1999, **17**:287–291.
24. Miura K, Furumoto T: **Cold signaling and cold response in plants.** *Int J Mol Sci* 2013, **14**:5312–5337.
25. Gao M-J, Allard G, Byass L, Flanagan AM, Singh J: **Regulation and characterization of four CBF transcription factors from *Brassica napus*.** *Plant Mol Biol* 2002, **49**:459–471.
26. Xue G-P: **Characterisation of the DNA-binding profile of barley HvCBF1 using an enzymatic method for rapid, quantitative and high-throughput analysis of the DNA-binding activity.** *Nucl Acids Res* 2002, **30**:e77.
27. Tong Z, Hong B, Yang Y, Li Q, Ma N, Ma C, Gao J: **Overexpression of two chrysanthemum DgDREB1 group genes caused delayed flowering or dwarfism in Arabidopsis.** *Pl Mol Biol* 2009, **71**:115–129.
28. Seki M, Narusaka M, Ishida J, Nanjo T, Fujita M, Oono Y, Kamiya A, Nakajima M, Enju A, Sakurai T, Satou M, Akiyama K, Taji T, Yamaguchi-Shinozaki K, Caminci P, Kawai J, Hayashikazi Y, Shinozaki K: **Monitoring the expression profiles of 7000 Arabidopsis genes under drought, cold and high-salinity stresses using a full-length cDNA microarray.** *Plant J* 2002, **31**:279–292.
29. Maruyama K, Sakuma Y, Kasuga M, Ito Y, Seki M, Goda H, Shimada Y, Yoshida S, Shinozaki K, Yamaguchi-Shinozaki K: **Identification of cold-inducible downstream genes of Arabidopsis DREB1A/CBF3 transcriptional factor using two microarray systems.** *Plant J* 2004, **38**:982–993.
30. Xiao H, Siddiqua M, Braybrook S, Nassuth A: **Three grape CBF/DREB1 genes are regulated by low temperature, drought and abscisic acid.** *Plant Cell Environ* 2006, **29**:1410–1421.
31. Xiao H, Tattersall E, Siddiqua M, Cramer GR, Nassuth A: **CBF4 is a unique member of the CBF transcription factor family of *Vitis vinifera* and *Vitis riparia*.** *Plant Cell Environ* 2008, **31**:1–10.
32. Wang S, Yang S, Yin Y, Guo X, Wang S, Hao D: **An *in silico* strategy identified the target gene candidates regulated by dehydration responsive element binding proteins (DREBs) in Arabidopsis genome.** *Plant Mol Biol* 2009, **69**:167–178.
33. Horstmann V, Huether CM, Jost W, Reski R, Decker EL: **Quantitative promoter analysis in Physcomitrella patens: a set of plant vectors activating gene expression within three orders of magnitude.** *BMC Biotech* 2004, **4**:13.
34. Narumi O, Mor S, Boku S, Tsuji Y, Hashimoto N, Nishikawa S-I, Yokata Y: **OUT, a novel basic helix–loop–helix transcription factor with an Id-like inhibitory activity.** *J Biol Chem* 2000, **275**:3510–3521.
35. Ohta S, Mita S, Hattori T, Nakamura K: **Construction and expression in tobacco of a β-glucuronidase (GUS) reporter gene containing an intron within the coding sequence.** *Plant Cell Physiol* 1990, **31**:805–814.
36. Vancanneyt G, Schmidt R, O'Connor-Sanchez A, Willmitzer L, Rocha-Sosa M: **Construction of an intron-containing marker gene: splicing of the intron in transgenic plants and its use in monitoring early events in Agrobacterium-mediated plant transformation.** *Mol Gen Genet* 1990, **220**:245–250.
37. Jefferson R, Kavanagh T, Bevan M: **GUS fusions: β-glucuronidase as a sensitive and versatile gene fusion marker in higher plants.** *EMBO J* 1987, **6**:3901–3907.
38. Samac DA, Tesfaye M, Dornbusch M, Saruul P, Temple SJ: **A comparison of constitutive promoters for expression of transgenes in alfalfa (*Medicago sativa*).** *Transgenic Res* 2004, **13**:349–361.
39. Chalfie MM, Tu Y, Euskirchen G, Ward WW, Prasher DC: **Green fluorescent protein as a marker for gene expression.** *Science* 1994, **263**:802–805.
40. Leffel S, Mabon S, Stewart N: **Applications of green fluorescent protein in plants.** *Biotechniques* 1997, **23**:912–918.
41. Ow D, Wood K, DeLuca M, DeWet J, Helsinki D, Howell S: **Transient and stable expression of the firefly luciferase gene in plant cells and transgenic plants.** *Sci* 1986, **234**:856–859.
42. Mayerhofer R, Langridge W, Cormier M, Szalay A: **Expression of recombinant Renilla luciferase in transgenic plants results in high levels of light emission.** *Plant J* 1995, **7**:1031–1038.
43. Mankin S, Allen G, Thompson W: **Introduction of a plant intron into the luciferase gene of *Photinus pyralis*.** *Plant Mol Biol Rep* 1997, **15**:186–196.
44. Cazzoneli CI, Christopher I, Velten J: **Construction and testing of an intron-containing luciferase reporter gene from *Renilla reniformis*.** *Plant Mol Biol Rep* 2003, **21**:271–280.
45. Behre G, Smith LT, Tenen DG: **Use of a promoterless Renilla luciferase vector as an internal control plasmid for transient co-transfection assays of Ras-mediated transcription activation.** *Biotechniques* 1999, **26**:24–26.
46. Odell JT, Nagy F, Chua NH: **Identification of DNA sequences required for activity of the cauliflower mosaic virus 35S promoter.** *Nature* 1985, **313**:810–812.
47. Benfey P, Chua N: **The cauliflower mosaic virus 35S promoter: combinatorial regulation of transcription in plants.** *Science* 1990, **250**:959–966.
48. Cazzonelli CI, Burke J, Velten J: **Functional characterization of the geminiviral conserved late element (CLE) in uninfected tobacco.** *Plant Mol Biol* 2005, **58**:465–481.
49. Eini O, Yang N, Pyvovarenko T, Pillman K, Bazanova N, Tikhomirov N, Eliby S, Shirley N, Sivasankar S, Tingey S, Langridge P, Hrmova M, Lopato S: **Complex regulation by apetela2 domain-containing transcription factors revealed through analysis of the stress responsive TdCor410b promoter from Durum wheat.** *Plos One* 2013, **8**:e58713.
50. Siddiqua M, Nassuth A: ***Vitis* CBF1 and *Vitis* CBF4 differ in their effect on Arabidopsis abiotic stress tolerance, development and gene expression.** *Plant Cell Env* 2011, **34**:1345–1359.

51. Sakuma Y, Liu Q, Dubouzet JG, Abe H, Shinozaki K, Yamaguchi-Shinozaki K: **DNA-binding specificity of the ERF/AP2 domain of Arabidopsis DREBs, transcription factors involved in dehydration- and cold-inducible gene expression.** *Biochem Biophys Res Comm* 2002, **290**:998–1009.
52. Hao D, Ohme-Takagi M, Sarai A: **Unique mode of GCC box recognition by the DNA-binding domain of ethylene-responsive element-binding factor (ERF domain) in plant.** *J Biol Chem* 1998, **273**:26857–26861.
53. Tuller T, Waldman YY, Kupiec M, Ruppin E: **Translation efficiency is determined by both codon bias and folding energy.** *Proc Natl Acad Sci U S A* 2010, **107**:3645–3650.
54. Höfgen R, Willmitzer L: **Storage of competent cells for Agrobacterium transformation.** *Nucleic Acids Res* 1988, **16**:9877.
55. Bendahmane A, Kanyuka K, Baulcombe DC: **The Rx gene from potato controls separate virus resistance and cell death responses.** *Plant Cell* 1999, **11**:781–791.
56. Vaucheret H: **Promoter-dependent trans-inactivation in transgenic tobacco plants; kinetic aspects of gene silencing and gene reactivation.** *C R Acad Sci* 1994, **317**:310–323.
57. Bradford M: **A rapid and sensitive method for the quantitation of microgram quantities of protein utilizing the principle of protein-dye binding.** *Anal Biochem* 1976, **72**:248–254.

Measuring the distribution of cellulose microfibril angles in primary cell walls by small angle X-ray scattering

Friederike Saxe[1*], Michaela Eder[1], Gunthard Benecke[1], Barbara Aichmayer[1], Peter Fratzl[1], Ingo Burgert[1,2,3] and Markus Rüggeberg[1,2,3]

Abstract

Background: X-ray scattering is a well-established method for measuring cellulose microfibril angles in secondary cell walls. However, little data is available on the much thinner primary cell walls. Here, we show that microfibril orientation distributions can be determined by small angle X-ray scattering (SAXS) even in primary cell walls. The technique offers a number of advantages: samples can be analyzed in the native hydrated state without any preparation which minimizes the risk of artifacts and allows for fast data acquisition. The method provides data averaged over a specimen region, determined by the size of the used X-ray beam and, thus, yields the microfibril orientation distribution within this region.

Results: Cellulose microfibril orientation distributions were obtained for single cells of the alga *Chara corallina,* as well as for the multicellular hypocotyl of *Arabidopsis thaliana*. In both, Chara and Arabidopsis, distributions with a broad scattering around mean microfibril angles of approximately 0° and 90° towards the longitudinal axis of the cells were found.

Conclusions: With SAXS, the structure of primary cell walls can be analysed in their native state and new insights into the cellulose microfibril orientation of primary cell walls can be gained. The data shows that SAXS can serve as a valuable tool for the analysis of cellulose microfibril orientation in primary cell walls and, in consequence, add to the understanding of its mechanical behaviour and the intriguing mechanisms behind cell growth.

Keywords: X-ray scattering, Primary cell wall, Cellulose microfibril orientation, *Chara corallina*, *Arabidopsis thaliana*

Background

Cellulose microfibril orientation plays an important role for plant cell wall mechanics and determines cell shape during growth. A number of complementary techniques can be used to show the arrangement of cellulose fibrils: transmission electron microscopy provides details with very high resolution; the field of view, however, is small [1,2]. In field emission scanning electron microscopy and atomic force microscopy, only the newly synthesised microfibrils at the inner face of the cell walls are accessible [3-6]. Polarised Fourier transform infrared microscopy yields one predominant microfibril orientation and can be extended to show a distribution of microfibril orientations [7]. Nevertheless, the method is not well suited for the application on primary cell walls as little wall material is available and oblique cellulose microfibril orientations can further impede the experiment. Fluorescent cellulose dyes like Pontamine Fast Scarlet 4B can be used to visualize the cellulose architecture in cell walls by confocal microscopy. While the dyes easily penetrate into root cells, it remains a challenge to apply the technique to intact hypocotyls [8]. Chemical and genetic analyses can shed light on composition and assembly [9,10], but a method to analyze the bulk of microfibril orientation of intact primary plant cell wall tissues is lacking.

SAXS allows the characterization of structures within the size range of 0.5 nm to 100 nm; this also applies to cellulose microfibrils. Only minor sample preparations

* Correspondence: friederike.saxe@mpikg.mpg.de
[1]Department of Biomaterials, Max Planck Institute of Colloids and Interfaces, 14476 Potsdam-Golm, Germany
Full list of author information is available at the end of the article

are required and high statistical accuracy results from averaging over a large sample volume. Scanning the sample with a microbeam enables spatially resolved measurements of the local microfibril orientation.

The SAXS-signal in plant tissues originates from differences in electron densities between the cellulose microfibrils and the surrounding matrix. It appears in close proximity of the primary beam at $q < 5\ nm^{-1}$, q being the modulus of the scattering vector $\vec{q}$, which is the difference between the vector of the incoming and the scattered X-ray beam. For wet samples, the SAXS signal is dominated by the scattering of the cellulose fibrils; contributions of pores can be neglected [11,12]. In SAXS, the long and rod shaped cellulose fibrils appear as flat disks in the scattering pattern, corresponding to the Fourier transform in reciprocal space. Hence, the scattering image emanating from a collection of fibrils in the cell wall represents the superposition of all disks originating from the different cellulose fibrils within the irradiated sample volume. The image appearing on a two-dimensional detector is a cut through the three-dimensional figure corresponding to these superposed disks (illustrated in Figure 1, [13]). The azimuthal intensity distribution of the resulting streaks is used to extract information on the distribution of microfibril orientation [14-16].

Measurement of cellulose microfibril orientation through small angle X-ray scattering (SAXS) is well established for the comparably thick secondary cell walls in which cellulose fibrils are mainly oriented in parallels [14,17,18]. This architecture leads to a strongly anisotropic SAXS signal from which mean orientations of the cellulose fibrils in secondary cell walls can be calculated [14].

For primary cell walls, Bayley et al. [19] obtained oriented SAXS patterns from the thick cell walls of *Avena* coleoptiles and deduced preferential orientations. Hydration dependency of microfibril spacing was investigated by SAXS on primary cell walls of celery collenchyma [20]. However, the primary cell walls of the model organism *Arabidopsis thaliana* are much thinner [21], resulting in weak scattering signals. To our knowledge, only one X-ray scattering study on the structure of secondary cell walls of *Arabidopsis* [22] exists. To date, no X-ray data on structure and orientation distribution of microfibrils in primary cell walls of *Arabidopsis* are available.

In this paper, we present SAXS analysis of primary cell walls from *Chara corallina* and *Arabidopsis thaliana.* The large cells of *Chara* enable the analysis of a single primary cell wall whereas for *Arabidopsis* data are obtained from the complex, multicellular hypocotyl. We deduce a microfibril angle distribution for both plants. The measurements were performed at the μ-spot beam line of the BESSY II synchrotron radiation facility (see Figure 2 for the setup). To obtain the microfibril angle distribution we adapted the simulation procedure of Rueggeberg et al. [23] which was developed for WAXD experiments on the basis of the simulation of Entwistle et al. [24]. Cell geometry, peak broadening and cellulose microfibril orientation distributions (not exclusively Gaussians) can be taken into account [23]. In addition, an analytical solution assuming cylindrical cell geometry was implemented to verify the results [15,16].

Results

Typical SAXS scattering patterns and the corresponding radial and azimuthal intensity distributions are shown in Figure 3 for *Chara corallina* and *Arabidopsis thaliana.* The contribution of background scattering from amorphous cellulose and other cell wall components can be

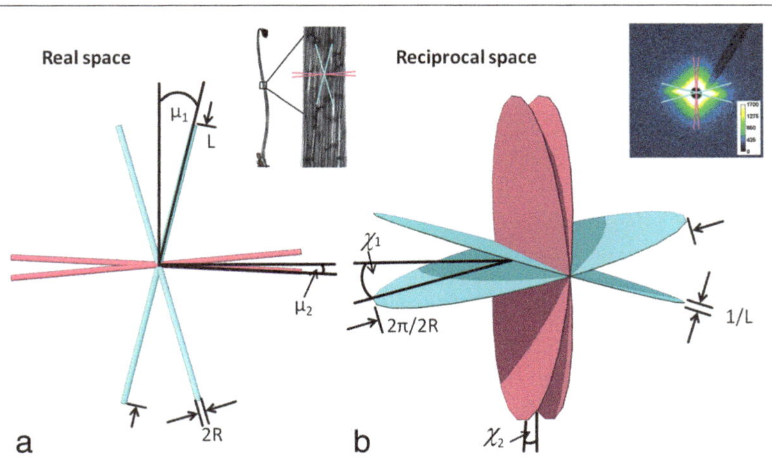

Figure 1 Principle of small angle X-ray scattering of cellulose microfibrils. a) Cellulose microfibrils in real space. Orientations are shown on the sample image at the upper right. Scale bar: 2 mm **b)** Fourier transform of the microfibrils are flat disks in reciprocal space that appear as two-dimensional scattering pattern on the detector (upper right). R: microfibril radius, L: microfibril length, μ: microfibril angle, α: cell wall orientation, χ: azimuthal peak positions in reciprocal space. Scalebar 2 mm. For simplicity only the calculated mean orientations are shown. Drawing adapted from [13].

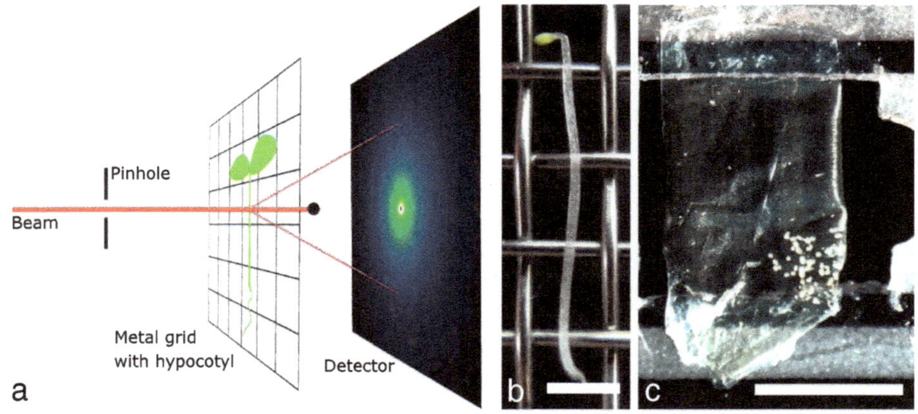

Figure 2 Experimental setup for small angle X-ray scattering. a) Setup at the µ-spot beamline (BESSY II). Beam size is adjusted by the pinhole before it hits the sample. The scattered signal is collected on a 2D-detector while the primary beam is kept back by a beam-stop. **b)** Sample preparation: 6d old, dark grown *Arabidopsis thaliana* hypocotyls are mounted onto a metal grid in their hydrated state. **c)** *Chara corallina* cell is cut open and a single cell wall is placed in the beam. White spots at the lower right of the sample are calciumcarbonate depositions (not relevant for the microfibril measurement). Scalebars: 2 mm.

accounted for by setting a baseline at the minimum scattering intensity at a q-value of approximately 4 nm^{-1} in the radial profile (Figure 3c and h). In the azimuthal intensity distribution (Figure 3b and g), which is used for orientation analysis, the area underneath the corresponding baseline (marked by the red line in Figure 3d and i) is not considered in the simulation. A theoretical intensity profile is calculated based on cell geometry and then fitted to the measured intensity distribution by varying the contribution of one or several cellulose orientation distributions according to Rueggeberg et al. [23]. In the present case, two Gaussian distributions were necessary to obtain appropriate fits.

A single unfolded cell wall of *Chara corallina* was analysed (Figure 3a). Two preferential orientations at ~0° and ~90° with respect to the longitudinal axis of the cell were found (Figure 3e and k). The peak areas correspond to the relative contribution of the microfibril angles to the total scattering signal. The distribution with its mean value at 90° (transversely oriented microfibrils) covered about 72% of the total area. Only about 28% of the scattering signal originated form the second microfibril distribution with a mean angle of around 0° with respect to the longitudinal axis of the cell. At intermediate microfibril angles the distributions overlap. This is indicated by the colour gradient in Figure 3e and j.

For *Arabidopsis thaliana* a section in the lower middle region of the hypocotyl was chosen for analysis (Figure 3f-i). As the beam diameter was approximately half the diameter of the hypocotyl, the results of the measurements represent an average of all the cells in the irradiated section. About 38% of the microfibrils were oriented with a mean angle of approximately 0° towards the cells axis while the remaining 62% were oriented transversely with a mean angle of approximately 90° towards the cells axis (Figure 3j and k).

For both organisms, the distributions of the cellulose microfibrils were bimodal and fairly broad. Contributions were detected from all angles within a range of 0 - 90° with respect to the longitudinal axis of the cell wall.

The orientation analysis through the simulation procedure under the assumption of cylindrical cells was compared to the calculated analytical solution for *Arabidopsis thaliana*. Both methods yielded a good quality of the fits (Figure 4a). At small angles however, the two distributions show differences in the microfibril angle distribution (Figure 4b). To further investigate the differing results at small microfibril angles, two theoretical scattering patterns were analysed. For equal scattering intensities, the obtained microfibril angle distributions for both methods, simulation procedure and analytical solution, corresponded well (Additional file 1: Figure S1a and b). For the theoretical intensity profile of a uniform distribution of microfibrils, the results of both methods deviate strongly from the theoretical, uniform distribution at small microfibril angles (Additional file 1: Figure S1c and d).

Discussion

SAXS provides the possibility to obtain data on the orientation distributions of cellulose microfibrils on native plant material. Depending on the chosen size of the X-ray beam, it is possible to either scan the sample with a spatial resolution down to the µm-range or to obtain an average over a larger sample volume containing many cells. However, the benefit of averaging over several cells requires some knowledge of the system at hand, as the signal represents an overlay of possibly

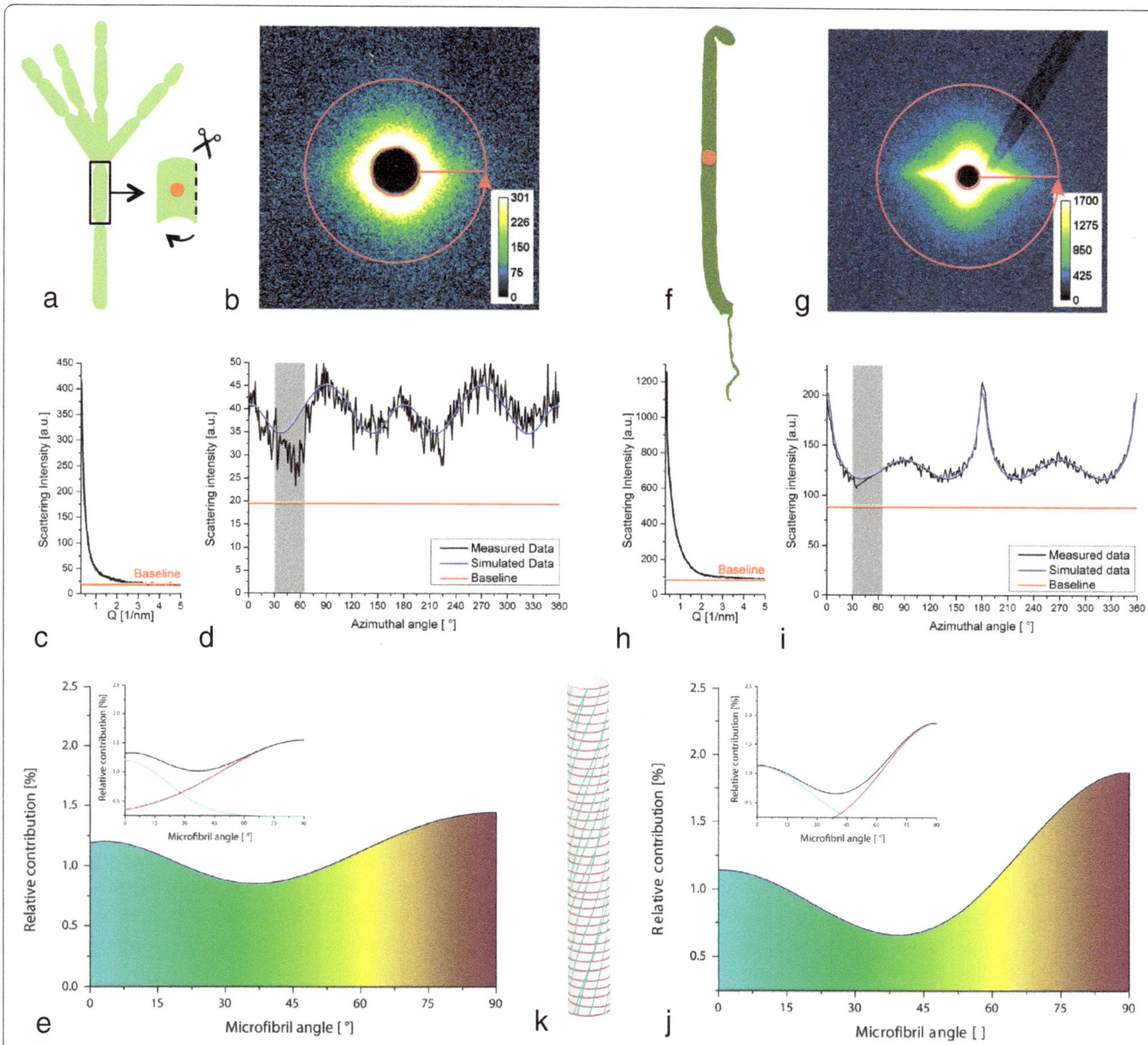

Figure 3 SAXS analysis.a-e: *Chara corallina*. a) One cell was cut open and a single cell wall was analysed. **f-j**: Primary cell walls of *Arabidopsis thaliana*. **a** and **f**) 6 day old dark grown hypocotyl is analysed. The beam diameter is close to the hypocotyl diameter. **b** and **g**) 2D scattering image: 360° azimuthal integration is marked. The scattering signal originates from electron density contrast between the cellulose microfibrils and water. **c** and **h**) The radial scattering profile is used to assign the background of the measurement. **d** and **i**) Integrated scattering signal (measured data) can be fitted by microfibril angle simulation (simulated data). The area beneath the baseline as well as the data range influenced by the glass capillary on which the beamstop was mounted (grey area) are not considered in the simulation. **e** and **j**) Relative contribution of each microfibril angle to the simulated scattering signal. The colour gradient represents the microfibril orientations from 0° (longitudinal towards the cell axis, indicated in blue) to 90° (transverse to the longitudinal axis of the cell, indicated in red). The insets show the two underlying distributions. At intermediate microfibril angles there is an overlap in which the two distributions cannot be distinguished. **k)** Schematic of a possible arrangement of the microfibrils in a cell with the colour scheme used in **e** and **j**.

different distributions of microfibril angles. The method cannot distinguish between two cell types with different distributions and one cell type showing two distributions. Therefore, alterations in cellulose microfibril orientation in the cell walls of different cell types or in different cell wall layers cannot be resolved. In multi-scale systems consisting of different cell types like the *Arabidopsis* hypocotyl the origin of the scattering signal can only be assigned by additional measurements e.g. after separation of different tissue types or by microscopy. The scattering image is not only averaged spatially, but also temporally, through the measurement time. Due to radiation damage it thus remains difficult to capture highly dynamic processes.

Scattering intensities of the presented model systems are low compared to intensities obtained from wood sections due to the small amount of material in primary cell

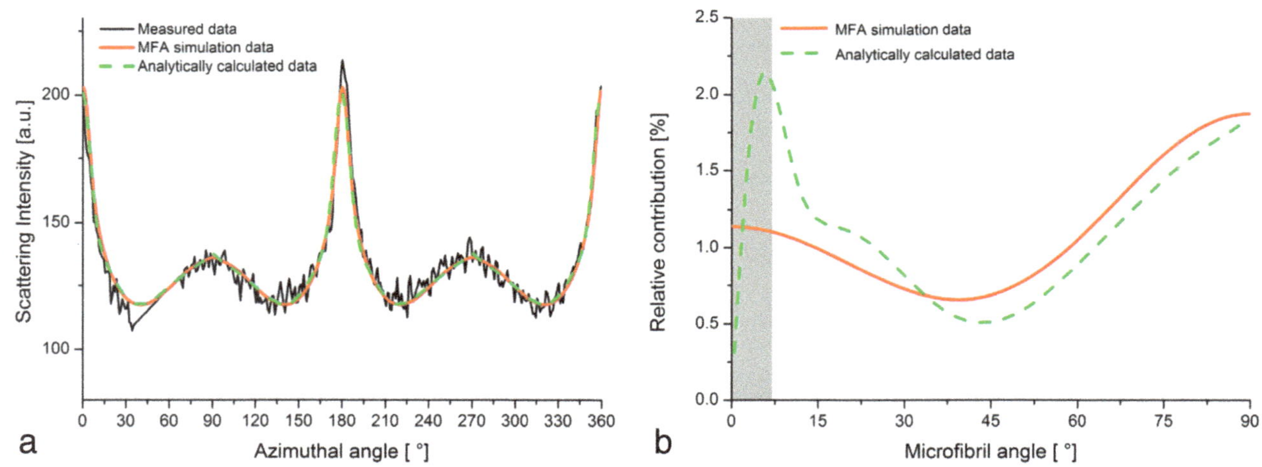

Figure 4 Comparison of the methodologies for *Arabidopsis thaliana* hypocotyl. a) Measured and calculated data for the simulation procedure and the analytically calculated solution. Measured data and fits are in good accordance for both methods. **b)** The resulting microfibril angle distributions. Deviations can be found especially at small microfibril angles where the real microfibril angle distribution can be approximated with less accuracy.

walls. However, the fitting procedure used to calculate cellulose orientation distributions tolerates a low signal to noise ratio (Figure 3d). Using synchrotron radiation has the advantage of a high brilliance of the beam and thus shorter measurement times. In principle though, the experiments can also be performed on a laboratory instrument.

In SAXS measurements, the scattering contrast is linked to the electron density difference between cell wall components. Hence, samples should be kept in the hydrated, native state. The reason being that the method is only applicable when the scattering of the microfibrils dominates the signal and additional contributions (e.g. arising from cracks or pores) are suppressed. In the case of hydrated plant cell walls, the highest electron density contrast exists between crystalline cellulose fibrils and the surrounding water-saturated matrix. Thus, a two-phase system is a good approximation and can be used as a basis for further analysis [11,12]. Upon drying, air-filled voids or cracks might appear and blur the scattered signal due to their strong electron density difference with respect to hydrated matrix.

For this study, two methods of data analysis were compared. The analytical solution is valid under the assumption of cylindrical cells, while the simulation procedure offers the advantage that other cell shapes can be considered as well. In the simulation procedure, the microfibril angle distribution is approximated by two Gaussian peaks to limit the number of fit parameters to the necessary minimum. When using the analytical solution, the number of degrees of freedoms is generally lower and more details on the microfibril angle distribution can be included into the fitting procedure. To increase the quality of the fit in the simulation procedure, it is possible to increase the number of Gaussian peaks approximating the microfibril angle distribution. Yet, a balance between the detailed representation of features of the measured data by the fit and the increase in fitting parameters has to be found. Both methods become imprecise in the range of very small microfibril angles. The analytical calculations of the microfibril angle distributions are valid for spherical cells and the results are then transferred to a cylindrical geometry. As the radius of a sphere in the direction normal to the vertical axis becomes infinitely small at the poles, the intensity of the scattering originating from the microfibrils in this region should in theory be concentrated in a point and thus become very high. Thus, in the case of small microfibril angles, the scattering pattern contains narrow peaks. However, due to instrumental broadening of the beam, these features are less pronounced in the measured data and depending on the chosen setup, the first degrees of the microfibril angle distribution – in this case , it is the first 5-7° [25] - cannot be accurately determined.

The effect of large scattering contributions of small microfibril angles is strongest for an equal distribution of all microfibril angles to the total scattering signal. The deviation of the simulated and analytically calculated solutions from the theoretical solution gives the range within which the results can maximally vary. Decreasing the interval of the azimuthal angle used for calculating the intensity distribution in the analytical solution further enhances the deviation from the theoretical result. The reason for this being that the theoretical scattering pattern approaches infinite values for microfibril angles close to zero, which cannot accurately be represented by the cosine functions used to fit the data (Equation 4).

Conclusions

The methods described here, allow for analysis of the cellulose microfibril orientation distributions within primary cell wall systems in their native state. The primary cell walls of *Arabidopsis thaliana* and *Chara corallina* both show a bimodal microfibril angle distribution such that the bulk of the microfibrils is oriented either transversely or longitudinally with a broad scattering. However, the scattering signal does not contain information on the orientation of microfibrils at small microfibril angles (up to 7° towards the longitudinal axis of the cell), which has to be taken into consideration when applying the method. The SAXS data can supplement the existing information on cell wall structure at different growth stages and thus, help to understand the complex mechanisms behind cell wall formation and cell growth.

Methods

Material

Chara corallina Klein ex Willd., em. R.D.W.(= C. australis R. Br.) was grown in the laboratory in 120 dm^3containers in a solution of distilled water with 1.0 mM NaCl, 0.2 mM KCl, 0.2 mM $CaSO_4x2H_2O$ and 1.0 mM $NaHCO_3$titrated to pH 8.2 by NaOH according to [26].

Arabidopsis thaliana (L.) hypocotyls of the ecotype Col-0 were investigated in this study. The sterilized seeds were plated in 2 g/l Murashige and Skoog basal medium in agar and incubated in the dark at 22 °C for 6 days.

X-ray scattering

X-ray measurements were performed at the μ-spot beamline at the synchrotron radiation facility of the Helmholtz-Zentrum Berlin für Materialien und Energie (BESSY II). The setup consists of a beam-defining pinhole, a guard pinhole and a two-dimensional position-sensitive CCD-detector MarMosaic 225 (Mar USA Evanston, USA) [25]. Fresh, hydrated *Arabidopsis* hypocotyls were mounted onto a metal grid (mesh size 2.5 mm; Figure 2b) through capillary forces and placed in the beam (transmission mode) (Figure 2a). From *Chara corallina* a 3 cm long cell was chosen for analysis, cut open on one side, unfolded and taped to a foliar frame in the hydrated state (Figure 2c). During the measurement the samples were kept wet by a humidifier. To achieve an integration over several cell walls the X-ray beam diameter was adjusted to ~100 μm, which is about half of the average diameter of an *Arabidopsis* hypocotyl. The wavelength was set to 0.82656 Ångstrom (energy of 15 keV) and typical measuring times were 60 s. The sample-detector distance was set to 320 mm for *Arabidopsis* (simultaneous detection of the SAXS and WAXD signal) and to 840 mm for Chara (detection of the SAXS signal only) and the q-axis was calibrated against a Quartz (WAXD) or Silverbehenate (SAXS) standard. A background reference frame taken without sample allowed for subtracting air scattering. The signals were also corrected for the instrumental noise (dark current).

In principal, if wide angle X-ray diffraction (WAXD) is recorded simultaneously with SAXS it may also be used for analyzing cellulose orientation. However, the WAXD signal of primary cell walls is much weaker compared to the SAXS signal due to low cellulose crystallinity in primary walls. The SAXS signal contains fewer disturbances from other components like cuticular waxes and in samples analyzed in the hydrated state, the main WAXD signal of cellulose is masked by the water halo. Thus, the sample can only be analyzed in the dry state, where the drying process might have induced changes in the fragile cell wall.

Data analysis

Beam centre, detector tilt and the sample-detector distance were determined using the Quartz and Silverbehenate measurements by the calibrant routine implemented in Fit2D [27]. Integration of the sample data was automated using Fit2D and Autofit, a Python based software developed by Chenghao Li, Aurelien Gourrier and Gerald A. Zickler at the Max-Planck-Institute of Colloids and Interfaces in Potsdam. The SAXS signal was integrated in two ways: Azimuthal integration gives the scattering intensity as a function of the scattering vector q (Figure 3b and g) and can be used to obtain structural information. To obtain information on microfibril orientation, the scattering signal is integrated radially, which results in the intensity as a function of the azimuthal angle χ. The intensity of the scattering signal as a function of the scattering vector q is high in the q-region of 0.3 – 2 nm^{-1} and then drops quickly to run into a minimum at values of q around 3 nm^{-1}. This minimum was used to assign the baseline for the following simulation. The SAXS q-region is then integrated azimuthally from 0-360° as shown in Figure 3c and g. Oriented cellulose microfibrils, amorphous non-oriented cell wall components as well as the background contribute to the azimuthal signal.

Simulation procedure

The microfibril angle distribution could be calculated from the azimuthal signal similar to the procedure described for wide angle X-ray diffraction (WAXD) in Rüggeberg et al. [23]. However, the equation for calculating the position χ of the intensity maxima on the detector in dependency of μ and α differs from that for WAXD. It has been deduced by several authors [18,28] and is given in Equation 1.

$$\tan\chi = \tan\mu \cdot \cos\alpha \qquad (1)$$

α is the cell wall orientation and μ the microfibril angle with respect to the longitudinal axis of the cell (Figure 1). For a single cell wall, like in the case of *Chara* that is placed perpendicular to the beam, the cell wall orientation α is 0 and the microfibril angle can be deduced directly from the scattering image as $\chi = \mu$. For *Arabidopsis* cylindrical cells are assumed and the peak positions are calculated for every combination of μ and α in 1° steps. Fit parameter is the contribution of every microfibril angle μ to the scattering signal with the restriction that the resulting microfibril angle distribution can be described by two Gauss peaks.

Analytical solution

For cylindrical cells it is possible to obtain an analytical solution. The variation of spiral angles within a cylindrical cell, in which the microfibril angle μ varies between $-\pi/2$ and $\pi/2$, is called g(μ). The microfibril angle distribution on a cylinder h(μ) is because of the symmetry of the round cell determined as $h(\mu) = g(\mu) + g(-\mu)$ with μ varying between 0 and $\pi/2$. The function is normalized such that $\int_0^{\frac{\pi}{2}} h(\mu)d\mu = 1$. The probability p(μ) that a cellulose fibril has an angle μ on a sphere is related to the microfibril angle h(μ) on a cylindrical cell by $h(\mu) = p(\mu)\sin\mu$. To deduce the microfibril angle distribution the equation for the intensity distribution given by [16] is inverted:

$$I(\chi) = \frac{2}{\pi}\int_{\chi}^{\frac{\pi}{2}} \frac{p(\mu)\sin(\mu)}{\sqrt{\cos^2\chi - \cos^2\mu}}d\mu = \frac{2}{\pi}\int_{\chi}^{\frac{\pi}{2}} \frac{h(\mu)}{\sqrt{\cos^2\chi - \cos^2\mu}}d\mu. \quad (2)$$

The inversion of the function can be done numerically [16], yet since I(χ) can only be determined from the detector image with limited precision, the SAXS data can be fitted by:

$$I(\chi) = \sum_{(n=1)}^{m} a_n \cos^{2\gamma_n}\chi \; for \left(0 \leq \chi \leq \frac{\pi}{2}\right) \; and \; \gamma_n \geq 0. \quad (3)$$

Then the microfibril angle distribution p(μ) becomes:

$$p(\mu) = \sum_{n=1}^{m} b_n \cos^{2\gamma_n}\mu \; with \; b_n = \frac{2a_n\Gamma(\gamma_n+1)/\Gamma(\gamma_n+\frac{1}{2})}{\sum_{k\geq 0} a_k\Gamma(\gamma_k+1)/\Gamma(\gamma_k+\frac{3}{2})} \quad (4)$$

(see appendix of [15]).

The measured data is fitted using Equation 3 and due to the symmetry of the scattering signal, averaging the four sectors is possible. In the next step the microfibril angle distribution of a cylindrical cell h(μ) can be calculated from p(μ). The intensity distribution can be calculated for arbitrary intervals of the azimuthal angle. In this case intervals of 1° were chosen which is a reasonable compromise between precision and calculation time.

The simulation procedure has been written in Excel and Python and is available from the authors on request.

Additional files

Additional file 1: Figure S1. Analysis of calculated theoretical scattering profiles. **a)** Fit of analytical solution and simulation procedure for equal scattering intensities over the entire azimuthal profile. **b)** The resulting microfibril angle distributions from both methods are in very good accordance. **c)** Fit of analytical solution and simulation procedure for the theoretically calculated scattering intensities of round cells with an equal distribution of all microfibril angles. Inlet: The resulting azimuthal scattering profile (green) of the simulation procedure has two contributing distributions (light and dark blue curves) as two Gauss peaks were used to fit the microfibril angle distribution. **d)** The resulting microfibril angle distributions from both methods deviate from the theoretical, uniform distribution. For microfibril angles larger than 7°, minor deviations can be observed that are a measure for the achievable accuracy of the results. The grey areas indicate the region of small microfibril angles in which the orientation cannot accurately be determined due to the instrumental broadening of the incident beam.

Competing interests
The authors declare that they have no competing interests.

Authors' contributions
FS, GB and MR carried out the X-ray measurements, data analysis and drafted the manuscript. FS, ME, GB, BA, IB, PF and MR discussed analyses, interpretation, and presentation. FS, MR and IB designed and coordinated the study. All authors read and approved the final manuscript.

Acknowledgements
We thank Joachim Fisahn, Heidemarie Held and Clara Sánchez-Rodríguez from the Max Planck Institute of Molecular Plant Physiology in Potsdam for providing the plant material of *Chara corallina* and *Arabidopsis thaliana*. We thank Stefan Siegel and Chenghao Li for their support during the X-ray measurements and Tobias Baskin (Biology Department, University of Massachusetts Amherst, Amherst, MA01003, United States) as well as Daniel Cosgrove (Department of Biology, Pennsylvania State University, University Park, PA16802, United States) for helpful discussions. IB thanks BAFU and Lignum, Switzerland for support of the Wood Materials Science group. We thank Rebekka Lauer for her support in creating the figures and Matthew Harrington and Lisa O'Conner for linguistic revisions.

Author details
[1]Department of Biomaterials, Max Planck Institute of Colloids and Interfaces, 14476 Potsdam-Golm, Germany. [2]Swiss Federal Institute of Technology Zurich, Institute for Building Materials, 8093 Zurich, Switzerland. [3]Swiss Federal Laboratories for Materials Science and Technology, Applied Wood Materials Laboratory, 8600 Duebendorf, Switzerland.

References

1. Lang JM, Eisinger WR, Green PB: **Effects of ethylene on the orientation of microtubules and cellulose microfibrils of pea epicotyl cells with polylamellate cell walls.** *Protoplasma* 1982, **110**:5–14.
2. Emons AMC: **Methods for visualizing cell-wall texture.** *Acta Botanica Neerlandica* 1988, **37**:31–38.
3. Sugimoto K, Williamson RE, Wasteneys GO: **New techniques enable comparative analysis of microtubule orientation, wall texture, and growth rate in intact roots of Arabidopsis.** *Plant Physiol* 2000, **124**:1493–1506.
4. Davies LM, Harris PJ: **Atomic force microscopy of microfibrils in primary cell walls.** *Planta* 2003, **217**:283–289.
5. Ding SY, Himmel ME: **The maize primary cell wall microfibril: a new model derived from direct visualization.** *J Agric Food Chem* 2006, **54**:597–606.

6. Zhang T, Mahgsoudy-Louyeh S, Tittmann B, Cosgrove D: **Visualization of the nanoscale pattern of recently-deposited cellulose microfibrils and matrix materials in never-dried primary walls of the onion epidermis.** *Cellulose* 2014, **21**:853–862.
7. MacKinnon IM, Šturcová A, Sugimoto-Shirasu K, His I, McCann MC, Jarvis MC: **Cell-wall structure and anisotropy in procuste, a cellulose synthase mutant of Arabidopsis thaliana.** *Planta* 2006, **224**:438–448.
8. Anderson CT, Carroll A, Akhmetova L, Somerville C: **Real-time imaging of cellulose reorientation during cell wall expansion in arabidopsis roots.** *Plant Physiol* 2010, **152**:787–796.
9. Persson S, Paredez A, Carroll A, Palsdottir H, Doblin M, Poindexter P, Khitrov N, Auer M, Somerville CR: **Genetic evidence for three unique components in primary cell-wall cellulose synthase complexes in Arabidopsis.** *Proc Natl Acad Sci U S A* 2007, **104**:15566–15571.
10. Harris D, Bulone V, Ding S-Y, DeBolt S: **Tools for cellulose analysis in plant cell walls.** *Plant Physiol* 2010, **153**:420–426.
11. Jakob HF, Tschegg SE, Fratzl P: **Hydration dependence of the wood-cell wall structure in Picea abies: a Small-Angle X-ray scattering study.** *Macromolecules* 1996, **29**:8435–8440.
12. Jungnikl K, Paris O, Fratzl P, Burgert I: **The implication of chemical extraction treatments on the cell wall nanostructure of softwood.** *Cellulose* 2008, **15**:407–418.
13. Lichtenegger H, Reiterer A, Tschegg S, Fratzl P: **Determination of spiral angles of elementary fibrils in the wood cell wall: comparison of small-angle X-ray scattering and wide-angle X-ray diffraction.** In *Microfibril angle in wood. Proc. IAWA/IUFRO Intn. Workshop on the significance of microfibril angle to wood quality.* Christchurch, New Zealand: University of Canterbury; 1998:140–156.
14. Jakob HF, Fratzl P, Tschegg SE: **Size and arrangement of elementary cellulose fibrils in wood cells: a small-angle x-ray scattering study of picea abies.** *J Struct Biol* 1994, **113**:13–22.
15. Fratzl P, Paris O, Klaushofer K, Landis WJ: **Bone mineralization in an osteogenesis imperfecta mouse model studied by small-angle x-ray scattering.** *J Clin Invest* 1996, **97**:396–402.
16. Perret R, Ruland W: **Single and multiple x-ray small-angle scattering of carbon fibres.** *J Appl Crystallogr* 1969, **2**:209–218.
17. Wardrop AB: **The low-angle scattering of x-rays by Conifer Tracheids.** *Text Res J* 1952, **22**:288–291.
18. Kantola M, Kähkönen H: **Small-angle X-ray investigation of the orientation of crystallites in finnish coniferous and deciduous wood fibers.** *Ann Acad Sci Fenn A VI Phys* 1963, **137**:1–14.
19. Bayley ST, Colvin JR, Cooper FP, Martinsmith CA: **The structure of the primary epidermal cell wall of Avena Coleoptiles.** *J Biophys Biochem Cy* 1957, **3**:171–182.
20. Kennedy C, Šturcová A, Jarvis M, Wess T: **Hydration effects on spacing of primary-wall cellulose microfibrils: a small angle X-ray scattering study.** *Cellulose* 2007, **14**:401–408.
21. Derbyshire P, Findlay K, McCann MC, Roberts K: **Cell elongation in Arabidopsis hypocotyls involves dynamic changes in cell wall thickness.** *J Exp Bot* 2007, **58**:2079–2089.
22. Fujita M, Himmelspach R, Hocart CH, Williamson RE, Mansfield SD, Wasteneys GO: **Cortical microtubules optimize cell-wall crystallinity to drive unidirectional growth in Arabidopsis.** *The Plant Journal* 2011, **66**:915–928.
23. Ruggeberg M, Saxe F, Metzger TH, Sundberg B, Fratzl P, Burgert I: **Enhanced cellulose orientation analysis in complex model plant tissues.** *J Struct Biol* 2013, **183**:419–428.
24. Entwistle KM, Kong K, MacDonald MA, Navaranjan N, Eichhorn SJ: **The derivation of the microfibril angle in softwood using wide-angle synchrotron X-ray diffraction on structurally characterised specimens.** *J Mater Sci* 2007, **42**:7263–7274.
25. Paris O, Li CH, Siegel S, Weseloh G, Emmerling F, Riesemeier H, Erko A, Fratzl P: **A new experimental station for simultaneous X-ray microbeam scanning for small- and wide-angle scattering and fluorescence at BESSY II.** *J Appl Crystallogr* 2007, **40**:466–470.
26. Fisahn J, McConnaughey T, Lucas WJ: **Oscillations in extracellular current, external pH and membrane potential and conductance in the alkaline bands of Nitella and Chara.** *J Exp Bot* 1989, **40**:1185–1193.
27. Hammersley A: *FIT2D V9.129 Reference Manual V3.1.* ; 1998.
28. Reiterer A, Jakob HF, Stanzl-Tschegg SE, Fratzl P: **Spiral angle of elementary cellulose fibrils in cell walls of Picea abies determined by small-angle X-ray scattering.** *Wood Sci Technol* 1998, **32**:335–345.

Circumnutation Tracker: novel software for investigation of circumnutation

Maria Stolarz*, Maciej Żuk, Elżbieta Król and Halina Dziubińska

Abstract

Background: An endogenous, helical plant organ movement named circumnutation is ubiquitous in the plant kingdom. Plant shoots, stems, tendrils, leaves, and roots commonly circumnutate but their appearance is still poorly described. To support such investigations, novel software *Circumnutation Tracker* (*CT*) for spatial-temporal analysis of circumnutation has been developed.

Results: *CT* works on time-lapse video and collected circumnutation parameters: period, length, rate, shape, angle, and clockwise- and counterclockwise directions. The *CT* combines a filtering algorithm with a graph-based method to describe the parameters of circumnutation. The parameters of circumnutation of *Helianthus annuus* hypocotyls and the relationship between cotyledon arrangement and circumnutation geometry are presented here to demonstrate the *CT* options.

Conclusions: We have established that *CT* facilitates and accelerates analysis of circumnutation. In combination with the physiological, molecular, and genetic methods, this software may be a powerful tool also for investigations of gravitropism, biological clock, and membrane transport, i.e. processes involved in the mechanism of circumnutat on.

Keywords: Circumnutation, Time-lapse, Plant movement, Software, *Helianthus annuus*

Background

Circumnutations have been investigated for over 130 years [1-4] but the methods for analysis thereof are still time consuming and weakly standardized. During his investigations, Charles Darwin recorded circumnutations in many plant species simply using a glass plate [1,4-6]. More recently, circumnutations have been investigated using a photo-diode containing apparatus [7] and currently by time-lapse images [8-10]. Some investigations search the core mechanism [11-13] and ultradian pacemaker of circumnutation [14]. The circumnutation analysis also accompanies investigations of gravitropic responses [10,15,16], growth mechanisms [17-20], ethylene signalling [21], IP_3 signalling [22], and glutamate signalling pathways [23] as well as investigations of aluminium treatment [9] and circadian clock [24-26]. Other studies explore the function of circumnutation in plant life [27-29]. Circumnutation analyses also accompany the research of right- and left-handed symmetry of twining organs [6,30] and they should be taken into account in investigations of heliotropism of organs [31]. The geometry of circumnutation is dependent on the morphological traits of plants [7,32,33]. Software for tracking root growth and development [34-36] or measuring hypocotyls and leaf rosettes [37,38] are known, but there is no tool for analysis of circumnutation – a ubiquitous phenomenon in plants. The aim of our work was to design software for analysis of standard circumnutation parameters in relation to the geographical direction applicable in various plant species. To our knowledge, the *Circumnutation Tracker* (*CT*) software presented here is the first free and open source tool for analysis of circumnutation.

Implementation

The *CT* is based upon cross-platform solutions and runs under the Windows (XP, Vista, 7) and Linux environment. Automatic options use Basic Linear Algebra Subroutine (BLAS) for calculations. The following steps are necessary for software installation: download *Circumnutation Tracker* and *CT* user guide (pdf) to your computer from http://circumnutation.umcs.lublin.pl and run: Circumnutation

* Correspondence: maria.stolarz@poczta.umcs.lublin.pl
Department of Biophysics, Institute of Biology and Biochemistry, Maria Curie-Skłodowska University, Akademicka 19, Lublin 20-033, Poland

Tracker.exe. All details of work with *CT* are presented in the *CT* user guide.

The CT workflow

The scheme of workflow with *CT* is presented in Figure 1. The crucial steps of working with *CT* include time-lapse video loading, calibration of time and space, manual harvesting of coordinates x and y, automatic determination of circumnutation parameters, and data and graph exporting.

Time-lapse video input - parameters of time-lapse video files opened by CT

CT works on black/white or colour time-lapse video images (AVI Audio Video Interleave), with any resolution. The supporting video codec's include H.264 and Indeo Video 5.

Time and space calibration

Circumnutating plants are elaborated by *CT* on a time-lapse video and circumnutation parameters are expressed in minutes and millimetres. The organ apex is usually chosen for tracing the circumnutations and their coordinates x and y changing during the experiment are determined. The centre of the coordinated system can be set to a preferred location e.g. in a place such as organ origin, by the *setup origin* option. Coordinates x and y harvested from the time-lapse video can be previously calibrated by the *frame timestamp* and *perspective correction* options. All technical details of work with *CT* are described in *CT* user guide.pdf.

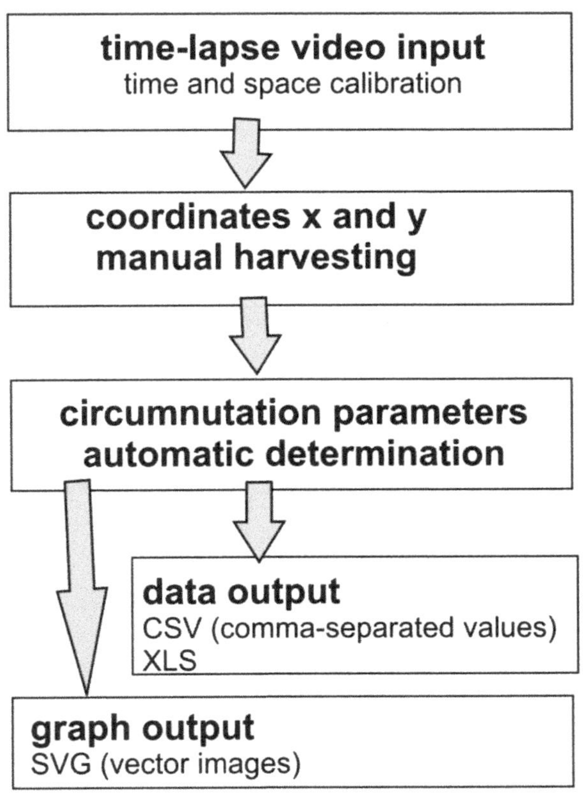

Figure 1 The workflow with *Circumnutation Tracker*.

Timestamp

Coordinates x and y are labelled with time stamps: year, month, hour, minutes, and seconds by the *frame timestamp* option.

Calibration and perspective correction

Conversion of pixels to millimetres and elimination of the illusion of an enlarged image when the plant grows toward the camera are performed in the *perspective correction* option using: frame width at the ground level in a pot, distance of the camera from the ground, plant height at the start and end of the experiment. Plant height at the end of our experiment was different for distilled-water and nutrient-solution growing seedlings (Table 1). Thus, the calibration and perspective corrections were made in different database files: Sample 1 (distilled water.db) and Sample 2 (nutrient solution.db), downloaded with the *CT* software.

Geographical directions

In many papers, the circumnutation trajectories are presented as viewed from above [15,18,28,39]. For a better understanding of the geometry of circumnutation in relation to plant morphology, we propose to set a top-view camera corresponding to the geographical plane and thus the coordinates x, y simultaneously correspond to the geographical east–west (EW) and north–south (NS), respectively. The seeds, seedlings, or older plants can be set to the experiment by juxtaposing their symmetry to the NS-EW directions (Additional file 1: Video 1, Figure 2C). We have an unquestionable point of reference for the circumnutation trajectory (especially circumnutation shape, angle, and direction) to the morphology of plants and environmental geographical directions by such settings of equipment and plants. This could also contribute to future studies on the role of circumnutation in morphogenesis including a study of right- and left-handing and symmetry of development. Additionally, linking the camera settings and plant symmetry with geographical directions will be useful for studying circumnutation together with heliotropism and phototropism [31]. The benefit of *CT* is that we can study every single circumnutation in an objective NS-EW plane, which is an advantage over Fourier and autocorrelation analysis as these provide information about frequency only.

Table 1 Comparison of circumnutation parameters of *Helianthus annuus* seedlings growing in distilled water and a nutrient solution

Seedling numero	Hypocotyl length (mm)	Period (min)	Length (mm)	Rate (mm min^{-1})	Shape	Angle (°)	cw%	ccw%	?%
			Distilled water						
1	85	270	18	0.07	0.43	-24	0	69	31
2	60	267	16	0.06	0.21	1	0	33	67
3	80	335	36	0.12	0.31	-3	40	30	30
4	90	293	19	0.07	0.21	-15	11	22	67
5	70	195	9	0.05	0.44	24	0	0	100
6	75	275	9	0.04	0.38	-43	8	0	92
7	65	400	11	0.03	0.54	-36	0	0	100
8	80	260	6	0.02	0.51	-20	0	0	100
Mean	76*****	287**	15**	0.06	0.38*	-15	7	19	73
SE	4	21	3	0.01	0.04	8	5	9	11
			Nutrient solution						
9	115	225	28	0.11	0.59	0	0	53	47
10	105	189	25	0.12	0.35	-17	21	16	63
11	112	188	97	0.48	0.57	3	11	74	16
12	116	217	42	0.19	0.61	-33	81	6	13
13	116	218	48	0.21	0.40	-40	69	6	25
14	113	175	73	0.40	0.56	-11	10	86	5
15	111	225	47	0.21	0.33	-30	6	81	13
16	121	209	99	0.47	0.55	6	76	18	6
Mean	114	206	57	0.27	0.49	-15	34	42	23
SE	2	7	10	0.05	0.04	6	12	12	7

The results were estimated for significance by a t-test at $p < 0.05$ *, $p < 0.01$ **, $p < 0.00001$ *****.

Coordinates x and y are manually harvested

Manual harvesting of the organ apex coordinates x and y by computer mouse clicks is controlled by the user. The coordinates x and y can be exported as a CSV (comma-separated values) file that can be opened in spreadsheet software (e.g. Microsoft Excel). The coordinates x and y presented as a time series are shown in Figure 3 for a *Helianthus annuus* hypocotyls. The manual harvesting is an advantage due to the independence of the background of circumnutating plants; therefore,

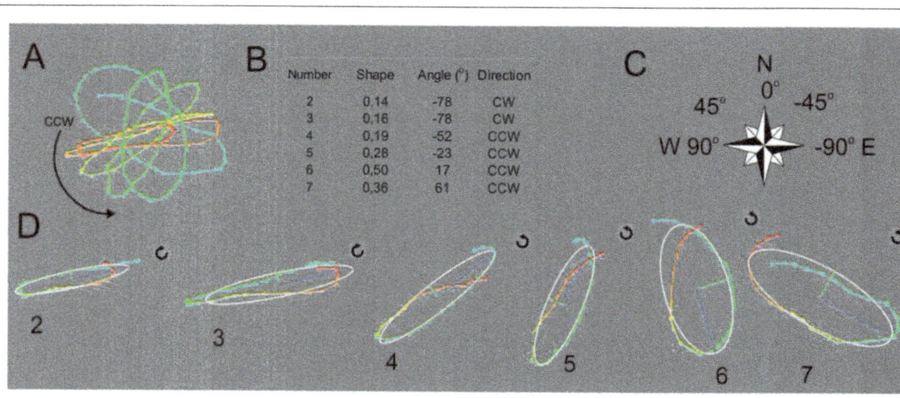

Figure 2 Fitting the shape, angle, and direction of circumnutations. A. The top view of the trajectory of circumnutation of the hypocotyls of one-week old *Helianthus annuus*. The example of seedling number 14 on Additional file 1: Video 1. The beginning and end of the circumnutation trajectory are marked by colour gradients from red to blue line **B**. Shape, angle, and direction of circumnutations number 2–7 (also in a Table 2). **C**. Arrangement of circumnutation angles in the geographical direction plane. **D**. Subsequent circumnutations with a marked shape (ellipse, white line), angle (blue line), and direction of circumnutations. In the geographic direction plane, the single circumnutation cycle is determined by two subsequent maximum northward bends of the hypocotyl.

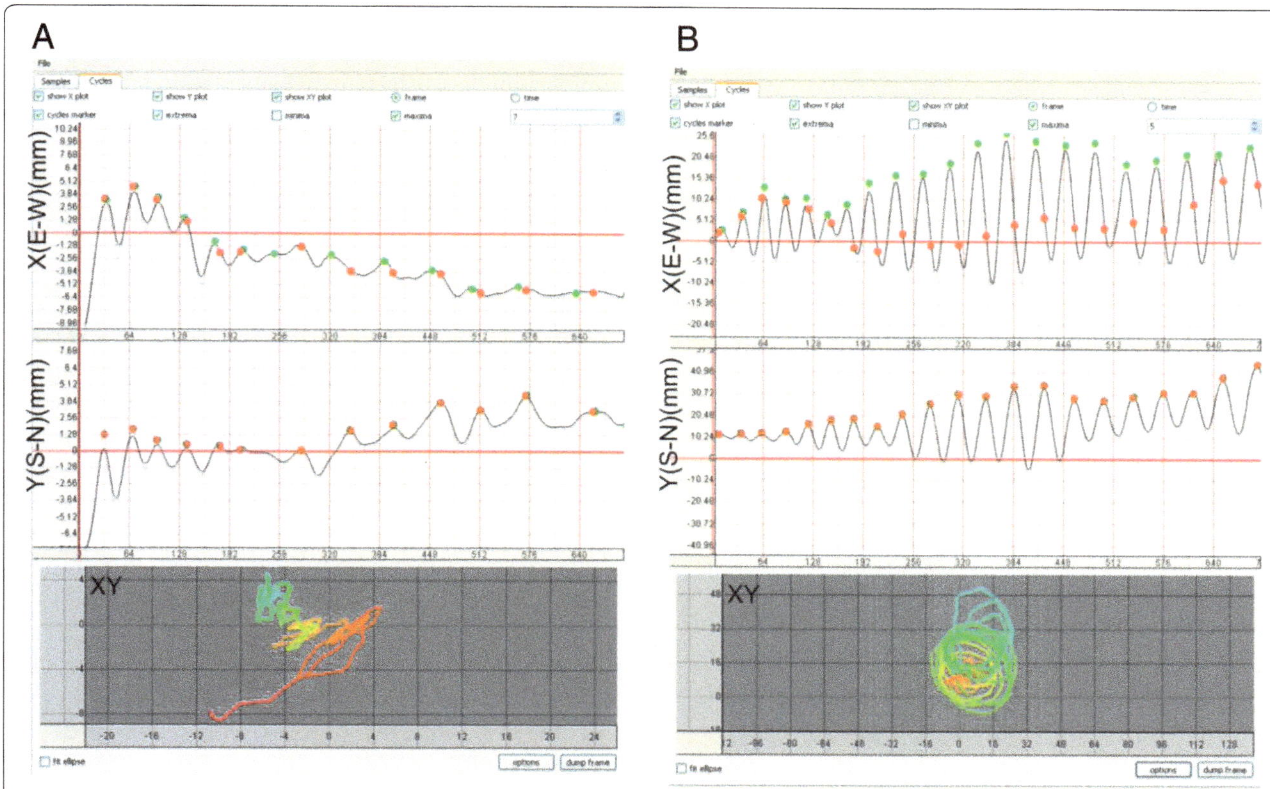

Figure 3 Screen shoots of the panel for determination of circumnutation cycles. Regular and irregular pattern of circumnutation can be elaborated successfully by *CT*. **A**. irregular (seedling 6 in Additional file 1: Video 1 and Figure 4A) and **B**. regular pattern (seedling 14 in Additional file 1: Video 1 and Figure 4B) of trajectory of circumnutation of *Helianthus annuus* hypocotyls lasting 63 hours. The black line is smoothed by a moving average filter; the green dots are maximum extremes calculated from coordinates x and y, the red dots are extremes confirmed by the user and its mark a single circumnutation between two maximum northward bends (more info in *CT* user guide.pdf).

CT is applicable in a wide range of filming conditions. We are also working on development of automatic harvesting of coordinates x and y as a prospective *CT* option.

Circumnutation cycles and parameters are automatically determined

Calculation of the extremes

After harvesting, the graphs of the time series of coordinates x and y are automatically plotted and an algorithm for calculation of extremes starts running (Figure 3). A single circumnutation cycle is determined by two subsequent extremes (maxima or minima). The extremes are detected on data smoothed using a moving average filter with a customizable filter size – the filter eliminates possible false extremes. Then, a discrete derivative is calculated to determine the slope of the curve. When the slope sign is changing – the extreme is detected, and when the slope changes from rising to falling, the extreme is the maximum (Figure 3; green dots) and otherwise – the minimum. We propose that in the geographic direction plane, the single circumnutation cycle is determined by two subsequent maximum northward bends of the organ (Figure 3; red dots). Figure 3 also shows that the regular and irregular patterns of circumnutation can be elaborated successfully by *CT*.

The following circumnutation parameters: period, length, rate, shape, and clockwise- and counter-clockwise directions are calculated by *CT* for every single circumnutation cycle.

- *period* (min) - the time that the organ apex needs to trace a single circumnutation cycle (time between two subsequent maximum northward bends of the organ).
- *length* (the length of the trajectory of circumnutation, mm) - the way that the organ apex covers during a single circumnutation cycle (between two subsequent maxima northward bends, Figure 3). This parameter corresponds to the term "amplitude" widely used in the literature.
- *rate* - the apex rate during a single circumnutation cycle (circumnutation length divided by period, mm min^{-1}).
- *shape (shape coefficient)* - a figure that is drawn in a horizontal plane by the organ apex during a single circumnutation, usually a wider or narrower ellipse or a circle. The shape is depicted by the shape

coefficient, which is expressed as a ratio of the length of the short axis to the length of the long axis of the ellipse; for a very narrow ellipse, it is close to 0.1, and 1 for a circular circumnutation. Determination of the algorithm for the shape coefficient of circumnutation is based on the ellipse fitting method described by Fitzgibbon A, Pilu M and Fisher RB [40]. Examples of narrow and wider circumnutations are shown in Figure 2.

- *angle* - the angle between the long axis of the ellipse and the line drawn by the direction of the geographic north–south (NS). The angle is determined with the arctangent of the long and short axis vector of an ellipse previously fitted into the circumnutation cycle. The N is a 0 degree, E −90 degree and W is a 90 degree (Figure 2C). The angle is not determined for the circular circumnutation. This parameter together with the shape coefficient could help in the study of a rosette-like trajectory, as reported by Baillaud L [41].
- *direction* - a clockwise (cw, clockwise, right-handed) or counter-clockwise (ccw, left-handed) direction of movement of the apex. Some circumnutations are indeterminate (marked as ?).

Table of circumnutation parameters – data output

The circumnutation parameter values can be exported as a CSV (comma-separated values) file that can be opened in spreadsheet software (e.g. Microsoft Excel) as shown in Table 2.

Trajectory of circumnutation – graph output

The circumnutation trajectory can be seen in *CT* (Figure 4) and exported to a SVG (vector image) type file that can be opened in graphics software (e.g. Corel Draw). Simply, PrintScreen of trajectory of circumnutation is also possible.

An example of *CT* use for analysis of circumnutation of *Helianthus annuus* seedlings is presented below.

Results and discussion

Circumnutation parameters vs. morphology of seedlings of *Helianthus annuus*

Dependence between plant morphology and the parameters of circumnutation was reported in *Avena* [7] and *Pisum* [32,33]. *Helianthus annuus* is a model plant for circumnutation and many studies have been carried out but no characteristics of circumnutations vs. seedlings morphology have been provided yet. Here, we have shown that circumnutations of seedlings growing in distilled water are different from those observed in plants growing in a nutrient solution (Additional file 1: Video 1). The parameters and trajectory of circumnutations determined using *CT* are shown in Table 1 and Figure 4. The distilled-water growing seedlings have a statistically significantly shorter hypocotyl ($p < 0.00001$) and length of circumnutation ($p < 0.01$) but a longer period ($p < 0.01$) than the nutrient-solution growing seedlings. The shape of distilled-water seedlings are slightly narrow ellipses ($p < 0.05$), and a dominating direction that is difficult to determine (73%). The nutrient-solution growing seedlings have wider ellipses, and *ccw* slightly dominated (42%) in relation to *cw* direction (34%). The long axis of the ellipses has the same angle arrangement on the geographical plane (−15°) in both groups and it is almost perpendicular to the axis of the cotyledon arrangement at the start of the experiment. Torsion of the hypocotyls was also observed and shown in Figure 4. The slowly growing distilled-water seedlings showed small torsion in relation to the faster growing nutrient-solution seedlings. The direction of torsion was usually counter to the dominating circumnutation direction as can be seen in seedlings no. 11, 12, 13, 14, 15, and 16 in Figure 4. Sometimes a rosette-like trajectory pattern was observed and its direction was usually ccw, as shown in Figure 5. The results presented above show the use of the *CT* software. The period and geometrical properties of subsequent circumnutation cycles are determined in irregularly and regularly circumnutating plants growing in different nutrient conditions. We expect that the analysis of circumnutation geometry will contribute to future studies of plant morphogenesis, including phyllotactic patterns and flower development [37,42].

Materials and methods

Experimental plants

Helianthus annuus L. seeds (PNOS, Ożarów Maz., Poland) were germinated on wet filter paper in a thermostated (25 ± 1°C) darkened chamber. Four-day (after 10 hours in light for hook opening) seedlings with 4.5 ± 0.5-cm long hypocotyls were cultivated hydroponically (eight plants per pot) in aerated distilled water and a nutrient solution (Additional file 1: Video 1). The nutrient solution contained 4 mM $Ca(NO_3)_2 \times 4H_2O$, 5 mM KNO_3, 1 mM $NH_4H_2PO_4$, 2 mM $MgSO_4 \times 7H_2O$; microelements: 0.085 mM Fe(III)citrate, 0.046 mM H_3BO_3, 0.0009 mM $MnCl_2 \times 4H_2O$, 0.0003 mM $CuSO_4 \times 5H_2O$, 0.0008 mM $ZnSO_4 \times 7H_2O$, 0.0001 mM $H_2MoO_4 \times 2H_2O$. The seedlings were set so that the cotyledons were arranged parallel to axis EW (Additional file 1: Video 1). The hydroponic culture was maintained for 63 hours under constant illumination, 40 μmol m^{-2} s^{-1} white light (Power Star HQT-T400 W/D OSRAM GmbH, Munich, Germany), at a temperature of 25 ± 1°C and relative humidity 50-70%.

Circumnutation measurements – time-lapse video

For circumnutation measurements, time-lapse video recordings started at 18:00 p.m. on the fourth day and

Table 2 Circumnutations parameters determined automatically; example of seedling number 14 on Additional file 1: Video 1

Numero of cycle	Start of cycle	End of cycle	Middle of cycle	Period (min)	Length (mm)	Rate (mm min^{-1})	Shape	Angle (°)	Direction
1	2013-07-30 18:35	2013-07-30 21:00	2013-07-30 19:47	145	17	0.12	0.13	-70	?
2	2013-07-30 21:00	2013-07-30 23:10	2013-07-30 22:05	130	32	0.25	0.14	-78	CW
3	2013-07-30 23:10	2013-07-31 01:45	2013-07-31 00:27	155	48	0.31	0.16	-78	CW
4	2013-07-31 01:45	2013-07-31 04:10	2013-07-31 02:57	145	40	0.28	0.19	-52	CCW
5	2013-07-31 04:10	2013-07-31 06:35	2013-07-31 05:22	145	39	0.27	0.28	-23	CCW
6	2013-07-31 06:35	2013-07-31 09:00	2013-07-31 07:47	145	45	0.31	0.50	17	CCW
7	2013-07-31 09:00	2013-07-31 11:30	2013-07-31 10:15	150	51	0.34	0.36	61	CCW
8	2013-07-31 11:30	2013-07-31 14:10	2013-07-31 12:50	160	56	0.35	0.66	86	CCW
9	2013-07-31 14:10	2013-07-31 17:10	2013-07-31 15:40	180	84	0.47	0.96	-15	CCW
10	2013-07-31 17:10	2013-07-31 20:10	2013-07-31 18:40	180	96	0.54	0.78	26	CCW
11	2013-07-31 20:10	2013-07-31 23:05	2013-07-31 21:37	175	103	0.59	0.73	45	CCW
12	2013-07-31 23:05	2013-08-01 02:05	2013-08-01 00:35	180	118	0.66	0.93	-87	CCW
13	2013-08-01 02:05	2013-08-01 05:15	2013-08-01 03:40	190	120	0.63	0.79	20	CCW
14	2013-08-01 05:15	2013-08-01 08:30	2013-08-01 06:52	195	104	0.54	0.62	35	CCW
15	2013-08-01 08:30	2013-08-01 11:40	2013-08-01 10:05	190	82	0.43	0.59	71	CCW
16	2013-08-01 11:40	2013-08-01 14:45	2013-08-01 13:12	185	64	0.35	0.83	79	CCW
17	2013-08-01 14:45	2013-08-01 18:00	2013-08-01 16:22	195	73	0.37	0.73	-88	CCW
18	2013-08-01 18:00	2013-08-01 21:10	2013-08-01 19:35	190	69	0.36	0.58	-84	CCW
19	2013-08-01 21:10	2013-08-02 00:20	2013-08-01 22:45	190	76	0.40	0.55	-52	CCW
20	2013-08-02 00:20	2013-08-02 04:00	2013-08-02 02:10	220	98	0.44	0.62	-41	CCW
21	2013-08-02 04:00	2013-08-02 08:00	2013-08-02 06:00	240	108	0.45	0.60	-4	CCW

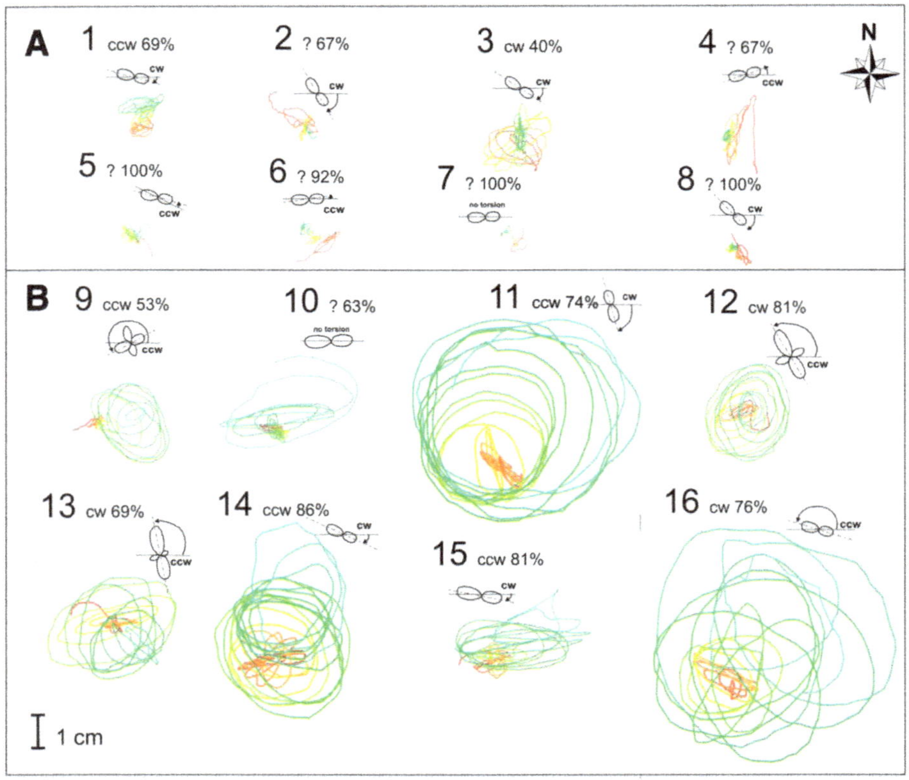

Figure 4 The top view of the trajectory of hypocotyl circumnutation and cotyledon torsion of *Helianthus annuus* seedlings. A. Distilled water. **B**. Nutrient solution. Cotyledon arrangement at the end of filming. Torsion in relation to beginning East–west cotyledon arrangement. All circumnutations during 63 hours of experiment. Numbers from 1 … 16 are numbers of seedlings in Additional file 1: Video 1.

ended at 09:00 a.m. on the seventh day of seedling growth (Additional file 1: Video 1). A monochromatic camera (Mintron MTV-1368CD, Mintron Enterprise Co. Ltd, Taipei, Taiwan) was used to record the circumnutation trajectory of the hypocotyl apex. The camera parameters (focus, aperture, and exposure time) remained constant during the experiment. The plants were filmed from the top and the camera was oriented corresponding to geographical plane and thus the coordinates x and y were simultaneously east–west (EW) and north–south (NS) of the geographical direction, respectively. Time-lapse images were recorded one frame per 5 minute by Gotcha! Multicam software (Prescient System Inc., West Chester, PA, USA). The system was calibrated by

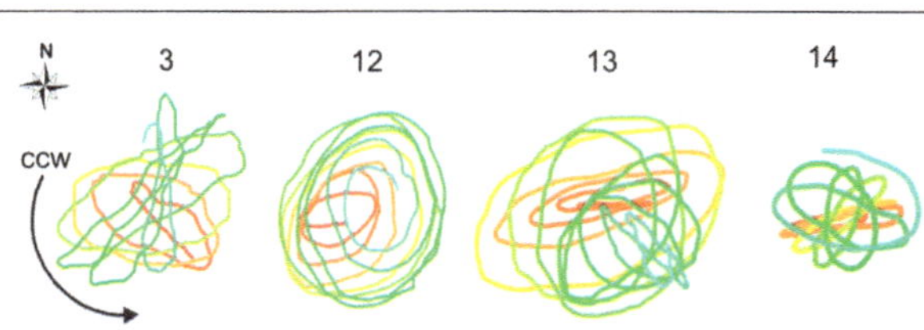

Figure 5 The top view of the rosette-like trajectory of circumnutation of *Helianthus annuus* hypocotyls. The seedlings (number 3, 12, 13, 14) are examples from Additional file 1: Video 1.

filming the line with a millimetre scale at the level of the organ origin (ground level). The time-lapse images were digitized using *Circumnutation Tracker* and Microsoft Excel programs. Experimental points (coordinates x, y of the stem apex on the horizontal plane) were determined at 5-min intervals.

Statistical analysis

The results obtained are presented as the mean ± SE in each experimental group. The results were estimated for significance by a t-test at $p < 0.05$ and $p < 0.01$, $p < 0.00001$.

Conclusions

We think that the *CT* software could be a useful tool for future research of circumnutation behaviour and may allow finding movement phenotypes [43]. The *CT* is an easy tool facilitating the circumnutation research, which could be helpful for plant physiology researchers and students. Therefore, the *CT* is available on http://circumnutation.umcs.lublin.pl. The user guide, explanation of installation, and samples are accessible. Nowadays, given the fast development of digital image recording [44,45], the time-lapse method of recording plant movement will develop rapidly; therefore, suitable software for time-lapse video analysis is required. Our software requires a minimal custom-made video input and can be adapted to different low-budget time-lapse imaging setups. In future, we will also work on fully automated harvesting of coordinates x y and 3D circumnutation-growth modelling.

Availability and requirements

- Project name: *Circumnutation Tracker*
- Project home page: http://circumnutation.umcs.lublin.pl
- Operating system(s): Windows (XP, Vista, 7), Linux
- Programming language: C++
- Other requirements: Qt, libavcodec, blas, lapack, armadillo

CT is freely available from the authors' web pages and source code are freely available on request. *CT* can be used, modified and distributed freely as long as this publication and the original authors are acknowledged. If research projects benefited much from *CT*, this publication should be cited in arising papers.

Additional file

Additional file 1: Video 1. A time-lapse video of circumnutation of *Helianthus annuus* seedlings growing in distilled water and nutrient solution.

Abbreviations

CT: Circumnutation Tracker; cw: Clock-wise; ccw: Counterclock-wise.

Competing interests

The authors declare that they have no competing interests.

Authors' contributions

MS wrote of this paper, designed the software architecture and methods. MŻ programming a software. EK and HD revised the final manuscript. All authors read and approved the final manuscript.

Acknowledgments

We thank Professor Kazimierz Trębacz for useful discussions and comments on the manuscript and Monika Stelmachowicz, Agnieszka Bara, Dariusz Buk for their extensive use of *CT* during its development.

References

1. Darwin C, Darwin F: *The Power of Movement in Plants.* London: John Murray; 1880.
2. Johnsson A: **Circumnutations: results from recent experiments on Earth and in space.** *Planta* 1997, **203**:147–158.
3. Kiss JZ, Choat B, Pittermann J, Strullu-Derrien C, Rioult J-P, Strullu D-G: **Plants circling in outer space.** *New Phytol* 2009, **182**:555–557.
4. Whippo CW, Hangarter RP: **The "sensational" power of movement in plants: a Darwinian system for studying the evolution of behavior.** *Am J Bot* 2009, **96**:2115–2127.
5. Edwards W, Moles AT: **Re-contemplate an entangled bank: the Power of Movement in Plants revisited.** *Bot J Linn Soc* 2009, **160**:111–118.
6. Edwards W, Moles AT, Franks P: **The global trend in plant twining direction.** *Global Ecol Biogeogr* 2007, **16**:795–800.
7. Israelsson D, Johnsson A: **A theory for circumnutations in Helianthus annuus.** *Physiol Plant* 1967, **20**:957–976.
8. Buda A, Zawadzki T, Krupa M, Stolarz M, Okulski W: **Daily and infradian rhythms of circumnutation intensity in Helianthus annuus.** *Physiol Plant* 2003, **119**:582–589.
9. Hayashi Y, Nishiyama H, Tanoi K, Ohya T, Nihei N, Tanioka K, Nakanishi TM: **An aluminum influence on root circumnutation in dark revealed by a new super-harp (high-gain avalanche rushing amorphous photoconductor) camera.** *Plant Cell Physiol* 2004, **45**:351–356.
10. Johnsson A, Solheim BGB, Iversen TH: **Gravity amplifies and microgravity decreases circumnutations in Arabidopsis thaliana stems: results from a space experiment.** *New Phytol* 2009, **182**:621–629.
11. Millet B, Badot PM: **The revolving movement mechanism in Phaseolus; New approaches to old questions.** In *Vistas on Biorhythmicity.* Edited by Greppin H, Degli Agosti R, Bonzon M. Geneva: University of Geneva; 1996:77–98.
12. Shabala SN, Newman IA: **Proton and calcium flux oscillations in the elongation region correlate with root nutation.** *Physiol Plant* 1997, **100**:917–926.
13. Shabala SN, Newman IA: **Root nutation modelled by two ion flux-linked growth waves around the root.** *Physiol Plant* 1997, **101**:770–776.
14. Koukkari WL, Bingham C, Hobbs JD, Duke SH: **In search of a biological hour.** *J Plant Physiol* 1997, **151**:352–357.
15. Hatakeda Y, Kamada M, Goto N, Fukaki H, Tasaka M, Suge H, Takahashi H: **Gravitropic response plays an important role in the nutational movements of the shoots of Pharbitis nil and Arabidopsis thaliana.** *Physiol Plant* 2003, **118**:464–473.
16. Yoshihara T, Iino M: **Circumnutation of rice coleoptiles: its occurrence, regulation by phytochrome, and relationship with gravitropism.** *Plant Cell Environ* 2005, **28**:134–146.
17. Care AF, Nefed'ev L, Bonnet B, Millet B, Badot PM: **Cell elongation and revolving movement in Phaseolus vulgaris L. twining shoots.** *Plant Cell Physiol* 1998, **39**:914–921.
18. Schuster J, Engelmann W: **Circumnutations of Arabidopsis thaliana seedlings.** *Biol Rhythm Res* 1997, **28**:422–440.
19. Spurny M: **Elongation and circumnutation oscillations of hypocotyl of pine seedlings (Pinus silvestris L.).** *Biol Plantarum* 1975, **17**:43–49.
20. Stolarz M, Krol E, Dziubinska H, Zawadzki T: **Complex relationship between growth and circumnutations in Helianthus annuus stem.** *Plant Signal Behav* 2008, **3**:376–380.
21. Binder BM, O'Malley RC, Wang WY, Zutz TC, Bleecker AB: **Ethylene stimulates nutations that are dependent on the ETR1 receptor.** *Plant Physiol* 2006, **142**:1690–1700.
22. Zachariassen E, Johnsson A: **Effects of lithium ions on the circumnutations of Helianthus hypocotyls.** *Physiol Plant* 1988, **72**:147–152.
23. Stolarz M, Krol E, Dziubinska H, Kurenda A: **Glutamate induces series of action potentials and a decrease in circumnutation rate in Helianthus annuus.** *Physiol Plant* 2010, **138**:329–338.
24. Charzewska A, Zawadzki T: **Circadian modulation of circumnutation length, period, and shape in Helianthus annuus.** *J Plant Growth Regul* 2006, **25**:324–331.
25. Niinuma K, Someya N, Kimura M, Yamaguchi I, Hamamoto H: **Circadian rhythm of circumnutation in inflorescence stems of Arabidopsis.** *Plant Cell Physiol* 2005, **46**:1423–1427.
26. Spurny M, Konecna D: **Diurnal and ultradiurnal oscilations of growing organs within framework of information-system of plant.** *Biol Plantarum* 1978, **20**:241–247.
27. Inoue N, Arase T, Hagiwara M, Amano T, Hayashi T, Ikeda R: **Ecological significance of root tip rotation for seedling establishment of Oryza sativa L.** *Ecol Res* 1999, **14**:31–38.
28. Kosuge K, Iida S, Katou K, Mimura T: **Circumnutation on the water surface: female flowers of Vallisneria.** *Sci Rep* 2013, **3**:1133.
29. Larson KC: **Circumnutation behavior of an exotic honeysuckle vine and its native congener: influence on clonal mobility.** *Am J Bot* 2000, **87**:533–538.
30. Migliaccio F, Piconese S: **Spiralizations and tropisms in Arabidopsis roots.** *Trends Plant Sci* 2001, **6**:561–565.
31. Vandenbrink JP, Brown EA, Harmer SL, Blackman BK: **Turning heads: The biology of solar tracking in sunflower.** *Plant Sci* 2014, **224**:20–26.
32. Britz SJ, Galston AW: **Physiology of movements in stems of seedling Pisum sativum L. cv Alaska. 2. The role of the apical hook and of auxin in nutation.** *Plant Physiol* 1982, **70**:1401–1404.

33. Tepper HB, Yang RL: **Influence of the shoot tip and leaves on circumnutation in green pea seedlings.** *Bot Acta* 1996, **109:**502–505.
34. Clark RT, MacCurdy RB, Jung JK, Shaff JE, McCouch SR, Aneshansley DJ, Kochian LV: **Three-dimensional root phenotyping with a novel imaging and software platform.** *Plant Physiol* 2011, **156:**455–465.
35. Lobet G, Pages L, Draye X: **A novel image-analysis toolbox enabling quantitative analysis of root system architecture.** *Plant Physiol* 2011, **157:**29–39.
36. Russino A, Ascrizzi A, Popova L, Tonazzini A, Mancuso S, Mazzolai B: **A novel tracking tool for the analysis of plant-root tip movements.** *Bioinspir Biomim* 2013, **8:**025004. doi:10.1088/1748-3182/8/2/025004 (http://iopscience.iop.org/1748-3190/8/2/025004)
37. De Vylder J, Vandenbussche F, Hu Y, Philips W, Van der Straeten D: **Rosette Tracker: an open source image analysis tool for automatic quantification of genotype effects.** *Plant Physiol* 2012, **160:**1149–1159.
38. Wang L, Uilecan IV, Assadi AH, Kozmik CA, Spalding EP: **HYPOTrace: image analysis software for measuring hypocotyl growth and shape demonstrated on arabidopsis seedlings undergoing photomorphogenesis.** *Plant Physiol* 2009, **149:**1632–1637.
39. Kitazawa D, Miyazawa Y, Fujii N, Nitasaka E, Takahashi H: **Characterization of a novel gravitropic mutant of morning glory, weeping2.** *Adv Space Res* 2008, **42:**1050–1059.
40. Fitzgibbon A, Pilu M, Fisher RB: **Direct least square fitting of ellipses.** *Ieee T Pattern Anal* 1999, **21:**476–480.
41. Baillaud L: **Mouvements autonomes des tiges, vrilles et autres organes à l'exception des organes volubiles et des feuilles.** In *Physiology of Movements/Physiologie der Bewegungen. Volume 17/2.* Edited by Aletsee L, Anker L, Baillaud L, Banbury GH, Brauner L, Crombie WM, Drews G, Girbardt M, Haupt W, Hoffmann-Berling H, Kaldewey H, Kamiya N, Larsen P, Nultsch W, Pohl R, Rufelt H, Snow R, Straka H, Umrath K, Ziegler H, Zurzycki J, Bünning E. Berlin Heidelberg: Springer; 1962:562–634. Encyclopedia of Plant Physiology / Handbuch der Pflanzenphysiologie.
42. Takahashi T, Matsuhara S, Abe M, Komeda Y: **Disruption of a DNA topoisomerase I gene affects morphogenesis in Arabidopsis.** *Plant Cell* 2002, **14:**2085–2093.
43. Kolukisaoglu U, Thurow K: **Future and frontiers of automated screening in plant sciences.** *Plant Sci* 2010, **178:**476–484.
44. Lobet G, Draye X, Perilleux C: **An online database for plant image analysis software tools.** *Plant Methods* 2013, **9:**38.
45. Spalding EP, Miller ND: **Image analysis is driving a renaissance in growth measurement.** *Curr Opin Plant Biol* 2013, **16:**100–104.

Protocol: a simple method for extracting next-generation sequencing quality genomic DNA from recalcitrant plant species

Adam Healey*, Agnelo Furtado, Tal Cooper and Robert J Henry

Abstract

Next-generation sequencing technologies rely on high quality DNA that is suitable for library preparation followed by sequencing. Some plant species store large amounts of phenolics and polysaccharides within their leaf tissue making genomic DNA extraction difficult. While many DNA extraction methods exist that contend with the presence of phenolics and polysaccharides, these methods rely on long incubations, multiple precipitations or commercially available kits to produce high molecular weight and contaminant-free DNA. In this protocol, we describe simple modifications to the established CTAB- based extraction method that allows for reliable isolation of high molecular weight genomic DNA from difficult to isolate plant species *Corymbia* (a eucalypt) and *Coffea* (coffee). The simplified protocol does not require multiple clean up steps or commercial based kits, and the isolated DNA passed stringent quality control standards for whole genome sequencing on Illumina HiSeq and TruSeq sequencing platforms.

Keywords: DNA extraction, Next-generation sequencing, CTAB, *Corymbia*, *Coffea*

Introduction

With the advent of next-generation sequencing (NGS) technologies, investigation into the genomes of important industrial plant species has never been easier or more economical. Traditionally, genome analysis has relied on relatively small amounts of DNA of moderate purity for the purpose of restriction enzyme fingerprinting or polymerase chain reaction (PCR) amplification. However, NGS technologies, such as the Illumina HiSeq and TruSeq platforms, allow for the investigation of the entire genome of plants and as such require the input of several micrograms of high quality DNA. In the context of NGS, high quality DNA is characterized as DNA that is predominantly high molecular weight with an A260/280 ratio between 1.8 and 2.0 and without contaminating substances, such as polysaccharides or phenolics [1,2], which impede or inhibit DNA library preparation for NGS. The quality and quantity requirements for plant DNA sequencing by NGS often confine extraction methods to using leaf material, which is problematic due to the accumulation of high amounts of phenolics and polysaccharides within a variety of species [3,4]. Polysaccharides, due to their chemical properties, coprecipitate with genomic DNA, giving solutions a viscous, glue-like appearance [5] and are known to inhibit the NGS library preparation (Nock C, personal communication). Phenolics, such as terpenoids and tannins, undergo rapid oxidation upon their release from leaf tissue and irreversibly bind to the phosphate backbone of DNA, characterized by the browning of leaf material [6,7]. Both contaminants prevent the use of DNA for molecular biology purposes, such as PCR, restriction digests, or sequencing by inhibiting the action of polymerases or endonucleases [8,9]. Forest trees, such as *Corymbia*, and species belonging to the *Coffea* genus also accumulate these contaminants in their leaves, limiting the study of their genomes [10-12].

The majority of DNA extraction methods from plant leaf tissue are derived from the original hexadecyltrimethylammonium bromide (CTAB) based method, described by Doyle and Doyle in 1987 [13]. To contend with the problems associated with phenolics and polysaccharides, the protocol has been modified to include polyvinylpyrrolidone

* Correspondence: adamhealey65@gmail.com
Queensland Alliance for Agriculture and Food Innovation, University of Queensland, Brisbane 4072, Australia

(PVP) and high salt solutions to isolate genomic DNA [5,6,8,9,14,15]. Unfortunately, these methods can be time consuming, either relying on long incubation steps, nuclei pre-extraction that increases handling time, or requiring multiple DNA washes and precipitations that decrease overall yield. For NGS library preparation, DNA from difficult-to-isolate plant species often require commercial kit based methods to supplement CTAB based extractions to generate genomic DNA of high enough quality to pass stringent conditions for library preparation [4,16]. Kit based extraction methods are intended to easily remove contaminants, but are often expensive, particularly when many samples are required for analysis. The problem of losing DNA through subsequent column washes or precipitations can be exacerbated when only small amount of leaf tissue is available for collection.

NGS quality control requirements are often very stringent and require preparation of DNA that is of high molecular weight with little evidence of band shearing, containing no evidence of contamination from protein, RNA or polysaccharides, and has a 260/280 nm absorbance ratio of approximately 1.8-2.0. A fast, simple, and reliable DNA extraction method that does not rely on long incubations, multiple DNA precipitations, or supplementation of commercial supplies or reagents to meet next-generation library preparation requirements will be invaluable to plant research. The method described below illustrates how the addition of PVP alone to an established CTAB based method does not necessarily translate to an effective DNA extraction protocol, and demonstrates how subtle manipulations to an extraction protocol can isolate high quality genomic DNA from recalcitrant plant species, free of contamination and suitable for NGS library preparation.

Materials and methods

Consumables

50 mL Falcon Tubes

RNAse A (Sigma Cat No. R6513)

Polyvinylpyrrolidone (PVP) (Sigma Cat No. PVP10) (not required)

Chloroform: Isoamyl alcohol 24:1 (Sigma Cat No. C0549)

Liquid nitrogen

β-mercaptoethanol (Sigma Cat No. 63689)

Trizma base (Sigma Cat No. 1503)

Ethylenediaminetetraacetic acid disodium salt dihydrate (EDTA) (Chem Supply Cat No. EA023)

Agarose (Amresco Cat No. 0710)

Sodium Chloride (NaCl) (Ajax Finechem Cat No. 1103414)

Hexadecyltrimethylammonium bromide (CTAB) (Sigma Cat No. 52365)

*Eco*RI (not required- used for quality assurance) (NEB Cat No. R0101S)

*Hin*dIII-HF (not required- used for quality assurance) (NEB Cat No. R3104S)

Reagents

Extraction Buffer: 100 mM Tris-HCl (pH 7.5), 25 mM EDTA, 1.5 M NaCl, 2% (w/v) CTAB, and 0.3% (v/v) β-mercaptoethanol- added immediately before use

RNAse A stock solution (10 mg/mL)

5 M NaCl

95% ethanol (v/v)

70% ethanol (v/v)

TE Buffer: 10 mM Tris-HCl (pH 7.6), 0.1 mM EDTA

CutSmart Buffer (NEB Cat No. B7204S)

NEBuffer EcoRI (NEB Cat No. B0101S)

Equipment

Mortar and Pestle

Water Baths (65°C and 37°C)

Centrifuge (capable of spinning 50 mL centrifuge tubes at 5000 × g)

-20°C Freezer

Gel electrophoresis system (e.g. Jordan Scientific JP-250)

NanoDrop UV/Vis Spectrophotometer (e.g. NanoDrop 8000, Thermo Scientific)

Plant material and tissue collection

Leaf tissue of *Corymbia citriodora* subsp. *variegata, Corymbia henryi, Corymbia torelliana, and Corymbia citriodora* subsp. *citriodora* was obtained from Queensland Department of Agriculture, Fisheries and Forestry in Gympie, Australia. Leaf tissue of *Coffea brassii* was obtained from the Australian Tropical Herbarium in Cairns, Australia. Leaf material after harvesting was transported on ice and stored at -80°C until subjected to DNA extraction.

Protocol

Preparatory steps

Before grinding, pre-chill the mortar and pestle (to minimize frozen tissue thawing) and 95% ethanol solution at -20°C. Pre-heat water baths (65°C and 37°C) before beginning the extraction. Once pre-heated, prepare 10 mL (per 1 g of leaf tissue) extraction buffer by adding 0.3% (v/v) β-mercaptoethanol in a 50 mL Falcon tube, and preheat in the 65°C water bath. PVP can also be added at this point, but is not required.

Grinding and tissue disruption

Using liquid nitrogen, grind 1 g of frozen leaf tissue into a fine powder. Place the powder into a new 50 mL Falcon tube and mix in the pre-heated extraction buffer. Put the sample into the 65°C water bath and mix by inversion every 10 min for 30 min- 1 h. After incubation, centrifuge the sample tube for 5 min at 5000 × g (to pellet and remove un-lysed leaf tissue) and decant the supernatant into a new 50 mL Falcon tube.

Protein Extraction and RNAse treatment

Add 1 volume of chloroform: isoamyl alcohol to the solution and mix by inversion for 5 min. Centrifuge the sample for 10 min at 5000 × g and pipette the upper aqueous phase into a new Falcon tube, taking care to avoid the aqueous/organic layer interface. Add 5 µL of RNAse A (10 mg/mL) to the solution and incubate at 37°C for 15 min with periodic, gentle mixing. After incubation, add 1 volume of chloroform: isoamyl alcohol to the solution and mix by inversion for 5 min. Centrifuge the solution for 10 min at 5000 × g and pipette the aqueous phase into a new Falcon tube, again taking care to avoid the organic layer.

Precipitation

Add ½ volume of 5 M NaCl to the sample and mix gently by inversion. Then, add 3 volumes of cold 95% ethanol and mix gently by inversion. Place the tubes into a -20°C freezer and incubate for 1 h. NOTE: do not leave the sample at -20°C for more than 1 h as both the CTAB and NaCl can precipitate from solution, preventing DNA isolation.

After incubation, centrifuge the Falcon tube for 10 min at 5000 × g to pellet the DNA. Carefully decant away the supernatant and wash the DNA pellet with 3 mL of 70% ethanol. Gently swirl the solution and centrifuge again for 10 min at 5000 × g. Carefully decant the supernatant and air-dry DNA pellet for 15 min at room temperature. Once dried, suspend DNA in 200 µL of TE buffer.

DNA quality and quantity assessment

Assess the quality of the extracted DNA using a NanoDrop UV/Vis spectrophotometer and 0.7% (w/v) agarose gel, looking for a single absorbance peak at 260 nm, a 260/280 absorbance ratio of 1.8-2.0, and no evidence of substantial band shearing or contamination (either RNA or polysaccharide).

Comments

Since the advent of the CTAB-based extraction method from plant leaves by Doyle and Doyle in 1987, many different iterations have been published, each with modifications to contend with the co-extractives of polyphenolics and polysaccharides present in the leaves of many plant species [3,5-8,15]. While having demonstrated their effectiveness for isolating DNA that is suitable for PCR amplification or restriction digests, all methods currently published in the literature require long incubations, and multiple precipitation steps and ethanol washes to produce RNA-free genomic DNA of high purity. As next-generation sequencing requires large amounts of high quality DNA, each additional precipitation and wash increases handling time and lowers overall yield. Commercial column based extraction kits, such as DNeasy (Qiagen, Australia) or Wizard (Promega, Australia), are effective for isolating contaminant free DNA from recalcitrant plant species, including eucalypts [4,16]. However, commercial kits can be expensive and carry the risk of losing DNA on the column, which in turn necessitates several extractions followed by pooling of DNA.

To test the modifications made to the extraction method (NGS protocol) against the well-established original CTAB method (used routinely in our laboratory to reliably extract high quality DNA from rice, sugarcane, barley and wheat for sequencing [17,18]), six grams of frozen *Corymbia citriodora* subsp. *variegata* leaf tissue was ground and aliquoted evenly into the extractions described below. The quality of DNA from each extraction was verified spectrophotometrically using a NanoDrop instrument and agarose gel electrophoresis. The NanoDrop absorbance profile is useful for detecting contamination such as protein, salts or polysaccharides, all of which can inhibit NGS library preparation. High quality DNA is characterized as having a 260/280 nm absorbance ratio of approximately 1.8, with a single absorbance peak at 260 nm. The spectrophotometric profile is also useful for detecting phenolic oxidation, as the aromatic structure will absorb at 230 and 270 nm [1]. If oxidation is suspected to have occurred, endonuclease digestion can be used to further assess DNA quality before library preparation as phenolics, which inhibit polymerases, also inhibit restriction enzymes [8,9].

Visualization of DNA on an agarose gel provides evidence of band shearing and RNA and polysaccharide contamination. Mechanical disruption, such as vortexing, causes DNA strands to shred apart, indicated by a wide DNA band with poor resolution. NGS library submission requires intact, high molecular weight genomic DNA, so all solution mixing steps were done by gentle inversion. Gel electrophoresis is also beneficial for visualizing RNA and polysaccharides, both of which contaminate sequencing reactions. RNA is evident as a distinct banding pattern at various sizes throughout the gel, whereas polysaccharides will migrate quickly and conglomerate at the bottom of the gel as a non-distinct fluorescent structure. Yield was determined through relative band intensity approximation with 100 and 200 ng λ DNA standards, as the NanoDrop concentration readings can inflate yield of genomic DNA.

Traditional CTAB extraction method

Using the original CTAB protocol, we were unsuccessful in isolating DNA from *Corymbia* leaves. During incubation at 65°C, the extraction solution began darkening, eventually turning brown. Upon precipitation, despite the observation of a small, brown pellet, the agarose gel failed to reveal any DNA (Figure 1, lane 4). The UV/Vis spectrophotometer absorbance peaks at 220-230 nm and 270-280 nm (Figure 2A) are likely due to polysaccharide, phenol and aromatic co-extractives [1]. The browning of

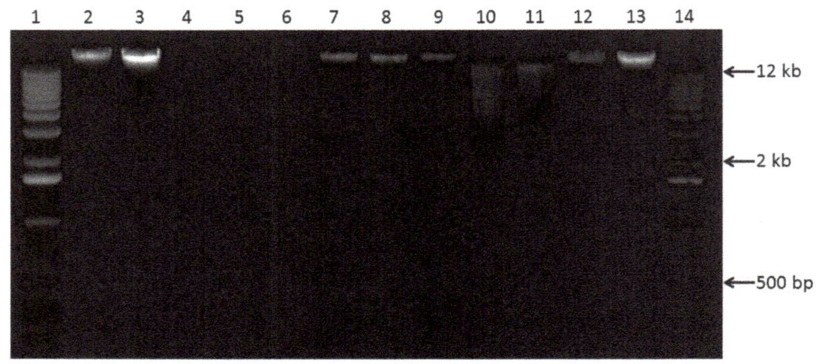

Figure 1 Genomic DNA preparation of *Corymbia citriodora* subsp. *variegata* resolved by electrophoresis. 1 kb DNA ladder (1, 14), 100 and 200 ng λ DNA standards (2, 3 and 12, 13) respectively. DNA extractions using the traditional CTAB-based method with no PVP (4), 1% PVP (5), and 4% PVP (6). DNA extractions using the NGS protocol with no PVP (7), 1% PVP (8), and 4% PVP (9). Endonuclease digestions of DNA extracted without PVP with *Eco*RI (10) and High-Fidelity *Hin*dIII (11). Results from six grams of leaf tissue finely ground using a mortar and pestle, then aliquoted (1 g) for each extraction. DNA was resolved by electrophoresis in a 0.7% agarose gel and visualized using SYBR Safe DNA gel stain. Percentages are represented as w/v. CTAB: hexadecyltrimethylammonium bromide; PVP: polyvinylpyrrolidone.

solution has been attributed to the oxidation of phenolic secondary metabolites in plant leaves [4,6,19,20], a known problem with *Corymbia* [11,12] and *Coffea* [10].

The addition of PVP into CTAB based extractions to absorb phenolics, preventing their oxidation that renders DNA unusable for downstream application, has been used successfully for other recalcitrant plant species [4,5,7,9], typically at a concentration of 1-2% (w/v). The addition of 1% and 4% PVP to the traditional CTAB extraction method failed to isolate any useable DNA from *Corymbia citriodora* subsp. *variegata*. Again, the browning of solution occurred, and upon precipitation, a minute brown

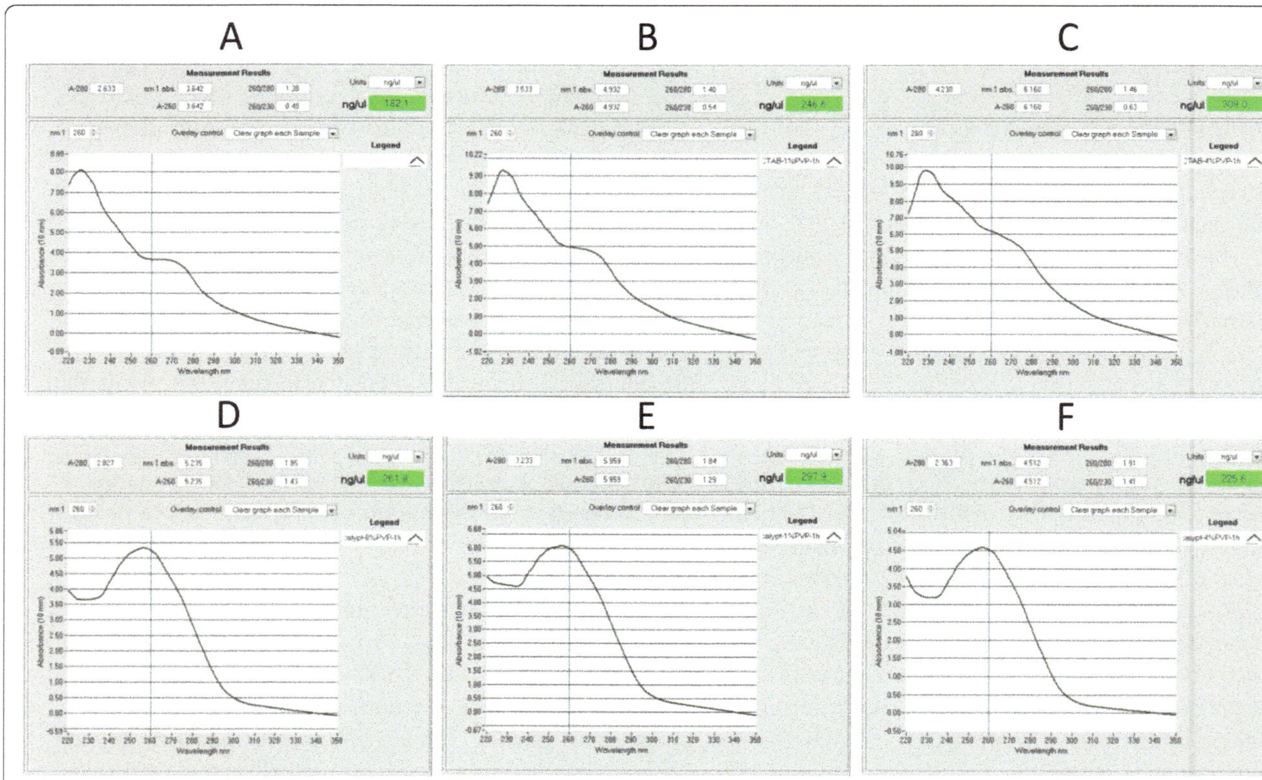

Figure 2 NanoDrop measurement profile of genomic DNA extractions from *Corymbia citriodora* subsp. *variegata*. DNA extractions using a traditional CTAB-based method with **(A)** no PVP, **(B)** 1% PVP and **(C)** 4% PVP. DNA extractions using the NGS protocol with **(D)** no PVP, **(E)** 1% PVP, and **(F)** 4% PVP. Results from six grams of leaf tissue finely ground using a mortar and pestle, then aliquoted (1 g) for each extraction. Percentages are represented as w/v. CTAB: hexadecyltrimethylammonium bromide; PVP: polyvinylpyrrolidone.

pellet was observed for each extraction. NanoDrop measurements revealed the persistence of the contamination absorbance peaks of 220-230 nm and 270-280 nm (Figure 2B-C). When resolved on a 0.7% agarose gel, no DNA was observed (Figure 1, lanes 5-6).

NGS extraction protocol

The NGS protocol allowed for the isolation of high quality DNA from *Corymbia citriodora* subsp. *variegata*. The NanoDrop spectrophotometer measurement profile showed a single absorbance peak at 260 nm, and a 260/280 ratio of 1.85 (Figure 2D) Gel electrophoresis revealed a single, high molecular weight DNA band with little evidence of shearing and no RNA or polysaccharide contamination (Figure 1 lane 7). To further evaluate the quality of the extracted genomic DNA, approximately 1 μg (per reaction) was digested overnight at 37°C with restriction enzymes *Eco*RI and High-Fidelity *Hin*dIII (New England BioLabs, Ipswich Massachusetts). Resolution of the digests on the agarose gel revealed efficient endonuclease activity of both enzymes (Figure 1 lanes 10 and 11, respectively).

The spectrophotometric profile and yield varied little when increasing amounts of PVP (1% and 4% w/v) were added to the extraction buffer. Each DNA extraction had a 260/280 absorbance ratio of 1.84 and 1.91 respectively (Figure 2E-F), and high molecular weight DNA band with little shearing or contaminants (Figure 1 lanes 8-9). Based on relative band intensity of the 2 μL of sample resolved on the gel with the 100 ng λ DNA standard, the method consistently yielded approximately 5 μg of DNA per gram of leaf tissue. Although the A260/230 ratios (a secondary measure of DNA quality) [1] for the extractions were lower than expected (1.41, 1.29, and 1.43 respectively), these results in combination with the endonuclease digestions suggested that PVP was not required to prevent phenolic oxidation, and the protocol was suitable for the isolation of DNA for whole genome NGS library preparation and sequencing. The modifications and considerations for the protocol are discussed below.

Modifications

The modifications to the previously cited methods were designed to simplify the protocol and maximize DNA yield by reducing the number of handling steps, DNA precipitations, and washes required, and eliminating the need for long incubations or supplementation with commercial based kits and reagents.

Phenolic oxidation

As shown (Figure 1 lanes 7-11 and Figure 2D-F), PVP was not required to prevent phenolic oxidation which renders DNA unsuitable for use. This is likely due to the presence of β-mercaptoethanol, a reducing agent [2], and the centrifugation step after 65°C incubation. The centrifugation and pelleting of un-lysed leaf material for removal was included to reduce the continued leeching of leaf phenolics into solution. Also, as un-lysed leaf tissue settles at the interface between the aqueous and organic phases during the first protein extraction step, its early removal increases the clarity between the two phases allowing easier pipetting of the aqueous portion. After centrifugation, chloroform: isoamyl alcohol was added as quickly as possible for further separate phenolics from the aqueous portion. If PVP has been added for phenolic absorption, the first protein extraction step will remove the majority, while the remainder is removed by the second extraction.

RNAse treatment

RNAse A treatment, a requirement for the isolation of high quality genomic DNA, is traditionally added after the DNA has been precipitated, washed and dissolved into a stabilizing buffer which necessitates additional steps to remove the enzyme and re-precipitate and wash the DNA. Each additional handling step and precipitation may produce DNA of higher quality, but decreases overall yield as typically, the simplest method for extraction will provide the most reliable result [21]. In this protocol, RNAse A was added between the two chloroform: isoamyl alcohol solvent extractions to allow for a single DNA precipitation step at the end of the protocol. As two washes using chloroform: isoamyl alcohol are required for high quality DNA extraction, the addition of the RNAse and the 15 minute incubation at 37°C after the first solvent extraction efficiently digests RNA, while the second solvent extraction removes the enzyme. This eliminates the need for further treatments, precipitations and washes once the DNA was re-suspended in TE buffer during the final step of the procedure.

Precipitation

Included with the protocol was the addition of a high salt solution before DNA was precipitated with 95% cold ethanol. Polysaccharides have a similar solubility to DNA and co-precipitate in either isopropanol or ethanol, inhibiting downstream molecular application [4]. The addition of a high salt buffer increases their solubility in ethanol, allowing their removal once the DNA has been precipitated and pelleted [22]. During the -20°C precipitation step, the 1 h incubation time should not be exceeded as NaCl and CTAB will eventually precipitate, preventing the DNA pellet from forming during centrifugation.

NGS library submission, preparation and sequencing

Based on the success of the method, the NGS protocol was applied to other samples of intended for NGS library preparation and sequencing. Due to limited amounts of leaf tissue available, 4% PVP was included with the

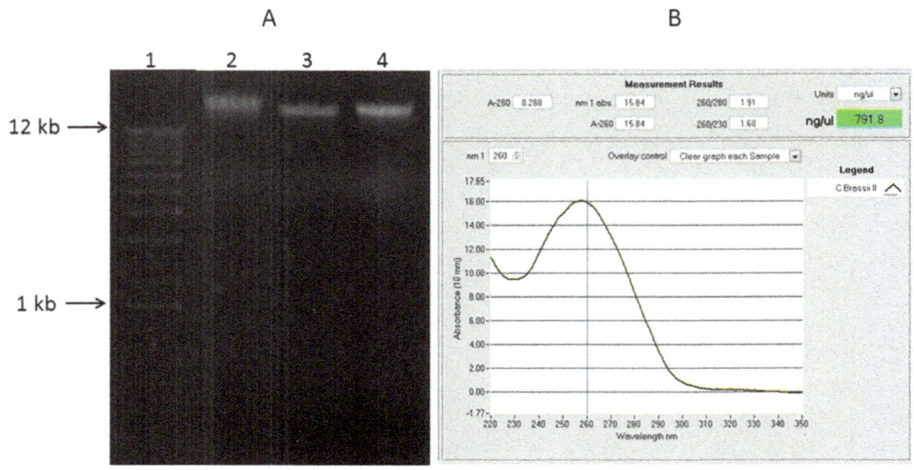

Figure 3 DNA quality and yield assessment for *Coffea brassii* genomic DNA using the NGS extraction protocol. A) Genomic DNA preparation of *Coffea brassii* resolved by electrophoresis. 1 kb DNA ladder (1), 100 ng and 200 ng λ DNA standards respectively (3, 4), and DNA extraction using the modified NGS extraction protocol (2). DNA was separated by electrophoresis in a 0.7% agarose gel and visualized using SYBR Safe DNA gel stain. The gel image was cropped to exclude an unrelated sample. **B)** NanoDrop measurement profile of *Coffea brassii* leaf extracted genomic DNA using the modified NGS extraction protocol.

extraction for full confidence that no phenolic oxidation would occur. High quality DNA was extracted from *Corymbia citriodora* subsp. *citriodora*, *Corymbia henryi*, *Corymbia citriodora* subsp. *variegata* and *Corymbia torelliana* for NGS library preparation and sequencing. The *Corymbia* genomic DNA samples were submitted to the Joint Genome Institute (JGI) for library preparation and sequencing on the Illumina HiSeq 2500 platform and passed their quality control measures which require: high molecular weight genomic DNA free of polysaccharide, RNA and protein contamination, and a 260/280 nm absorbance ratio between 1.6 and 2.2.

The robustness of the NGS protocol was demonstrated with another recalcitrant plant genus, *Coffea* [10], to isolate high quality DNA from *Coffea brassii* for sequencing. As only 0.1 grams of *C. brassii* leaf material was available, the protocol was modified to use 5 mL of extraction buffer and 40 μL of TE buffer to resuspend the DNA pellet. The DNA extraction was successful, the NanoDrop spectrophotometer profile showing a single 260 nm absorbance peak and a 260/280 nm absorbance ratio of 1.91 (Figure 3B). Resolution of 2 μL of DNA by gel electrophoresis revealed high molecular weight DNA with little evidence of shearing and no observable contamination (Figure 3A). Based on relative band intensity with the 100 ng λ DNA standard, approximately 1.5-2 μg of DNA was isolated, indicating a theoretical yield of 15-20 μg of DNA per gram of leaf tissue. The *C. brassii* DNA sample was submitted to the Australian Genome Research Facility (AGRF) for library preparation and paired end sequencing on the Illumina TruSeq platform.

A summary of the DNA extractions for NGS library preparation and sequencing by JGI and AGRF is provided in Table 1.

Sequencing quality

As sequencing is influenced by the quality of DNA provided, raw Illumina reads from JGI and AGRF were assessed using read quality distributions, generated by CLC Bio Genomics WorkBench, software version 5.5.2 (CLC Bio, Denmark). Read quality distributions are based upon PHRED quality

Table 1 Summary of *Corymbia* and *Coffea* genomic DNA extractions and sequencing results from JGI and AGRF

Species	Absorbance 260/280 nm	Quantity of DNA submitted (μg)	NGS quality control results	Number of sequencing reads
Corymbia henryi	1.78	19.1	Pass (JGI)	206,959,160 (HiSeq)
Corymbia citriodora subsp. *citriodora*	1.81	4.4	Pass (JGI)	169,513,988 (HiSeq)
Corymbia citriodora subsp. *variegata*	1.82	5.5	Pass (JGI)	234,021,522 (HiSeq)
Corymbia torelliana	1.93	6.3	Pass (JGI)	213,411,194 (HiSeq)
Coffea brassii	1.91	1.5	Pass (AGRF)	145,197,482 (TruSeq)

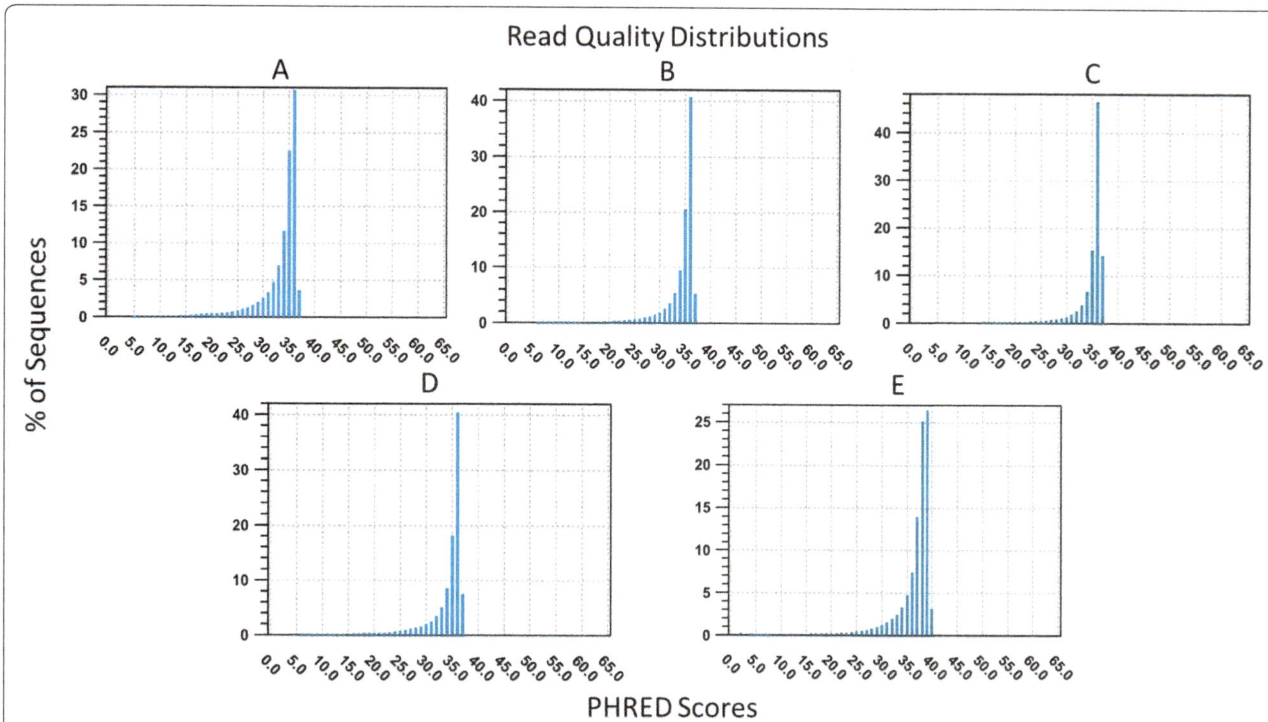

Figure 4 Paired end read quality distributions of *Corymbia* and *Coffea* samples from Illumina HiSeq and TruSeq library preparations. **(A)** *Corymbia citriodora* subsp. *citriodora*, **(B)** *Corymbia henryi*, **(C)** *Corymbia torelliana*, **(D)** *Corymbia citriodora* subsp. *variegata*, and **(E)** *Coffea brassii*. The x-axis represents the PHRED quality score and the y-axis represents the percentage of sequences with a particular score, normalized to the total number of sequences. The distribution graph was generated using CLC Bio software (Version 5.5.2).

scores, which estimate the probability of error per base call [23]. The read quality distribution visualizes this data for the entire sequencing library, normalized to the total number or sequences.

Each *Corymbia* sequencing library (prepared by JGI, filtered above a PHRED score of 5) and the *C. brassii* sequencing library (prepared by AGRF) produced over 100 million reads per library, with a modal PHRED quality score of 36 and 39, respectively (Figure 4A-E). This represents a base call accuracy of approximately 99.999%, providing high confidence in the quality of DNA submitted. Despite lower A260/230 ratios for the submitted *Corymbia* (~1.4) and *Coffea* (1.68) samples, there were no observable differences between the library preparation and sequencing quality for the two species on either sequencing platform.

Conclusion

The described method, developed to improve genomic DNA extractions from leaf tissue of recalcitrant plant species, is a marked improvement over other methods as it does not require multiple clean up steps, precipitations, or commercial based kits or reagents. Using the protocol, high quality DNA was isolated from species of *Corymbia* and *Coffea* that passed stringent Illumina NGS library submission requirements, despite high amounts of high leaf phenolics and polysaccharides. The method was generated with the intent of using a single protocol for all plant species, regardless of the presence or absence of DNA co-extractive contaminants. With this robust protocol, whole-genome sequencing is possible from recalcitrant plant species using established DNA sequencing technologies for advanced bioinformatics investigations.

Competing interests

The authors declare they have no competing interests.

Authors' contributions

AH isolated genomic DNA from *Corymbia* and drafted the manuscript with sections written by AF. AH, AF, and RH developed and optimized the NGS protocol. TC extracted genomic DNA from *C. brassii* using the NGS protocol. All authors read and approved the final manuscript.

Acknowledgements

This work was supported by the University of Queensland and a postgraduate scholarship from the Queensland Alliance for Agriculture and Food Innovation. The authors would like to thank David Lee of the University of the Sunshine Coast and the Queensland Department of Agriculture, Fisheries and Forestry (QDAFF) for providing the *Corymbia* leaf material for extraction providing helpful comments for the manuscript, and Jason Lupoi of the Joint BioEnergy Institute for his comments. We thank Prof. Darren Crayn of Australian Tropical Herbarium for supplying the *C. brassii* leaves, Myrna Constantin for endonuclease digestions, and Joint Genome Institute and Australian Genome Research Facility for library preparation and sequencing of samples.

References

1. Thermo Scientific: **260/280 and 260/230 Ratios.** In *T009-Technical Bulletin.* 2008:1–2.
2. Kasem S, Rice N, Henry RJ: **DNA Extraction from Plant Tissue.** In *Plant Genotyping II: SNP Technology.* Edited by Henry RJ, CAB International. 2008:219–271.
3. Križman M, Jakše J, Baričevič D, Javornik B, Prošek M: **Robust CTAB-activated charcoal protocol for plant DNA extraction.** *Acta Agric Slov* 2006, **87:**427–433.
4. Tibbits JFG, Mcmanus LJ, Spokevicius AV, Bossinger G: **A rapid method for tissue collection and high-throughput isolation of genomic DNA from mature trees.** *Plant Mol Biol Report* 2006, **24:**81–91.
5. Porebski S, Bailey LG, Baum BR: **Modification of a CTAB DNA extraction protocol for plants containing high polysaccharide and polyphenol components.** *Plant Mol Biol Report* 1997, **15:**8–15.
6. Sahu SK, Thangaraj M, Kathiresan K: **DNA extraction protocol for plants with high levels of secondary metabolites and polysaccharides without using liquid nitrogen and phenol.** *ISRN Mol Biol* 2012, **2012:**1–6.
7. Maliyakal JE: **An efficient method for isolation of RNA and DNA from plants containing polyphenolics.** *Nucleic Acids Res* 1992, **20:**2381.
8. Kim CS, Lee CH, Shin JS, Chung YS, Hyung NI: **A simple and rapid method for isolation of high quality genomic DNA from fruit trees and conifers using PVP.** *Nucleic Acids Res* 1997, **25:**1085–1086.
9. Khanuja SPS, Shasany AK, Darokar MP, Kumar S: **Rapid isolation of DNA from dry and fresh samples of plants producing large amounts of secondary metabolites and essential oils.** *Plant Mol Biol Report* 1999, **17:**1–7.
10. Labat M, Augur C, Rio B, Perraud-Gaimé I, Sayadi S: **Biotechnological Potentialities Of Coffee And Similar With Olive, Two Models Of Agroindustrial Products Rich In Polyphenolic Compounds: A Review.** In *Coffee Biotechnology and Quality.* Edited by Sera T, Soccol CR, Pandey A, Roussos S, Kluwer Academic Publishers. 2000:517–531.
11. Asante KS, Brophy JJ, Doran JC, Goldsack RJ, Hibbert DB, Larmour JS: **A comparative study of the seedling leaf oils of the spotted gums: species of the *Corymbia* (Myrtaceae), section Politaria.** *Aust J Bot* 2001, **49:**55–66.
12. Brophy JJ, Forster PI, Goldsack RJ, Brynn Hibbert D: **The essential oils of the yellow bloodwood eucalypts (*Corymbia*, section Ochraria, Myrtaceae).** *Biochem Syst Ecol* 1998, **26:**239–249.
13. Doyle J, Doyle J: **A rapid procedure for DNA purification from small quantities of fresh leaf tissue.** *Phytochem Bull* 1987, **19:**11–15.
14. Carrier G, Santoni S, Rodier-Goud M, Canaguier A, De Kochko A, Dubreuil-Tranchant C, This P, Boursiquot J-M, Le Cunff L: **An efficient and rapid protocol for plant nuclear DNA preparation suitable for next generation sequencing methods.** *Am J Bot* 2011, **98:**e13–15.
15. Peterson DG, Boehm KS, Stack SM: **Isolation of milligram quantities of nuclear DNA from tomato (*Lycopersicon esculentum*), a plant containing high levels of polyphenolic compounds.** *Plant Mol Biol Report* 1997, **15:**148–153.
16. Shepherd M, Cross M, Stokoe RL, Scott LJ, Jones ME: **High-throughput DNA extraction from forest trees.** *Plant Mol Biol Report* 2002, **20:**425.
17. Furtado A: **DNA Extraction from Vegetative Tissue for Next-Generation Sequencing.** In *Cereal Genomics Methods and Protocols.* Edited by Henry RJ, Furtado A, Springer Protocols. 2014:1–5.
18. Furtado A, Henry RJ: **The wheat Em promoter drives reporter gene expression in embryo and aleurone tissue of transgenic barley and rice.** *Plant Biotechnol J* 2005, **3:**1–14.
19. Couch JA, Fritz PJ: **Isolation of DNA from plants high in polyphenolics.** *Plant Mol Biol Report* 1990, **8:**8–12.
20. Katterman FRH, Shattuck VI: **An effective method of DNA isolation from the mature leaves of *Gossypium* species that contain large amounts of phenolic terpenoids and tannins.** *Prep Biochem* 1983, **13:**347–359.
21. Henry RJ: **Plant DNA Extraction.** In *Plant Genotyping DNA Fingerprinting Plants.* Edited by Henry RJ, CAB International. 2001:239–250.
22. Fang G, Hammar S, Grumet R: **A quick and inexpensive method for removing polysaccharides from plant genomic DNA.** *Biotechniques* 1992, **13:**52–54.
23. Ewing B, Green P: **Base-calling of automated sequencer traces using Phred. II. Error probabilities.** *Genome Res* 1998, **8:**186–194.

Next generation sequencing and *de novo* transcriptomics to study gene evolution

Achala S Jayasena[1], David Secco[1], Kalia Bernath-Levin[1], Oliver Berkowitz[1,3], James Whelan[2] and Joshua S Mylne[1*]

Abstract

Background: Studying gene evolution in non-model species by PCR-based approaches is limited to highly conserved genes. The plummeting cost of next generation sequencing enables the application of *de novo* transcriptomics to any species.

Results: Here we describe how to apply *de novo* transcriptomics to pursue the evolution of a single gene of interest. We follow a rapidly evolving seed protein that encodes small, stable peptides. We use software that needs limited bioinformatics background and assemble four *de novo* seed transcriptomes. To demonstrate the quality of the assemblies, we confirm the predicted genes at the peptide level on one species which has over ten copies of our gene of interest. We explain strategies that favour assembly of low abundance genes, what assembly parameters help capture the maximum number of transcripts, how to develop a suite of control genes to test assembly quality and we compare several sequence depths to optimise cost and data volume.

Conclusions: *De novo* transcriptomics is an effective approach for studying gene evolution in species for which genome support is lacking.

Keywords: *De novo* transcriptomics, Gene evolution, PawS1, Cyclic peptides

Background

De novo transcriptomics

Next generation sequencing (NGS) technology and the accompanying drop in the per-base cost of the sequence data has enabled many new approaches in molecular biology such as whole genome sequencing [1], transcriptomics [2], metagenomics [3], epigenomics [4], proteomic applications such as ProteinSeq [5], and single-cell sequencing [6]. Before NGS, transcriptomics was largely limited to model species using oligo arrays hybridised to transcript libraries. NGS has allowed for transcriptomic studies leading to gene discovery [2,7], SNP detection [8], simple sequence repeat discovery [9], and gene pathway description [10]. The challenge with non-model transcriptomes is a bioinformatic one, they need to be assembled without any reference genomes *i.e. de novo*. One approach to avoid *de novo* assembly is to assemble the transcriptomes of a non-model species using a partial reference-based strategy with sequences from a closely related model species [11].

De novo transcriptome assembly will continue to become less challenging as the length of NGS reads increases [12] and as software develops better algorithms for unsupported assemblies. A wealth of open source and LINUX-based software developed for transcriptome assembly exist including Trinity [13], SOAPdenovo (http://soap.genomics.org.cn/soapdenovo.html), Velvet [14], CAP3 [15], and TGICL [16]. Commercial software for the mainstream operating systems also exists with examples including CLC Genomics Workbench (CLC bio), SeqMan Ngen (DNASTAR), and gsAssembler (Life Technologies).

The *PawS1* gene

De novo transcriptomics is being used to study the evolution of the *PawS1* (Preproalbumin with SFTI-1) genes, which were first discovered in sunflower (*Helianthus annuus*) [17]. Sunflower PawS1 (151-residue protein) and a close relative PawS2 (137-residue protein) have an unusual dual biosynthesis in that each are matured into two very different proteins; a 10.5 kDa napin-type seed storage albumin and an 'extra' 1.5 kDa cyclic peptide that is buried in

* Correspondence: joshua.mylne@uwa.edu.au
[1]The University of Western Australia, School of Chemistry and Biochemistry & ARC Centre of Excellence in Plant Energy Biology, 35 Stirling Highway, Crawley, Perth 6009, Australia
Full list of author information is available at the end of the article

the proalbumin sequence (Figure 1) [17]. To study *PawS1* evolution, genes have been amplified using PCR from related species with primers designed against genomic DNA sequence flanking the *PawS1* and *PawS2* genes. This approach identified *PawS1* genes from Asteraceae members related to sunflower and defined a new class of seed peptide termed "PawS-Derived Peptides" (PDPs) that is at least 18 million years old [18]. Seed storage proteins are already known to evolve rapidly [19], but the PDP region in *PawS1* genes evolved faster than the adjacent albumin [18]. The PCR-based amplification strategy also discovered some *PawS-like* (*PawL1*) genes that share many features with *PawS1*, but the PDP region does not produce a stable peptide and is proposed to be an ancestral form of *PawS1* [18].

The heterologous PCR approach is limited in that it cannot amplify *PawS1* genes from Asteraceae species that are more distantly related to sunflower. Very little sequence information is available for the species in which *PawS1* genes are present. The first draft of the sunflower genome is pending and has been challenging due to its large size (3,600 Mbp) and high (81%) content of transposable elements [22]. In Elliott *et al.* [18] *de novo* transcriptomes for sunflower and mountain arnica (also known as leopard's bane, wolf's bane, and mountain tobacco; *Arnica montana*) indicated NGS was a viable approach to study the evolution of *PawS1*. We have subsequently assembled *de novo* transcriptomes for dozens of seed RNA samples and have found by cloning the putative genes and finding supporting peptide evidence by mass spectrometry that NGS-based study of single genes is a viable approach. Among the species analysed, *Zinnia haageana* (also known as Persian carpet, Mexican Zinnia, dwarf Zinnia) was particularly challenging due to presence of a large number of closely-related *PawS1* genes.

Drawbacks of NGS-based gene discovery

NGS is not the best approach for all studies of gene evolution. If the gene sequence of interest is highly conserved, a PCR approach remains the most cost-effective. Quality control of NGS data is critical due to the potential for sequence or assembly errors. Cloning large number of genes, even with primers designed from the assembled sequences can be laborious, especially if the assembled transcript does not contain a complete open reading frame (ORF). RNA-seq is of course dependent on the gene being transcribed so it will not detect gene evolution that triggers a loss in gene expression. Finally for labs to adopt this approach there are the obvious bioinformatic barriers, the most obvious being the handling and processing of millions of reads which might require new computational skills or software in addition to new hardware to store the data - formidable considering the target gene constitutes only a very small proportion of the reads gathered.

Benefits of an NGS-based approach to studying gene evolution

Despite these drawbacks, applying NGS has its benefits; (i) The type of sequencing can be selected based on the objective. To assemble transcriptomes from higher organisms, high data volumes are required, so they are usually sequenced using either Illumina or SOLiD technologies

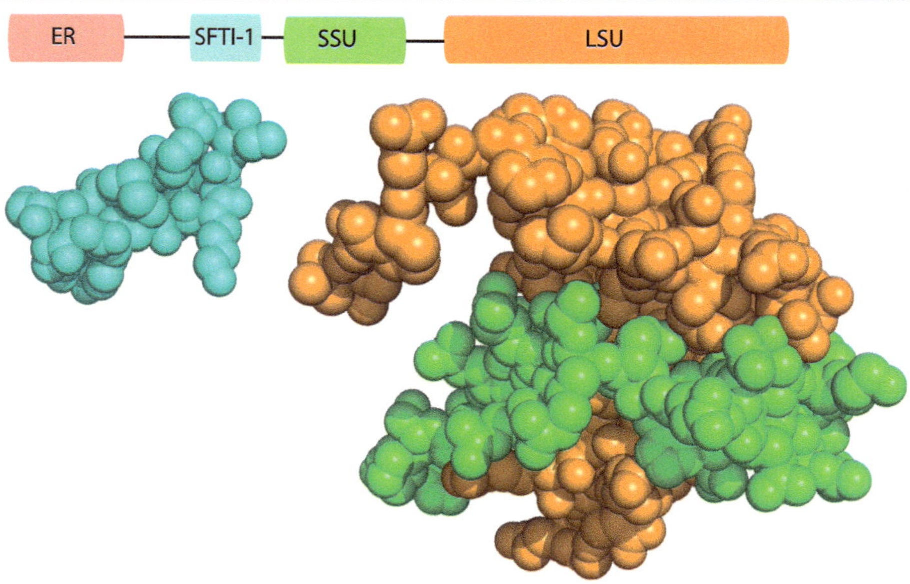

Figure 1 Structure of the sunflower preproalbumin PawS1 and its mature proteins. PawS1 is matured into two proteins [17] represented here as sphere-formatted structural models; a small peptide called SFTI-1 (cyan, PDB 1sfi [20]) and a heterodimeric seed storage albumin consisting of a small subunit (SSU, green) and a large subunit (LSU, orange), here represented by a *Brassica napus* napin structure (PDB Bnlb [21]).

(reviewed in [23]). The 454 platform creates longer reads, but the error rate is relatively high (reviewed in [24]). (ii) Software, such as CLC Genomics Workbench (CLC bio) is designed for users without extensive bioinformatics training. This software had been used by different groups to assemble *de novo* transcriptomes in non-model plants [25,26]. (iii) NGS-based methods can provide the full ORF and this is preferable for situations where PCR primers may only be designed within ORFs and must be followed by 3' and 5' RACE; (iv) NGS can discover transcripts in any species for which RNA can be purified and so permits studying fast-evolving genes; (v) A suite of control genes ranging in expression can indicate whether the absence of a specific transcript is due to either a poor assembly or poor data for assembly; (vi) RNA-seq by NGS provides the expression level of transcripts [27], and (vii) Unlike microarray based transcriptomic approaches, RNA-seq by NGS can also find splice variants if present [28].

Although we have an interest in the evolution of the *PawS1* gene [18], *PawS1* also serves as a good test case for the application of *de novo* transcriptomics to track gene evolution. PawS1 is a precursor for albumin and seed storage genes in general have been shown to evolve rapidly [19]. PawS1 also encodes a peptide whose mass can be detected readily using mass spectrometry, proving proteinacous evidence for assembled transcripts. Finally, the *PawS1* gene is found in sunflower, for which no genome is currently available and is in a plant family studied little with genomic approaches. Most species we have tested contain only one or two *PawS1* genes, but we found that *Z. haageana* has many and this offered an ideal dataset to test the ability to separate related genes and test the parameters for gene discovery.

Here, using four Illumina RNA-seq datasets (including the challenging *Z. haageana*) the bioinformatic strategies adopted to discover *PawS1* transcripts are outlined, the selection of a suite of control genes to evaluate the quality of seed-RNA assemblies are detailed to help explain why a *PawS1* gene might not be present in an assembled transcriptome. These strategies can be used to develop an NGS approach to studying the evolution of any gene in model and non-model species.

Results and discussion

De novo transcriptomics approach for *PawS1* genes

To demonstrate the use of NGS to study gene evolution, two recently published and publicly available datasets originating from dry-seed mRNA of *H. annuus* (SAMN02569067) [18] and *A. montana* (SAMN02569068) [18] were used. Additionally new sequence data from dry-seed mRNA of *Z. haageana* (SAMN02933922) and *Heliopsis helianthodes* (commonly known as smooth oxeye or false sunflower) (SAMN02933923) were also obtained. All sequence data were paired-end reads. The reads were assembled using CLC Genomics Workbench 6.5.1 (CLC bio).

Quality control of the reads

Although CLC Genomics can clean reads during import, the open source FASTX toolkit (http://hannonlab.cshl.edu/fastx_toolkit/) was used for read trimming and filtering as it offers more flexibility to define quality thresholds.

Table 1 shows the number of clean reads obtained after trimming and filtering. For *Z. haageana* and *H. helianthoides* datasets, trimming parameters were maintained as quality threshold of the read ends (t) at 30 and the minimum length after trimming (l) at 50. A quality threshold (t) of 30 means the probability of incorrect base call is 1 in 1,000 bases or the base call accuracy is 99.9%. A minimum length after trimming (l) of 50 means reads shorter than 50 bp after trimming will be discarded.

Filtering parameters for these two datasets were set as q30, p90. A quality threshold (q) of 30 means the base call accuracy is 99.9%. The minimum percentage of bases that must match (p) the specified quality threshold (in this case q30) is 90. For example, quality filtering of the *H. helianthoides* dataset to achieve a quality threshold of 22 (q22, p90) removed only 8% reads compared to 22% when q30, p90 is used (Table 1). A quality threshold of 20 (q20) equals to a base call accuracy of 99%. Similarly 15% of raw reads were removed when quality filtering parameters were set at q30, p90 in the *Z. haageana* dataset. Less stringent filtering parameters (*i.e.* q22, p90) removed only 5% of the raw reads.

Clean reads from both *H. helianthoides* and *Z. haageana* under the two different quality filtering parameters were assembled with the word size 60, paired method described below. The assemblies were queried for PawS1 using tBLASTn. Two different quality filtering parameters did not affect the number of *PawS/PawL* transcripts found in both species.

However, it was observed that quality filtering with less stringent parameters followed by assembly helped obtain longer *PawS/PawL* transcripts, and sometimes identified novel transcripts in some datasets.

Performing the assembly

To find the optimum CLC parameters to identify the maximum number of *PawS/PawL* transcripts, we assembled *H. helianthoides* and *Z. haageana* data using two different methods (Figure 2). Initially all forward and reverse paired-end reads were combined and assembled under CLC default settings where the word size (*k-mer*) is automatically set by the program, indicating the minimum contig length as 100. The word size automatically set by the program was 23 for both datasets. In this case paired information was not used for assembly. In the second method, paired information was used and a

Table 1 Assembly statistics

Species	Read length	Raw reads	Clean reads	Assembler	N50	Contig count
H. annuus	101	2 x 21,404,702	40,742,686	CLC (ws60,paired)	482	59,530
A. montana	101	2 x 14,458,043	27,516,042	CLC (ws60,paired)	485	45,194
Z. haageana	101	2 x 38,382,090	64,649,107	CLC (autows,non-paired)	308	205,324
Z. haageana	101	2 x 38,382,090	64,649,107	CLC (ws60,paired)	435	80,460
Z. haageana	101	2 x 38,382,090	72,756,408	CLC (ws60,paired)	629	40,764
H. helianthoides	101	2 x 109,627,594	169,128,716	CLC (autows,non-paired)	305	443,800
H. helianthoides	101	2 x 109,627,594	169,128,716	CLC (ws60,paired)	497	151,272
H. helianthoides	101	2 x 109,627,594	200,130,791	CLC (ws60,paired)	496	162,563

Clean reads were assembled using two methods; automatic word size (autows, 23), non-paired and word size 60 (ws60), paired method. Number of c ean reads when quality filtering was done to achieve a quality threshold (q) of 30 and 22 are shown for *Z. haageana* and *H. helianthoides* datasets. N50 refers to the contig length where 50% of the assembly is represented by contigs of this size or longer.

range of word sizes was tested as well as setting the minimum contig length to 300, instead of the default 200. The number of *PawS*/*PawL* transcripts found in assemblies with different word sizes varied greatly and among the word sizes tested, 60 gave the maximum number of transcripts from all four RNA-seq datasets.

Development of a suite of core transcripts

Traditional quality metrics such as N50 (N50 refers to the contig length where 50% of the assembly is represented by contigs of this size or longer) and contig count are often used to represent the quality of *de novo* assemblies, but such figures can be mis-leading [29,30]. Parra *et al.* [30] have described a more reliable method to assess the quality of *de novo* assembled genomes. The most important outcome from a genome or a transcriptome is their catalogue of genes. So Parra *et al.* [30] used a set of core genes which should be present in all eukaryotic genomes as controls and analysed how completely these core genes are assembled in eukaryotic genomes assembled using different methods.

We also used a similar approach to assess the quality of our *de novo* transcriptome assemblies. The CLC *de novo* assembly tool has the option to map reads back to contigs permitting an evaluation of abundance. This permits selection of transcripts with a range of expression levels from high to low.

We assembled the sunflower transcriptome using a word size of 60 with paired-end reads and mapped the reads back to contigs and selected subsets of contigs to represent high, medium, and low levels. To identify the transcripts represented by these contigs we used the BLAST at NCBI tool menu in CLC. Examining these BLAST hits identified a suite of six core genes (Table 2) expressed in seeds. Dozens of Asteraceae seed *de novo* transcriptomes were queried for these core genes to make sure they were conserved in this plant family. The six include transcripts expressed at different levels and included long and short transcripts. Ideal core transcripts for this purpose would be highly conserved single-copy genes. However, without genome support, it was challenging to find such transcripts for Asteraceae members.

The first core transcript we selected was a highly expressed gene belonging to the *LATE EMBRYOGENESIS ABUNDANT* (*LEA*) family. This *LEA* family transcript was found to be ubiquitous in the sunflower mature (dry) seed *de novo* transcriptome. LEA protein is generally formed during the latter part of seed development and was first described in cotton seeds [31], but later found in seeds of many other land plants, also in vegetative tissues. LEA proteins are considered to be involved in plant stress and dehydration tolerance (reviewed in [32]).

The second core transcript was an *OLEOSIN* (*OLE*) gene family member. Oil in seeds is normally stored in oil bodies composed of oleosin proteins that control their structure and lipid accumulation [33]. The majority of Asteraceae are oilseeds and we found this *OLE* to be ubiquitous and highly expressed.

The third core transcript was an *ASPARTIC PROTEINASE* (*AP*). *AP*s are widely distributed in the plant kingdom and found in many plant tissues including seeds [34]. Plant aspartic proteinases are thought to be involved in protein processing, protein degradation, senescence, stress responses, and sexual reproduction (reviewed in [35]). This particular *AP* was moderately expressed in sunflower.

The fourth core transcript was a *PATHOGENESIS RELATED* (*PR*) gene. PRs in plants are normally involved in defence signalling. PR proteins are induced by plant hormones such as jasmonic acid, ethylene, and salicylic acid (reviewed in [36]). *PR* has low expression in sunflower and is the most similar in expression to *PawS1*.

The fifth core transcript we selected was a *CYSTEINE PROTEASE-1* (*CP-1*). The closest member from *Arabidopsis* to this *CP-1 is RD21B* which is a plant papain-like cysteine protease. Plant papain-like cysteine proteases are involved in plant defence, development, and

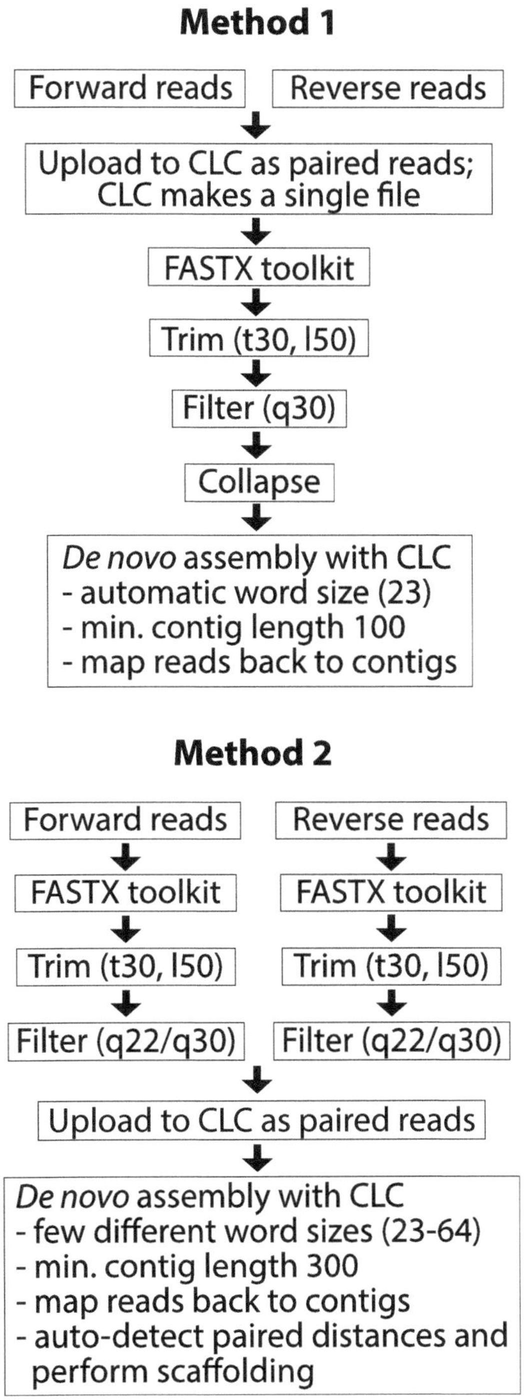

Figure 2 Flow diagram showing the two different assembly methods used. In the first method raw reads were combined before trimming, filtering, and collapsing. Then they were assembled using CLC Genomics using the automatic word size (23) set by the program indicating the minimum contig length as 100. In the second method forward and reverse reads were trimmed and filtered individually. High or low quality filtering parameters were tried. Trimmed and filtered reads were assembled using CLC Genomics. Several different word sizes were tried and the minimum contig length was set to 300. Paired distances were automatically detected by the program and scaffolding was performed.

senescence [37]. *CP-1* is a low abundance transcript in sunflower.

The sixth core transcript was a *SERINE/THREONINE PROTEIN KINASE* (*PK*). PKs are widely distributed in the plant kingdom and are found to be involved in many signalling cascades such as plant hormone signalling, defence responses or tolerance to stresses [38]. This *PK* had the lowest abundance among the six core transcripts.

Of the core transcripts *CP-1* and *PK* were typically expressed at a lower level than our target gene *PawS1* (Table 3) and therefore were the most valuable indicators for the quality of assembly. Presence of *CP-1* and *PK* contigs suggested a high likelihood that transcripts with similar expression levels such as *PawS1/PawL1*, would be included in our assemblies. This was especially important for species where assemblies lacked *PawS1/PawL1* genes.

Using tBLASTn we identified the corresponding core transcripts in each of our *de novo* assembled seed transcriptomes and calculated the percentage of ORF coverage of each core transcript compared to the query. PawS1 is a shorter protein of around 151 amino acids, but all the core transcripts except *LEA* encode proteins that are larger than PawS1 (Table 2). Mapping raw reads to each core transcript (as well as *PawS/PawL* transcripts if present) provided an indication of the relative abundance.

Caution must be used with core transcripts and it is preferable to rely on a suite rather than one or two transcripts. As an example in *Z. haageana* only 57% of *OLEOSIN* is assembled with the automatic word size, non-paired method despite average coverage being very high at 108,400 (Table 3). With word size of 60, paired method, 95% of the transcript is assembled, but the average coverage drops to 12,436 (Table 3). Subtly different gene duplicates and even alleles can confound the assemblies. We selected the transcript with the best matching score from the tBLASTn output to the GenBank ID in Table 2. Many of the control genes are members of multigene families (*e.g.* there are four *OLEOSIN* genes in *Arabidopsis*) which might explain the disparity of results we observed.

Discovering novel *PawS/PawL* transcripts

We used sunflower PawS1 protein sequence as a query to search all assemblies for novel *PawS/PawL* transcripts with the tBLASTn procedure. Precursors of napin-type seed storage albumins typically contain an endoplasmic reticulum (ER) signal and a spacer region which are discarded during the protein maturation process, followed by a polypeptide chain. The polypeptide chain is composed of two subunits separated by a spacer region (Figure 1). The small subunit typically has two conserved cysteines and the large subunit six conserved cysteines (reviewed in [39]). Typically in PawS1 proteins, a short stretch of amino acids with two conserved Cys residues are buried between the spacer region and

Table 2 Details of the core genes used to assess the quality of *de novo* transcriptome assemblies

Gene name	GenBank ID	Length (amino acid)	Average coverage
Late embryogenesis abundant (LEA)	X59700.1	104	174,360
Oleosin (OLE)	X78679.1	183	9,894
Aspartic proteinase (AP)	AB025359.2	509	1,561
Pathogenesis related (PR)	AB091075.1	158	503
Cysteine protease-1 (CP-1)	AB109186.1	461	126
Serine/threonine protein kinase (PK)	AB090881.1	439	50

The level of transcription in sunflower ranges from high (*LEA*) to low (*PK*). The GenBank ID for the amino acid sequence used to tBLASTn each *de novo* transcriptome is shown. The average coverage in a word size 60, paired method transcriptome provides an indication of their relative abundance. This control set was found to be appropriate for the Asteraceae. For reference, the most closely related sequences in *Arabidopsis thaliana* are *LEA4-5* (At5g06760), an *OLEOSIN* family member (At3g01570), *APA1* (At1g11910), *MLP423* (At1g24020), *RD21B* (At5g43060), and *Protein kinase (PK)* super family member (At5g15080).

the small subunit followed by a four amino acid tail (in most cases these four amino acids are Gly-Leu-Asp-Asn) [17,18]. Furthermore, most of these buried peptides (PDPs) starts with a Gly residue and ends with Asp [18]. PawL1 structure is very similar PawS1 except that the buried amino acid sequence does not have the two conserved Cys residues and does not produce a stable peptide [18]. We looked for these characteristic amino acid arrangements when defining novel *PawS/PawL* transcripts from our assemblies.

Elliott *et al.* [18] found *PawS1, PawS2,* and *PawL1* in sunflower and *PawL1* in *A. montana* seed *de novo* transcriptomes assembled using Trinity. The same transcripts were detected using our assembly, but no additional *PawS/PawL* by assembling the two datasets with CLC using word size 60, paired method.

For *H. helianthoides* the same three *PawS1* genes were identified as in Elliott *et al.* [18] amplified using PCR. *H. helianthoides PawS1a* and *PawS1b* gene sequences are very similar to each other and differ only by two nucleotides causing protein sequences to differ in two amino acids. In addition to those three *PawS1* genes, we identified a previously unknown *PawL1* transcript. These four transcripts were present in both the automatic word size, non-paired assembly as well as in the word size 60, paired assembly. This is a good indication that CLC assemblies are reliable in terms of identifying novel *PawS/PawL* transcripts.

In the *Z. haageana* transcriptome, 15 putative *PawS1* and *PawL1* transcripts were identified (Figure 3). However, only twelve transcripts were present in the automatic word size, non-paired assembly (Table 3). The word size 60, paired assembly identified ten transcripts (Table 3). *ZhPawS1g, ZhPawS1i,* and *ZhPawL1a* identified from the initial assembly were not found from the word size 60 assembly. Meantime *ZhPawS1h* and *ZhPawS1j* were only detected in the word size 60 assembly. *ZhPawS1l* could be detected when a word size of 62 was used. Similarly *ZhPawLc* was found only in word size 45 assembly.

Word size had a considerable impact on the outcome from different assemblies. Haznedaroglu *et al.* [40] also reported that each individual *k-mer* value assembly is missing certain biological information. Assembling with several different word sizes will ensure the capture of a maximum number of transcripts. A dataset of around 50 million reads could be assembled within 2–3 hours using CLC Genomics running on a high-speed desktop computer, meaning that several word sizes could be tested throughout a day before selecting the best assembly for a given dataset. In terms of other quality parameters of *de novo* transcriptomes such as contig length, hybrid assemblies, redundancy reduction, and error tolerance, it was shown that CLC Genomics and TGICL pipeline gave better assemblies when different software available for assembling short reads generated by NGS were compared [26].

When comparing the two methods used to assemble *Z. haageana* and *H. helianthoides* datasets we observed that automatic word size (23), non-paired assemblies where minimum contig length was set to 100 created a larger number of contigs when compared to the word size 60, paired assemblies with a minimum contig length of 300 (Table 1). This might be because assemblies with the first method keep all transcripts longer than 100 bp, while the latter removes transcripts shorter than 300 bp. Also the word size (23 vs 60) makes a difference in the number of contigs created. The number of contigs is an important consideration in a *de novo* transcriptome, but when the focus is on a single gene, it is worth retaining biological information to avoid possible loss of information. It is worth noting that our target gene is around 400–600 bp long. We can identify the PDP region which is typically around 14 amino acids, followed by its characteristic tail even in partially assembled shorter transcripts. So it is beneficial to set the minimum contig length to a lower value and also to try few different word sizes as mentioned above to minimize the loss of biological information for this type of study.

The *H. helianthoides* mRNA was sequenced in greater depth (>200 million raw reads) than the other three species (Table 1). Despite the greater depth, the same genes were identified as previously described by Elliott *et al.* [18]

Table 3 *De novo* transcriptome assembly quality statistics

Gene	*H. annuus* ws60, paired	*A. montana* ws60, paired	*Z. haageana* autows, non-paired	*Z. haageana* ws60, paired	*H. helianthoides* autows, non-paired	*H. helianthoides* ws60, paired
LEA	90%	80%	97%	79%	53%	73%
	165,041	23	25,957	5,584	162,273	176,283
OLE	100%	100%	57%	95%	100%	45%
	14,242	23,511	108,400	12,436	114,811	145,334
AP	78%	100%	63%	89%	97%	97%
	1,666	405	237	43	5,704	5,704
PR	100%	100%	89%	100%	79%	100%
	4,922	6,183	6	13	44,849	36,182
CP-1	51%	72%	100%	98%	100%	94%
	122	44	70	66	363	355
PK	73%	63%	83%	97%	75%	99%
	63	48	59	40	175	229
PawS1a	100%	-	100%	100%	70%	70%
	532		3,862	3,362	1,827	1,827
PawS1b	84%	-	79%	100%	32%	40%
	221		809	927	88	86
PawS1c	-	-	29%	58%	32%	41%
			15	58	798	782
PawS1d	-	-	70%	41%	-	-
			19,930	7,916		
PawS1e	-	-	33%	36%	-	-
			2,236	7,394		
PawS1f	-	-	33%	42%	-	-
			2,142	2,518		
PawS1g	-	-	44%	-	-	-
			7,071			
PawS1h	-	-	-	37%	-	-
				129		
PawS1i	-	-	30%	-	-	-
			550			
PawS1j	-	-	-	40%	-	-
				30		
PawS1k	-	-	22%	52%	-	-
			4	8690		
PawL1a	98%	100%	100%	-	100%	100%
	25	400	4		4,711	4,711
PawL1b	-	-	40%	87%	-	-
			4,125	4,736		

All four species were assembled with word size (ws) 60, paired method. In addition, *Z. haageana* and *H. helianthoides* transcriptomes were assembled with the automatic word size (autows, 23), non-paired method. The first line shows the percentage coverage of the identified transcripts when compared to the expected full length protein query. The second line shows the average coverage by mapping trimmed, filtered reads on identified full/partial transcripts. Average coverage means sum of the bases of the aligned part of all the reads divided by the length of the reference sequence.

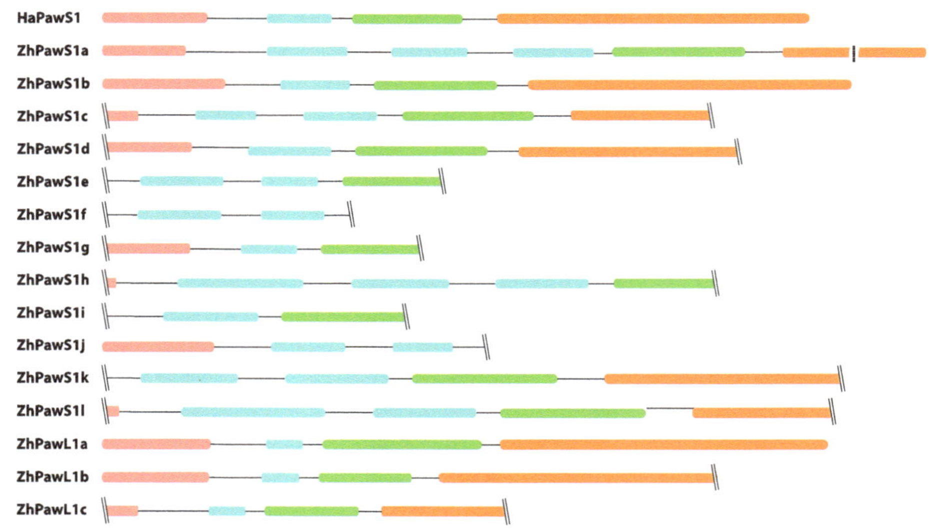

Figure 3 Putative *Z. haageana* PawS1 and PawL1 proteins identified by *de novo* transcriptomics. Colours indicate ER signal (pink), spacer regions (black), the peptide-coding regions (aqua), albumin small subunit (green), albumin large subunit (orange). Predictions of regions are based on comparison to sunflower PawS1. Partial transcripts are indicated by two lines. For the purpose of this figure, only part of the predicted large albumin subunit in the greatly expanded ZhPawS1a was displayed (dashed line).

in *H. helianthoides* as well as in sunflower and *A. montana* where sequencing depth was moderate (Table 1). In addition to these two paired-end assemblies described here, several other Asteraceae seed *de novo* transcriptomes have been assembled with single-end reads (unpublished). Moderate or even low depth, single-end read assemblies were not inferior to paired-end assemblies in terms of discovering novel *PawS*/*PawL* transcripts as well as assembling the suite of core transcripts selected.

Cloning and LC-MS

Z. haageana PawS/*PawL* genes have not been described in previous studies. To confirm the transcripts identified from the *Z. haageana* seed *de novo* transcriptome are not a result of mis-assembly, PCR was used to clone the full ORF of five of these sequences (*PawS1a, b, d, e, g*) and one *PawL1* sequence (*PawL1a*) from seed genomic DNA (Figure 4). Sequence similarity at the nucleotide level between the cloned products and their corresponding assembled transcript was 100% for *PawS1b*, *PawS1e*, *PawS1g*, *PawL1a*, 98% for *PawS1a*, and 91% for *PawS1d*. Partial, overlapping transcripts were identified for *PawS1d* from assemblies with two different word sizes. The full length sequence could be manually reconstructed by combining these partial, overlapping sequences. Primers for *PawS1d* were designed based on this contig. This might explain the relatively higher error rate observed between the cloned *PawS1d* sequence and the transcript from the assembly.

Furthermore, LC-MS was used to provide evidence for these newly discovered *Z. haageana* PDPs at the protein level. The *Z. haageana PawS1* transcripts appear to give rise to seventeen unique PDPs (certain PDPs were repeated in the same gene or in different genes). By extracting peptides from the same seeds and analysing the resulting peptide LC-MS profile we found matching masses for eleven of the seventeen peptides predicted from *PawS1* sequences (Table 4). Synthetic peptides for seven of these peptides were compared to native extracts confirming their existence in LC-MS by displaying the same mass and retention time (Figure 5).

Conclusions

Many groups have used *de novo* transcriptomic approaches for gene discovery in non-model species. Examples include identification of: genes encoding metabolites with biotechnological value [7], pigment pathway genes in spiders [41], phosphate starvation-responsive genes in wheat [2], genes differentially expressed during limb regeneration in salamanders [42], and genes responsible for agronomic traits in pennycress [25]. Comprehensive RNA-seq development resources are becoming available and include that of Wit *et al.* [43], which extends to population genomics. Here we employ a highly focused application for *de novo* transcriptomics to study gene evolution. With the cost of RNA-seq lowering, we believe it is an effective way to study even a single gene.

With the *H. annuus*, *A. montana*, and *H. helinathoides* data we showed that *PawS1* genes previously identified using the traditional PCR approach could be readily detected. Additionally when a *de novo* transcriptomics approach was used a novel *PawL1* transcript could be

ZhPawS1a MAVKVALVVLAVSAILAFVEVSGYRTSITTVTIQDNGICFKDPFGSTLCAPYGLGENPIEDNGICFKD
PFGSTLCAPDGLGDNPIADNGICFKDPFGSTLCAPDGLGDNPREQCDRQIPIQQLNHCQMHLTRSFGY
KLRTMAVENPIQQQHLSLCCNQLQQVEEQCQCKAVRQVAKQALKQLQQQPGGQQRIMKQMLKKAQNLP
NECSLKCSI*

ZhPawS1b MAVKVALVVHVLAVTAILAFLEASSYRTSITTVAIQDNGSCQPLLWGRFCFPDGLRDNPREQCDRQIS
ILQLNHCQMRLTTSFDYMMAVENPKQQEHLNLCCKQLQQVQEQCQCEAVRQVAKQALKQLQQQPGGQK
RIMAEQVVKTAQTLPNKCNLKCSI*

ZhPawS1d MAVKVALVVLAVTAILAFVEVSGYRTSITTVAIQDNGQCEILPWFPFPEACGPDGLGDNPREQCDRQI
PIQQLNHCQMHLTSSFGYKLRMPMENSKQQQHLSLCCKQLQQVEEQCQCKAIGQVAKQALKQLPQQAG
GQQQQQQMMAEQMLKKAQNLPNQCNLKCSI*

ZhPawS1e MAVKVALVVLAVTAILAFVEVSGYRTSITTVAIQDNGLCTILPWPPYLEVCGLDGLQDIQDNGPCYSY
QSCFPDGLGDNPREQCDRQIPIQQLNHCQMHLTSSFGYKLRMPMENSKQQQHLSLCCKQLQQVEEQCQ
CKAIGQVAKQALKQLPQQAGGQQQQQQQMMAEQMLKKAQNLPNQCNLKCSI*

ZhPawS1g MAVKVALVVLAVTAILAFVEVSGYRTSITTVAIQDNGPCYSYQSCFPDGLGDNPREQCDRQIPIQQLN
HCQMHLTSSFGYKLRMPMENSKQQQHLSLCCNQLQQVEEQCQCKAVRQVAKQALKQLPQQAGGQQQQQ
MMAEQMLKKAQNLPNQCNLKCSI*

ZhPawL1a MAKLALLLLTFTALVAFFSASAYKTTIITTTTTTTYDDNGLLPPILDGLDNPWEQCRSQIAIEKLNHCQ
MHLTQGVKRPFQQQHEHLKQCCSQLQNVNPQCQCDALNLVFNEARQKADVIKTRLMLIEAYELPNLCS
LQLQDCSIAAPRV*

Figure 4 Sequences encoded by *Z. haageana PawS1* and *PawL1* genes confirmed by PCR. Colour scheme as Figure 2. Region delimitation is inferred by comparison to sunflower PawS1.

captured from *H. annuus* [18] and in this work, a *PawL1* transcript from *H. helianthoides*. Seed mRNA for *Zinnia haageana* was also sequenced, and this species displayed extensive duplication of *PawS1* which could make assemblies challenging. This same complexity, however allowed us to demonstrate how parameters such as the stringency of quality filtering, word size, assembly method, and the depth of sequencing affected transcript identification. A method to assess the quality of an assembly using core transcripts was also outlined. A commercial software package was sufficient and it could parse closely related genes into separate contigs.

In summary, we outline a comprehensive protocol with commercial software ideal for users with minimal bioinformatics experience and explain how to take seed total RNA through to unsupported assemblies. We found that to improve the likelihood of finding a specific gene, it was beneficial to try several different word sizes during assembly. We also detail how we developed our own set of control genes that varied in expression level among Asteraceae seeds and how we used them to assess quality. Thus the approaches described here are adaptable for laboratories that traditionally do not have expertise with NGS large datasets. Our data illustrate

Table 4 *Zinnia haageana* PDPs matching masses observed in the LC-MS profile

Peptide sequence	Predicted mass (Da)	Observed mass (Da)	LC matches synthetic
GICFKDPFGSTLCAPY	858.88	858.86	Yes
cyclo-GICFKDPFGSTLCAPD	825.86	825.88	Yes
cyclo-GQCEILPWFPFPEACGPD	993.43	993.43	Yes
cyclo-GPCYSYQSCFPD	673.75	673.75	Yes
cyclo-GRPCYTLQSCFPD	733.81	733.83	Yes
cyclo-GLCTILPWPPYLEVCGLD	984.98	984.98	Yes
cyclo-GPCFPMGPWGPFCIPD	850.86	850.86	Yes
cyclo-GRGCFGFPPICFPD	746.83	746.83	ND
cyclo-SAACSHLPPGLREMCAAWSFD	1114.99	1114.97	ND
cyclo-SAACSHLQPVLREMCVARFD	1107.02	1107.01	ND
cyclo-GAACSHIEPGLREMCAAWSFFD	1189.51	1189.48	ND

The predicted mass is for the doubly charged [M^{+2}] ion and assumes the Cys residues form a disulfide bond (oxidised). For cyclic peptides the monoisotopic mass is reduced by 18 Da. The observed masses from LC-MS experiments are calculated from the doubly charged [M^{+2}] ions. Seven synthetic peptides were ordered, which allowed additional confirmation of the retention time of the mass during liquid chromatography (LC). The retention times for some masses were not confirmed using synthetic peptides (ND).

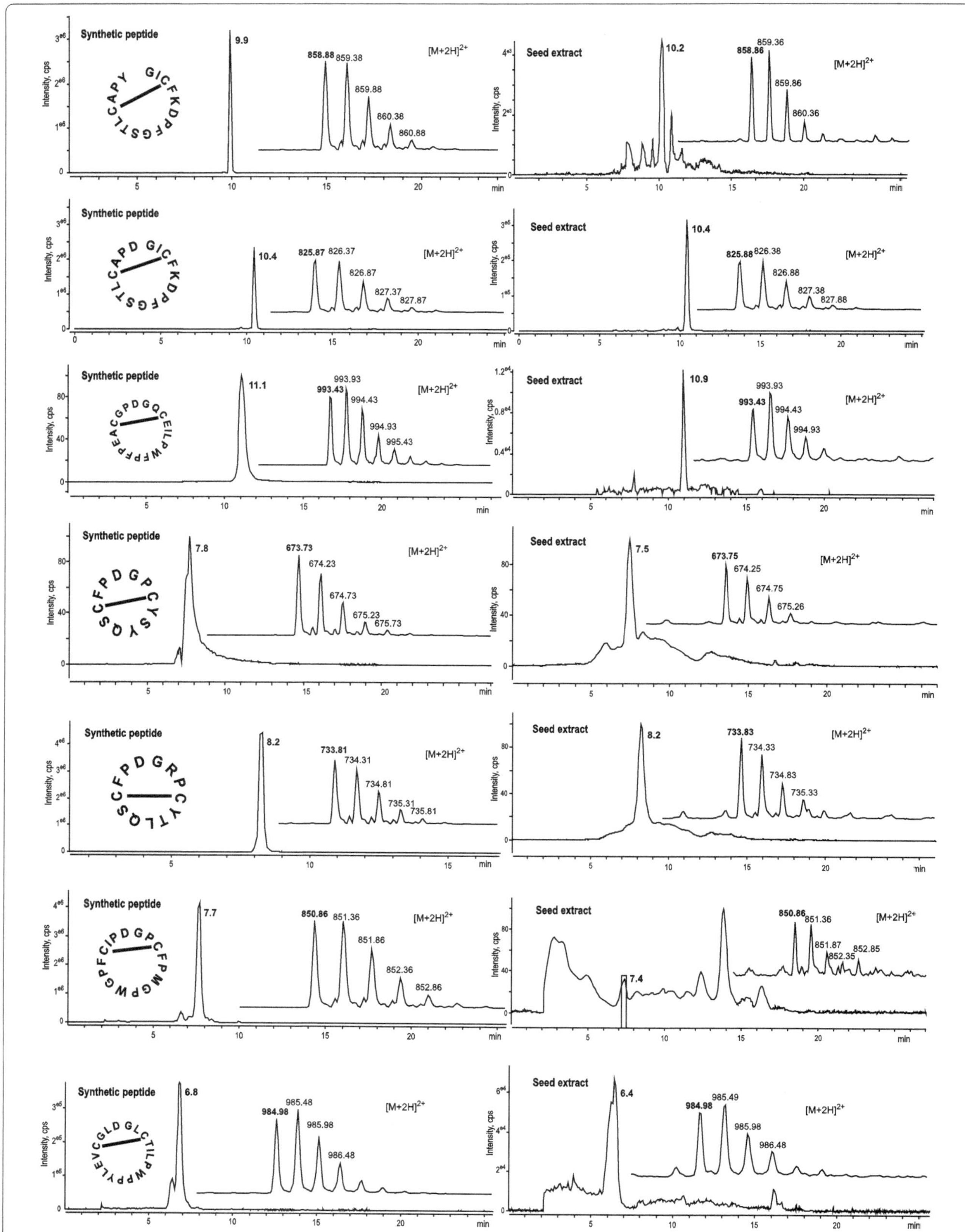

Figure 5 (See legend on next page.)

(See figure on previous page.)
Figure 5 *In planta* confirmation of PDPs in peptide extracts of *Zinnia haageana* dry seeds. Left side panel is the Extracted Ion Chromatograms (XICs) and average mass spectrums (inset) of the synthetic peptides showing the $[M + 2H]^{2+}$ ions. Average retention time for each peptide is shown in minutes next to the XIC. Right side panels are XICs and mass spectrums of corresponding native peptides identified from the *Z. haageana* seed extract run under the same conditions.

how *de novo* transcriptomics is a viable approach in non-model species for tracking the evolutionary history of a target gene.

Methods

RNA extractions

Total RNA was extracted from mature seeds of *Z. haageana* and *H. helianthodes* using the mini-hot phenol method [44], which is based on a method by Botella *et al.* [45]. RNA was DNAse treated followed by further purification using the NucleoSpin RNA Clean-up kit (Macherey-Nagel). The quality of the RNA was tested by measuring the A260/280 and A260/230 ratios using a NanoDrop Spectrophotometer and visualizing on a 1% agarose gel. When visualising total RNA extracted from dry seeds on a gel, note that the rRNA bands are not as prominent as they are for leaf total RNA, but if the RNA is intact it will similarly migrate with discrete banding. Following the above extraction and clean up procedure, some seed RNA samples will still contain pigments, but we do not find these to noticeably inhibit subsequent RNA-seq or reverse transcription.

Library construction

Sequencing libraries were generated using the TruSeq RNA sample prep kit (Illumina). To do so, 300–1000 ng of total mRNA from mature seeds was used according to the manufacturer's instructions. Sequencing was then performed on an Illumina HiSeq 1500, as a 2 × 101 paired-end run.

Read processing and *de novo* transcriptome assembly

We used a high-speed desktop computer for all data analysis. Specifically, the hardware was a Dell Precision T1650 desktop computer with Intel Xeon E3-1270 (Quad Core), 32GB DDR3 1600MHz RAM, and 2TB Hard Drive. Data was stored locally on a suite of network attached hard drives, specifically four Western Digital Caviar Red NAS WD30EFRX, 3TB capacity each with SATA 6Gb/s interface, 3.5" form factor, IntelliPower RPM, and 64MB cache enclosed in a QNAP TS-412 1.2GHz 4 Bay enclosure.

The quality of the raw sequences was visually inspected using the open-source software FastQC (http://www.bioinformatics.babraham.ac.uk/projects/fastqc/). The FASTX toolkit (http://hannonlab.cshl.edu/fastx_toolkit/) was used to remove poor quality reads. *De novo* transcriptome assembly was performed with CLC Genomic Workbench 6.5.1. Two assembly methods were applied (Figure 2).

In the automatic word size, non-paired method, raw forward and reverse sequences were uploaded to CLC software indicating they were paired reads. CLC sorted these reads and stored them together in a single file. Reads in this single file were trimmed using the FASTQ quality trimmer tool in the FASTX toolkit to obtain a quality threshold of the read ends of 30 by maintaining the minimum length after trimming at 50. FASTQ quality filter tool removed poor quality reads to maintain a quality threshold at 30 and minimum percentage of bases that match the quality threshold 30 as 90. Then the reads were collapsed using the FASTX collapser tool to remove identical duplicates and uploaded to CLC Genomics as a FASTA file. This was assembled using the CLC *de novo* assembly under default settings (Graph parameters : Automatic word size - yes, Automatic bubble size - yes. Mapping options: Map reads back to contigs; Mismatch cost 2, Insertion cost 3, Deletion cost 3, Length fraction 0.8, Similarity fraction 0.8) except that the minimum contig length was set to 100 (default is 200). The word size and the bubble size set by the program in this case were 23 and 50 respectively for all datasets.

For the paired method, raw forward and reverse sequences were trimmed and filtered using the FASTX toolkit maintaining the same parameters as above. Filtered forward and reverse reads were uploaded into CLC Genomics indicating that they are paired reads. Different word sizes (*i.e.* 20, 35, 40, 45, 50, 60, 62, 63, and 64) as well as the automatic word size set by the program were tried. CLC default settings were used except the following changes. Minimum contig length: 300, Auto detect paired distances: yes, Perform scaffolding: yes.

Each assembly was searched for the presence of *PawS1* and *PawL1* genes using the tBLASTn procedure using the 151-residue sunflower PawS1 protein sequence as the query.

Selecting core gene for quality assessment

Sunflower seed *de novo* transcriptome was assembled with the word size 60, paired method. Then reads were mapped back to contigs maintaining the parameters as below. Mismatch cost: 2, Insertion cost: 3, Deletion cost: 3, Length fraction: 0.8, Similarity fraction: 0.8, and Update contigs: Yes. Contigs were ordered based on their average coverage values. Six main subsets of contigs were selected to include contigs with average coverage value above 100,000, in the range of 10,000, 1,000, 500 (*PawS1* is in this range), 100, and contigs with average coverage values less than 100. These subsets of contigs were extracted and searched for their representative transcripts using the BLAST at NCBI tool. BLAST outputs for each subset were carefully examined and six core transcripts were selected to assess the quality of transcriptomes (Table 2). Care was taken to include transcripts coding for common plant proteins and mRNA that are expressed especially in seeds, also long and short transcripts. Dozens of Asteraceae *de novo* seed transcriptome assemblies were searched for these core genes and they appear to be conserved among all species tested.

Each transcriptome was queried for these core genes using the tBLASTn procedure. If several closely related transcripts were observed in the output the one with the best matching score was selected as the target gene. Coverage of identified core transcript was calculated as a percentage relatively to the length of the query protein. To evaluate the abundance of selected transcripts in each species at the mRNA level, trimmed and filtered raw reads were mapped to the ORF/partial ORF using the map reads to reference tool in CLC. Mapping parameters were maintained as Mismatch cost: 2, Insertion cost: 3, Deletion cost: 3, Length fraction: 0.8, Similarity fraction: 0.8.

Peptide synthesis, peptide extract and LC-MS confirmation

Synthetic peptides used as standards in LC-MS were produced by Wuxi Nordisk Biotech Co., Ltd, China. Peptides were extracted from mature *Z. haageana* seeds as described [17]. Peptides were resuspended in 5% acetonitrile (v/v), 0.1% formic acid (v/v) and loaded onto a C18 high capacity nano LC chip trapping column (160 nL) (Agilent) in 95% Buffer A (0.1% formic acid) and 5% Buffer B (0.1% formic acid in acetonitrile) using a 1200 series capillary pump (Agilent). After loading samples, the trapping column was switched in-line with a 1200 series nano pump (Agilent) and a C18 analytical column (75 mm × 150 mm) and then using a gradient, 5% B to 60% B in 15 min, peptides were eluted into a 6510 Series QTOF mass spectrometer (Agilent). The QTOF was operated in MS-mode only. MS data were collected at eight spectra per second.

Cloning *Z. haageana PawS1* and *PawL1* genes

Genomic DNA was extracted from mature *Z. haageana* seeds using the DNeasy Plant Mini Kit (Qiagen). Full length ORFs of *Z. haageana PawS1a* and *PawS1b* were specifically amplified using primers AJ58 (5′ CATC GAT CCT AGA AGA CAA TGG CAG TT 3′) and AJ59 (5′ GTG CGT AAG TGC GTA CAT TAA CCT 3′) and AJ60 (5′ GAT CCG ACT ACG ATG GCA GTT AAA G 3′) and AJ61 (5′ GAG TTA AAC ACA GAC CAC ACG 3′) respectively using *PfuUltra* high fidelity DNA polymerase (Agilent Technologies). In addition, the full length *PawL1a* was amplified with above genomic DNA with primers AJ134 (5′ TCC ACA ATG ACG AAA CTC ACA CTC 3′) and AJ56 (5′ ATC TTT ATT ACA AAT ACA TAC ATA GGC 3′) using *Taq*-DNA polymerase. Amplified fragments were cloned into pGEM-Teasy vector (Promega). Three independent clones were sequenced for each transcript to determine if there were errors introduced by *Taq* DNA polymerase.

ZhPawS1d was assembled only partially in two assemblies with different word sizes, but they were overlapping. The full ORF could be reconstructed by combining these overlapping partial sequences. Based on that reconstructed full length ORF, primers were designed to amplify *ZhPawS1d* gene. PCR with *Z. haageana* seed genomic DNA and primers AJ65 (5′ ATG GCA GTT AAA GTT GCA CTT G 3′) and AJ66 (5′ CAA GCC TAA TAG CTT GAG ACA G 3′) produced a broadly-migrating band on an agarose gel indicating multiple products ranging in size had been amplified. The products of this PCR reaction were cloned into pGEM-Teasy vector (Promega) and 41 independent clones were sequenced. Among these 41 clones *ZhPawS1a*, *ZhPawS1b*, *ZhPawS1d*, *ZhPawS1e*, and *ZhPawS1g* were present.

Accession codes

Zinnia haageana genes have been deposited in GenBank under the accession codes [GenBank: KM243336-KM243341]. The sequence data used to assemble the *Z. haageana* and *H. helianthodes* transcriptomes were deposited in the National Centre for Biotechnology Information under BioSample accession numbers [SRA: SAMN02933922 and SAMN02933923] respectively.

Competing interests

The authors declare that they have no competing interests.

Authors' contributions

ASJ designed and performed the research and analysed the data. DS and JW constructed sequencing libraries and generated sequencing data. DS, OB, and JW helped develop control genes. KB-L performed peptide LC-MS experiments and analysed the data. JSM extracted RNA. ASJ and JSM wrote the manuscript with input and final approval from all authors.

Acknowledgements

The authors would like to thank Mr Hayden Walker, ARC Centre of Excellence in Plant Energy Biology, for IT support. A.S.J. is supported by an International Postgraduate Research Scholarship (IPRS) and an Australian Postgraduate Award (APA). D.S. was supported by an ARC Super Science Fellowship FS100100022. K.B-L. was supported by Australian Research Council (ARC) grant DP120103369. O.B. is supported by an ARC Linkage grant LP0990355. J.W. is supported by an ARC Centre of Excellence grant CE140100008. J.S.M. is supported by an ARC Future Fellowship FT120100013. This work was supported by ARC grant DP130101191.

Author details

[1]The University of Western Australia, School of Chemistry and Biochemistry & ARC Centre of Excellence in Plant Energy Biology, 35 Stirling Highway, Crawley, Perth 6009, Australia. [2]La Trobe University, Department of Botany, School of Life Sciences & ARC Centre of Excellence in Plant Energy Biology,

AgriBio, the Centre for AgriBioscience, 5 Ring Road, Melbourne, Bundoora Victoria 3086, Australia. [3]The University of Western Australia, School of Plant Biology, 35 Stirling Highway, Crawley, Perth 6009, Australia.

References

1. Walker TM, Ip CLC, Harrell RH, Evans JT, Kapatai G, Dedicoat MJ, Eyre DW, Wilson DJ, Hawkey PM, Crook DW, Parkhill J, Harris D, Walker AS, Bowden R, Monk P, Smith EG, Peto TEA: **Whole-genome sequencing to delineate *Mycobacterium tuberculosis* outbreaks: a retrospective observational study.** *Lancet Infect Dis* 2013, **13:**137–146.
2. Oono Y, Kobayashi F, Kawahara Y, Yazawa T, Handa H, Itoh T, Matsumoto T: **Characterisation of the wheat (*triticum aestivum* L.) transcriptome by *de novo* assembly for the discovery of phosphate starvation-responsive genes: gene expression in Pi-stressed wheat.** *BMC Genomics* 2013, **14:**1–14.
3. Qin J, Li R, Raes J, Arumugam M, Burgdorf KS, Manichanh C, Nielsen T, Pons N, Levenez F, Yamada T, Mende DR, Li J, Xu J, Li S, Li D, Cao J, Wang B, Liang H, Zheng H, Xie Y, Tap J, Lepage P, Bertalan M, Batto J, Hansen T, Le Paslier D, Linneberg A, Nielsen HB, Pelletier E, Renault P, *et al*: **A human gut microbial gene catalogue established by metagenomic sequencing.** *Nature* 2010, **464:**59–65.
4. Cruickshanks HA, McBryan T, Nelson DM, VanderKraats ND, Shah PP, van Tuyn J, Singh Rai T, Brock C, Donahue G, Dunican DS, Drotar ME, Meehan RR, Edwards JR, Berger SL, Adams PD: **Senescent cells harbour features of the cancer epigenome.** *Nat Cell Biol* 2013, **15:**1495–1506.
5. Darmanis S, Nong RY, Vänelid J, Siegbahn A, Ericsson O, Fredriksson S, Bäcklin C, Gut M, Heath S, Gut IG, Heath S, Gut IG, Wallentin L, Gustafsson MG, Kamali-Moghaddam M, Landegren U: **ProteinSeq: High-performance proteomic analyses by proximity ligation and next generation sequencing.** *PLoS One* 2011, **6:**e25583.
6. Navin N, Kendall J, Troge J, Andrews P, Rodgers L, McIndoo J, Cook K, Stepansky A, Levy D, Esposito D, Muthuswamy L, Krasnitz A, McCombie WR, Hicks J, Wigler M: **Tumour evolution inferred by single-cell sequencing.** *Nature* 2011, **472:**90–94.
7. Xiao M, Zhang Y, Chen X, Lee E, Barber CJS, Chakrabarty R, Desgagné-Penix I, Haslam TM, Kim Y, Liu E, MacNevin G, Masada-Atsumi S, Reed DW, Stout JM, Zerbe P, Zhang Y, Bohlmann J, Covello PS, De Luca V, Page JE, Ro DK, Martin VJ, Facchini PJ, Sensen CW: **Transcriptome analysis based on next-generation sequencing of non-model plants producing specialized metabolites of biotechnological interest.** *J Biotechnol* 2013, **166:**122–134.
8. Novaes E, Drost DR, Farmerie WG, Pappas GJ Jr, Grattapaglia D, Sederoff RR, Kirst M: **High-throughput gene and SNP discovery in *Eucalyptus grandis*, an uncharacterized genome.** *BMC Genomics* 2008, **9:**1–14.
9. Zhang J, Liang S, Duan J, Wang J, Chen S, Cheng Z, Zhang Q, Liang X, Li Y: ***De novo* assembly and characterisation of the transcriptome during seed development, and generation of genic-SSR markers in Peanut (*Arachis hypogaea* L.).** *BMC Genomics* 2012, **13:**90.
10. Wan L, Han J, Sang M, Li A, Wu H, Yin S, Zhang C: ***De novo* transcriptomic analysis of an oleaginous microalga: pathway description and gene discovery for production of next-generation biofuels.** *PLoS One* 2012, **7:**e35142.
11. Franssen S, Shrestha R, Brautigam A, Bornberg-Bauer E, Weber A: **Comprehensive transcriptome analysis of the highly complex Pisum sativum genome using next generation sequencing.** *BMC Genomics* 2011, **12:**227.
12. Quail M, Smith M, Coupland P, Otto T, Harris S, Connor T, Bertoni A, Swerdlow H, Gu Y: **A tale of three next generation sequencing platforms: comparison of Ion Torrent, Pacific Biosciences and Illumina MiSeq sequencers.** *BMC Genomics* 2012, **13:**341.
13. Grabherr MG, Haas BJ, Yassour M, Levin JZ, Thompson DA, Amit I, Adiconis X, Fan L, Raychowdhury R, Zeng Q, Chen Z, Mauceli E, Hacohen N, Gnirke G, Rhind N, di Palma F, Birren BW, Nusbaum C, Lindblad-Toh K, Friedman N, Regev A: **Full-length transcriptome assembly from RNA-Seq data without a reference genome.** *Nat Biotech* 2011, **29:**644–652.
14. Zerbino DR, Birney E: **Velvet: Algorithms for *de novo* short read assembly using de Bruijn graphs.** *Genome Res* 2008, **18:**821–829.
15. Huang X, Madan A: **CAP3: A DNA sequence assembly program.** *Genome Res* 1999, **9:**868–877.
16. Pertea G, Huang X, Liang F, Antonescu V, Sultana R, Karamycheva S, Lee Y, White J, Cheung F, Parvizi B, Tsai J, Quackenbush J: **TIGR Gene Indices clustering tools (TGICL): a software system for fast clustering of large EST datasets.** *Bioinformatics* 2003, **19:**651–652.
17. Mylne JS, Colgrave ML, Daly NL, Chanson AH, Elliott AG, McCallum EJ, Jones A, Craik DJ: **Albumins and their processing machinery are hijacked for cyclic peptides in sunflower.** *Nat Chem Biol* 2011, **7:**257–259.
18. Elliott AG, Delay C, Liu H, Phua Z, Rosengren KJ, Benfield AH, Panero JL, Colgrave ML, Jayasena AS, Dunse KM, Anderson MA, Schilling EE, Ortiz-Barrientos D, Craik DJ, Mylne JS: **Evolutionary Origins of a Bioactive Peptide Buried within Preproalbumin.** *Plant Cell* 2014, **26:**981–995.
19. Kreis M, Shewry PR: **Unusual features of cereal seed protein structure and evolution.** *Bioessays* 1989, **10:**201–207.
20. Luckett S, Garcia RS, Barker JJ, Konarev AV, Shewry PR, Clarke AR, Brady RL: **High-resolution structure of a potent, cyclic proteinase inhibitor from sunflower seeds.** *J Mol Biol* 1999, **290:**525–533.
21. Rico M, Bruix M, González C, Monsalve RI, Rodríguez R: **^{1}H NMR assignment and global fold of napin BnIb, a representative 2S albumin seed protein.** *Biochemistry* 1996, **35:**15672–15682.
22. Natali L, Cossu R, Barghini E, Giordani T, Buti M, Mascagni F, Morgante M, Gill N, Kane N, Rieseberg L, Cavallini A: **The repetitive component of the sunflower genome as shown by different procedures for assembling next generation sequencing reads.** *BMC Genomics* 2013, **14:**686.
23. Martin JA, Wang Z: **Next-generation transcriptome assembly.** *Nat Rev Genet* 2011, **12:**671–682.
24. Liu L, Li Y, Li S-L, Hu N, He Y, Pong R, Lin D, Lu L, Law M: **Comparison of next-generation sequencing systems.** *J Biomed Biotechnol* 2012, **2012:**11.
25. Dorn KM, Fankhauser JD, Wyse DL, Marks MD: ***De novo* assembly of the pennycress (*Thlaspi arvense*) transcriptome provides tools for the development of a winter cover crop and biodiesel feedstock.** *Plant J* 2013, **75:**1028–1038.
26. Bräutigam A, Mullick T, Schliesky S, Weber APM: **Critical assessment of assembly strategies for non-model species mRNA-Seq data and application of next-generation sequencing to the comparison of C3 and C4 species.** *J Exp Bot* 2011, **62:**3093–3102.
27. Mortazavi A, Williams BA, McCue K, Schaeffer L, Wold B: **Mapping and quantifying mammalian transcriptomes by RNA-Seq.** *Nat Meth* 2008, **5:**621–628.
28. Wang ET, Sandberg R, Luo S, Khrebtukova I, Zhang L, Mayr C, Kingsmore SF, Schroth GP, Burge CB: **Alternative isoform regulation in human tissue transcriptomes.** *Nature* 2008, **456:**470–476.
29. O'Neil S, Emrich S: **Assessing *de novo* transcriptome assembly metrics for consistency and utility.** *BMC Genomics* 2013, **14:**465.
30. Parra G, Bradnam K, Ning Z, Keane T, Korf I: **Assessing the gene space in draft genomes.** *Nucleic Acids Res* 2009, **37:**289–297.
31. Dure L III, Croudh M: **Developmental biochemistry of cotton seed embryogenesis, and termination: changing messenger ribonucleic and populations as shown by *in vitro* and *in vivo* protein synthesis.** *Biochemistry* 1981, **20:**4162–4168.
32. Hong-Bo S, Zong-Suo L, Ming-An S: **LEA proteins in higher plants: structure, function, gene expression and regulation.** *Colloids Surf B Biointerfaces* 2005, **45:**131–135.
33. Siloto RMP, Findlay K, Lopez-Villalobos A, Yeung EC, Nykiforuk CL, Moloney MM: **The accumulation of oleosins determines the size of seed oilbodies in arabidopsis.** *Plant Cell* 2006, **18:**1961–1974.
34. Chen X, Pfeil JE, Gal S: **The three typical aspartic proteinase genes of *Arabidopsis thaliana* are differentially expressed.** *Eur J Biochem* 2002, **269:**4675–4684.
35. Simões I, Faro C: **Structure and function of plant aspartic proteinases.** *Eur J Biochem* 2004, **271:**2067–2075.
36. van Loon LC, Rep M, Pieterse CMJ: **Significance of inducible defense-related proteins in infected plants.** *Annu Rev Phytopathol* 2006, **44:**135–162.
37. Richau KH, Kaschani F, Verdoes M, Pansuriya TC, Niessen S, Stüber K, Colby T, Overkleeft HS, Bogyo M, Van der Hoorn RAL: **Subclassification and biochemical analysis of plant papain-like cysteine proteases displays subfamily-specific characteristics.** *Plant Physiol* 2012, **158:**1583–1599.
38. Hardie DG: **Plant protein serine/threonine kinases: classification and functions.** *Annu Rev Plant Physiol Plant Mol Biol* 1999, **50:**97–131.
39. Shewry PR, Napier JA, Tatham AS: **Seed storage proteins: structures and biosynthesis.** *Plant Cell* 1995, **7:**945–956.
40. Haznedaroglu BZ, Reeves D, Rismani-Yazdi H, Peccia J: **Optimization of *de novo* transcriptome assembly from high-throughput short read**

sequencing data improves functional annotation for non-model organisms. *BMC Bioinformatics* 2012, **13**:170.

41. Croucher P, Brewer M, Winchell C, Oxford G, Gillespie R: **De novo characterization of the gene-rich transcriptomes of two color-polymorphic spiders, *Theridion grallator* and *T. californicum* (Araneae: Theridiidae), with special reference to pigment genes.** *BMC Genomics* 2013, **14**:862.
42. Wu C-H, Tsai M-H, Ho C-C, Chen C-Y, Lee H-S: **De novo transcriptome sequencing of axolotl blastema for identification of differentially expressed genes during limb regeneration.** *BMC Genomics* 2013, **14**:434.
43. De Wit P, Pespeni MH, Ladner JT, Barshis DJ, Seneca F, Jaris H, Therkildsen NO, Morikawa M, Palumbi SR: **The simple fool's gide to population genomics via RNA-Seq: an introduction to high-throughput sequencing data analysis.** *Mol Ecol Resour* 2012, **12**:1058–1067.
44. Mylne JS, Chan LY, Chanson AH, Daly NL, Schaefer H, Bailey TL, Nguyencong P, Cascales L, Craik DJ: **Cyclic peptides arising by evolutionary parallelism via asparaginyl-endopeptidase-mediated biosynthesis.** *Plant Cell* 2012, **24**:2765–2778.
45. Botella JR, Arteca JM, Schlagnhaufer CD, Arteca RN, Phillips AT: **Identification and characterization of a full-length cDNA encoding for an auxin-induced 1-aminocyclopropane-1-carboxylate synthase from etiolated mung bean hypocotyl segments and expression of its mRNA in response to indole-3-acetic acid.** *Plant Mol Biol* 1992, **20**:425–436.

Using monoclonal antibodies to label living root hairs: a novel tool for studying cell wall microarchitecture and dynamics in *Arabidopsis*

Emily R Larson[1,4], Mary L Tierney[1,2], Berke Tinaz[3] and David S Domozych[3*]

Abstract

Background: The *Arabidopsis* root hair represents a valuable cell model for elucidating polar expansion mechanisms in plant cells and the overall biology of roots. The deposition and development of the cell wall is central to the root hair expansion apparatus. During this process, incorporation of specific wall polymers into the growing wall architecture constitutes a critical spatio-temporal event that controls hair size and growth rate and one that is closely coordinated with the cell's endomembrane, cytoskeletal and signal transduction apparatuses.

Results: In this study, the protocol for live cell labeling of roots with monoclonal antibodies that bind to specific wall polymers is presented. This method allows for rapid assessment of root hair cell wall composition during development and assists in describing changes to cell wall composition in transgenic mutant lines. Enzymatic "unmasking" of specific polymers prior to labeling allows for refined interpretation of cell wall chemistry. Live cell immunofluorescence data may also be correlated with transmission electron microscopy-based immunogold labeling.

Conclusions: Live *Arabidopsis* root hairs may be labeled with cell wall polymer-specific antibodies. This methodology allows for direct visualization of cell wall dynamics throughout development in stable transgenic plant lines. It also provides an important new tool in the elucidation of the specific interactions occurring between membrane trafficking networks, cytoskeleton and the cell wall deposition/remodeling mechanism.

Keywords: Arabidopsis, Root hairs, Live cell labeling, Immunocytochemistry, Polar growth, Cell wall

Background

The expansion dynamics of a plant cell are directly controlled by the microarchitecture of its cell wall. Modulations to cell wall constituents via new polymer deposition and/or remodeling of pre-existing polymers create loosened or "softened" zones with less tensile strength throughout the wall or at specific sites therein. Internal turgor pressure generates the non-vectorial force against the wall that drives the expansion process at these softened zones [1,2]. Many plant cells exhibit *diffusive* growth whereby growth is roughly equivalent on all faces of the expanding cell. However, other cell types grow in a *polar* fashion where wall and cell expansion are focused at a specific point or front [3,4]. Tensile resistance of the wall to turgor is less at this front, that in turn, allows for a localized but controlled cell expansion. At other regions of the cell the wall retains sufficient tensile strength to resist turgor-driven pressure. This type of growth often leads to distinct tubular shapes, as exemplified by pollen tubes, root hairs and moss protonemata.

The most well studied polar expansion system in land plants is the pollen tube [5,6]. At the tip of a growing tube, pectin dynamics create a less-rigid wall that promotes polarized expansion driven by turgor [7,8]. In the shank areas adjacent to the tip, modulation of wall chemistry creates a rigid matrix organization that is resistant to turgor pressure. This includes de-esterification of the pectin followed by calcium (Ca^{2+}) complexing to yield a rigid gel and the addition of ß(1–3)-glucan (callose) and ß(1–4)-glucan (cellulose) to the wall.

Root hairs represent specialized extensions of the root epidermis that are also formed by polar expansion. These structures are profoundly important for the survival of a plant as they are critical for the uptake of

* Correspondence: ddomoz@skidmore.edu
[3]Department of Biology, Skidmore College, Saratoga Springs, NY, USA
Full list of author information is available at the end of the article

water and minerals and are involved in the establishment of symbiotic associations with resident microbiota of soils [9-11]. Surprisingly, far less is known about wall composition and architecture, and its role in polar expansion of root hairs than for other polar expanding plant cells. This is due to the exceptionally fragile nature of the hairs, especially when handling during experimental manipulation, and the difficulty in obtaining sufficient amounts of root hair wall material for biochemical studies. Presently, it is thought that cellulose microfibril arrangement is random at the growing root hair tip, which consequently creates a softened zone to promote unidirectional expansion [12,13]. Lateral expansion along the shank of the hair is restricted by the production of a secondary cell wall containing an organized helicoid arrangement of microfibrils that make the wall resistant to turgor pressure [14-17]. To date, only limited information is available concerning the non-cellulosic components that play key roles in tethering microfibrils, and that form the matrix in which the cellulose is embedded [18,19]. This results in an incomplete understanding of the root hair cell wall composition/organization and its required role in polar expansion.

Over the past two decades, the use of monoclonal antibodies (mAbs) with specificity toward epitopes of various cell wall polymers has greatly enhanced our understanding of cell wall chemistry. Employment of mAbs in high throughput microarrays allows for rapid screening of large numbers of polymers in different taxa, tissues, cell types and cell wall fractions [20,21]. Similarly, mAbs have been important in mapping specific polymers in various plants and parts therein using light microscopy (LM)-based immunofluorescnce and transmission electron microscopy (TEM)-based immunogold labeling [22-24]. For virtually all of these studies, cells and tissues are typically fixed using chemical or cryofreezing methods, dehydrated and embedded in a plastic or wax matrix that allows for subsequent sectioning prior to imaging [25]. While these protocols have yielded valuable results, they may cause extraction of polymers, artefact formation and wall damage during preparation. More importantly, these techniques only provide snapshots of the cell wall and limited developmental information dealing with the dynamics of the wall in a live plant. Recently, we have devised live-labeling strategies for green algae using mAbs and other molecular probes [26,27]. These have provided detailed information about wall architecture and wall development during cell expansion in live cells. In this paper, we report on the development of live cell imaging protocol for root hair development in *Arabidopsis thaliana*. This methodology allows for rapid mapping of cell wall polymers using mAbs with live root hairs, and provides a valuable new technique when coupled with transgenic cell lines for detailed studies dealing with wall deposition, secretory dynamics and cell expansion.

Results

The protocol developed here for live labeling of root hairs is both simple and convenient, and is outlined in Figure 1. The elimination of fixation, dehydration and embedding is advantageous especially for handling such fragile specimens as root hairs. This methodology also does not require sectioning or dewaxing protocols and allows for 3-dimensional image acquisition via confocal laser scanning microscopy (CLSM). The use of multi-welled petri dishes allowed for assessment of large numbers of roots in small volumes of labeling solutions of 500 μL or less. Root hairs could also be treated with wall-degrading enzymes to remove and unmask specific polymers prior to labeling [28,29]. We also tested root hair viability, labeling quality/intensity with and without the inclusion of a detergent (0.85 mM Triton-X100) in the labeling protocol and noted no discernable difference in labeling results. Control experiments also showed that the labeling was specific.

Table 1 provides a summary of antibodies used and labeling results that were obtained for the following phenotypes: wild-type (wt), the *vti13* single mutant, *prp3* single mutant, the *prp3 vti13* double mutant, *csdl2* single mutant, the *csdl3-2* double mutant and the *xxt1*, *xxt2* and *xxt5* single mutants. In wt root hairs, intense labeling was noted for mAbs specific for epitopes of xyloglucan, xylan and arabinogalactan protein with moderate labeling for extensin and homogalacturonan (HG), specifically low esterified HG. Similar results were observed for the *prp3* mutant walls. In *vti13* seedlings, xyloglucan- and xylan-epitope labeling was absent in root hairs. For the double mutant, *prp3 vti13*, xyloglucan and xylan labeling was present in root hairs. The *csld2*-1 mutant produced notably small root hairs [30] while the *csld3-2* mutant produced no root hairs [30]. The *csld2-1* mutant had similar labeling to the wt except for the absence of LM19-probed HG and LM11-probed ß(1–4)-xylan/arabinoxylan. The *xxt* mutants yielded similar results except for the lack of LM10 labeling in the *xxt1* line. Most notable though was the complete lack of labeling with the xyloglucan probes, LM15 and LM25 in all three mutant lines. These results provide evidence that differences in root hair cell wall organization or composition in different genetic backgrounds can be used to identify root hair mutant phenotypes in live seedlings.

For detailed analysis of imaging, we chose the xylosylated-xyloglucan specific mAb, LM15 [29]. LM15 labeled the root hairs and epidermis of the wt root (Figure 2A, B) with a subtle gradient of higher labeling at the tip and lesser intensity at the base (Figure 2C).

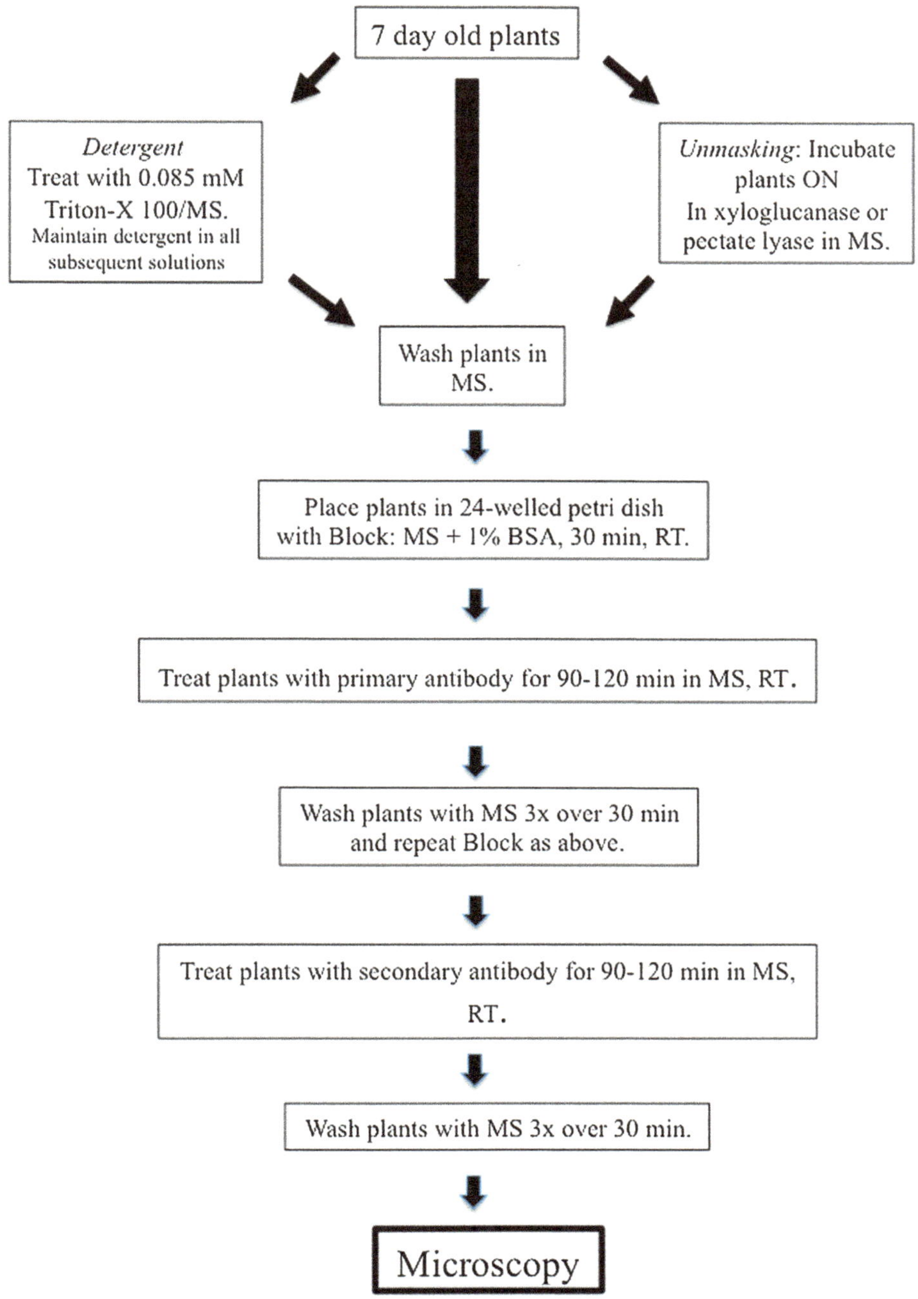

Figure 1 Schematic outline of live cell labeling protocol.

No discernable difference was noted in labeling when the protocol included the detergent-containing base, MS-Tri-X100 (Figure 2E, F). When pre-treated with 5U of xyloglucanase for 90 min, the root hairs did not label with LM15 (Figure 2G, H). When pre-treated with 3U of pectate lyase for 90 min, LM15 labeling was unaltered (Figure 2I, J). TEM immunogold labeling supplemented our CLSM imaging and demonstrated that LM15 labeling was indeed localized at the cell wall (Figure 2K). A control experiment whereby the primary antibody was eliminated from the protocol showed no labeling (Figure 2L, M). LM15 labeled the cell walls of the root hairs of the *prp3* mutant (Figure 3A, B). LM15 did not label the root hairs of the *vti13* mutant (Figure 3C, D) but did label the double mutant, *prp3 vti13* (Figure 3E, F).

LM10, a mAb with specificity toward xylan, labeled the root hairs of the WT, *prp3* and double mutant, *prp3 vti13*, but not the *vti13* mutant (Figure 4A, D, G, H). Labeling of pectin with relative low levels of esterification using LM19 demonstrated that labeling was present in all

Table 1 Summary of antibodies employed and results of immunocytochemical labeling

mAb	Specificity/Reference	WT	vti13	prp3	prp3/vti13	csld2	csld3-2	xxt1	xxt2	xxt5
PECTINS										
JIM7	**High DE HG** [31]	-	-	-	-	-	NA	-	-	-
LM19	**Low DE HG** [23]	++	++	+	++	-	NA	+	+	+
LM5	**(1–4)-galactan** [32]	-	-	-	-	-	NA	-	-	-
LM6	**(1–5)-arabinan** [33]	-	-	-	-	-	NA	-	-	-
INRA-RU1	**Rhamnogalacturonan-I** [34]	-	-	-	-	-	NA	-	-	-
XYLOGLUCAN										
LM15	**xylosylated XyG** [29]	++	-	++	++	+	NA	-	-	-
LM25	**galactosylated XyG epitope** [20]	++	-	+	++	+	NA	-	-	-
MANNAN										
BS-400-4	**ß(1–4)-mannan/galacto-ß(1–4)-mannan** [35]	-	-	-	-	-	NA	-	-	-
XYLAN										
LM10	**(1–4)-xylan** [36]	++	-	++	++	++	NA	-	−+	++
LM11	**ß(1–4)-xylan/arabinoxylan** [36]	+	-	+	+	-	NA	-	-	-
AGP										
LM2	**AGP** [37]	-	-	-	-	-	NA	-	-	-
JIM13	**AGP** [38]	+	+	+	+	+	NA	+	+	+
JIM8	**AGP** [39]	+	+	+	+	+	NA	+	+	-
EXTENSIN										
LM1	**Extensin** [40]	++	++	+	++	+	NA	+	−	+
JIM20	**Extensin** [41]	+	+	+	+	+	NA	+	−	+

Key: ++ = intense labeling, + = moderate labeling, − = ss no label, NA = not applicable, no root hairs formed.

genotypes (Figure 4B, E, I; Table 1). Interestingly, JIM7, a mAb with specificity toward high-esterified HG, did not label the root hairs of any of the tested genetic backgrounds (e.g. Figure 4C; Table 1). Previous studies, using fixed sections of plant roots have shown that both JIM5 and JIM 7 label the cell walls of root hairs [42]. These results suggest that the pectin epitopes recognized by JIM5 and JIM7 may not be accessible to the antibody in growing root hairs when using live seedlings. JIM20, a mAb with specificity toward extensin, labeled the root hairs of all genotypes as exemplified by *vti13* (Figure 4F; Table 1). These results help define the dynamics of the root hair cell wall and how organization or composition is altered in root hair mutants, *prp3* and *vti13*.

These methods were then used to investigate the root hairs of mutant lines defective in the synthesis of structural carbohydrates within the cell wall. The *csld2-1* mutant produced small root hairs that labeled inconsistently with LM15 or LM25. The developing root hairs near the root tip labeled but no label was noted in fully expanded hairs further removed from the root tip (Figure 5A-C). No hairs were formed in the *csld3-2* line as has been previously described [31,43-45], but it is important to note that LM15/LM25 labeling was noted in the epidermis (Figure 5D, E). For the *xxt* mutant lines, root hairs were present but did not label with either LM15 or 25, as illustrated for *xxt1* (Figure 5F, G). In summary, the results of this study highlight a convenient and simple means to identify other cell wall mutant lines and better define changes in their wall microarchitecture.

Discussion

Live cell labeling of root hairs of Arabidopsis provides a valuable new tool in elucidating wall microarchitecture and development. This simple and relatively fast method of mapping specific wall polymers offers much potential in adding detail to cell wall deposition/organization and cell expansion dynamics. In this study, we demonstrated labeling of xyloglucans in the root hairs of WT, *prp3* single mutant and the double mutant, *prp3 vti13* using the xylosylated-xyloglucan specific mAb, LM15, and the galactosylated xyloglucan specific mAb, LM25. These observations supplement recent research where a galacturonic acid-containing xyloglucan is found in *Arabidopsis* root hair cell walls [18]. Likewise, we show a higher intensity labeling at the root tip in our study with LM15 in WT and *prp3 vti13* double mutant root hairs compared to the labeling in the *vti13* single mutant, which has a branching root hair phenotype under our growth conditions [46]. This observation suggests that xyloglucan

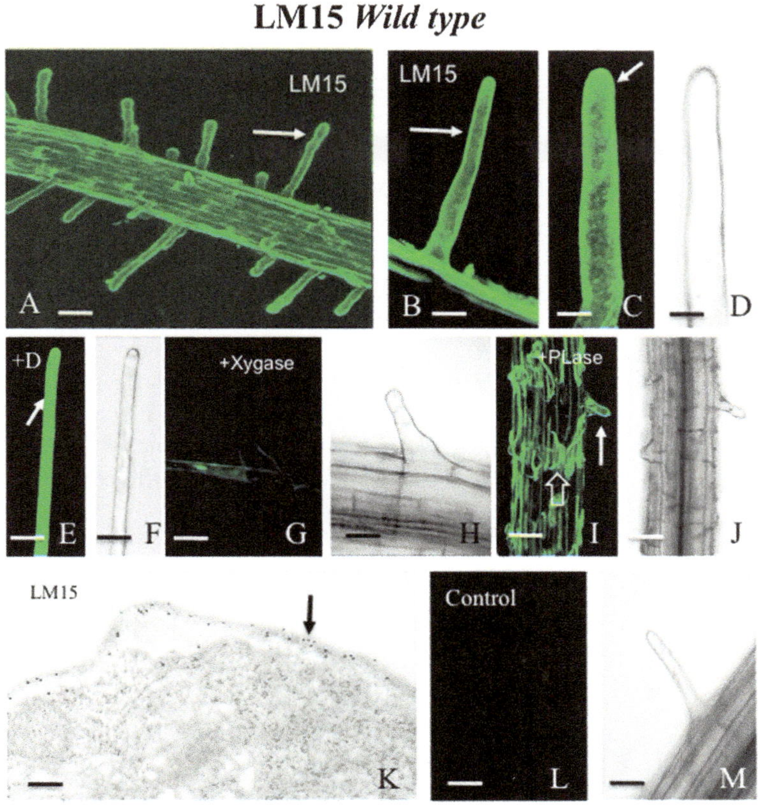

Figure 2 Live cell LM15 labeling of wild type (wt) Arabidopsis root hairs. A: Overview of live root demonstrating strong LM15 labeling of root hairs (arrows) and epidermis (hollow arrow). Bar = 100 μm. **B**: Magnified view of a root hair labeled with LM15 (arrow). Bar = 19 μm. **C** and **D**: LM15 labeling of root hair highlighting a slight but notable gradient with more intense labeling at the root hair tip (C, arrow). D is the DIC profile of C. Bars for C and D = 12 μm. **E** and **F**: LM15 labeling of root hair (E, arrow) after incubation of root in 0.085 mM Triton-X detergent (+D), during labeling process. Note labeling of root hair is comparable to that observed with non-detergent labeling (A-C). F is the DIC profile of E. Bars for E and F = 30 μm. **G** and **H**: LM15 labeling after root was pre-incubated overnight in 5U of xyloglucanase (+Xygase). Note the lack of labeling of root and hairs (G). H is the DIC profile of G. Bars for G and H = 18 μm. **I** and **J**: LM15 labeling after root was pre-incubated overnight in 3U of pectate lyase (+PLase). Note that the root hair labels (I, solid arrow). The epidermal cells labeling appears in strips (I, hollow arrow) but still labels. J is the DIC profile of I. Bars for I and J = 50 μm. **K**: Immunogold profile of root hair cell wall labeled with LM15 (arrow), Bar = 50 nm. **L** and **M**: Control profile whereby the primary antibody was eliminated during labeling. M is the DIC profile of L. Bars for L and M = 30 μm.

polymers may play a key role in providing controlled tensile resistance to turgor pressure at the root hair tip during growth. This tensile support might be especially important in this zone where less organized cellulose microfibril organization has been noted [12]. Our study has also shown that wall glycoproteins such as extensin and arabinogalactan proteins can be detected in the root hair wall of live seedlings, supporting recent work that also showed the presence of these polymers in the wall [19,47,48].

In the *Arabidopsis* root hair, xyloglucan and other hemicellulosic polymers like xylan may also be critical to regulating wall extensibility. The absence of labeling of certain epitopes in the *vti13* single mutant and the recovery of these labeling patterns in the *prp3 vti13* double mutant suggests that there is a reorganization of the cell wall matrix in *vti13* root hairs, rather than a loss of these components from the cell wall. Therefore, the *vti13* mutant has an altered deposition or development of the cell wall such that the epitopes are not available to the mAb labeling. The reemergence of the epitopes in the *prp3 vti13* is evidence that the cell wall has gone through another reorganization, not a loss of specific constituents of the matrix.

Our results also show that this methodology can also be used for studying wall mutant lines, especially those deficient in the synthesis of specific wall polymers. The *xxt* mutants did not label with LM15 or LM25, antibodies specific for xyloglucan epitopes. Genetic analysis of *xxt1* and *xxt2* mutants has shown that xyloglucan synthesis is required for root hair growth [42,49]. In addition, similar amounts of xyloglucan can be detected within the walls of wt and single *xxt1* and *xxt2* mutant seedlings using OLIMP analysis, suggesting that these two loci compensate for one other during xyloglucan synthesis [42]. The lack of labeling of root hairs in *xxt* mutant seedlings with LM15/LM25 in this study

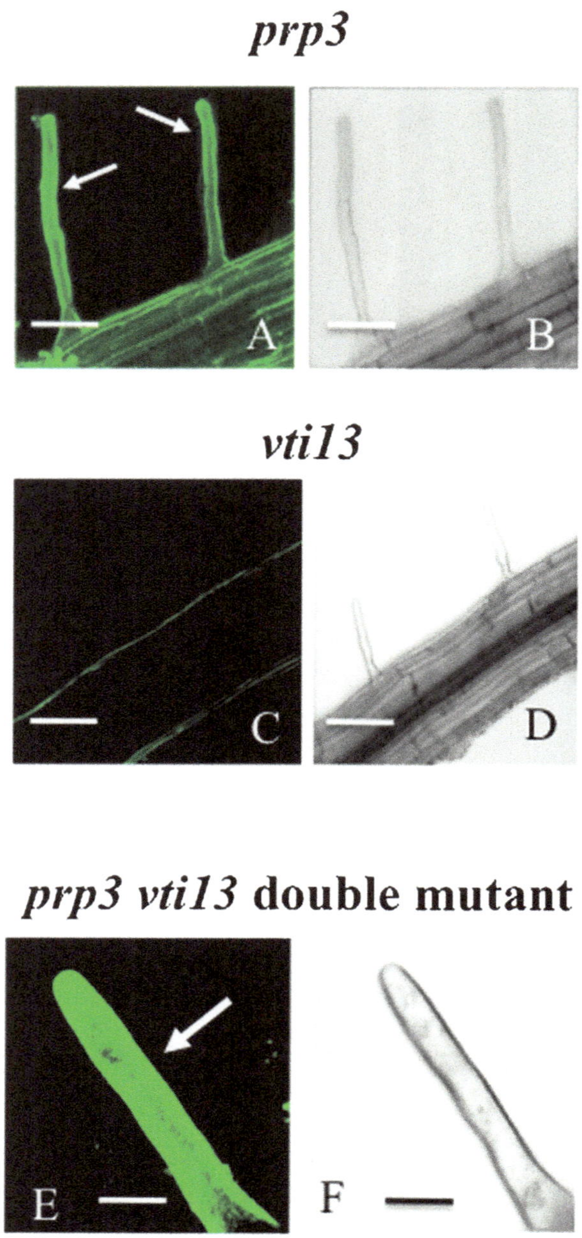

Figure 3 LM15 labeling of live cells of *prp3, vti13,* and *prp3/vti13* double mutant seedlings. **A** and **B**: LM15 labels the hairs of the *prp3* mutant. B is the DIC profile of A. Bars for A and B = 50 μm. **C** and **D**: LM15 does not label the root hairs of the *vti13* mutant. D is the DIC profile of C. Bars for C and D = 100 μm. **E** and **F**: In the *prp3/vti13* double mutant, LM15 labels the root hair (E, arrow). F is the DIC profile of E. Bars for E and F = 20 μm.

suggests subtle changes in xyloglucan orientation within the wall may eliminate the availability of the LM15 and LM25 epitopes using live seedlings. For the *csld* mutants, lines deficient in the synthesis of ß-glucans (e.g. mannose, cellulose; [30,50,51]), the results were limited as the two lines either did not produce root hairs or yielded small hairs with inconsistent labeling. Overall, the ability to discern these nuances of cell wall metabolism makes this live labeling protocol a powerful tool in defining phenotypes in cell wall and root hair mutants. A next step in expanding use of this method will be labeling with two or more wall-specific mAbs.

Conclusion

Live cell labeling of cell wall epitopes allows for dynamic imaging of wall structure during development as well as important application for co-labeling studies. For example, using this methodology along with fluorescent protein-fusion constructs of various components of the secretory and cytoskeletal machinery of the hair offer powerful and specific mechanisms for deciphering the coordinated interactions of specific subcellular systems in the development of the cell wall and constituent polymers therein.

Methods

General

Seeds of wt and various Arabidopsis mutants (all in a Columbia background) were surface sterilized with a 20% (v/v) bleach solution and sown on MS medium (1X Murashige and Skoog salts, 1% (w/v) sucrose, 1X Gamborg's vitamin solution, 5 mM MES pH6) and solidified with 1.3% (w/v) agarose as described in Larson et al. [36]. The *vti13* mutant (SALK_075261) was obtained from the Arabidopsis Biological Research Center (ABRC) and confirmed to be a null mutant. The *prp3* and *vti13* single mutants were crossed using standard procedures and the homozygous double mutant was identified in the F2 generation using genomic PCR (described in [36]). CSLD- and XXT-mutant lines were a generous gift of Dr. Ken Keegstra. Murashige-Skoog (MS) medium was obtained from Sigma Chemical (St. Louis, MO, USA) and antibodies were obtained from Plant Probes (Leeds, UK), Sigma Chemical (St. Louis, MO, USA) and a generous gift from Dr. Marie Christine-Ralet (INRA, Nantes, FR).

Immunofluorescence labeling

All labeling was performed in a 24-welled petri dish (non-tissue culture treated dishes; Fisher Scientific, Pittsburg, PA, USA). Arabidopsis seedlings were grown vertically on MS-containing media for 7 days after which they were gently removed from the agarose surface, placed in the wells of the petri dish and incubated for 30 min in 500 μL of liquid MS medium containing 1.0% (w/v) non-fat Carnation Instant milk. This represented the blocking solution for immunolabeling. In an initial test of labeling quality and root hair viability, we compared samples processed in MS as the buffer for labeling with samples processed in MS and detergent (0.85 mM Triton-X100). No difference was noted between the two solutions and the inclusion of detergent was

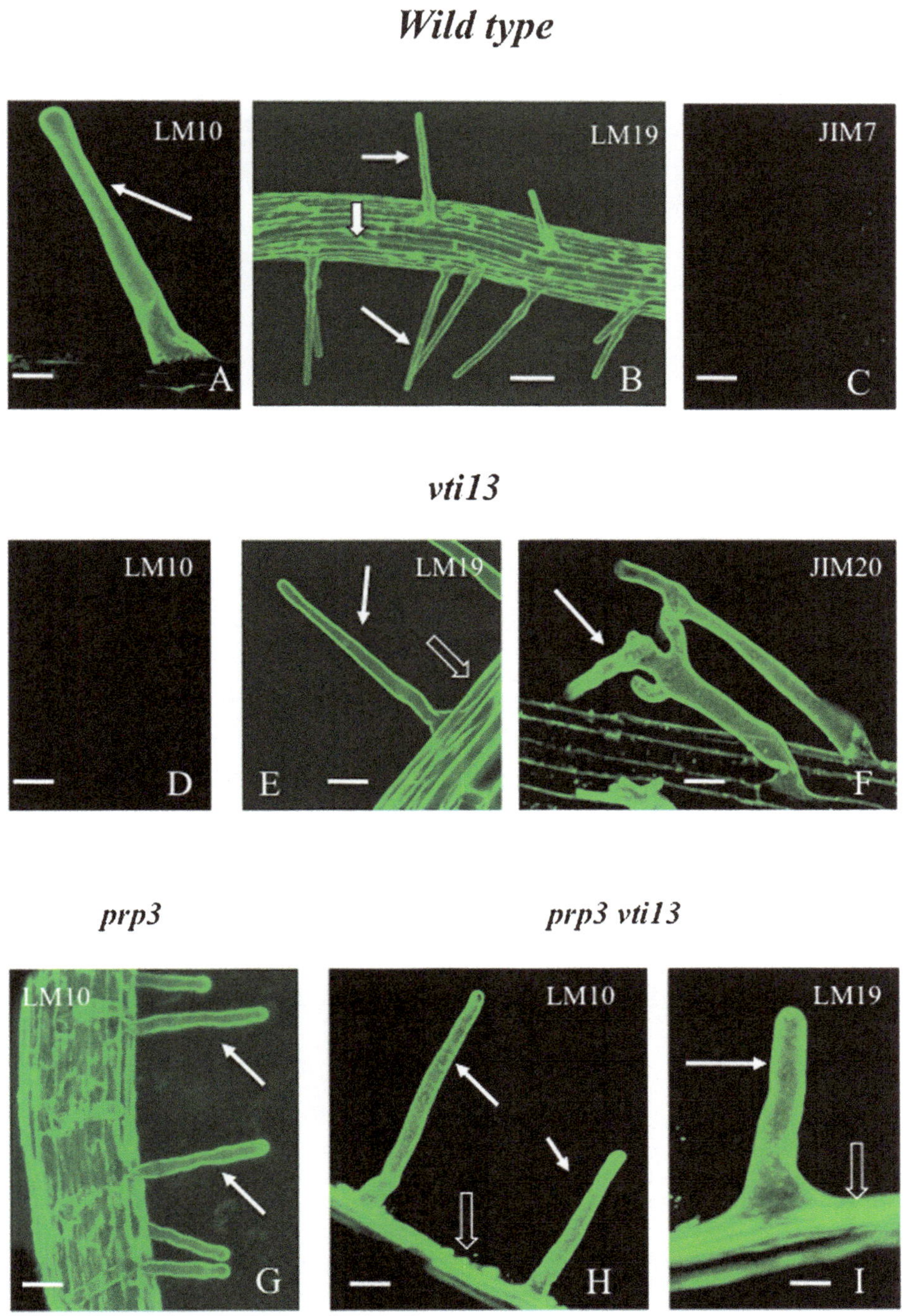

Figure 4 Live cell profiles of immunolabeled root hairs. **A**: LM10 labeling of a wt root hair (arrow). Bar = 10 μm. **B**: LM19 labeling of wt root hairs (solid arrows) and root epidermis (hollow arrow). Bar = 80 μm. **C**: Lack of JIM7 labeling of wt root hairs. Bar = 20 μm. **D**: The root hairs of the *vti13* mutant do not label with LM10. Bar = 30 μm. **E**: LM19 labeling of *vti13* root hairs (solid arrow) and root epidermis (hollow arrow). Bar = 35 μm. **F**: JIM 20 labeling of root hairs of the *vti13* (arrow). Bar = 15 μm. **G**: LM10 labeling of root hairs (arrows) of *prp3* mutant. Bar = 15 μm. **H**: LM10 labeling of root hairs of *prp3/vti13* (solid arrows) and root epidermis (hollow arrow). Bar = 15 μm. **I**: LM19 labeling of root hairs of *prp3/vti13* (solid arrows) and root epidermis (hollow arrow). Bar = 10 μm.

maintained for all subsequent labeling and is noted as MS-Tri-X100 below. After blocking, the seedlings were washed 3X over 30 min with liquid MS medium and then placed in 500 μL of primary antibody solution that consisted of a 1/10 (v/v) dilution of wall polymer-specific antibody in MS-Tri-X100. Plates were gently shaken on a laboratory rotator for 90 min in the dark at Room Temperature (RT). Seedlings were again washed 3X with MS-Tri-X100, blocked (as described above) and washed again. Seedlings were incubated and gently

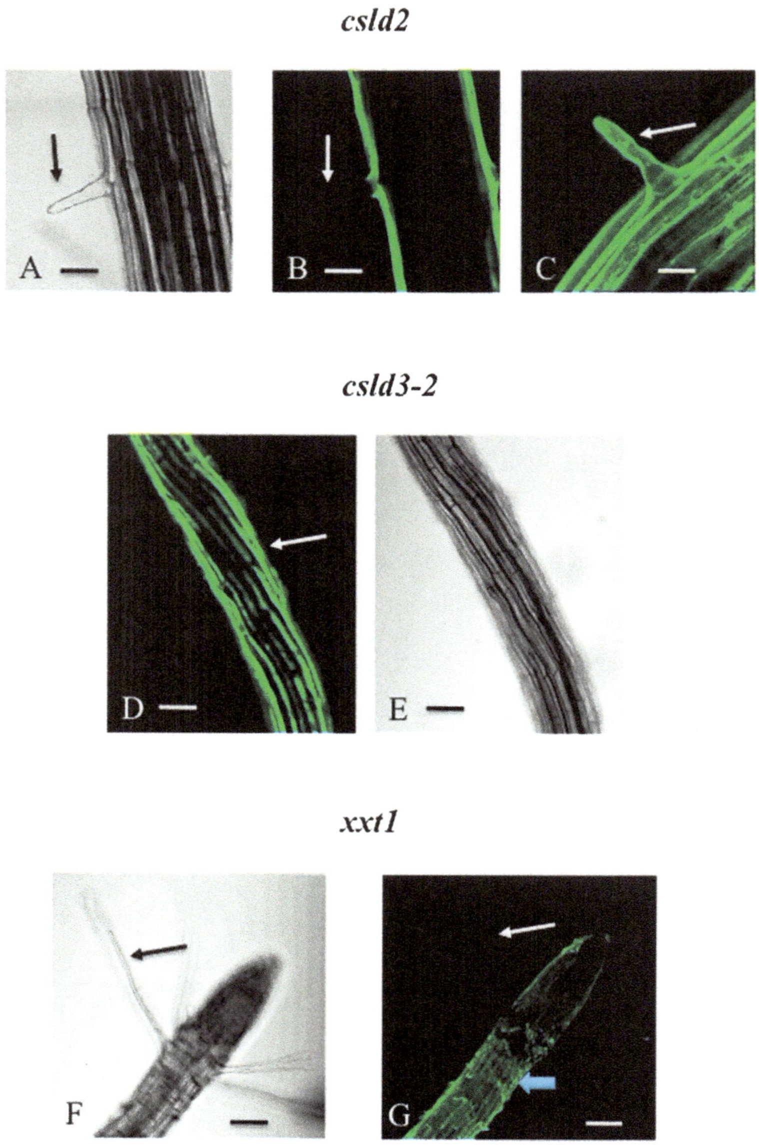

Figure 5 Live cell profiles of immunolabeled root hairs of cell wall mutant lines. A: A small root hair of the *csld-2* line located approximately 2 mm from the root tip. Bar = 20 μm. **B**: LM15 labeling of the root hair identified in "A". Note the absence of labeling (arrow). Bar = 20 μm. **C**: LM15 labeling of a small root hair (arrow) located just behind the root tip of the *csld2* mutant. Bar = 12 μm. **D**: No root hairs were found in the *csld3-2* mutant but the epidermis labeled with LM15 (arrow). Bar = 20 μm. **E**: DIC image of the root shown in "D". Bar = 20 μm. **F**: The root hairs of the *xxt1* mutant line (arrow). Bar = 30 μm. **G**: LM15 labeling of the root hairs (arrow) observed in "F". Bar = 15 μm.

shaken for 90 min in the dark at RT in 500 μL of a 1/75 (v/v) dilution of secondary antibody (anti-rat-TRITC; Sigma) in MS-Tri-X100. They were then washed with liquid MS 3X over 30 min. The seedlings were left in MS liquid medium until microscopic viewing. For control, primary antibody labeling was eliminated from the protocol. For each primary antibody tested at least 10–15 root hairs of 3 separate seedlings were observed. This process was also performed three times.

For unmasking experiments, some seedlings were gently shaken in solutions of 5U xyloglucanase (Megazyme, IR) or 3U pectolyase (Megazyme) in MS for 90 min at RT. The seedlings were then removed, washed 3X with MS and processed for labeling (see above).

The labeled seedlings were placed in a 200 μL drop of liquid MS in the well of a single-welled immunoslide (EMS, Ft. Washington, PA, USA). A glass coverslip was placed gently over the seedling. The depression of the immunoslide prevented crushing of the root and hairs. For some slides, the coverslip was affixed to the slide with small drops of nail polish. Root hairs were then observed using an Olympus BX-61 microscope equipped with a Fluoview 300 confocal system. The fluorescence signals were also observable using wide field Olympus

BX-60 or IX70 fluorescence microscopes with 50 watt mercury lamp[s and TRITC filter sets.

Transmission Electron Microscopy (TEM)

Roots were fixed in 0.5% (v/v) glutaraldehyde (EMS, Ft. Washington, PA, USA) in Sorensen's Phosphate buffer (EMS; pH 7.2) for 40 min at 4°C. They were washed 3X with Sorensen's buffer and post-fixed in 0.5% (w/v) OsO_4 (EMS) in Sorensen's buffer for 1 h at 4°C. After washing with buffer (see above), the seedlings were dehydrated in acetone and embedded in Spurrs Low Viscosity plastic. 60–80 nm sections were obtained using a Reichert Ultracut ultramicrotome and collected on formvar coated nickel grids. Immunogold labeling was performed using previously developed techniques [52].

Competing interests

The authors declare that they have no competing interests.

Authors' contributions

All authors participated in the design of this study as well as immunofluorescence labeling of the root hairs. DSD performed the TEM. All authors read and approved the final manuscript.

Acknowledgements

This work was supported by grants the National Science Foundation to DSD (NSF-MCB 0919925 and NSF-DBI 0922805).

Author details

[1]Cellular, Molecular, and Biomedical Science Program, University of Vermont, Burlington, VT, USA. [2]Department of Plant Biology, University of Vermont, Burlington, VT, USA. [3]Department of Biology, Skidmore College, Saratoga Springs, NY, USA. [4]Laboratory of Plant Physiology and Biophysics, Institute of Molecular Cell and Systems Biology, Bower Building, University of Glasgow, Glasgow G12 8QQ, UK.

References

1. Geitmann A: **How to shape a cylinder: Pollen tube as a model system for the generation of complex cellular geometry.** *Sex Plant Reprod* 2010, **23**:63–71.
2. Chebli Y, Geitmann A: **Mechanical properties governing pollen tube growth.** *Funct Plant Sci Biotech* 2007, **1**:232–245.
3. Ivakov A, Persson S: **Plant cell shape: modulators and measurements.** *Front Plant Sci* 2013. doi:10.3389/fpls.2013.00439.
4. Domozych DS, Fujimoto C, LaRue T: **Polar expansion dynamics in the plant kingdom: A diverse and multifunctional journey on the path to pollen tubes.** *Plants* 2013, **2**:148–173.
5. Hepler PK, Rounds CM, Winship LJ: **Control of cell wall extensibility during pollen tube growth.** *Mol Plant* 2013, **6**:998–1017.
6. Chebli Y, Kaneda M, Zerzour R, Geitmann A: **The cell wall of the Arabidopsis thaliana pollen tube- Spatial distribution, recycling and network formation of polysaccharides.** *Plant Physiol* 2012, **160**:1940–1955.
7. Palin R, Geitmann A: **The role of pectin in plant morphogenesis.** *BioSyst* 2012, **109**:397–402.
8. Mollet J-C, Leroux C, Dardelle F, Lehner A: **Cell wall composition, biosynthesis and remodeling during pollen tube growth.** *Plants* 2013, **2**:107–147.
9. Leitner D, Klepsch S, Ptashnyk M, Marchant A, Kirk GJD, Schnepf A, Roose T: **A dynamic model of nutrient uptake by root hairs.** *New Phytol* 2010, **185**:792–802.
10. Schmidt W, Schikora A: **Different pathways are involved in phosphate and iron stress-induced alterations of root epidermal cell development.** *Plant Physiol* 2001, **125**:2078–2084.
11. Bates TR, Lynch JP: **The efficiency of *Arabidopsis thaliana* (Brassicaceae) root hairs in phosphorus acquisition.** *Am J Bot* 2000, **87**:964–970.
12. Akkerman M, Franssen-Verheijen MAW, Immerzeel P, Hollander LDEN, Schel JHN, Emons AMC: **Texture of cellulose microfibrils of root hair cell walls of *Arabidopsis thaliana*, *Medicago truncatula*, and *Vicia sativa*.** *J Microsc* 2012, **247**:60–67.
13. Dumais J, Shaw SL, Steele CR, Long SR, Ray PM: **An anisotropic-viscoplastic model of plant cell morphogenesis by tip growth.** *Int J Dev Biol* 2006, **50**:209–222.
14. Emons AMC, Ketelaar T: **Intracellular organization: A prerequisite for root hair elongation and cell wall deposition.** In *Plant Cell Monographs: Root Hairs.* Edited by Emons AMC, Ketelaar T. Berlin: Springer; 2009:27–44.
15. Galway ME: **Root hair cell walls: filling in the framework.** *Can J Bot* 2006, **84**:613–621.
16. Nielsen E: **Plant cell wall biogenesis during tip growth in root hair cells.** In *Plant Cell Monographs: Root Hairs.* Edited by Emons AMC, Ketelaar T. Berlin: Springer; 2009:85–102.
17. Shaw SL, Dumais J, Long SR: **Cell Surface Expansion in Polarly Growing Root Hairs of *Medicago truncatula*.** *Plant Physiol* 2000, **124**:959–970.
18. Peña MJ, Kong Y, York WS, O'Neill MA: **A galacturonic acid-containing xyloglucan is involved in Arabidopsis root hair tip growth.** *Plant Cell* 2012, **24**:4511–4524.
19. Velasquez SM, Ricardi MM, Dorosz JG, Fernandez PV, Nadra AD, Pol-Fachin L, Egelund J, Gille S, Harholt J, Ciancia M, Verli H, Pauly M, Bacic A, Olsen CE, Ulvskov P, Petersen BL, Somerville C, Iusem ND, Estevez JM: **O-Glycosylated Cell Wall Proteins Are Essential in Root Hair Growth.** *Science* 2011, **332**:1401–1403.
20. Pedersen HL, Fangel JU, McCleary B, Ruzanski C, Rydahl MG, Ralet M-C, Farkas V, von Schantz L, Marcus SE, Andersen MCF, Field R, Ohlin M, Knox JP, Clausen MH, Willats WGT: **Versatile high-resolution oligosaccharide microarrays for plant glycobiology and cell wall research.** *J Biol Chem* 2012, **287**:39429–39438.
21. Moller I, Marcus SE, Haeger A, Verhertbruggen Y, Verhoef R, Schols H, Mikklesen JD, Knox JP, Willats W: **High-throughput screening of monoclonal antibodies against plant cell wall glycans by hierarchial clustering of their carbohydrate microarray binding profiles.** *Glycoconj J* 2008, **25**:37–48.
22. Hervé C, Marcus SE, Knox JP: **Monoclonal antibodies, carbohydrate-binding modules, and the detection of polysaccharides in plant cell walls.** In *The Plant Cell Wall: Methods and Protocols, Methods in Molecular Biology, vol. 715.* Edited by Popper ZA. New York: Springer/Humana Press; 2011:103–113.
23. Verhertbruggen Y, Marcus SE, Haeger A, Ordaz-Ortiz JJ, Knox JP: **An extended set of monoclonal antibodies to pectic homogalacturonan.** *Carbohydr Res* 2009, **344**:1858–1862.
24. Talboys PJ, Zhang HM, Knox JP: **ABA signalling modulates the detection of the LM6 arabinan cell wall epitope at the surface of *Arabidopsis thaliana* seedling root apices.** *New Phytol* 2011, **190**:618–626.
25. Lee KJD, Knox JP, Žárský V, Cvrckova F: **Resin embedding, sectioning and immunocytochemical analyses of plant cell walls in hard tissues.** In *Plant Cell Morphogenesis: Methods and Protocols, Methods in Molecular Biology, Vol. 1080.* New York: Springer / Humana Press; 2013:41–52.
26. Domozych DS, Brechka H, Britton A, Toso M: **Cell wall growth and modulation dynamics in a model unicellular green alga – *Penium margaritaceum*: live cell labeling with monoclonal antibodies.** *J Bot* 2011. doi:10.1155/2011/632165.
27. Domozych DS, Lambiasse L, Kiemle SN, Gretz MR: **Cell wall development and bipolar growth in the desmid *Penium margaritaceum*. Asymmetry in a symmetric world.** *J Phycol* 2009, **45**:879–893.
28. Hervé C, Rogowski A, Gilbert HJ, Knox JP: **Enzymatic treatments reveal differential capacities for xylan recognition and degradation in primary and secondary plant cell walls.** *Plant J* 2009, **58**:413–422.
29. Marcus SE, Verhertbruggen Y, Herve C, Ordaz-Ortiz JJ, Farkas V, Pedersen HL, Willats WGT, Knox JP: **Pectic homogalacturonan masks abundant sets of xyloglucan epitopes in plant cell walls.** *BMC Plant Biol* 2008, **8**:60–71.
30. Bernal AJ, Yoo CM, Mutwil M, Jensen JK, Hou G, Blaukopf C, Sørensen I, Blancaflor EB, Scheller HV, Willats WGT: **Functional analysis of the cellulose synthase-like genes Csld1, Csld2, and Csld4 in tip-growing *Arabidopsis* cells.** *Plant Physiol* 2008, **148**:1238–1253.
31. Clausen MH, Willats WGT, Knox JP: **Synthetic methyl hexagalacturonate hapten inhibitors of anti-homogalacturonan monoclonal antibodies LM7, JIM5 and JIM7.** *Carbohydr Res* 2003, **338**:1797–1800.
32. Jones L, Seymour GB, Knox JP: **Localization of pectic galactan in tomato cell walls using a monoclonal antibody specific to (1->4)-beta-D-galactan.** *Plant Physiol* 1997, **113**:1405–1412.

33. Willats WGT, Marcus SE, Knox JP: **Generation of a monoclonal antibody specific to (1- > 5)-alpha-L-arabinan.** *Carbohydr Res* 1998, **308:**149–152.
34. Ralet M-C, Tranquet O, Poulain D, Moïse A, Guillon F: **Monoclonal antibodies to rhamnogalacturonan I backbone.** *Planta* 2010, **231:**1373–1383.
35. Pettolino FA, Hoogenraad NJ, Ferguson C, Bacic A, Johnson E, Stone BA: **A (1–4)-beta-mannan-specific mono- clonal antibody and its use in the immunocytochemical location of galactomannans.** *Planta* 2001, **214:**235–242.
36. McCartney L, Marcus SE, Knox JP: **Monoclonal antibodies to plant cell wall xylans and arabinoxylan.** *J Histochem Cytochem* 2005, **53:**543–546.
37. Smallwood M, Yates EA, Willats WGT, Martin H, Knox JP: **Immunochemical comparison of membrane associated and secreted arabinogalactan-proteins in rice and carrot.** *Planta* 2006, **198:**452–459.
38. Yates EA, Valdor J-F, Haslam SM, Morris HR, Dell A, Mackie W, Knox JP: **Characterization of carbohydrate structural features recognized by anti-arabinogalactan-protein monoclonal antibodies.** *Glycobiol* 1996, **6:**131–139.
39. McCabe PF, Valentine TA, Forsberg LS, Pennell RI: **Soluble signals from cells identified at the cell wall establish a developmental pathway in carrot.** *Plant Cell* 1997, **9:**2225–2241.
40. Smallwood M, Beven A, Donovan N, Neill SJ, Peart J, Roberts K, Knox JP: **Localization of cell wall proteins in relation to the developmental anatomy of the carrot root apex.** *Plant J* 1994, **5:**237–246.
41. Xu C, Takac T, Burbach C, Menze D, Samaj J: **Developmental localization and the role of hydroxyproline rich glycoproteins during somatic embryogenesis of banana (*Musa* spp. AAA).** *BMC Plant Biol* 2011, **11:**38.
42. Cavalier DM, Lerouxel O, Neumetzler L, Yamauchi K, Reinecke A, Freshour G, Zabotina OA, Hahn MG, Burgert I, Pauly M, Raikhel NV, Keegstra K: **Disrupting two *Arabidopsis thaliana* xylosyltransferase genes results in plants deficient in xyloglucan, a major primary cell wall component.** *Plant Cell* 2008, **20:**1519–1537.
43. Galway ME, Eng RC, Schiefelbein JW, Wasteneys GO: **Root hair-specific disruption of cellulose and xyloglucan in AtCSLD3 mutants, and factors affecting the post-rupture resumption of mutant root hair growth.** *Planta* 2011, **233:**985–999.
44. Favery B, Ryan E, Foreman J, Linstead P, Boudonck K, Steer M, Shaw P, Dolan L: ***KOJAK* encodes a cellulose synthase-like protein required for root hair cell morphogenesis in Arabidopsis.** *Genes Dev* 2001, **15:**79–89.
45. Wang X, Cnops G, Vanderhaeghen R, De Block S, Van Montagu M, Van Lijsebettens M: **AtCSLD3, a cellulose synthaselike gene important for root hair growth in Arabidopsis.** *Plant Physiol* 2001, **126:**575–586.
46. Larson ER, Domozych DS, Tierney ML: **SNARE VTI13 plays a unique role in endosomal trafficking pathways associated with the vacuole and is essential for cell wall organization and root hair growth in Arabidopsis.** *Ann Bot* 2014. in press.
47. Baumberger N, Ringli C, Keller B: **The chimeric leucine rich repeat/extensin cell wall protein LRX1 is required for root hair morphogenesis in *Arabidopsis thaliana*.** *Genes Dev* 2001, **15:**1128–1139.
48. Nguema-Ona E, Coimbra S, Vicré-Gibouin M, Mollet J-C, Driouich A: **Arabinogalactan proteins in root and pollen-tube cells: distribution and functional aspects.** *Ann Bot* 2012, **110:**383–404.
49. Zabotina OA, Wilhelmina TG, Van De V, Freshour G, Drakakaki G, Cavalier D, Mouille G, Hahn MG, Keegstra K, Raikhe NV: **Arabidopsis *XXT5* gene encodes a putative α-1,6-xylosyltransferase that is involved in xyloglucan biosynthesis.** *Plant J* 2008, **56:**101–115.
50. Park S, Szumlanski AL, Gu F, Guo F, Nielsen E: **A role for CSLD3 during cell-wall synthesis in apical plasma membranes of tip-growing root-hair cells.** *Nat Cell Biol* 2011, **13:**973–980.
51. Yo C-M, Quan L, Blacaflor EB: **Divergence and redundancy in *CSLD2* and *CSLD3* function during *Arabidopsis thaliana* root hair and female gametophyte development.** *Front Plant Sci* 2011. doi: 10.3389/fpls.2012.00111.
52. Baylson FA, Stevens BW, Domozych DS: **Composition and synthesis of pectin and protein components of the cell wall of *Closterium acerosum* (Chlorophyta).** *J Phycol* 2001, **37:**796–809.

11

A novel approach to the generation of seamless constructs for plant transformation

Remy Kronbak, Christina Rønn Ingvardsen, Claus Krogh Madsen and Per Langkjær Gregersen*

Abstract

Background: When creating plant transformation vectors, full control of nucleotides flanking the insert in the final construct may be desirable. Modern ligase-independent methods for DNA-recombination are based on linearization by classical type II restriction endonucleases (REs) alone or in combination with nicking enzymes leaving residual nucleotides behind in the final construct. We here explore the use of type IIS and type IIB REs for vector linearization that combined with sequence and ligase-independent cloning (SLIC) overcomes this problem and promotes seamless gene-insertion in vectors. Providing the basis for a collection of biolistic plant transformation vectors ready to be cloned with different genes-of-interest, we present two vectors, where promoter and terminator are joined by a spacer. During spacer-removal linearization (SRL), type IIS and type IIB REs remove their own recognition sequences from the vector leaving no undesired, short sequences behind.

Results: We designed two plant transformation vectors prepared for SRL in combination with SLIC, pAUrumII and pAUrumIII, harboring a spacer with recognition sites for a type IIS and IIB RE, respectively. The gene for a green fluorescent protein, *gfp*, was successfully cloned into both vectors; traces of pAUrumIII, however, contaminated the transformation due to incomplete linearization, an issue not encountered with the type IIS linearized pAUrumII. Both constructs, pAUrumII-*gfp* and pAUrumIII-*gfp*, were functional, when tested *in vitro* on wheat and barley endosperm cells for transient *gfp* expression.

Conclusions: All nucleotides flanking an insert in a biolistic plant transformation vector can be customized by means of SRL in combination with SLIC. Especially type IIS REs promote an efficient cloning result. Based on our findings, we believe that the SRL system can be useful in a series of plant transformation vectors, favoring the presence of functional sequences for optimal expression over redundant cloning-site remnants.

Keywords: Plant transformation vector, linearization, Type IIS restriction endonucleases, Type IIB, In-Fusion™, Seamless cloning, Biolistics, Cereals, Transient expression

Background

Combining DNA from various sources in single constructs typically for the purpose of over-expressing genes-of-interest is essential to modern research in fields like genetics, bioinformatics and biotechnology. Classical type II restriction endonucleases (REs) recognize and cleave short, palindromic sequences creating blunt or sticky ends (5'- or 3'-overhangs typically of one to a few nucleotides in length) and for decades they have been employed in a cleaving-and-ligating oriented manner. A range of limitations, however, accompanies this method. First, direction of the inserted fragment cannot be controlled unless the cloning is based on two REs creating different overhangs. Second, cloning of more than one fragment is not efficient and direction and order are not controllable using a single classical type II RE. Third, recognition sequences for the RE in question can only be present at the desired cloning location of the vector and flanking the insert, not elsewhere in the vector backbone or in the insert sequence. Fourth, ligase treatment of the inserted fragment and the backbone vector is required before transformation to *E. coli*. This is due to the fact that matching overhangs are most often not of more than four nucleotides in length. Such overhangs will likely dissociate at transformation temperatures, when non-ligated.

* Correspondence: per.gregersen@agrsci.dk
Science and Technology, Department of Molecular Biology and Genetics, Aarhus University, Forsøgsvej 1, DK-4200 Slagelse, Denmark

In recent years methods have been developed, which overcome most of the obstacles mentioned [1]. These methods are in general based on longer matching overhangs than most classical type II REs offer, and include ligase-independent cloning (LIC) [2], sequence and ligase-independent cloning (SLIC) [3] (equivalent to the commercially available In-Fusion™ cloning system, Clontech Laboratories, [4,5]), uracil specific excision reagent (USER™, New England Biolabs) cloning [6,7], circular polymerase extension cloning (CPEC) [8] and one-step isothermal *in vitro* recombination (Gibson Assembly™, Synthetic Genomics) [9]. The overhangs are typically at least 8–15 nt in size and remain sticky at transformation temperatures of 37–42°C independent of ligation. Several fragments can be assembled in the correct order and direction in a single reaction as long as all the sequences of homology involved (from which the matching overhangs are generated) are unique. In the In-Fusion™ system, which we use here, a poxvirus DNA polymerase displays 3'-5' exonuclease activity against duplex DNA, when dNTP levels are low, due to its proofreading capacity. When the entire matching overhangs are created from 15 nt sequences of homology, spontaneous annealing of these overhangs occur, inhibiting further exonuclease activity of the polymerase. Thus, the assembled hybrid DNA only contains nicks or short gaps around the assembled sequences, which will be repaired and ligated by *E. coli* after transformation [4,5].

Meeting agricultural challenges, such as disease protection and drought tolerance, with the usage of genetically modified (GM) crops has not received great public acceptance because of the skepticism associated with the introduction of foreign DNA to an organism; an issue that intragenesis and cisgenesis have been developed to counteract (reviewed in [10,11]). For these two techniques, the gene pool, from which DNA can be introduced into an organism, is limited to crossable species and thus identical to the conventional breeder's gene pool. Regardless of the GM technique employed, one might consider to avoid the presence of any unnecessary DNA sequence. Also in connection with gene fusion, there must be a seamless junction between a vector with an existing open reading frame (ORF) and an ORF-continuing insert if it is important to exclude additional amino acids between the fused proteins [12]. So for several reasons, it may be desired to fully control the nucleotides flanking the insert/inserts in the final construct, e. g. to have the gene-of-interest start codon at the exact location as the start codon of the native gene for the promoter in question, or to have only a Kozak consensus sequence in front of the gene for optimal mRNA translation or a tag sequence right after for tracking the expression of a gene, which has a high level of similarity to others. None of the mentioned cloning methods directly offer this, when the cloning is based on linearization by classical type II REs or a combination of REs and nicking enzymes (USER™ cassette) leaving residual nucleotides behind in the final construct. We explore here the use of type IIS and type IIB REs (reviewed in [13]) in the vector linearization process that, when combined with SLIC, overcomes this problem and creates seamless junctions. Type IIS REs are enzymes that recognize specific DNA sequences, but whereas classic type II REs cleave the DNA strands within their recognition sequence, type IIS cleave at a certain distance, most often a few base pairs away, from a non-palindromic recognition sequence. The type IIS RE *Eam1104I* has previously been used with LIC and SLIC for seamless gene fusion in *E. coli* expression vectors [14,15]. Dependent on the presence of S-adenosyl methionine (SAM), type IIB REs have the capability to cleave the DNA strands at both sides of the recognition sequence at defined distances. By using type IIS and type IIB REs for linearization, the recognition sequence can be separated from the cleaved sequence and in that way not reappear in the final construct.

Providing the basis for a collection of prepared plant transformation vectors, we here present two vectors, where promoter and terminator are joined by a spacer that during linearization is removed by either a type IIS or type IIB RE. During this spacer-removal linearization (SRL) these endonucleases remove their own recognition sequences from the vector leaving no undesired nucleotides in the final construct. SRL is combined with In-Fusion™, which is simple and efficient.

Results

Creating plant transformation vectors

As the first two of a collection of plant transformation vectors prepared for SRL used with In-Fusion™ cloning, pAUrumII and pAUrumIII were designed (Figure 1). They were prepared for gold particle bombardment and have thus no active elements in connection with plant cell entry like e.g. border sequences of binary vectors for *Agrobacterium tumefaciens*-mediated transformation. The 35S promoter of Cauliflower Mosaic Virus was chosen as it is constitutively active throughout most plant tissues. For transcription termination, the nopaline-synthase (NOS) terminator of *A. tumefaciens* was selected.

Between the promoter and the terminator, pAUrumII has a spacer of 833 bp containing a recognition site for the type IIS RE *LguI* (5'–GCTCTTCN$^{\vee}$NNN$_{\wedge}$–3') at both ends with the cleaving sites being the last 3 bp of 35S and first 3 bp of NOS, respectively. So when pAUrumII is linearized with this enzyme, it creates 3 nt 5'-overhangs at these location, while it removes its own recognition sites from the vector (Figure 2A). The nucleotides of these 5'-overhangs are included in the 15 nt overhangs created during In-Fusion™ enzyme treatment.

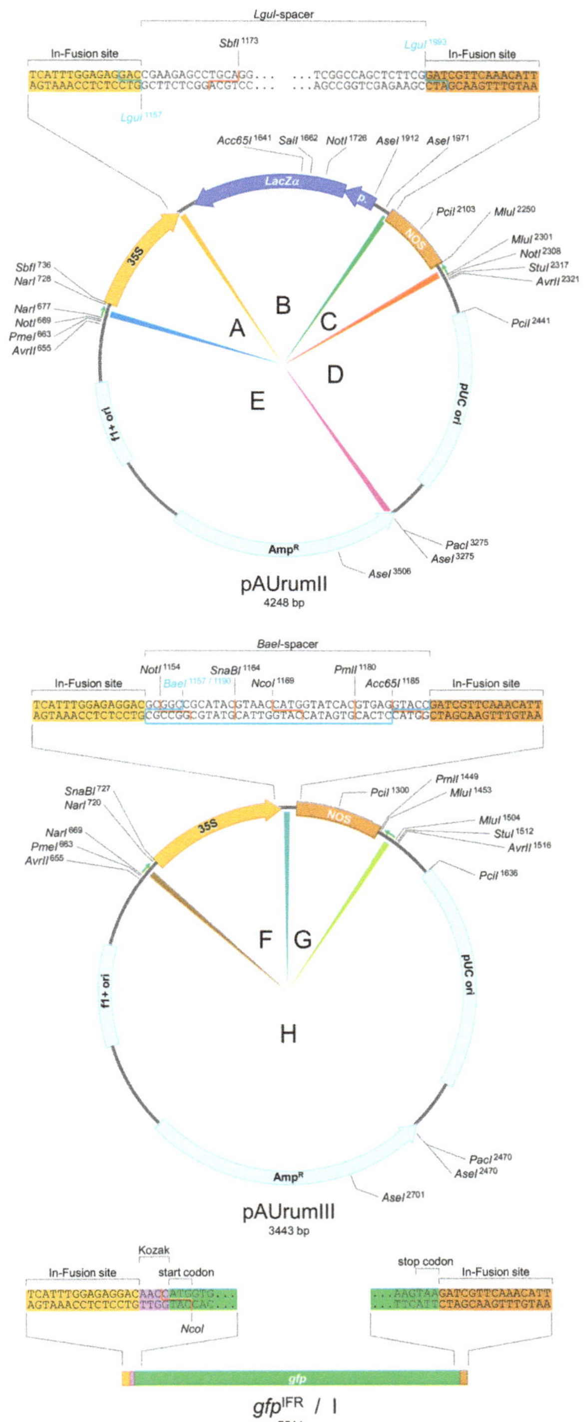

Figure 1 Maps of vectors and insert. pAUrumII prepared from the fragments A–E, pAUrumIII from F–H and the In-Fusion™-ready gene for green fluorescent protein, *gfp*^{IFR} (the I fragment) are mapped with selected RE cleaving sites and nucleotide sequences. Colors of the fragment separators of the two plasmid maps correspond to the indications of sequence homology in Figure 3. Small green arrows indicate randomly generated sequences for optimized primer annealing (not used here) and the sequences can be removed with *NarI* and *MluI*.

We placed the *LacZα* coding sequence with promoter between the *LguI* sites for detecting carry-over of non-linearized vector, when cloning genes-of-interest.

Five PCR fragments (see Methods and Figure 1), approx. 60 fmol of each, were In-Fusion™ cloned to form pAUrumII. Most of the appearing colonies were blue due to the presence of the *LacZα* coding sequence and 6 of 20 analyzed blue colonies contained a correctly assembled construct, when analyzed with *AseI*, *NotI* and *SalI* (not shown); we did not look into what the remaining 14 colonies contained.

The spacer used in pAUrumIII is much shorter since cleavage at both the promoter and the terminator locations are performed physically by the same type IIB RE, *BaeI*. This enzyme is associated with the sequence $5'\text{–}_{\wedge}NNNNN^{\vee}N_{10}ACN_4GTAYCN_{7\wedge}NNNNN^{\vee}\text{–}3'$ (Y = C or T) with the recognition site being located between two cleaving sites. Therefore, when linearizing with *BaeI*, the enzyme creates 5 nt 3'-overhangs both at the 35S 3'-end and the NOS 5'-end, while it removes its own recognition sequence from the construct (Figure 2B). The nucleotides of these 3'-overhangs will be chewed back by the In-Fusion™ enzyme, which then continues chewing back, thus creating the 15 nt 5'-overhangs for assembly. Obviously, the size of a type IIB spacer does not allow the presence of the *LacZα* coding sequence. Among the many optional nucleotides between the two cleaving sites of the spacer, however, we placed a recognition site for *SnaBI* for future exchange of promoter (a similar site is present upstream of the 35S) and a site for *PmlI* for the terminator to be exchanged (downstream of the NOS a similar site is present). Conventional cloning using *NotI* + *Acc65I* can also be performed for direction controlled but not seamless insertion.

Approx. 60 fmol of each of three fragments (see Methods and Figure 1) were In-Fusion™ cloned to form pAUrumIII and since they were all PCR amplified from pAUrumII, which holds the *LacZα* coding sequence in its spacer, any template carry-over would lead to colonies being blue. Few blue colonies were formed and among the many white colonies we found 16 out of 16 analyzed with *AvrII* and *PciI* (not shown) to contain the correct construct.

SRL and In-Fusion™ cloning of *gfp*

Primers were designed to amplify *gfp*(S65T) [16] and incorporate the monocotyledon Kozak consensus sequence, AACC [17], as well as the 15 bp extensions necessary to facilitate In-Fusion cloning (Figures 2C and 3).

The vectors pAUrumII and pAUrumIII were linearized overnight. From the agarose gel we found no indications of an incomplete *LguI*-digestion of pAUrumII, whereas a small amount of non-linearized (supercoiled) pAUrumIII was left by *BaeI* (Figure 4A). Also an amount of some

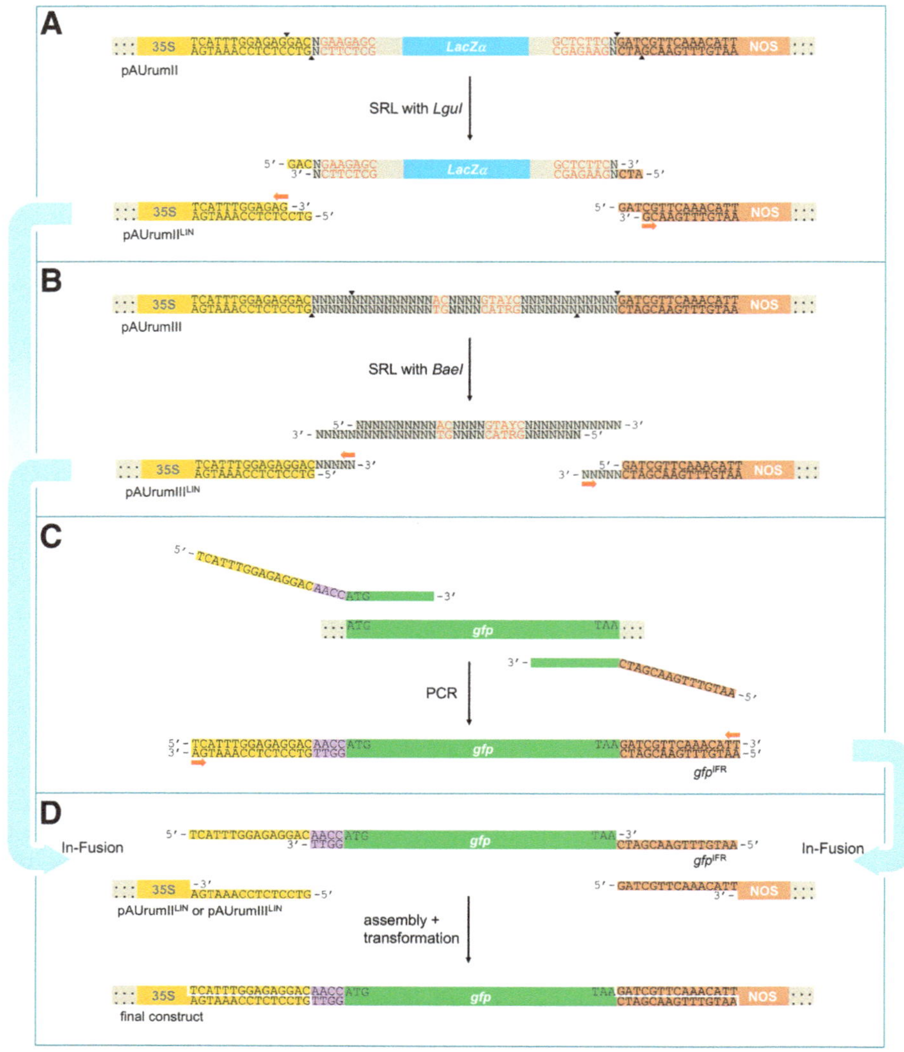

Figure 2 Overview of SRL used with In-Fusion™ cloning. Red letters indicate recognition sites and black triangles cleaving sites of *LguI* and *BaeI*. Red arrows indicate where In-Fusion™ enzyme exonuclease activity will occur. **A**. Spacer-removal linearization (SRL) of pAUrumII (to pAUrumII^{LIN}) with type IIS RE *LguI*. The spacer holds the *LacZα* coding sequence with its promoter. **B**. SRL of pAUrumIII (to pAUrumIII^{LIN}) with type IIB RE *BaeI*. **C**. Extended primers used in PCR amplification of *gfp*. The In-Fusion™-ready product, *gfp*^{IFR}, has 15 nt of homology with the 35S promoter 3'-end and 15 nt of homology with the NOS terminator 5'-end. Moreover, it has the monocot Kozak consensus sequence AACC in front of the start codon. **D**. In-Fusion™ reaction and assembly of pAUrumII^{LIN} or pAUrumIII^{LIN} with *gfp*^{IFR} and subsequent transformation of non-ligated construct to *E. coli*, where repair and ligation will occur. There are no unwanted nucleotides in the final construct.

relaxed, probably nicked, state of circular pAUrumIII was observed.

Insert and linearized vector were equimolarly mixed for In-Fusion™ reaction, and the mixture of non-ligated, assembled fragments was transformed into *E. coli* (Figure 2D). Among colonies from the pAUrumII-*gfp* transformation (n = 299, distributed on three plates) no blue were found. Plasmid mini-preparations of liquid overnight cultures of 15 randomly selected colonies from one plate were analyzed with *NotI* (Figure 4B) and *AseI* + *NcoI* (not shown), and all digestions gave the expected fragments. Blue/white screening was not an option, when selecting colonies from the pAUrumIII-*gfp* transformation (n = 367, distributed on three plates). 15 randomly selected colonies from one plate were analyzed with *AvrII* (Figure 4C) and *AseI* + *NcoI* (not shown), and 11 gave the expected fragments. In four colonies the analysis revealed the presence of pAUrumIII; in one case we found indications that both pAUrumIII and pAUrumIII-*gfp* were present, suggesting that the colony was developed from two single cells.

Transient *gfp* expression

The collection of plant transformation vectors to be built is intended mainly for crops like barley and wheat. For this reason pAUrumII-*gfp* and pAUrumIII-*gfp* were

pAUrumII	A-for	5'-AGAAGCGTCAATTGCCTACCCAGTCGAACCCACATAGGGCGCCTGCAGGGTCAACATGGTGGAGCACG-3'
	A-rev	5'-CCTGCAGGCTCTTCGGTCCTCTCCAAATGAAATGAACTTCC-3'
	B-for	5'-CGAAGAGCCTGCAGGAAAGGGCGAAAAACCGTCTATC-3'
	B-rev	5'-GATCCGAAGAGCTGGCCGATTCATTAATGCAGC-3'
	C-for	5'-CCAGCTCTTCGGATCGTTCAAACATTTGGCAATAAAGTT-3'
	C-rev	5'-AGACAACTTTCAGGCTTTGCAGGAATGGAGGGTATCGAACGACGCGTCCCGATCTAGTAACATAGATGAC-3'
	D-for	5'-GCCTGAAAGTTGTCTCGCGACGCGTGCGGCCGCAGGCCTCCTAGGCTTCCTCGCTCACTGACTC-3'
	D-rev	5'-GCCTCATTAATTAAGCATTGGTAACTGTCAGACC-3'
	E-for	5'-CTTAATTAATGAGGCACCTATCTCAGC-3'
	E-rev	5'-GCAATTGACGCTTCTATCCGAGCGGCGCCGCGGCCGCGTTTAAACCTAGGCCAATTCGCCCTATAGTGAGT-3'
pAUrumIII	F-for	5'-AGAAGCGTCAATTGCCTACCCAGTCGAACCCACATAGGGCGCCTACGTAGTCAACATGGTGGAGCACG-3'
	F-rev	5'-GATACCATGGTTACGTATGCGGCCGCGTCCTCTCCAAATGAAATGAACTTCC-3'
	G-for	5'-CGTAACCATGGTATCACGTGAGGTACCGATCGTTCAAACATTTGGCAATAAAGTT-3'
	G-rev	5'-AGACAACTTTCAGGCTTTGCAGGAATGGAGGGTATCGAACGACGCGTCACGTGCCCGATCTAGTAACATAGATGAC-3'
	H-for	5'-GCCTGAAAGTTGTCTCGCGACGCGTAGGCCTCCTAGGCTTCCTCGCTCACTGACTC-3'
	H-rev	5'-GCAATTGACGCTTCTATCCGAGCGGCGCCGTTTAAACCTAGGCCAATTCGCCCTATAGTGAGT-3'
gfp[IFR]	I-for	5'-TCATTTGGAGAGGACAACCATGGTGAGCAAGGGCGAG-3'
	I-rev	5'-AATGTTTGAACGATCTTACTTGTACAGCTCGTCCATGC-3'

Figure 3 PCR primer sequences. The fragment to be PCR amplified (A–I) and the orientation, forward (for) or reverse (rev), are indicated in the primer name in front of the primer sequence. Sequence with a single, black underline represents the part of the primer that anneals to the template. Sequence in bold, italic represents the 15 nt for In-Fusion™ cloning and pairwise coloring (A–H primers) represents homology; the same colors are used in Figure 1, indicating that the fragments are assembled from these homologous sequences, thus forming the two plasmids. Sequence with double underline represents recognition sites for *LguI* (A-rev, C-for) or *BaeI* (G-for), whereas sequence with dashed, red underline represents the monocotyledon Kozak consensus sequence (I-for).

tested *in vitro* on wheat and barley endosperm cells. Two days after bombardment, the material was evaluated under UV light in a stereo microscope and it was observed that both constructs worked in both species; the wheat endosperms, however, contained more green glowing cells than those of barley, and for both species not all endosperms contained *gfp*-expressing cells due to uneven quality of the endosperms and/or varying positioning of these in relation to the bombardment (Figure 5).

Discussion

Today's cloning of vectors is to a large extent based on the creation of long matching overhangs between the fragments to be assembled. We investigated here the possibility of starting a collection of biolistic plant transformation vectors based on SLIC (specifically In-Fusion™) ready to be cloned with different genes-of-interest, while having the full control of any single nucleotide present in the final construct. For this we used a spacer occupying the cloning site. This spacer is removed during the linearization process by a type IIS or IIB RE, and along with the spacer, the enzymes remove their own recognition sequence as well. Our demonstration that 35S–(Kozak)*gfp*–NOS was efficiently cloned and fully operational in a transient expression study on cereal endosperms, indicates that there is a basis for starting such a collection.

A method that leaves out undesired, short sequences in final transformation vectors may be relevant in the creation of genetically modified crops. Recombining DNA in a way that reduces the presence of redundant sequences should in our opinion be strived for. Especially for intragenesis [10,11], SRL combined with In-Fusion™ cloning has relevance. A range of genes-of-interest could be inserted in a spacer-removal-linearized vector with an appropriate promoter and terminator, all sequences being available within crossable relatives to the species to be transformed. When it comes to cisgenesis [10,11], however, the gene-of-interest must be flanked by its own natural promoter and terminator, and it is, therefore, not possible to prepare promoter–spacer–terminator vectors suitable for different genes. By means of the flanking sites for RE *SbfI*, the 35S promoter can easily be exchanged with a range of different promoters, e.g. a native gene promoter. Such promoter can easily be inserted through In-Fusion™ cloning, while ensuring that the *LguI* and *SbfI* sites are restored from extended PCR primers prior to this step. If, however, one or more sites for *LguI* are present in the new promoter, the spacer needs to be recreated for the use of another type IIS RE (it is irrelevant that a chosen type IIS RE may also cleave within the spacer sequence).

Type IIS REs have successfully been used in connection with seamless genetic recombination (reviewed in [12]). In a sophisticated variant of classical cleaving-ligating cloning, the type IIS RE *BsaI* has been employed [18]. Subcloning a fragment from one vector, leaving recognition sites for *BsaI* in the backbone, into another vector, linearized with the same type IIS RE by removing a fragment holding the recognition sites, has several advantages. Besides having no *BsaI* recognition sites in the final construct, the advantages include the full control of insert orientation, as well as simultaneous ligation in

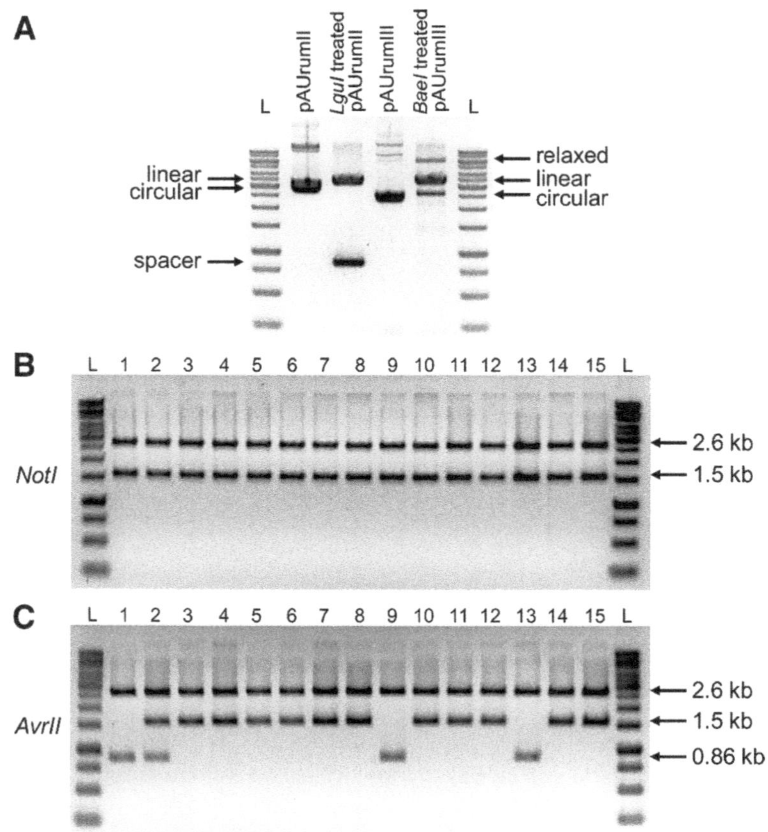

Figure 4 Vector linearization and RE analysis of cloned *gfp*. A. pAUrumII in its circular (supercoiled) and *LguI* treated, linear state with the *LguI*-spacer removed next to pAUrumIII in its circular (supercoiled) and *BaeI* treated, linear state. The *BaeI* treated pAUrumIII also appears in a relaxed, circular state. The short *BaeI*-spacer is not visible. **B**. Restriction analysis for pAUrumII-*gfp* with *NotI* on plasmid DNA from 15 colonies. The correct construct is separated in two fragments of 2.6 kb and 1.5 kb, respectively. **C**. Restriction analysis for pAUrumIII-*gfp* with *AvrII*. The correct construct is separated in two fragments of 2.6 kb and 1.5 kb, respectively. The construct separated in a 2.6 kb and a 0.86 kb fragment is pAUrumIII. For all gels, L indicates the O'GeneRuler 1 kb DNA Ladder (Thermo Scientific).

the digestion solution. The method, referred to as Golden Gate, has been further developed to shuffle several DNA fragments in single reactions [19]. Also the type IIB RE *BaeI* has previously been used in an advanced technique for shuffling DNA named sequence-independent site-directed chimeragenesis (SISDC) [20]. In addition to these ligase-dependent methodologies, we demonstrate here the type IIS and IIB REs as a useful means for SRL used with In-Fusion™ cloning, which has the advantage of being ligase-independent and thus relatively rapid. The applications of this method are in our opinion not limited to plant science, but can also be relevant in connection with expression vectors for *E. coli* and yeast. The principle was shown useful in connection with protein expression in *E. coli* [15].

Omitting undesired nucleotides in In-Fusion™-cloned final constructs could be performed in different ways [12]. In the In-Fusion™ cloning, the insert could be received by a vector, which is not linearized from a circular state, but instead prepared as a linear PCR product from a circular or linear template ([21], and as proposed by the In-Fusion™ manufacturer). The primers used to generate this linear vector will determine, which nucleotides will be flanking the insert, and a construct with similar seamlessly joined promoter, gene-of-interest (insert) and terminator, as we demonstrate by using SRL with In-Fusion™ cloning, can be achieved. PCR amplifying the insert-receiving vector, however, may lead to PCR mistakes in the vector backbone, and this risk is increased with the vector size. The linearization option may lead to contamination with carry-over of non-linearized vector, when transforming *E. coli* with the In-Fusion™-cloning mixture; creating a linear vector from PCR amplification, however, may similarly lead to contamination with carry-over of template-vector. Template-vector may be degraded prior to transformation with a type IIM RE like *DpnI*, which only digests methylated DNA and thus spares PCR products, but addition of this step to the procedure contributes to the over-all working time. So, from this point of view it is not clear, which is preferable;

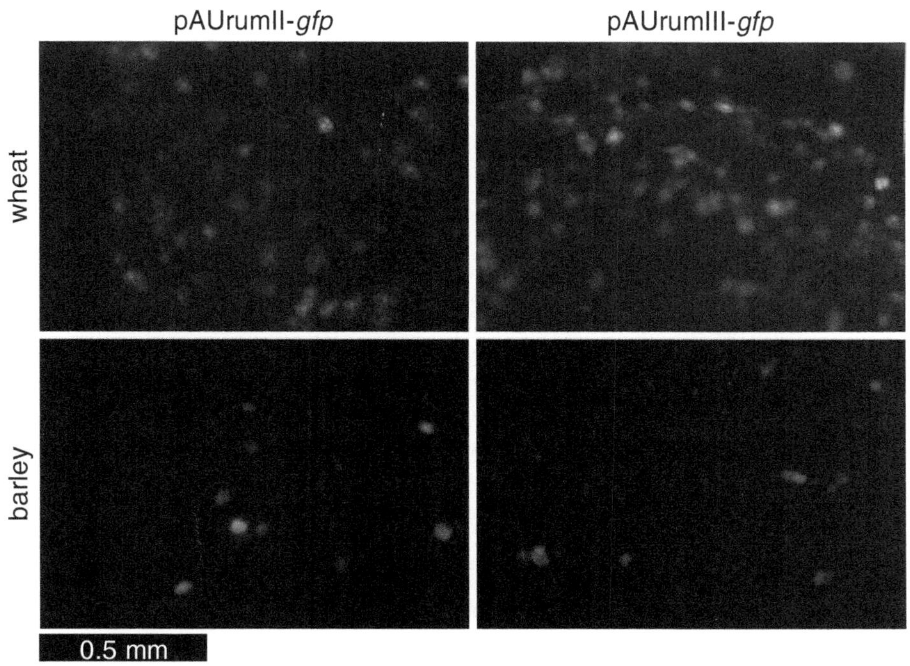

Figure 5 Transient *gfp* expression in endosperms. Wheat (upper) and barley (lower) immature endosperms two days after bombardment with pAUrumII-*gfp* (left) or pAUrumIII-*gfp* (right) evaluated under UV light. The bar length represents 0.5 mm. Handling parameters such as positioning of the endosperms in the gene gun influence the outcome of the experiment. Endosperms with *gfp* expression were therefore preferentially selected for the illustration.

linearization or PCR amplification of the vector. Another way of cloning a gene-of-interest seamlessly between a promoter and a terminator, possibly with few extra desired nucleotides flanking it, is to insert promoter and terminator as separate fragments simultaneously with the gene-of-interest, *i.e.* a 4-fragment In-Fusion™ cloning. SLIC is, due to the long, unique overhangs, suitable for cloning more than one fragment; the efficiency of the cloning, however, most likely decreases with every fragment added to the cloning, and the risk of contamination with carried-over template increases. Besides, the risk of introducing PCR errors in the promoter and terminator is also present.

SRL takes advantage of the special characteristics of type IIS and IIB REs, having their recognition site separated from their cleaving site/sites. *LguI* and *BaeI* were chosen among several other type IIS and IIB REs and the main criteria was that recognition sites for the enzyme in question were not present outside the spacer sequence. It seems that *BaeI* does not digest the vector to the same level as *LguI*, for which no visible non-linearized vector was present in the agarose gel (Figure 4A). The possibility that this holds true for other type IIB REs, as a possible consequence of the cleaving process being more complicated than for type IIS REs, or if the choice of *BaeI* was just unfortunate, was not examined.

The use of two recognition sites for a type IIS RE rather than one for a type IIB gives a size-independent spacer. We took advantage of this and placed the coding sequence of *LacZα* with its promoter in the *LguI*-spacer. In this way it is possible to detect colonies transformed with non-linearized vector, which may be present even after an overnight digestion. The type IIB spacer does not allow such a screening system. For both pAUrumII and pAUrumIII, the linearized and the circular (supercoiled) vector run close to each other on an agarose gel (Figure 4A), and because *BaeI* and possibly other type IIB REs do not digest completely, it would especially here be an advantage to have the possibility of blue/white screening.

SLIC and the similar methods that rely on the creation of long matching overhangs allow the use of unpurified PCR products in the assembly process. But the relatively easy step of purifying linearized vector and PCR products is probably worth doing in order to reduce the level of false positives. The time saved by skipping these steps is easily used on doing more plasmid mini-preparations and RE analysis to find the right clone.

Conclusions

All nucleotides flanking an insert in a plant transformation vector can be controlled from the combination of spacer-removal linearization (SRL) of the insert-receiving vector and sequence and ligation-independent cloning. We demonstrate that both type IIS and IIB restriction endonucleases can be used to remove the spacer occupying the cloning site of the vector; it seems, however, that the use of a type IIS has more advantages for an efficient cloning outcome. Based on our findings, we believe that the SRL system can be useful for the generation of expression vectors such as plant transformation vectors.

Methods

PCR

All PCRs were carried out with Herculase II Fusion DNA polymerase (Agilent) or Velocity DNA polymerase (Bioline) in agreement with general protocol guidelines. PCR primers were purchased from Invitrogen, Germany. As many of the primers had a relatively long sequence attached to the template-matching sequence (Figure 3), 6–8 initial cycles with 58°C annealing temperature were performed followed by an increase in temperature to 63°C for an additional 30 cycles.

Purification

To avoid carry-over of template DNA in the cloning reactions, all PCR fragments were purified from a 1.2–1.5% (w/v) agarose gel with ethidium bromide or GelStar (Lonza) staining and subsequent spin column recovery (NucleoSpin Gel and PCR Clean-up, Macherey Nagel). Likewise, linearized vectors were purified from agarose gel followed by spin column recovery to eliminate the presence of non-linearized vectors in the *E. coli* transformation solution.

Creating plant transformation vectors

The plant transformation vector pAUrumII was created from five PCR produced fragments (Figures 1 and 3). Fragment A contained the Cauliflower Mosaic Virus (CaMV) 35S promoter and fragment C contained the nopaline synthase terminator (NOS) of *Agrobacterium tumefaciens*. Promoter and terminator were linked by fragment B holding the *LacZα* coding sequence with its promoter. Fragment B was flanked by recognition sequences for the type IIS RE *LguI* with cleaving sites pointing outward. Fragments B, D and E were amplified from a pBluescriptII SK + (Agilent) template, and D combined with E forms the vector backbone. We wanted to introduce a recognition site for the RE *PacI* in the beta-lactamase (ampicillin resistance) gene for having the possibility of linearizing the final vector opposite the expression cassette if desired (not relevant here); the *PacI* site, however, does not disturb the ORF of the beta-lactamase gene. In the 15 common base pairs (for In-Fusion™ cloning) of fragments D and E the *PacI* site was introduced. Fragment D, closest to the terminator fragment C, was amplified so that the existing recognition site for *LguI* of pBluescriptII SK + was not included. The 35S–*LguI*-spacer–NOS cassette was flanked by two short, randomly generated sequences (R package version 2.14.0) with optimized PCR-primer sites prepared for tracking full-lengths inserted expression cassettes in the genome of regenerated plants; hence the great length of some of the PCR primers for this construct. In the present study, however, these sequences were not utilized, as we tested the construct only transiently.

All five fragments, ranging in size from 314 to 1,683 bp, were joined in equimolar amounts (approx. 60 fmol of each) in 10 μL 1X In-Fusion™ enzyme mix (Clontech) for 15 minutes reaction at 50°C before assembly on ice. 50 μL competent Stellar (Clontech) *E. coli* cell solution in 2.2 mL microcentrifuge tubes on ice were transformed with 2.5 μL of the In-Fusion™ reaction solution according to the manufacturer's recommendations before spreading on LB agar plates (100 mg/L ampicillin, 100 mg/L X-gal) for overnight incubation at 37°C.

Plasmid mini-preparations (FastPlasmid Mini Kit, 5Prime) of overnight LB cultures (100 mg/L ampicillin) of randomly selected blue colonies were analyzed with the REs *AseI* (New England Biolabs), *NotI* and *SalI* (both Fermentas).

Three fragments, F, G and H, for the transformation vector pAUrumIII were PCR amplified using pAUrumII as template (Figures 1 and 3). F holds the 35S promoter and G the NOS terminator. The relatively short type IIB RE spacer, with a recognition site for *BaeI*, was introduced between F and G from the PCR primers. Fragment H corresponds to D + E and when combined with the promoter and terminator fragments, only few nucleotides make pAUrumIII differ from pAUrumII outside the 35S–*BaeI*-spacer–NOS cassette; these are RE recognition sites irrelevant for construct functionality.

The In-Fusion™ cloning was carried out as for pAUrumII and plasmid mini-preparations of overnight cultures of randomly selected white colonies were analyzed with *AvrII* and *PciI* (both New England Biolabs).

A plasmid midi-preparation (NucleoBond Xtra Midi/Maxi, Macherey Nagel) of a selected positive clone of each of pAUrumII and pAUrumIII were verified by sequencing at Eurofins MWG Operon, Germany.

Linearizing plant transformation vectors

Five micrograms of pAUrumII was digested 15 h at 37°C using 6 U/μg of the type IIS RE *LguI* (*SapI*) (Fermentas) in a 100 μL reaction solution. In another 100 μL reaction solution containing 20 μM S-adenosyl methionine, 5 μg pAUrumIII was digested 15 h at 25°C using 10 U/μg of the type IIB RE *BaeI* (New England Biolabs). Both digestions were terminated at 65°C for 20 min, and to further reduce the risk of transforming with empty vectors, both the linearized vectors, pAUrumII^{LIN} or pAUrumIII^{LIN}, were treated with 0.6 U/μg calf intestinal alkaline phosphatase (Invitrogen) for 30 min at 37°C before purification.

Preparing *gfp* for insertion

To evaluate the constructs, the gene for a green fluorescent protein, *gfp*, was chosen for insertion (we use the gene for sGFP(S65T) [16], originating from plant transformation vector pVec8-GFP [GenBank: FJ949107.1]). Using the I-primers (Figure 3), we prepared an In-Fusion™-ready fragment by PCR, gfp^{IFR}, *i.e.* the ORF of *sgfp*(S65T), with the monocotyledon Kozak consensus sequence AACC in front of the ATG start codon, extended at both ends with nucleotide sequences corresponding to the last 15 nucleotides of the 35S promoter, TCATTTGGAGAGGAC, and the first 15 of the NOS terminator, GATCGTTCAAACATT, respectively (Figures 1 and 2C).

In-Fusion™ cloning of *gfp*

Approx. 60 fmol of gfp^{IFR} were joined in equimolar amounts with pAUrumII^{LIN} or pAUrumIII^{LIN} in 10 μL 1X In-Fusion™ enzyme mix, and reaction and transformation occurred as described for the vectors above. After the 1 h shaking of the transformed bacteria, a 1:100 dilution was prepared in S.O.C. medium and 100 μL were spread on each of three LB agar plates (100 mg/L ampicillin, 100 mg/L X-gal) before incubation overnight at 37°C.

Plasmid mini-preparations of overnight LB cultures (100 mg/L ampicillin) of randomly selected single white colonies were analyzed with REs *NotI* for pAUrumII-*gfp*, *AvrII* for pAUrumIII-*gfp* and *AseI* + *NcoI* for both constructs (all New England Biolabs).

A plasmid midi-preparation of a selected positive clone of each of pAUrumII-*gfp* and pAUrumIII-*gfp* was made and correct gene insertion was verified by sequencing at Eurofins MWG Operon, Germany.

Preparation of plant material

Immature seeds of barley, *Hordeum vulgare* cv. 'Gunhild' (NordGen) and wheat, *Triticum aestivum* cv. 'AvocetYr10' from green house were collected and sterilized 10 min in a sodium hypochlorite solution of about 1.5% active chlorine (VWR) and 0.05% (v/v) Tween 20 (Sigma) before three times wash in sterile Milli-Q water. Sterilely handled, the seeds were cut open with scissors at the embryo end and endosperms were squeezed out and placed 3 x 5 in the center of Petri dishes containing 1X Murashige & Skoog medium including modified vitamins (Duchefa Biochemie, pH = 5.8) supplied with 36.5 g/L mannitol, 20 g/L maltose, 3.5 g/L phytagel, 1.5 g/L ammonium nitrate, 1 g/L casein hydrolysate, 690 mg/L L-proline and 150 mg/L myo-inositol. One plate each of barley and wheat endosperms was prepared for each of the two *gfp*-constructs. The endosperms were stored for about an hour in darkness until bombardment.

Gold particle bombardment for transient *gfp* expression

The gold preparation and coating procedure below was based on available procedures [22–24]. Twenty milligrams 0.6 μm gold particles (Bio-Rad) were heated overnight at 180°C. After cooling, the particles were suspended in 1 mL 2-propanol and transferred to a microcentrifuge tube, ultra-sonicated for 2 min and incubated at room temperature for 15 min before another minute of ultra-sonication followed by 1 min of centrifugation at 13,000 g. The supernatant was discarded and the particles were washed three times in 500 μL sterile Milli-Q water with ultra-sonication for 2 min and centrifugation at 13,000 g for 1 min each time. Gold particles were resuspended in 670 μL sterile Milli-Q water, vortexed for 2 min and stored at −20°C in 50 μL aliquots in microcentrifuge tubes.

For each of the two plasmids, a 50 μL gold particle aliquot was thawed and ultra-sonicated for 2 min. With the tube lid open, the particles were vortexed gently, while 5 μg plasmid in aqueous solution and subsequently 50 μL 2.5 M calcium chloride premixed with 20 μL 0.1 M spermidine were added. The vortexing was continued 1 min followed by centrifugation at 13,000 g for 5 sec. The DNA coated particles were washed in 140 μL 2-propanol and spun again. Finally, they were suspended in 50 μL 2-propanol and kept few minutes on ice until bombardment.

Bombardment was performed at 25 inHg vacuum and 1,100 psi helium pressure using a Bio-Rad PDS-1000/He Biolistic gene gun in a sterile bench. The DNA-gold mixture was ultra-sonicated for 3 x 1 sec and 10 μL droplets were left on macrocarriers, which were then kept in closed Petri dishes on a non-vibrating surface, *i.e.* outside the sterile bench, for slow 2-propanol evaporation to occur, ensuring an equal, smooth distribution of gold particles. The bombarded endosperms were kept in darkness in a 23°C incubator for two days before *gfp* expression was evaluated under UV light in a stereo microscope (Leica WILD MZ8).

Competing interests

The authors declare that they have no competing interests.

Authors' contributions

RK developed SRL, conducted the experimental part and wrote the manuscript. CRI contributed to the development of the gold coating procedure, for which she also found the relevant literature. CKM suggested the use of the *LacZα* coding sequence in the *LguI*-spacer. All authors agreed on the final appearance of the manuscript after careful review.

Acknowledgements

We would like to thank Toni Wendt and Inger B Holme for the plasmid pVec8-GFP, from which the CaMV 35S promoter and the *gfp* gene originated, and Kim H Hebelstrup for the plasmid $\text{pUCE}_{\text{Ubi:USER:NOS}}$, from where the *Agrobacterium* NOS terminator was derived.

References

1. Wang T, Ma X, Zhu H, Li A, Du G, Chen J: **Available methods for assembling expression cassettes for synthetic biology.** *Appl Microbiol Biotechnol* 2012, **93:**1853–1863.
2. Aslanidis C, de Jong PJ: **Ligation-independent cloning of PCR products (LIC-PCR).** *Nucleic Acids Res* 1990, **18:**6069–6074.
3. Li MZ, Elledge SJ: **Harnessing homologous recombination in vitro to generate recombinant DNA via SLIC.** *Nat Methods* 2007, **4:**251–256.
4. Hamilton MD, Nuara AA, Gammon DB, Buller RM, Evans DH: **Duplex strand joining reactions catalyzed by vaccinia virus DNA polymerase.** *Nucleic Acids Res* 2007, **35:**143–151.
5. Zhu B, Cai G, Hall EO, Freeman GJ: **In-Fusion™ assembly: seamless engineering of multidomain fusion proteins, modular vectors, and mutations.** *Biotechniques* 2007, **43:**354–359.
6. Nour-Eldin HH, Hansen BG, Nørholm MHH, Jensen JK, Halkier BA: **Advancing uracil-excision based cloning towards an ideal technique for cloning PCR fragments.** *Nucleic Acids Res* 2006, **34:**e122.
7. Bitinaite J, Rubino M, Varma KH, Schildkraut I, Vaisvila R, Vaiskunaite R: **USER™ friendly DNA engineering and cloning method by uracil excision.** *Nucleic Acids Res* 2007, **35:**1992–2002.
8. Quan J, Tian J: **Circular polymerase extension cloning of complex gene libraries and pathways.** *PLoS One* 2009, **4:**e6441.
9. Gibson DG, Young L, Chuang R-Y, Venter JC, Hutchison CA, Smith HO: **Enzymatic assembly of DNA molecules up to several hundred kilobases.** *Nat Methods* 2009, **6:**343–345.
10. Holme IB, Wendt T, Holm PB: **Intragenesis and cisgenesis as alternatives to transgenic crop development.** *Plant Biotechnol J* 2013, **11:**395–407.
11. Jacobsen E, Schouten H: **Cisgenesis: an important sub-invention for traditional plant breeding companies.** *Euphytica* 2009, **170:**235–247.
12. Lu Q: **Seamless cloning and gene fusion.** *Trends Biotechnol* 2005, **23:**199–207.
13. Pingoud A, Jeltsch A: **Structure and function of type II restriction endonucleases.** *Nucleic Acids Res* 2001, **29:**3705–3727.
14. Wyborski DL, Bauer JC, Zheng C-F, Felts K, Vaillancourt P: **An *Escherichia coli* expression vector that allows recovery of proteins with native N-termini from purified calmodulin-binding peptide fusions.** *Protein Expr Purif* 1999, **16:**1–10.
15. Lu Q, Burns MC, McDevitt PJ, Graham TL, Sukman AJ, Fornwald JA, Tang X, Gallagher KT, Hunsberger GE, Foley JJ, Schmidt DB, Kerrigan JJ, Lewis TS, Ames RS, Johanson KO: **Optimized procedures for producing biologically active chemokines.** *Protein Expr Purif* 2009, **65:**251–260.
16. Chiu W-l, Niwa Y, Zeng W, Hirano T, Kobayashi H, Sheen J: **Engineered GFP as a vital reporter in plants.** *Curr Biol* 1996, **6:**325–330.
17. Lukaszewicz M, Feuermann M, Jérouville B, Stas A, Boutry M: **In vivo evaluation of the context sequence of the translation initiation codon in plants.** *Plant Sci* 2000, **154:**89–98.
18. Engler C, Kandzia R, Marillonnet S: **A one pot, one step, precision cloning method with high throughput capability.** *PLoS One* 2008, **3:**e3647.
19. Engler C, Gruetzner R, Kandzia R, Marillonnet S: **Golden gate shuffling: a one-pot DNA shuffling method based on type IIs restriction enzymes.** *PLoS One* 2009, **4:**e5553.
20. Hiraga K, Arnold FH: **General method for sequence-independent site-directed chimeragenesis.** *J Mol Biol* 2003, **330:**287–296.
21. Benoit RM, Wilhelm RN, Scherer-Becker D, Ostermeier C: **An improved method for fast, robust, and seamless integration of DNA fragments into multiple plasmids.** *Protein Expr Purif* 2006, **45:**66–71.
22. Sawant S, Singh PK, Tuli R: **Pretreatment of microprojectiles to improve the delivery of DNA in plant transformation.** *BioTechniques* 2000, **29:**246–248.
23. Kikkert JR, Vidal JR, Reisch BI: **Stable transformation of plant cells by particle bombardment/biolistics.** In *Methods in Molecular Biology, Volume 286: Transgenic Plants: Methods and Protocols.* Edited by Peña L. Totowa: Humana Press Inc; 2004:61–78.
24. Sparks CA, Jones HD: **Biolistics transformation of wheat.** In *Methods in Molecular Biology, Transgenic Wheat, Barley and Oats: Production and Characterisation., Volume 478.* Edited by Jones HD, Shewry PR. New York: Humana Press; 2009:71–92.

12

High-throughput phenotyping of plant resistance to aphids by automated video tracking

Karen J Kloth[1,2,3*], Cindy JM ten Broeke[1], Manus PM Thoen[1,2,3], Marianne Hanhart-van den Brink[1], Gerrie L Wiegers[1,3], Olga E Krips[4], Lucas PJJ Noldus[4], Marcel Dicke[1] and Maarten A Jongsma[3]

Abstract

Background: Piercing-sucking insects are major vectors of plant viruses causing significant yield losses in crops. Functional genomics of plant resistance to these insects would greatly benefit from the availability of high-throughput, quantitative phenotyping methods.

Results: We have developed an automated video tracking platform that quantifies aphid feeding behaviour on leaf discs to assess the level of plant resistance. Through the analysis of aphid movement, the start and duration of plant penetrations by aphids were estimated. As a case study, video tracking confirmed the near-complete resistance of lettuce cultivar 'Corbana' against *Nasonovia ribisnigri* (Mosely), biotype Nr:0, and revealed quantitative resistance in Arabidopsis accession Co-2 against *Myzus persicae* (Sulzer). The video tracking platform was benchmarked against Electrical Penetration Graph (EPG) recordings and aphid population development assays. The use of leaf discs instead of intact plants reduced the intensity of the resistance effect in video tracking, but sufficiently replicated experiments resulted in similar conclusions as EPG recordings and aphid population assays. One video tracking platform could screen 100 samples in parallel.

Conclusions: Automated video tracking can be used to screen large plant populations for resistance to aphids and other piercing-sucking insects.

Keywords: Aphids, Arabidopsis, Automated video tracking, Host plant resistance, *Lactuca sativa*, Phenotyping, Piercing-sucking insects, *Arabidopsis thaliana*

Background

More than 100 aphid species (Aphididae) are economically significant pest insects and most crops are host to at least one species [1]. Aphids feed on phloem sap, and to reach the phloem they move their stylets between plant cells towards a sieve element, making short punctures in cells along the way. Most probes are prematurely interrupted in the epidermis and mesophyll. When, however, a phloem vessel is reached, aphids can ingest phloem sap continuously for many hours or even days [2]. Although aphids inflict little tissue damage, they transmit plant viruses and deplete host plants of photoassimilates and free amino acids [3,4]. In wild plant populations aphids rarely constitute pests due to effective natural defence strategies, such as epicuticular waxes, protease inhibitors, and induced production of secondary metabolites [4-12]. After generations of domestication many of these defence traits have been diminished or lost in cultivated plants, making them vulnerable targets of herbivorous insects [13,14]. The genetic backgrounds of resistance mechanisms still remain largely elusive and genomics studies strongly depend on the capacity for phenotyping large panels of plants. Few high-throughput methods have been established for assessing plant resistance to insect herbivores, such as aphids or other piercing-sucking insects [15-20]. Generally, two approaches are used to quantify the level of plant defence against aphids; either assessment of aphid population development or investigation of aphid feeding behaviour. Aphid population assays are generally the most demanding in terms of time and space, since they require the availability of a climate-controlled compartment for 1 or 2 weeks and extensive

* Correspondence: karen.kloth@wur.nl
[1]Laboratory of Entomology, Wageningen University, P.O. Box 16, 6700, AA Wageningen, The Netherlands
[2]Laboratory of Plant Physiology, Wageningen University, P.O. Box 16, 6700, AA Wageningen, The Netherlands
Full list of author information is available at the end of the article

manual work [21-23]. On the contrary, aphid feeding behaviour can be measured within a couple of hours via the Electrical Penetration Graph (EPG) technique. EPG recording delivers electrical waveforms comprising information on the plant tissue that is penetrated (phloem vessel, xylem vessel or other cells) and the stylet penetration activity (cell puncture, salivation, ingestion, penetration difficulties) [24,25]. EPG studies have shown that aphids prolong phloem ingestion on suitable host plants and delay and reduce feeding on resistant or non-host plants [26-33]. The high specificity of the information about plant tissue and key components of aphid behaviour, makes this methodology appealing for exploring defence mechanisms. A drawback of EPG is, however, the restricted capacity, generally 8 plants per setup [34], and the labour-intensive nature of wiring aphids and annotating electrical signals.

Here, we present the methodology and validation of image-based tracking of aphid feeding behaviour. Automated video tracking was introduced in the early 1990s and has since been used in many animal behaviour studies [35-41]. Video tracking involves software-engineered pattern analysis of grids of pixels in order to quantify the location and movement of subjects over time. In this study, we used movement patterns of the central body point of aphids to estimate the duration of plant penetrations made by the aphid's mouth parts. Previous EPG studies showed that probes shorter than approximately 3 minutes represent penetrations in the epidermis and/or mesophyll [26], and that probes involving phloem uptake last on average at least 25 min [27,42,43]. This allowed us to discriminate test probes from putative phloem uptake events in video observations in order to identify plants that are resistant to aphids. We benchmarked the performance of the high-throughput video tracking platform against EPG recordings and aphid population development assays, using natural accessions of *Arabidopsis thaliana*, and lettuce cultivars, *Lactuca sativa*, in combination with the green peach aphid, *Myzus persicae* (Sulzer), and the black-currant lettuce aphid, *Nasonovia ribisnigri* (Mosely) (Hemiptera: Aphididae), respectively.

Results

Tracking aphid feeding behaviour

Automated video tracking of aphid feeding behaviour was performed using video tracking software and a stationary camera mounted above 20 no-choice arenas. We introduced one aphid onto each arena, consisting of an agar substrate almost completely covered by a leaf disc, and recorded 20 arenas simultaneously with a frame rate of 25 frames s^{-1} (Figure 1, Additional file 1: Figure S1). Because the aphid's mouthparts were not visible in the multi-arena setup, we made the assumption that when the aphid's centre point was located on the leaf disc and did not move, the aphid was penetrating the leaf tissue with its stylets. By assessing video images by eye, we defined velocity thresholds for the start and end of probing events of two aphid species, *M. persicae* and *N. ribisnigri* (Figure 2, Additional file 1: Figure S2). According to our observations, the software was more vulnerable to premature probe endings of *N. ribisnigri* due to body movements during probing (such as event γ in Figure 2). As this aphid species is somewhat larger (±1.9 mm body length, versus ± 1.7 mm for *M. persicae*), movements around the fixated mouth resulted in a higher tangential velocity, and therefore required a higher velocity threshold.

Accuracy

To test the accuracy of the platform, we performed automated video tracking and human observations simultaneously. A camera was attached to a stereo microscope to

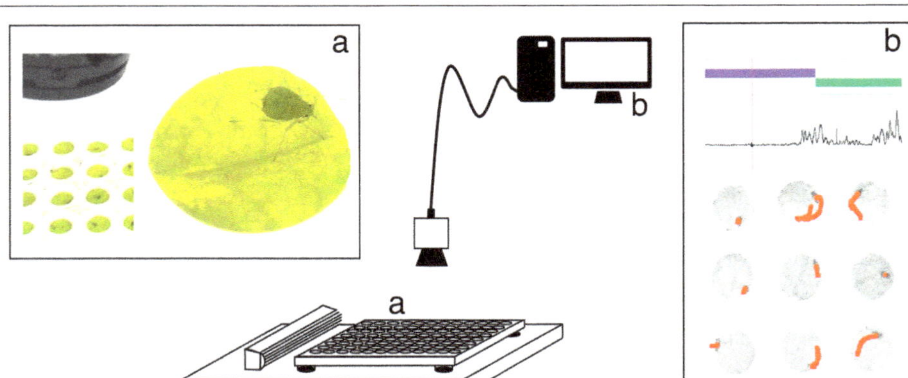

Figure 1 Video tracking platform. A stationary camera is mounted above a microtitre plate which is placed on top of a backlight unit with ventilation at the left. Wells in the microtitre plate contained a leaf disc and an aphid **(a)**. Cling film was wrapped around the plate to prevent aphids from escaping. The camera was connected to a computer with EthoVision® XT video tracking software **(b)**. Movements of the aphid's centre point were automatically tracked (red track shows movements across 30 seconds). With this information the software calculated aphid velocity (line graph) and estimated probing (purple bar) and non-probing events (green bar).

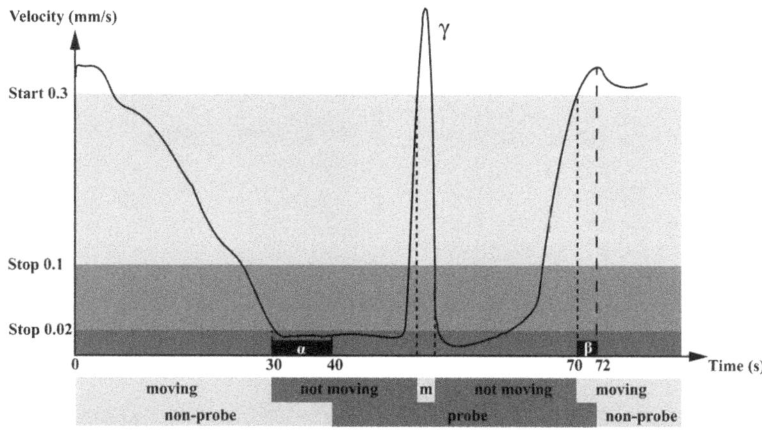

Figure 2 Velocity thresholds for registration of probes. An example of how aphid feeding behaviour was measured using a resolution of 275 pixels per mm^2. Subject states can be defined as 'moving' or 'not moving' by means of two thresholds: the start velocity at which the subject begins to move, and the stop velocity at which the state changes from moving to not moving. Probe starts were recorded if the velocity of the aphid's centre point dropped below 0.02 mm/s for at least 10 seconds (α). Probe stops were recorded if the velocity of *M. persicae* aphids exceeded 0. 3 mm/s for at least 2 seconds (β), or 0.35 mm/s for at least 2 seconds in the case of winged *N. ribisnigri* aphids. To avoid premature probe endings due to short movements during probing (event γ), probe stops were only recorded when velocity increased above 0. 1 mm/s for more than 2 seconds. Figure adjusted from the EthoVision XT Reference Manual (version 8) [44].

deliver a side-view on the arena for manual scoring of probes (Additional file 2). Among a total of 139 probes of 16 different *M. persicae* aphids scored by hand, 88% was detected with video tracking (Figure 3a). Undetected and false positive probes involved only short events (<3 min). Of the detected probes, 19% was either underrated (multiple 'true' probes were considered as one probe), or overrated (one 'true' probe was translated into multiple probes by the software). Underrated samples were caused by undetected probe stops due to slow movements below the velocity threshold. Overrated samples were caused by false probe stops when, for example, the aphid was immobile on the edge of the leaf disc and the assigned position continuously switched between an "on the leaf disc" and "off the leaf disc" status (Figure 3b). Three times this incident occurred, leading to 17 redundant probes of which 10 were filtered out automatically (see Methods, section Software settings). Other reasons for premature probe stops were abdominal movements during probing related to e.g. reproduction or honeydew excretion. The longer probes lasted, the higher the risk was of encountering such incidents. Indeed automatically tracked probes were in general biased to end 73 to 12 seconds too early (Figure 3c), and the total duration of probing was underestimated (on average 46 min ± 2.5 min standard error, versus 50 min ± 1.9, $P = 0.01$, Mann–Whitney U test, total observation duration: 55 min). Nevertheless, the recorded number and duration of probes were highly correlated to human observations (Figure 4, average $r^2 = 0.7$ with 275 pixels per mm^2). Other parameters, such as distance moved, were also highly correlated with feeding behaviour in general, but were less informative with regard to long probes (Figure 4l). Although automated video tracking did not achieve a precision as high as manual scoring, it enabled observing multiple arenas simultaneously. In the above described tests, we used 275 pixels per mm^2, equal to a coverage of 20 arenas with our 768 × 576 pixels camera. To determine whether the capacity could be increased, we repeated the experiment with only 155 pixels per mm^2, equal to a coverage of 35 arenas, but found that reduced resolution resulted in decreased correlations with human observations (average $r^2 < 0.5$).

Benchmarking against EPG recording with Arabidopsis

To validate whether automated video tracking delivered a reliable proxy for plant resistance, feeding behaviour of *M. persicae* was measured during 8 hours continuous recording on two natural accessions of Arabidopsis, Co-2 and Sanna-2 (Additional file 3). These accessions were selected from a population of hundreds of accessions based on preliminary video tracking data. Automated video tracking showed that *M. persicae* aphids walked larger distances on Co-2 and reduced the mean duration of long probes (Table 1). EPG recordings on intact plants confirmed shorter durations of (sustained) phloem ingestion, and additionally revealed more short probes, non-probing behaviour and a delayed phloem uptake on Co-2 (Table 1). This behaviour is an indication of both epidermis/mesophyll-located and phloem-located resistance in Co-2 against *M. persicae*. All aphids ingested phloem, but quantitative differences in feeding

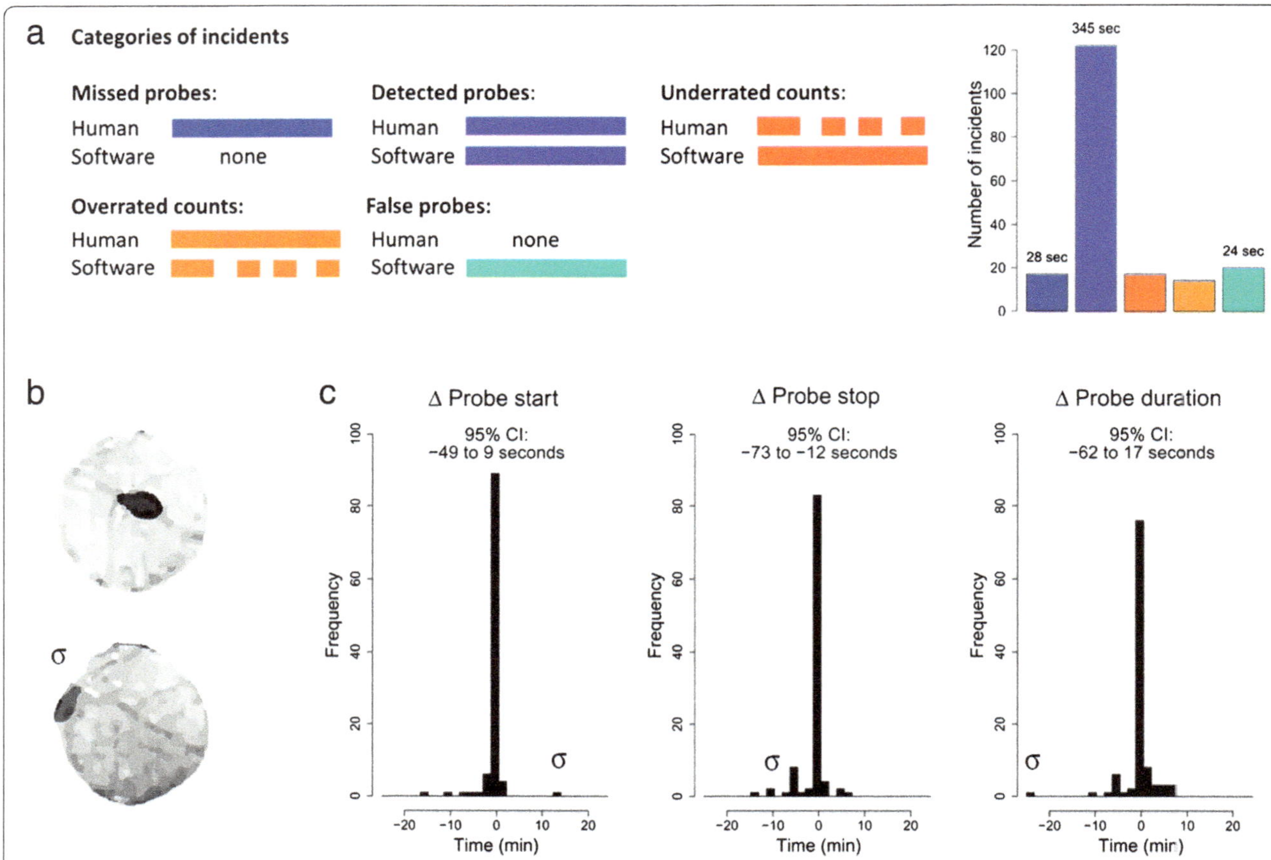

Figure 3 Accuracy of automated tracking in comparison to human observations. *M. persicae* feeding behaviour was measured on Arabidopsis leaf discs by automated video tracking and human observations simultaneously. **(a)** Out of 139 probes of 16 aphids scored by hand, 88% was detected by automated video tracking. Probes were considered a match when their duration overlapped at least partially. Some of the detected probes were matched by too few (underrated) or too many (overrated) probes. For these situations, the amount of missed or redundant probes is shown. 17 Probes went undetected and 20 false probes were recorded. Mean duration per probe is shown above the bars. **(b)** Screenshots of the top-view video used for automated tracking. The lower image (σ) shows an aphid positioned on the edge of the leaf disc for more than 20 min, causing overrated probe counts by the software due to continuous switching between an "on the leaf disc" and "off the leaf disc" status. **(c)** Differences between software and human observations per matched probe. 95% Confidence Intervals are shown above the histograms. Negative values correspond to too early probe starts, too early probe endings, resp. too short duration of probes compared to the human observations. In case of overrated probe counts, the probe with the most similar duration as the manually scored probe was included. The outlier caused by the example in **(b)** is annotated with σ.

behaviour between aphids on Co-2 and on Sanna-2 were already apparent in the first hour (Figure 5). A reproduction assay on intact plants confirmed that Co-2 was indeed more resistant than Sanna-2, although the resistance was not absolute. Depending on plant age, aphids either started reproduction later or produced fewer offspring (Figure 6). Although we had been able to correctly identify a quantitative difference in resistance with automated video tracking, the effects were smaller than in EPG recordings on intact plants. To verify whether the plant line effects in the video tracking assay were attenuated due to the use of excised plant tissue, the EPG experiment was repeated with leaf discs. Particularly for the resistant accession, aphid feeding behaviour was different and involved more phloem uptake and fewer short probes on leaf discs compared to intact plants (Additional file 1: Table S2). The only significant difference between the accessions that remained was a reduced duration of phloem uptake events on Co-2 (Table 1). In addition, contribution of salivation to the phloem phase, required to suppress (callose-mediated) sieve-plate occlusion [45], was equal on leaf discs but higher on intact plants of Co-2 (Figure 7). This indicates that the resistance mechanisms in intact plants were partially lost in leaf discs.

Benchmarking against EPG recording with lettuce

Apart from a study system with partial resistance, an example of near-complete resistance was tested with the video tracking platform. The behaviour of black-currant lettuce aphids, *N. ribisnigri*, biotype Nr:0 was recorded on two near-isogenic lettuce cultivars, the resistant 'Corbana' and susceptible 'Terlana'. Previous studies showed

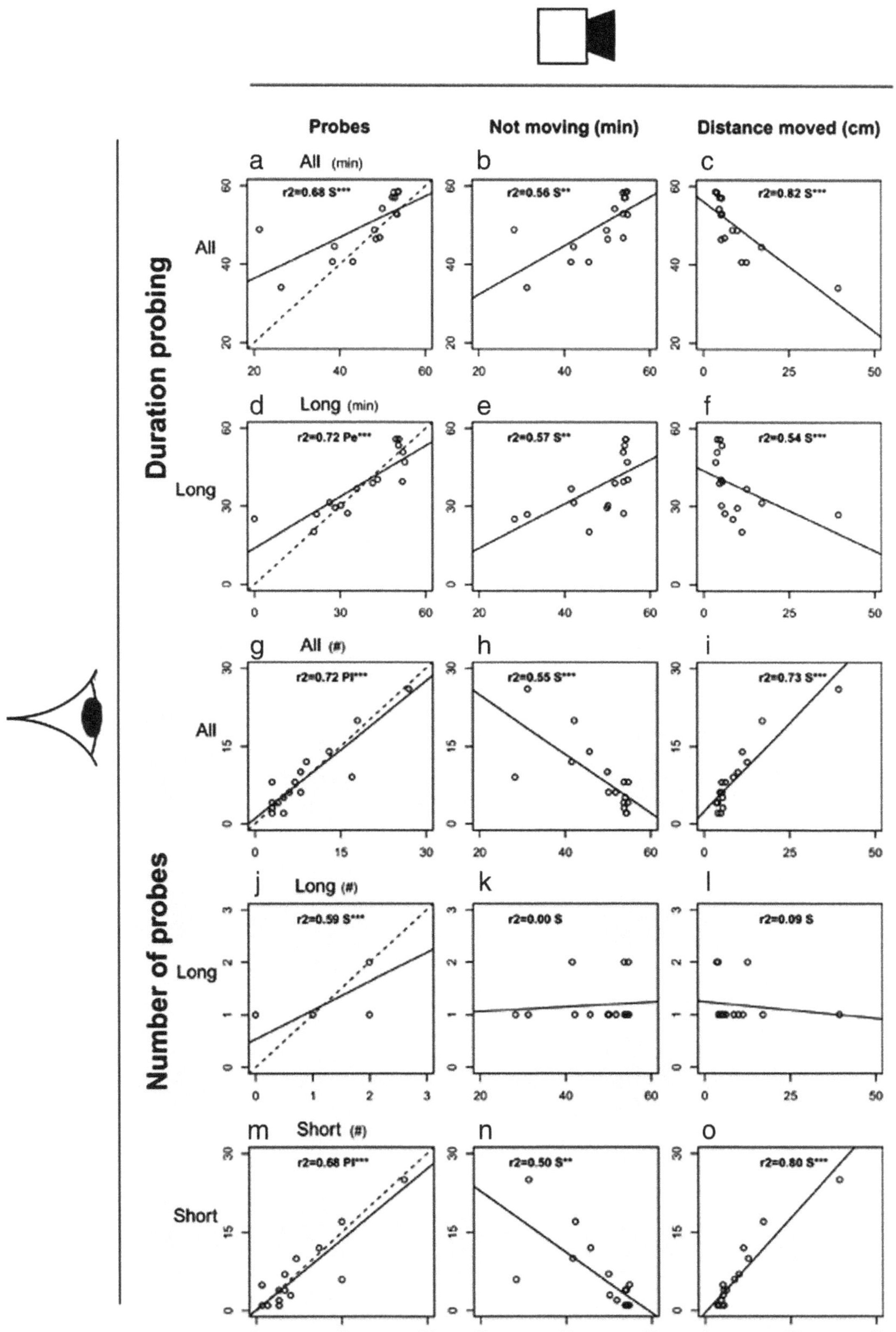

Figure 4 (See legend on next page.)

(See figure on previous page.)

Figure 4 Correlation between automated video tracking and human observations. *M. persicae* behaviour was measured by automated video tracking (x-axes) and human observations simultaneously (y-axes). Three categories of probes were distinguished: All probes, Long probes (>15 min), and Short probes (<3 min). The duration (min) and number of probes measured by human observations were compared to: **(a,d,g,j,m)** the duration (min) and number of probes (all, long, and short probes) measured by video tracking, **(b,e,h,k,n)** the total time not moving (min), and **(c,f,i,l,o)** the distance moved by the aphids (cm) (*P < 0.05; **P < 0.01; ***P < 0.001, Pe = Pearson correlation test, Pl = Pearson correlation test on log transformed data, S = Spearman correlation test, dashed lines represent a hypothetical $r^2 = 1$, n = 16 recordings of 1 aphid for 55 min, 275 pixels per mm^2).

Table 1 Feeding behaviour of *M. persicae* on two Arabidopsis accessions and *N. ribisnigri* on two lettuce cultivars

		Arabidopsis – *M. persicae*		Lettuce – *N. ribisnigri*	
		Co-2 R	**Sanna-2 S**	**Corbana R**	**Terlana S**
EPG intact plants	Total duration non-penetration	139 ± 15	60.2 ± 8.1 Ts***	121 ± 16	29.2 ± 5.8 Tl***
	Total duration phloem feeding	110 ± 13	296 ± 31 M***	2.34 ± 1.68	399 ± 17 M***
	Total duration phloem feeding (>10 min)	27.7 ± 7.9	259 ± 32 M***	0.00 ± 0.00	399 ± 17 M***
	Total duration other penetration activities	231 ± 15.8	124 ± 28 M**	357 ± 16	51.5 ± 14 T***
	Number of non-penetrations	57.7 ± 6.2	22.4 ± 3.1 M***	27.8 ± 7.58	7.58 ± 1.33 T***
	Number of short probes (<3 min)	49.8 ± 5.8	18.4 ± 2.4 M***	13.7 ± 2.0	4.53 ± 1.17 M***
	Number of phloem feeding events	6.37 ± 0.55	4.28 ± 0.70 T*	0.11 ± 0.07	1.16 ± 0.12 M***
	Mean duration of phloem feeding events	17.6 ± 1.4	117 ± 26 M***	22.2 ± 6.16	379 ± 26 M*
	Latency to first phloem feeding event	188 ± 25	111 ± 26 M*	228 ± 94	106 ± 17 M
	Contribution salivation to phloem phase (%)	13.3 ± 2.2	2.45 ± 1.3 M***	98.7 ± 1.0	1.15 ± 0.3 M***
EPG leaf discs	Total duration non-penetration	164 ± 48	173 ± 59 M	-	-
	Total duration phloem feeding	220 ± 42	201 ± 48 T	-	-
	Total duration phloem feeding (>10 min)	158 ± 46	177 ± 47 T	-	-
	Total duration other penetration activities	95.3 ± 25.7	106 ± 17 T	-	-
	Number of non-penetrations	19.1 ± 5.7	14.4 ± 3.0 Ts	-	-
	Number of short probes (<3 min)	16.6 ± 4.8	8.63 ± 2.0 Ts	-	-
	Number of phloem feeding events	8.11 ± 1.53	2.63 ± 0.84 M*	-	-
	Mean duration of phloem feeding events	34.2 ± 10.8	99.2 ± 31.7 Ts*	-	-
	Latency to first phloem feeding event	144 ± 51	250 ± 53 T	-	-
	Contribution salivation to phloem phase (%)	3.03 ± 0.6	3.63 ± 2.2 M	-	-
Video leaf discs	Total duration non-probing	61.5 ± 9.7	42.6 ± 9.4 T	243 ± 27	148 ± 25 T*
	Total duration long probes (>25 min)	338 ± 20	377 ± 17 T	153 ± 26	283 ± 25 T***
	Total duration sust. probes (>35 min)	276 ± 27	353 ± 19 M*	132 ± 24	260 ± 25 T***
	Total duration other probes (<= 25 min)	80.6 ± 12.4	60.5 ± 9.9 T	84.1 ± 7.0	48.4 ± 9.3 M***
	Number of non-penetrations	33.2 ± 3.9	22.1 ± 3.8 T	31.1 ± 3.1	23.0 ± 2.6 T
	Number of short probes (<3 min)	20.3 ± 3.1	12.4 ± 3.4 T	22.4 ± 3.1	15.6 ± 2.3 M
	Number of long probes (> = 25 min)	5.95 ± 0.36	4.71 ± 0.36 M*	2.19 ± 0.33	3.25 ± 0.30 T*
	Mean duration of long probes (> = 25 min)	62.8 ± 7.0	90.0 ± 9.8 M*	72.4 ± 9.5	99.8 ± 11.2 M*
	Latency to first long probe (> = 25 min)	18.5 ± 5.8	27.1 ± 7.1 M	142 ± 27	65.1 ± 14.2 M*
	Total duration not moving (min)	445 ± 7.3	465 ± 3.5 M	257 ± 27	350 = 25 T*
	Total distance moved (cm)	46.9 ± 4.6	34.0 ± 4.6 T*	204 ± 47	103 = 23 T
	Max velocity (mm/s)	0.72 ± 0.12	0.53 ± 0.05 T	2.78 ± 0.54	1.88 ± 0.20 M

Within each EPG and video tracking experiment it was tested whether aphid behaviour differed between the two plant lines. The mean duration and latency were only calculated if the corresponding events did occur (all samples of *M. persicae*; *N. ribsinigri*: EPG n = 2 Corbana and n = 20 Terlana, video tracking n = 21 Corbana and n = 26 Terlana).

Means ± standard error, *P < 0.05; **P < 0.01; ***P < 0.001, T = Student's *t*-test, Ts = Student's t-test on square root transformed data, Tl = Student's t-test on log transformed data, M = Mann–Whitney *U* test, R = resistant plant line, S = susceptible plant line, time is represented in min.

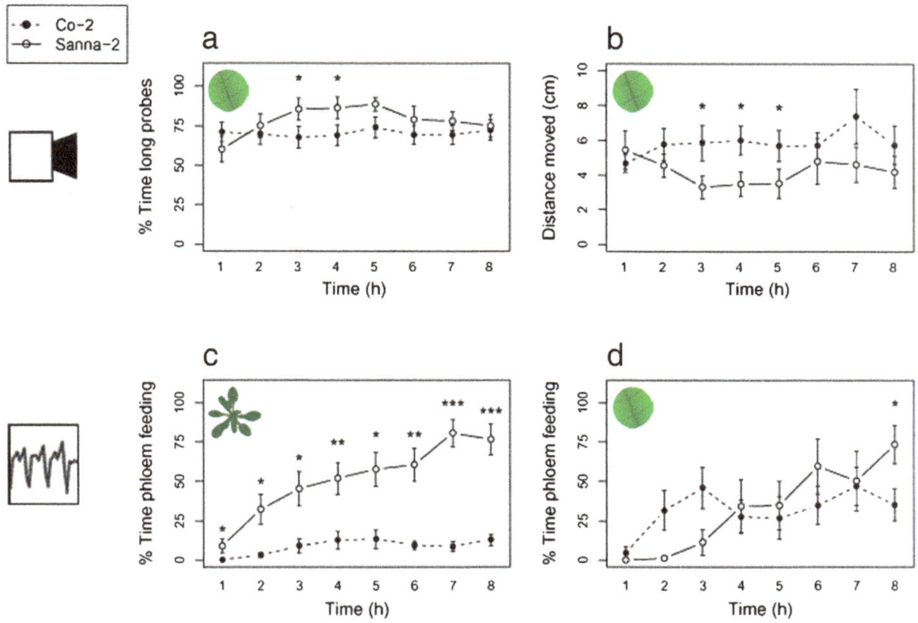

Figure 5 Behavioural parameters of *M. persicae* on two natural Arabidopsis accessions, Co-2 (resistant) and Sanna-2 (susceptible). **(a)** Percentage of the time spent on long probes (>25 min), and **(b)** distance moved (cm) were measured by automated video tracking. Percentage of the time spent on phloem feeding (waveform 5) were measured by **(c)** EPGs on intact plants, and **(d)** EPGs on leaf discs (Mann–Whitney U test, *$P < 0.05$; **$P < 0.01$; ***$P < 0.001$, video tracking: Co-2 n = 20, Sanna-2 n = 17, EPG recording intact plants: n = 19, EPG recording leaf discs: Co-2 n = 9, Sanna-2 n = 8, error bars represent standard error).

that the *Nr* gene is responsible for near-complete resistance in Corbana against this biotype of aphids, mainly due to a phloem-located mechanism [34,46]. Our video tracking observations on leaf discs were compared to EPG recording data by ten Broeke et al. [47]. Seven out of nine video tracking variables confirmed that cultivar Corbana was more resistant than cultivar Terlana (Table 1). Aphids on Corbana spent less time on long probes and more time on shorter probes and other activities. In addition, aphids increased their walking activity over

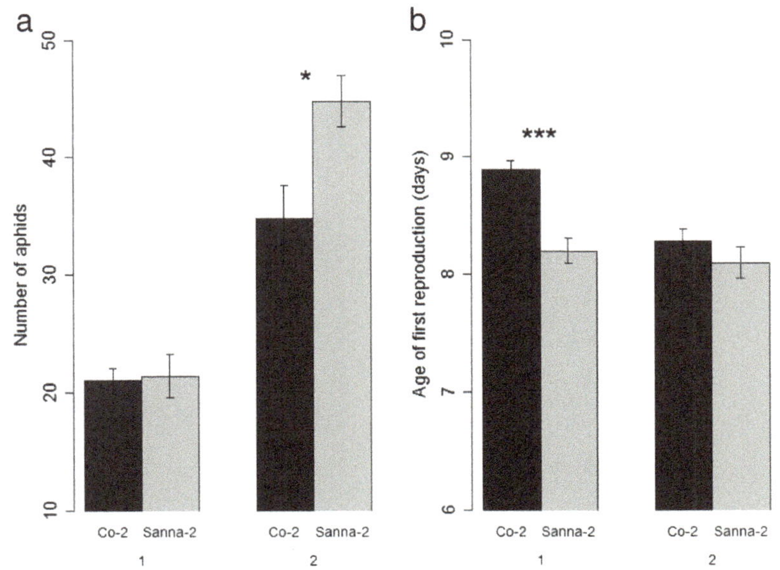

Figure 6 Reproduction of *M. persicae* on two Arabidopsis accessions. One neonate aphid was introduced to a 2.5-week-old plant (assay 1) or a 3.5-week-old plant (assay 2). **(a)** Total number of aphids per plant 2 weeks after infestation. **(b)** Days until the first nymph was produced by the aphid (Mann–Whitney U test, *$P < 0.05$, ***$P < 0.001$, assay 1: Co-2 n = 19, Sanna-2 n = 15, assay 2: Co-2 n = 14, Sanna-2 n = 13, error bars represent standard error).

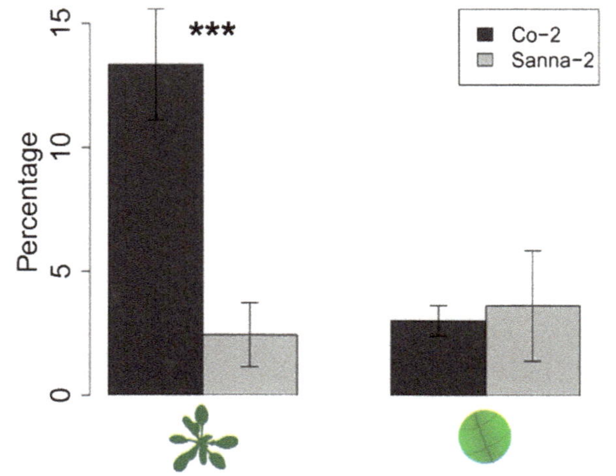

Figure 7 Contribution of salivation to phloem ingestion. Percentage of time spent salivating in the phloem compared to the total phloem phase (salivation + ingestion) of *M. persicae* aphids on Arabidopsis accessions Co-2 (resistant) and Sanna-2 (susceptible) (Mann–Whitney U test, *P < 0.05; **P < 0.01; ***P < 0.001, left bars: EPG recording intact plants: n = 19, right bars: EPG recording leaf discs: Co-2 n = 9, Sanna-2 n = 8, error bars represent standard error).

time on both cultivars, but generally covered larger distances on Corbana leaf discs (mixed linear model: time effect: P = 0.00, cultivar effect: P = 0.03, time × cultivar interaction: P = 0.77, Figure 8). Yet, the resistance effect was less pronounced in video tracking compared to EPG recording on intact plants: only 11% of the aphids in EPG recordings showed phloem ingestion on Corbana plants, while 78% of the aphids in the video assay performed long probes on Corbana. These long probing events could include other activities, such as water ingestion from xylem vessels, since EPGs showed that on Corbana plants more aphids penetrated xylem sieve elements (12 aphids on Corbana versus 2 aphids on Terlana).

Throughput

Using simulated data with a similar plant line effect as the data sets from the plant-aphid assays described here, we assessed the required sample size and recording duration for automated video tracking (Table 2). With 20 replicates of 8-hour observations, significant resistance was detected in more than 80% of the cases for the Arabidopsis plant line effect on *M. persicae* (2 response variables tested per simulated data set, Bonferroni correction: P < 0.025). The near-complete resistance of Corbana lettuce against *N. ribisnigri* biotype Nr:0 was detected in more than 80% of the cases with 15 replicates of 4 hours of video observations. Subtle differences in resistance in Arabidopsis were more difficult to detect when video observations were shorter than 8 hours (Table 2). On the other hand, reducing the video duration to the first 4 hours improved the detection of near-complete resistance, as with *N. ribisnigri* biotype

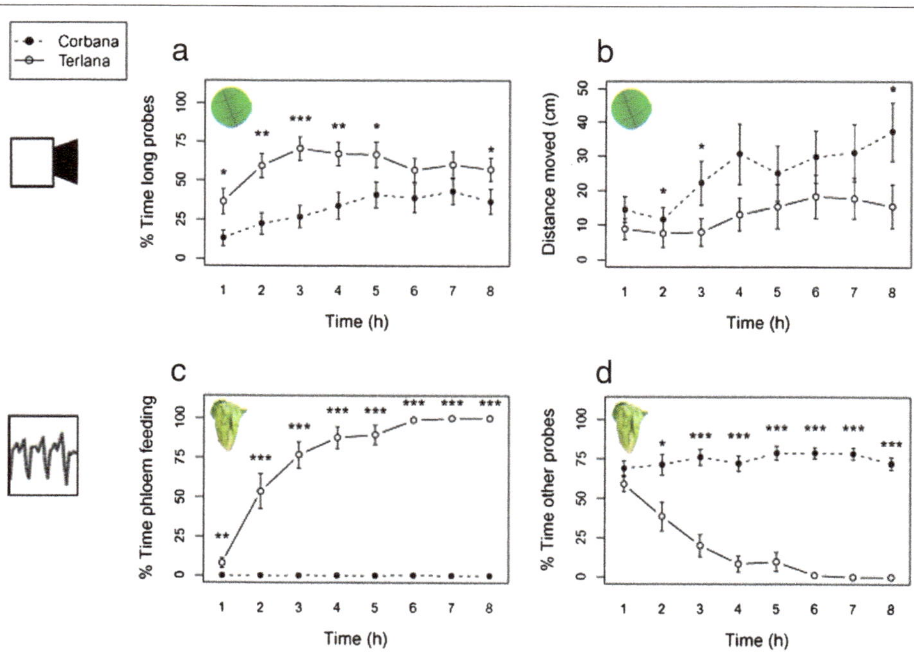

Figure 8 Behavioural parameters of *N. ribisnigri* on two lettuce cultivars, Corbana (resistant) and Terlana (susceptible). **(a)** Percentage of the time spent on long probes (>25 min), and **(b)** distance moved (cm) were measured by automated video tracking on leaf discs. **(c)** Percentage of the time spent on phloem feeding (waveform 5), and **(d)** percentage of time spent on other probes (pathway, phloem salivation and xylem feeding) were measured by EPGs on intact plants (Mann–Whitney U test per time bin, *P < 0.05; **P < 0.01; ***P < 0.001, video tracking: Corbana n = 27, Terlana n = 28, EPG recording: n = 19).

Table 2 Required video duration and number of replicates for identifying a significant effect

Plant-aphid system	Duration	Replicates	Detection rate
Arabidopsis - *M. persicae*	8 h	20	> 80%
	6 h	25	> 80%
Lettuce – *N. ribisnigri*	8 h	20	> 80%
	4 h	15	> 80%

Student's t-tests have been applied to subsamples of two simulated data sets with a similar mean and standard deviation as two response variables from the video tracking assays. The percentage of tests with a significant outcome in at least one of the two variables is represented as the detection rate (Bonferroni correction: $P < 0.025$).

Nr:0 on the Corbana lettuce cultivar. Apparently, in this case, the precision of video tracking decreased over time. While the EPG recording with lettuce did not reveal an increase in aphid activities in the xylem or mesophyll over time (Figure 8), the last stretch of the video observation was likely confounded by sessile behaviour other than probing. The risk of falsely rejecting the null hypothesis was limited to 1% (*M. persicae* on Arabidopsis accession Col-0). Overall, video tracking required similar observation durations as EPG recording, but a larger sample size to detect significant plant effects (Table 3). The required amount of replicates was, however, compensated by screening many samples simultaneously and automated data annotation.

Discussion

Leaf discs

The effect sizes measured in video tracking with leaf discs were substantially smaller compared to EPG recording on intact plants. EPG recording on leaf discs confirmed that the application of excised plant tissue partially impaired plant resistance [47,48], possibly due to the interrupted supply of ions and metabolites in the phloem, or due to the interference by jasmonic acid and ethylene mediated wound responses [49]. Furthermore, aphids can be disturbed by the decrease in pressure in the sieve elements of excised plant tissue, although they are well capable of active uptake of sap [50,51]. The increase of coagulating proteins and cellular debris in the phloem after plant wounding may plug sieve plates and the aphid's food canal in the stylets [28,52,53]. To prevent such potential clogging of sieve elements, aphids might increase the injection of watery saliva into the phloem or shorten their feeding events, but neither of these effects were observed consistently. To maintain turgor better the use of leaves still connected to intact plants would be favourable, but this is currently not feasible in view of poor detection of aphids in more complex environments. Arenas designed to hold entire detached leaves or seedlings on agar could, however, be a feasible alternative to leaf discs.

Table 3 Comparison of automated video tracking and EPG recording characteristics

	EPG	Video
Plant material	Intact plants	Leaf discs
Min. observation duration	4-8 h	4-8 h
Min. number of replicates (identification rate ±80%)	3-5	15-20
Maximum sample size per set up	± 8	± 100
Preparation time per sample	± 5 min	± 2 min
Annotation of electrical patterns/video images	± 15 min	Automated

Minimum duration of observations and minimum sample sizes were estimated with simulations. Preparation time per replicate reflects a rough estimation of time required for an experienced person to prepare one arena, resp. plant-aphid individual, for a recording.

Application

High-throughput phenotyping techniques of sucking insect species are urgently needed in view of functional genomics studies aiming to find subtle allelic differences in plant populations measuring many hundreds of plants. Conventional methods, like EPG and population studies, are less scalable for this purpose and carry much higher investments in terms of time (labour, duration) and costs (equipment, greenhouses). In this study, automated video tracking was used to study aphid feeding behaviour, but it could as well be applied to track the behaviour of other piercing-sucking insects. We recommend to validate the velocity thresholds for each species first, by checking several video files by hand. As shown here with two aphid species, size and velocity can differ and will affect the accuracy of probe estimations. When studying plants with thick or dark leaves, increased resolution, better (macro) lenses, and lateral light sources instead of backlight can help to improve the detection of insects. We worked with EthoVision XT video tracking and analysis software, but other programs or programming environments, such as MatLab and ImageJ, could as well serve as robust video tracking tools [37,40,54].

Conclusions

The aim of this study was to develop a high-throughput method to screen large plant populations for resistance to aphids and other piercing-sucking insects. For the first time it is shown that automated video tracking of aphid body movement can be used to estimate how often the insects are penetrating plant tissue and are reaching the vascular bundle. The use of leaf discs instead of intact plants enhanced the throughput of the video tracking platform, but EPG recording illustrated that resistance effects were partially lost in leaf discs. Nevertheless, we could identify both intermediate and extreme levels of resistance with video tracking. In Arabidopsis accession Co-2, we found a quantitative resistance level. This was confirmed in additional bioassays, suggesting the involvement of constitutive or rapidly activated resistance mechanisms in both epidermis/mesophyll and phloem, resulting in a small detrimental effect on the aphid population. The video tracking platform also confirmed the near-complete resistance of the lettuce cultivar Corbana to *N. ribisnigri* biotype Nr:0. Although video tracking requires more replicates to identify resistant plants than the conventional EPG technique, it can screen many samples simultaneously in a confined space. In addition, computerized data acquisition reduces laborious exercises, such as annotation of electrical patterns or counting of aphid populations, and only little plant material is required which can be advantageous when studying segregating populations with only one plant per genotype. These features make automated video tracking a valuable phenotyping method for screening large plant populations for resistance to piercing-sucking insects that are serious pests in our crops.

Methods

Plants and insects

Arabidopsis, *Arabidopsis thaliana* (L.) Heynh., plants were grown for 4–5 weeks in pots (5 cm diameter) with pasteurized potting soil (4 h at 80°C; Lentse potgrond, Lent, The Netherlands) in a climate room at 21 ± 1°C, 50-70% relative humidity, an 8/16 h day/night cycle, and a light intensity of 200 μmol m^{-2} s^{-1}. Four natural accessions of Arabidopsis were used throughout this study: Col-0 (CS76113), Van-0 (CS76297), Co-2 (CS28163) and Sanna-2 (CS76223). Seeds were acquired from the European Arabidopsis Stock Centre and propagated by the Laboratory of Genetics, Wageningen University.

Lettuce, *Lactuca sativa* (L.), cultivars Corbana (resistant) and Terlana (susceptible) were grown for 3 to 4 weeks in a greenhouse compartment at a temperature of 20 ± 3°C during the day and 18 ± 3°C during the night, 50-70% relative humidity and a 14/10 h day/night cycle using artificial lighting. Seeds were acquired from Enza Zaden bv. *Myzus persicae* (Sulzer) aphids were reared in a climate room on radish plants at 19°C, 50-70% relative humidity and a 16/8 h day/night cycle. *Nasonovia ribisnigri* (Mosely) biotype Nr:0 aphids were reared on the susceptible lettuce cultivar Terlana in a greenhouse compartment at a temperature of 20 ± 3°C during the day and 18 ± 3°C during the night, 50-70% relative humidity, and a 14/10 h day/night cycle.

Video tracking platform

Aphid behaviour was recorded with an analogue, monochrome camera (Ikegami, model: I CD-49E, type: REV, 768 × 576 pixels) with a varifocal lens (Computar H3Z4512 CS-IR, 4.5-12.5 mm F1.2) mounted above the arenas (Figure 1). An arena consisted of a well in a 96-well microtitre plate, having a 6.5 mm inner diameter (Sarstedt, sterile flat bottom suspension cells. No. 831835500), containing a leaf disc with the abaxial side up on a substrate of 1% agar (technical agar no.3, Oxoid). One aphid was introduced per arena and cling film was tightly wrapped around the plate to prevent aphids from escaping. The microtitre plate was placed on a platform, 1 cm above a backlight unit (FL tubes, 5000 K). A fan was attached between the platform and backlight unit to prevent water condensation inside the arenas. Room temperature was controlled at 21-22°C.

Software settings

EthoVision XT 8.5 video tracking and analysis software (Noldus Information Technology bv, Wageningen, The Netherlands) was used for automated video tracking of aphid feeding behaviour in multiple arenas simultaneously [41,55]. Subject detection was achieved with grey scaling (Additional file 1: Table S1). Arenas contained two zones: the leaf disc (zone 1) and the space surrounding the leaf disc (zone 2) (Additional file 1: Figure S1). Zone 1 had a diameter of approximately 5 mm, excluding the outer edges of the leaf disc to prevent aphids on the arena wall to be falsely assigned to the leaf disc. Because zone 1 and zone 2 required different grey scale thresholds, optimal thresholds for zone 1, the leaf disc, were chosen. Consequently, only behavioural data acquired in zone 1 were used throughout this study. Velocity and time thresholds appropriate to starting and ending a probe were fine-tuned using simultaneous observations of the top-view camera (275 pixels per mm^2) and a side-view camera attached to a stereo microscope (20-40 × magnification), capturing close-up recording of proboscis and antennae movements of *M. persicae* aphids (Additional file 2). A probe start was automatically recorded when the aphid was positioned on the leaf disc and its velocity dropped below 0.02 mm/s and did not exceed 0.3 mm/s for at least 10 seconds (Figure 2, Additional file 1: Figure S2). A probe stop was recorded when aphid velocity exceeded 0.3 mm/s for the relatively small wingless *M. persicae* or 0.35 mm/s for the larger winged *N. ribisnigri* and did not decrease below 0.1 mm/s for at least 2 seconds. Confounding movements during probing were generally characterized by a repetitive pattern of short movements. The 2 seconds time delay prevented that these movements resulted in false probe stops. Zone-transition problems, which occurred when aphids were positioned exactly on the edge of zone 1 and zone 2, were filtered from the data set after acquisition in EthoVision XT, with the statistical computing program R (Additional file 4). These incidences, characterized by a train of consecutive short probes in the output, were filtered out by excluding probes with a duration of less than 3 seconds that were preceded by a very short non-probe bout of maximally 15 seconds. These thresholds were selected by hand using some examples of zone transition problems in this study.

Video recording versus human observations

To validate automated tracking of probes with manual scoring, we used a camera mounted on a stereo microscope (20-40×) with a side-view on a single arena (n = 16) (Additional file 2). Each replicate consisted of a 55 min continuous recording of one arena with a single adult *M. persicae* aphid and an Arabidopsis leaf disc, by both the top-view and side-view camera. Aphids were starved between 30 minutes and three hours before the experiment. Recordings with the top-view camera were performed at two distances: capturing 20 arenas with 275 pixels per mm^2, and capturing 35 arenas with 155 pixels per mm^2. Leaf discs of 6 mm in diameter were cut just below the leaf apex of 4–5 week old Col-0 and Van-0 plants. The Observer® XT 10 software (Noldus Information Technology bv, Wageningen, The Netherlands) was used for manual scoring of probes. Probe starts were manually recorded when body movement stopped, the proboscis was touching the leaf and antennae moved backwards. If the aphid's proboscis was obscured, body arrestment on the leaf disc with subsequent backward movement of antennae was defined as a probe start [56,57]. Probe endings were manually recorded when antennae moved upward and the aphid removed its proboscis from the leaf, or, when the latter was not visible, when the antennae moved upwards followed by locomotion. Apart from probe estimations, we also tracked the "total time not moving" across the whole observation, using a start velocity of 0.3 mm/s and a stop velocity of 0.02 mm/s. Velocities were averaged across 5 frames, using a sample rate of 5 frames per second.

Video tracking assays

In each recording twenty arenas were tracked simultaneously for 8 hours, with a frame rate of 25 s^{-1}, and a resolution of 275 pixels per mm^2 (Additional file 3). All arenas consisted of a different plant and aphid individual and within each recording the 2 involved plant lines were equally represented. For Arabidopsis accessions Co-2 and Sanna-2, automated video tracking was performed with 7 to 8 day old wingless *M. persicae* aphids (Co-2 n = 20, Sanna-2 n = 17). Leaf discs of 6 mm in diameter were made just below the apex of intermediately aged leaves. Aphid survival was checked the day after recording. Subject detection was checked after data acquisition on 6 time points across the video. Three samples with no or low quality detection were excluded from the analysis. Video tracking of winged *N. ribisnigri* biotype Nr:0 on lettuce cultivars Terlana and Corbana was performed with 4 mm leaf discs (Corbana n = 27, Terlana n = 28). In view of the large contour of winged *N. ribisnigri* aphids, we used arenas with leaf discs of 4 mm diameter and a 3–4 mm leaf edge-to-wall distance in order to have a clear distinction between aphids on the leaf disc and aphids on agar or the arena wall. Leaf discs were made near the leaf base of the third oldest leaf, next to the mid vein. None of the aphids had died the day after recording. Five samples with no or low quality detection were excluded from the analysis. The response variable "duration not moving" was measured using a start velocity of 0.3 mm/s and a stop velocity of 0.02 mm/s. Velocities were averaged across 5 frames, using a sample rate of 5 frames per second.

EPG recording

Feeding behaviour of the green peach aphid, *M. persicae*, was analysed with EPG recording on two natural accessions of Arabidopsis, Co-2 and Sanna-2, during 8-hour observations. EPG recording was made on both intact plants (Co-2 n = 19, Sanna-2 n = 18) and leaf discs (Co-2 n = 9, Sanna-2 n = 8), using direct currents (DC) according to the methodology of ten Broeke et al. [34]. An electrode was inserted in the potting soil or agar respectively, and a thin gold wire (1.5 cm length for intact plants, 1 cm length for leaf discs) was gently attached to the dorsum of 8 to 11 day old wingless aphids with silver glue. The electrical circuit was completed when the aphid's piercing-sucking mouthparts penetrated the plant cuticle and the electrical signals, correlated to stylet activities, were recorded instantly [25]. Each replicate consisted of a different aphid and plant individual, employing one leaf disc per plant. Leaf discs of 9 mm in diameter were processed just below the apex of intermediately aged Arabidopsis leaves and placed abaxial side up in a Petri dish on a 1% agar substrate. A transparent plastic sheet covered the agar surrounding the leaf disc to prevent aphids to get stuck or make probes in the agar. Aphids that did not start probing within the first 3 hours of the observation were excluded from the analysis. EPG recording of winged *N. ribisnigri* biotype Nr:0 on lettuce cultivars Corbana and Terlana has been made in a previous study by ten Broeke et al. [47] (8-hour recording, n = 19).

Aphid population development

One *M. persicae* neonate (0 to 24 h old) was transferred to each Arabidopsis plant in a climate chamber (21 ± 1°C, 50-70% relative humidity, an 8/16 h day/night cycle, light intensity of 200 μmol m^{-2} s^{-1}). In the first assay 2.5-week-old plants were infested, in the second assay 3.5-week-old plants. A soap-diluted water barrier prevented aphids from moving between plants. Six, seven, and eight days after introduction the presence of the aphid and its offspring was checked. None of the aphids developed wings.

14 Days after infestation the number of aphids was counted per plant. Plants without an adult aphid 8 days after introduction, and plants without any adults or neonates 14 days after introduction were excluded from the analysis (assay 1: Co-2 n = 19, Sanna-2 n = 15; assay 2: Co-2 n = 14, Sanna-2 n = 13).

Simulations

In simulations, 10^4 random draws were taken from a normal distribution with the mean and standard deviation of a response variable of the Arabidopsis-*M. persicae* and lettuce-*N. ribisnigri* data sets (Additional file 1: Table S3). For video observations data was simulated with two probing variables: the mean duration of long probes and the total duration of sustained probes. For EPGs the total duration of phloem ingestion was simulated. Random draws were excluded when values were below zero, below the minimum duration of the probe category, or above the maximum recording duration. The generated data sets were subsampled with 1000 iterations without replacement for several replicate levels (n = 10, 15, 20, 25, 30, 35, 40). Student's t-tests were executed for each iteration and the percentage of significant p-values per replicate level was calculated. Video tracking simulation tests were defined significant if they had a P-value below $\alpha = 0.025$ for at least one of the two probing variables (Bonferroni correction: $\alpha = 0.05/2$). For EPG simulations one variable and P-values below $\alpha = 0.05$ already delivered maximum detection rates. This process was performed on complete data sets of EPG and video recording (8 h observations) and on data sets rescaled to shorter durations (6 and 4 hour observations). The proportion of tests where the null hypothesis is incorrectly rejected, was calculated with simulations based on a data set of 8-hour video recording of *M. persicae* on Arabidopsis accession Col-0 (data set n = 53, replicate levels n = 15 and n = 20, two variables, $P < 0.025$, Additional file 1: Table S3).

Statistical analysis

An R script was written to calculate response variables of video tracking, such as the total number and total duration of short and long probes in each observation and for each hour (Additional file 4). For EPG recording, the start time and duration of waveforms were analysed with the EPG PROBE 3.0 software (EPG-Systems, Wageningen, The Netherlands). Further calculations and analyses of EPG data were performed with the statistical computing program R. The duration of phloem ingestion events in EPG recording were calculated as the sum of three subsequent waveforms: (a) inter- and intracellular penetrations followed by (b) phloem salivation and (c) phloem ingestion. Bar graphs were produced with the R package sciplot version 1.1-0 (Morales 2012) [58]. Data distributions and homogeneity of variances were tested with a Shapiro test and a Levene's test. In case data transformations (square root, log, logit, arcsine) did not result in a distribution that approaches a normal distribution, non-parametric tests were applied. Human observations were compared to video tracking parameters with a paired *t*-test or, when data were not normally distributed with a Wilcoxon signed ranks matched pairs test. Correlations were tested with a Pearson correlation test or, when data were not normally distributed, with a Spearman correlation test. For benchmarking of video tracking against EPGs with susceptible and resistant Arabidopsis and lettuce lines and for the reproduction assay, response variables were tested with a Student's *t*-test, or when the data were not normally distributed with a Mann–Whitney *U* test. Walking activity of aphids was tested across 8 time bins of 1 hour. The distance moved was not normally distributed and, therefore, transformed to ranks ranging from the lowest to highest value within the complete data set. A mixed linear model was applied on the ranks, using plant line, time bin, and plant line x time bin interaction as fixed effects and plant/aphid individual as a random effect.

Additional files

Additional file 1: Figure S1. Arena settings. **Figure S2.** Trial Control Settings. **Table S1.** Subject detection settings for *M. persicae*. **Table S2.** Absolute differences between EPG recording and automated video tracking. **Table S3.** Video tracking variables used to generate simulations.

Additional file 2: Movie of aphid feeding behaviour measured by manual annotations and automated video tracking simultaneously. Events were estimated where the *M. persicae* aphid penetrated the Arabidopsis leaf disc with its piercing-sucking mouth parts. Manual scoring of probes was achieved with a side-view camera mounted on a stereo microscope and The Observer XT software. Automated video tracking was performed with a top view on the arena (275 pixels/mm) and EthoVision XT software (Noldus Information Technology bv, Wageningen, The Netherlands). EthoVision XT recorded a probe when the aphid's centre point was located on the leaf disc and did not move. Other motion parameters, such as the distance moved (cm) and velocity (cm/s), were also acquired with automated tracking. The movie is displayed at double speed.

Additional file 3: Movie of automated video tracking of 20 aphids simultaneously. *M. persicae* aphids were tracked on Arabidopsis leaf discs using EthoVision XT video tracking software (Noldus Information Technology bv, Wageningen, The Netherlands). The aphid's centre point (red) was automatically detected and tracks of the previous 30 seconds are visualised in red. Arenas are numbered from 1 (top left) to 20 (bottom right). The movie is displayed at 16x speed.

Additional file 4: R script for feature extraction from EthoVision XT output.

Abbreviation

EPG: Electrical Penetration Graph.

Competing interests

Two authors, OK and LN, contributed to the development of the video tracking platform and are affiliated with Noldus Information Technology bv. All involved authors confirm that the contents of this manuscript have not been affected by any competing interest whatsoever.

Authors' contributions

The hardware setup was designed by KK, MT, OK, LN and MJ. Software settings were optimized by KK, MT and OK. KK and GW performed the manual observations of aphid behaviour, automated video tracking experiments and aphid population assays. CtB and MHvdB performed the EPG recording and annotation of electrical signals. KK wrote the R script, analysed the data and prepared the figures. The movies in Additional files 2 and 3 were made by KK, OK and LN. The manuscript was written by KK, MD and MJ. All authors read and approved the final manuscript.

Acknowledgements

This work was supported by The Netherlands Organization for Scientific Research (NWO) through the Technology Foundation Perspective Programme 'Learning from Nature' [STW10989]. We would like to thank Lia Hemerik and Freddy Tjallingii for their valuable advice during this study.

Author details

[1]Laboratory of Entomology, Wageningen University, P.O. Box 16, 6700, AA Wageningen, The Netherlands. [2]Laboratory of Plant Physiology, Wageningen University, P.O. Box 16, 6700, AA Wageningen, The Netherlands. [3]Plant Research International, Wageningen University and Research Center, P.O. Box 16, 6700, AA Wageningen, The Netherlands. [4]Noldus Information Technology bv, P.O. Box 268, 6700, AG Wageningen, The Netherlands.

References

1. Blackman RL, Eastop VF. Aphids on the world's herbaceous plants and schrubs. London: John Wiley & Sons, Ltd. & Natural History Museum; 2006.
2. Tjallingii WF. Regulation of phloem sap feeding by aphids. In: Chapman RF, De Boer G, editors. Regulatory mechanisms in insect feeding. US: Springer; 1995. p. 190–209.
3. Minks AK, Harrewijn P. World crop pests. Aphids. Their biology, Natural enemies and control. Amsterdam, the Netherlands: Elsevier Science Publishers; 1989.
4. De Vos M, Van Oosten VR, Van Poecke RMP, Van Pelt JA, Pozo MJ, Mueller MJ, et al. Signal signature and transcriptome changes of *Arabidopsis* during pathogen and insect attack. MPMI. 2005;18:923–37.
5. Eigenbrode SD, Espelie KE. Effects of plant epicuticular lipids on insect herbivores. Annu Rev Entomol. 1995;40:171–94.
6. Wang E, Wang R, DeParasis J, Loughrin JH, Gan S, Wagner GJ. Suppression of a P450 hydroxylase gene in plant trichome glands enhances natural-product-based aphid resistance. Nat Biotechnol. 2001;19:371–4.
7. Ceci LR, Volpicella M, Rahbé Y, Gallerani R, Beekwilder J, Jongsma MA. Selection by phage display of a variant mustard trypsin inhibitor toxic against aphids. Plant J. 2003;33:557–66.
8. Mewis I, Appel HM, Hom A, Raina R, Schultz JC. Major signaling pathways modulate *Arabidopsis* glucosinolate accumulation and response to both phloem-feeding and chewing insects. Plant Physiol. 2005;138:1149–62.
9. Smith CM, Boyko EV. The molecular bases of plant resistance and defense responses to aphid feeding: current status. Entomol Exp Appl. 2007;122:1–16.
10. Pegadaraju V, Louis J, Singh V, Reese JC, Bautor J, Feys BJ, et al. Phloem-based resistance to green peach aphid is controlled by Arabidopsis PHYTOALEXIN DEFICIENT4 without its signaling partner ENHANCED DISEASE SUSCEPTIBILITY1. Plant J. 2007;52:332–41.
11. Du Y, Poppy GM, Powell W, Pickett JA, Wadhams LJ, Woodcock CM. Identification of semiochemicals released during aphid feeding that attract parasitoid *Aphidius ervi*. J Chem Ecol. 1998;24:1355–68.
12. Will T, van Bel AJE. Physical and chemical interactions between aphids and plants. J Exp Bot. 2006;57:729–37.
13. Kollner TG, Held M, Lenk C, Hiltpold I, Turlings TCJ, Gershenzon J, et al. A maize (E)-beta-caryophyllene synthase implicated in indirect defense responses against herbivores is not expressed in most American maize varieties. Plant Cell. 2008;20:482–94.
14. Wink M. Plant breeding: importance of plant secondary metabolites for protection against pathogens and herbivores. Theor Appl Genet. 1988;75:225–33.
15. Eenink AH, Dieleman FL. Screening *Lactuca* for resistance to *Myzus persicae*. Neth J Plant Pathol. 1977;83:139–51.
16. Smyrnioudis IN, Harrington R, Katis NI. A simple test for evaluation of transmission efficiency of barley yellow dwarf virus by aphids. Phytoparasitica. 2002;30:535–8.
17. Chen X, Vosman B, Visser RGF, van der Vlugt RAA, Broekgaarden C. High throughput phenotyping for aphid resistance in large plant collections. Plant Methods. 2012;8:33.
18. Stelinski L, Tiwari S. Vertical T-maze choice assay for arthropod response to odorants. J Vis Exp. 2013;72:e50229.
19. Kloth KJ, Thoen MPM, Bouwmeester HJ, Jongsma MA, Dicke M. Association mapping of plant resistance to insects. Trends Plant Sci. 2012;17:311–9.
20. Smith CM, Khan ZR, Pathak MD. Techniques for evaluating insect resistance in crop plants. Florida: CRC Press LLC; 1994.
21. Broekgaarden C, Poelman EH, Steenhuis G, Voorrips RE, Dicke M, Vosman B. Responses of *Brassica oleracea* cultivars to infestation by the aphid *Brevicoryne brassicae*: an ecological and molecular approach. Plant Cell Environ. 2008;31:1592–605.
22. Mewis I, Tokuhisa JG, Schultz JC, Appel HM, Ulrichs C, Gershenzon J. Gene expression and glucosinolate accumulation in *Arabidopsis thaliana* in response to generalist and specialist herbivores of different feeding guilds and the role of defense signaling pathways. Phytochemistry. 2006;67:2450–62.
23. Moran PJ, Thompson GA. Molecular responses to aphid feeding in Arabidopsis in relation to plant defense pathways. Plant Physiol. 2001;125:1074–85.
24. McLean DL, Kinsey MG. A technique for electronically recording aphid feeding and salivation. Nature. 1964;202:1358–9.
25. Tjallingii WF. Electrical recording of stylet penetration activities. In: Minks AK, Harrewijn P, editors. Aphids, their biology, natural enemies and control, vol. 2B. Amsterdam: Elsevier; 1988. p. 95–108.
26. van Emden HF, Harrington R. Aphids as crop pests. Oxfordshire, UK: CABI; 2007.
27. van Helden M, Tjallingii WF. Tissue localisation of lettuce resistance to the aphid N. ribisnigri using electrical penetration graphs. Entomol Exp Appl. 1993;68:269–78.
28. Tjallingii WF. Salivary secretions by aphids interacting with proteins of phloem wound responses. J Exp Bot. 2006;57:739–45.
29. Will T, Tjallingii WF, Thönnessen A, Van Bel AJE. Molecular sabotage of plant defense by aphid saliva. Proc Natl Acad Sci U S A. 2007;104:10536–41.
30. Boquel S, Giordanengo P, Ameline A. Divergent effects of PVY-infected potato plant on aphids. Eur J Plant Pathol. 2011;129:507–10.
31. Nalam VJ, Keeretaweep J, Sarowar S, Shah J. Root-derived oxylipins promote green peach aphid performance on arabidopsis foliage. Plant Cell. 2012;24:1643–53.
32. ten Broeke CJM, Dicke M, van Loon JJA. Resistance in a *Lactuca virosa* accession to a new biotype of *Nasonovia ribisnigri*. Euphytica. 2013;193:265–75.
33. Alvarez AE, Tjallingii WF, Garzo E, Vleeshouwers V, Dicke M, Vosman B. Location of resistance factors in the leaves of potato and wild tuber-bearing *Solanum* species to the aphid *Myzus persicae*. Entomol Exp Appl. 2006;121:145–57.
34. ten Broeke CJM, Dicke M, van Loon JJA. Performance and feeding behaviour of two biotypes of the black currant-lettuce aphid, *Nasonovia ribisnigri*, on resistant and susceptible *Lactuca sativa* near-isogenic lines. Bull Entomol Res. 2013;103:511–21.
35. Bakchine E, Pham-Delegue MH, Kaiser L, Masson C. Brief communication: computer analysis of the exploratory behavior of insects and mites in an olfactometer. Physiol Behav. 1990;48:183–7.
36. Shcherbakov D, Schill RO, Brümmer F, Blum M. Movement behaviour and video tracking of *Milnesium tardigradum* Doyère, 1840 (Eutardigrada, Apochela). Contrib Zool. 2010;79:33–8.
37. Pinkiewicz TH, Purser GJ, Williams RN. A computer vision system to analyse the swimming behaviour of farmed fish in commercial aquaculture facilities: a case study using cage-held atlantic salmon. Aquac Eng. 2011;45:20–7.
38. Moreno-Delafuente A, Garzo E, Moreno A, Fereres A. A plant virus manipulates the behavior of its whitefly vector to enhance its transmission efficiency and spread. PLoS ONE. 2013;8:e61543.
39. Spitzen J, Spoor CW, Grieco F, ter Braak C, Beeuwkes J, van Brugge SP, et al. A 3D analysis of flight behavior of *Anopheles gambiae* sensu stricto malaria mosquitoes in response to human odor and heat. PLoS ONE. 2013;8:e62995.
40. Dell AI, Bender JA, Branson K, Couzin ID, de Polavieja GG, Noldus LPJJ, et al. Automated image-based tracking and its application in ecology. Trends Ecol Evol. 2014;29:417–28.

41. Noldus LPJJ, Spink AJ, Tegelenbosch RAJ. Computerised video tracking, movement analysis and behaviour recognition in insects. Comput Electron Agric. 2002;35:201–27.
42. Tjallingii WF. Sieve element acceptance by aphids. Eur J Entomol. 1994;91:47–52.
43. Prado E, Tjallingii WF. Behavioral evidence for local reduction of aphid-induced resistance. J Insect Sci. 2007;7:1–8.
44. Grieco F, Loijens L, Krips OE, Smit G, Spink AJ, Zimmerman P. EthoVision XT reference manual. Wageningen: Noldus Information Technology; 2010.
45. Will T, Kornemann SR, Furch ACU, Tjallingii WF, van Bel AJE. Aphid watery saliva counteracts sieve-tube occlusion: a universal phenomenon? J Exp Biol. 2009;212:3305–12.
46. Reinink K, Dieleman FL. Comparison of sources of resistance to leaf aphids in lettuce (lactuca sativa L.). Euphytica. 1989;40:21–9.
47. Ten Broeke CJM. Unravelling the resistance mechanism of lettuce against *Nasonovia ribisnigri*. Wageningen: Wageningen University; 2013.
48. Liu Y, McCreight JD. Responses of *Nasonovia ribisnigri* (Homoptera: Aphididae) to susceptible and resistant lettuce. J Econ Entomol. 2006;99:972–8.
49. Leon J, Rojo E, Sanchez-Serrano JJ. Wound signalling in plants. J Exp Bot. 2001;52:1–9.
50. Will T, Hewer A, Van Bel AJE. A novel perfusion system shows that aphid feeding behaviour is altered by decrease of sieve-tube pressure. Entomol Exp Appl. 2008;127:237–45.
51. Mittler TE. Applications of artificial feeding techniques for aphids. In: Minks AK, Harrewijn P, editors. Aphids, their biology, natural enemies ano control, vol. 2B. Amsterdam: Elsevier; 1988. p. 145–70.
52. Turgeon R. The puzzle of phloem pressure. Plant Physiol. 2010;154:578–81.
53. Martin B, Rahbé Y, Fereres A. Blockage of stylet tips as the mechanism of resistance to virus transmission by *Aphis gossypii* in melon lines bearing the Vat gene. Ann Appl Biol. 2003;142:245–50.
54. Husson SJ, Steuer Costa W, Schmitt C, Gottschalk A. Keeping track of worm trackers. In: Community eTCeR, editor. Wormbook. 2012. p. 1–17. http://www.wormbook.org.
55. Noldus LPJJ, Spink AJ, Tegelenbosch RAJ. EthoVision: a versatile video tracking system for automation of behavioral experiments. Behav Res Methods Instrum Comput. 2001;33:398–414.
56. Hardie J, Holyoak M, Taylor NJ, Griffiths DC. The combination of electronic monitoring and video-assisted observations of plant penetration by aphids and behavioural effects of polygodial. Entomol Exp Appl. 1992;62:233–9.
57. Hardie J, Powell G. Close-up video combined with electronic monitoring of plant penetration and behavioral effects of an aphid (Homoptera: Aphididae) antifeedant. In: Walker PG, Backus EA, editors. Principles and applications of electronic monitoring and other techniques in the study of Homopteran feeding behavior. US: Entomological Society of America; 2000. p. 201–11.
58. R-Core-Team. R: a language and environment for statistical computing. Vienna, Austria: R Foundation for Statistical Computing; 2013.

Estimating rice chlorophyll content and leaf nitrogen concentration with a digital still color camera under natural light

Yuan Wang[1], Dejian Wang[1*], Peihua Shi[2] and Kenji Omasa[3]

Abstract

Background: The color of crop leaves is closely correlated with nitrogen (N) status and can be quantified easily with a digital still color camera and image processing software. The establishment of the relationship between image color indices and N status under natural light is important for crop monitoring and N diagnosis in the field. In our study, a digital still color camera was used to take pictures of the canopies of 6 rice (*Oryza sativa* L.) cultivars with N treatments ranging from 0 to 315 kg N ha^{-1} in the field under sunny and overcast conditions in 2010 and 2011, respectively.

Results: Significant correlations were observed between SPAD readings, leaf N concentration (LNC) and 13 image color indices calculated from digital camera images using three color models: RGB, widely used additive color model; HSV, a cylindrical-coordinate similar to the human perception of colors; and the $L^*a^*b^*$ system of the International Commission on Illumination. Among these color indices, the index b^*, which represents the visual perception of yellow-blue chroma, has the closest linear relationship with SPAD reading and LNC. However, the relationships between LNC and color indices were affected by the developmental phase. Linear regression models were used to predict LNC and SPAD from color indices and phasic development. After that, the models were validated with independent data. Generally, acceptable performance and prediction were found between the color index b^*, SPAD reading and LNC with different cultivars and sampling dates under different natural light conditions.

Conclusions: Our study showed that digital color image analysis could be a simple method of assessing rice N status under natural light conditions for different cultivars and different developmental stages.

Keywords: Digital still color camera, Leaf color, Image processing technology, Natural light, Nitrogen, Rice

Background

Nitrogen (N) is one of the most important nutrients essential for the growth of crops, and is a major component of chlorophyll and protein which are closely associated with leaf color, crop growth status and yield [1]. Insufficient N supply leads to smaller leaves, lower chlorophyll content and less biomass production, and consequently, reduced grain yield and quality [2,3]. Conversely, excessive N application can lead to the environmental problems of water and atmospheric pollution [4]. Hence, measuring crop N status timely is critical for increasing N use efficiency and environmental quality [5,6].

* Correspondence: djwang@issas.ac.cn
[1]Institute of Soil Science, Chinese Academy of Sciences, 71 East Beijing Road, Nanjing 210008, PR China
Full list of author information is available at the end of the article

The spectral reflectance of crop leaf or canopy is known to be correlated with N status [7,8]. The instruments for measuring spectral reflectance are the chlorophyll meter [9-11], multi-spectral sensor [12], hyper-spectral sensor [7,8] and commercial digital camera [13-15], are used in precision agriculture for growth monitoring, nitrogen diagnosis and site-specific crop management. The SPAD meter (SPAD-502, Minolta Camera Co., Osaka, Japan), measures leaf chlorophyll content nondestructively, has potential for improving N use efficiency without affecting grain yield in real-time nitrogen management (RTNM) experiments [9,11]. A disadvantage of the SPAD-502 for assessing crop N status is its small sampling area (6 mm^2). In addition, the measurements are subject to operator bias so a large number of repetitions are needed to obtain reliable results [16,17], and SPAD meter experiences difficulties in distinguishing chlorophyll levels when crops are near or above the optimum N supply [10,18]. In contrast, satellite or airborne-mounted hyperspectral sensors can obtain spectral information in a larger sampling area and record more spectral bands [7,8,12]. The high cost of images, infrequent satellite overpasses and risk of images being obscured by clouds limit the application of these platforms for commercial use [19]. The proximal sensors GreenSeeker® (NTech Industries Inc., USA) and Yara N-sensor (Yara International ASA, Germany) which measure red and near infra-red (NIR) reflectance, overcome some of the limitations of satellite or airborne sensors [20], but their accuracy is influenced by background soil interference [8].

Alternatively, images from digital still color cameras, which record spectral information of visible bands, have a low cost but very high image resolution (consumer cameras in 2014 record up to 40 megapixels per image). At a sensor height of about 1 meter above the canopy, high-spatial-resolution images separate crops from background soil or other interferences, which is important for accurate diagnosis of N status when the vegetation fraction was low [13,21]. Moreover, images from digital still cameras contain a large amount of information about the crop structure and leaf color, such as leaf orientation, plant height, biomass accumulation and leaf senescence [22-24], and these parameters are easy to obtain [13,25] with existing software, such as MatLab® (MathWorks Inc.), the free-ware package ImageJ [26]. Previous studies showed that canopy cover estimated from the images was not only highly correlated with leaf area index (LAI), aboveground biomass and N accumulation [13,27], but was also stable in varying environmental conditions [28,29]. In addition to canopy cover, color digital images provide spectral information in the visible bands which are closely related to the leaf N concentration (LNC) and SPAD readings [17]. Hunt *et al.* [30] found that the triangular greenness index (TGI), which was derived from red, green and blue bands of a digital still color camera, was sensitive to leaf chlorophyll content of a whole canopy.

The color-related indices from digital still cameras can diagnose crop N status [17,30,31]. Previous studies on the analysis of leaf color were mostly undertaken in controlled light conditions [17,32-34]. This approach could reduce the impact of light on the image color, and easily obtain a reliable relationship between N status and leaf color indices. However, the results derived from the controlled light cannot be completely applied to natural light because of the variable light conditions [15,35,36]. Besides, there are still many uncertainties in the use of digital still cameras for N diagnosis under natural light conditions, and further validation is necessary to ensure the application in the field.

In this study, experiments with different N application rates were carried out in the field under sunny and overcast conditions (1) to analyze the relationship between chlorophyll content, LNC and canopy color related indices in different cultivars and stage of phasic development, (2) to establish the possible models for the diagnosis of crop N status using image color indices, and (3) to validate the applicability of the models under different natural light conditions.

Results and discussion

Correlation of color related indices and crop nitrogen status

Correlations between two rice N parameters (leaf N concentration and SPAD reading) and 13 image-color related indices (Eq. 1–10) were calculated with individual and pooled cultivars in 2011. Similar results were obtained from the three rice cultivars, i.e., Liangyoupeijiu, Nanjing45 and Nanjing46. Therefore, only correlation coefficients from the Liangyoupeijiu dataset (n = 72) and the pooled dataset (n = 240) were displayed in Table 1. SPAD readings showed significant correlations with each color index except g in both datasets. Among these color indices, L^*, b^*, R, G, B, r and INT were negatively correlated with SPAD readings, while the other indices were positively correlated with SPAD readings. Most of the color indices were closely correlated with LNC, however, the magnitude and direction of the correlation coefficients were not consistent with those between color indices and SPAD readings. Overall, the indices derived from the CIE $L^*a^*b^*$ color model had relatively higher correlation coefficients with SPAD readings and LNC. In this color model, the index b^* represents the visual perception of yellow-blue chroma which is similar with the leaf color variation, and it has been used in many other studies for image color analysis [37,38]. Therefore, we select index b^* as a representative for further analysis.

Relationships between SPAD, LNC and the color index b^*

Regression analyses were performed between SPAD readings, LNC and the color index b^* using the 2011 dataset. Positive linear relationships were observed between LNC and SPAD readings, with the same trends in different sampling dates and cultivars (Figure 1). The determination coefficient (R^2) in different sampling dates varied from 0.61 to 0.88 along with the root mean square error (RMSE) from 1.81 to 2.64 g kg^{-1}. The LNC decreased with the rice development, while the maximum SPAD values increased with rice growth. Smaller RMSE was obtained in jointing and booting stages than in vegetative and tillering stages. Similarly, Xue et al. [39] reported that the ratio index of NIR/green (R_{810}/R_{560}) reached the best accuracy with LNC at jointing stage. When data were pooled across the sampling dates, there

Table 1 Correlation coefficients between SPAD readings, leaf nitrogen concentration (LNC, g kg^{-1}) and image color related indices (digital number from three color models: RGB, HSV and CIE $L^*a^*b^*$)

	SPAD	LNC	L^*	a^*	b^*	b^*/a^*	H	R	G	B	r	g	b	INT	VI_{Green}
SPAD	1.00	0.21	−0.67**	0.78**	−0.90**	0.76**	0.83**	−0.68**	−0.67**	−0.48**	−0.76**	0.01	0.41**	−0.63**	0.66**
LNC	0.32**	1.00	0.48**	−0.26*	−0.06	0.35**	0.13	0.47**	0.48**	0.63**	−0.37**	−0.68**	0.65**	0.52**	−0.28*
L^*	−0.62**	0.35**	1.00	−0.83**	0.73**	−0.43**	−0.63**	1.00**	1.00**	0.96**	0.38**	−0.64**	0.21	1.00**	−0.90**
a^*	0.61**	−0.04	−0.70**	1.00	−0.86**	0.50**	0.69**	−0.82**	−0.84**	−0.69**	−0.52**	0.18	0.17	−0.80**	0.62**
b^*	−0.73**	−0.35**	0.54**	−0.72**	1.00	−0.86**	−0.95**	0.74**	0.73**	0.51**	0.87**	0.04	−0.51**	0.68**	−0.73**
b^*/a^*	0.60**	0.48**	−0.29**	0.36**	−0.91**	1.00	0.96**	−0.45**	−0.43**	−0.18	−0.99**	−0.26*	0.72**	−0.37**	0.63**
H	0.61**	0.40**	−0.39**	0.42**	−0.91**	0.97**	1.00	−0.64**	−0.63**	−0.40**	−0.95**	−0.09	0.58**	−0.58**	0.75**
R	−0.63**	0.33**	1.00**	−0.68**	0.56**	−0.34**	−0.44**	1.00	1.00**	0.96**	0.39**	−0.64**	0.21	1.00**	−0.92**
G	−0.62**	0.35**	1.00**	−0.71**	0.54**	−0.29**	−0.39**	1.00**	1.00	0.96**	0.38**	−0.64**	0.21	1.00**	−0.90**
B	−0.37**	0.58**	0.91**	−0.47**	0.15*	0.09	−0.03	0.90**	0.91**	1.00	0.12	−0.82**	0.48**	0.98**	−0.83**
r	−0.56**	−0.54**	0.19**	−0.35**	0.89**	−0.99**	−0.95**	0.24**	0.19**	−0.20**	1.00	0.36**	−0.79**	0.31**	−0.56**
g	0.05	−0.66**	−0.62**	0.03	0.31**	−0.44**	−0.34**	−0.60**	−0.61**	−0.87**	0.56**	1.00	−0.86**	−0.69**	0.58**
b	0.27**	0.69**	0.27**	0.17**	−0.66**	0.79**	0.71**	0.23**	0.26**	0.63**	−0.87**	−0.90**	1.00	0.29*	−0.08
INT	−0.56**	0.42**	0.99**	−0.64**	0.45**	−0.20**	−0.31**	0.99**	0.99**	0.95**	0.10	−0.70**	0.37**	1.00	−0.89**
VI_{Green}	0.67**	−0.10	−0.85**	0.42**	−0.65**	0.62**	0.68**	−0.88**	−0.85**	−0.68**	−0.51**	0.43**	0.02	−0.83**	1.00

Numbers in the upper diagonal were calculated with the Liangyoupeijiu (n = 72) dataset and numbers in the lower diagonal were calculated with the combined data of Liangyoupeijiu, Nanjing45 and Nanjing46 in 2011 (n = 240).
**indicate the significance at $P < 0.01$, *indicate the significance at $P < 0.05$.

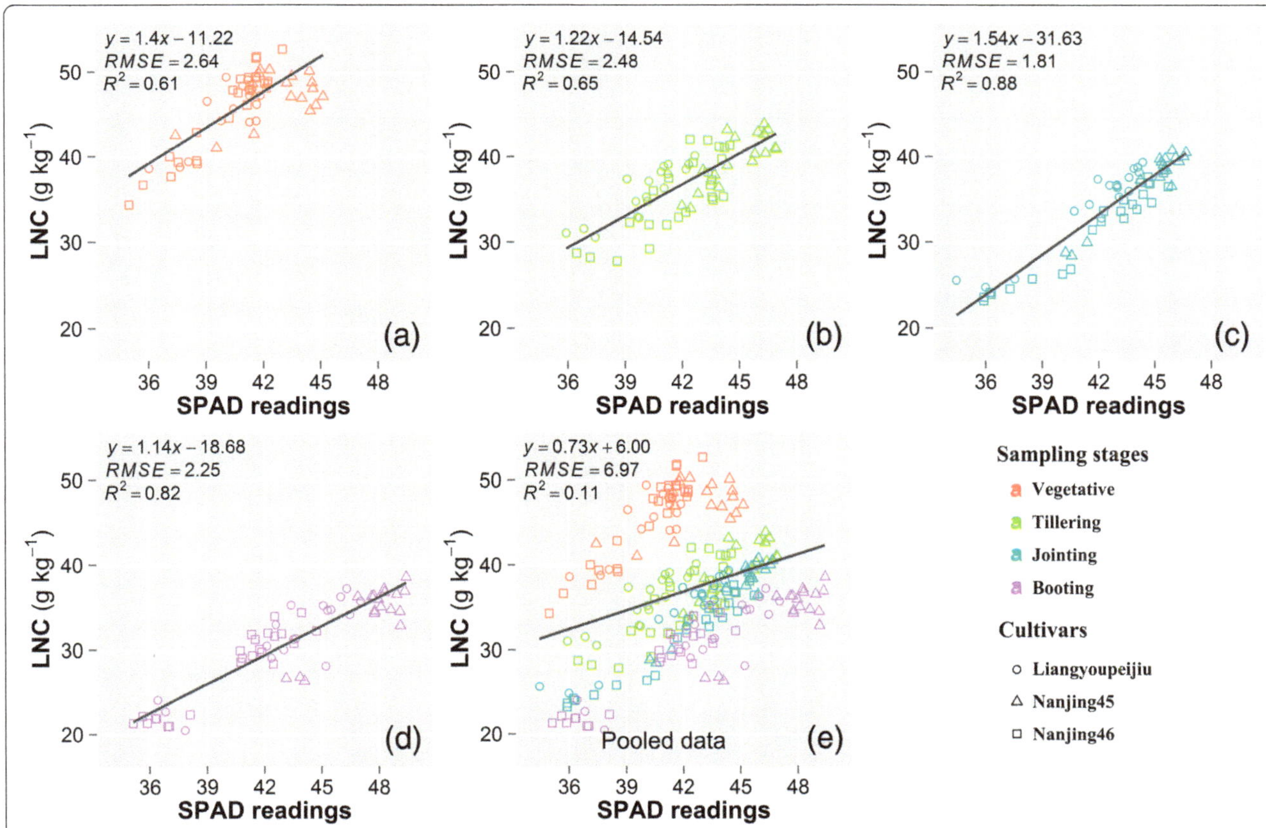

Figure 1 Relationships between leaf nitrogen concentration (LNC, g kg^{-1}) and SPAD readings in vegetative (a), tillering (b), jointing (c), booting (d) stages, and the pooled data of the four stages (e) in 2011.

was no significant trend observed between SPAD readings and LNC (Figure 1e).

Significant negative linear relationships were seen between LNC and color index b^*, with R^2 ranging from 0.58 to 0.86 in the four developmental stages (Figure 2). The R^2 between b^* and LNC at booting stage (Figure 2d) was less than that between SPAD readings and LNC. Same as the relationship between SPAD readings and LNC, no significant trend was observed between b^* and LNC when data pooled across the sampling dates (Figure 2e). In addition, there were negative linear relationships between b^* and SPAD readings (Figure 3). The R^2 in the vegetative stage was lower than that in the other stages. There were obvious differences among cultivars for the relationship between b^* and SPAD readings, especially the Liangyoupeijiu in jointing and booting stages (Figure 3c and d) and the Nanjing45 in vegetative stage (Figure 3a). In this case, the regression analysis was carried out with individual cultivars (Table 2). Overall, higher R^2 were observed from individual cultivars and sampling dates than that from the pooled dataset (Table 2). However, the intercepts and slopes of the linear relationship varied with rice cultivars, these differences might be partly caused by the different plant type among cultivars, with a loose shape and large mean leaf angle in hybrid indica rice (Liangyoupeijiu) [40,41] while tight shape and small mean leaf angle in japonica rice (Nanjing45, Nanjing46). The different plant type lead to different distribution of reflectance [42,43] and finally caused different image-derived indices.

In our experiments, image acquisition was carried out in the field under natural light near solar noon which was the period with the most stable illumination at the top of the atmosphere. This makes sure that the light intensity is not changing too much during image acquisition in a single day. The results from Table 2 indicated that reliable estimates of N status could be obtained from images taken under natural light. Considering the stability of N diagnosis in different locations and sampling dates, individual sampling dates could not meet the needs of crop monitoring and N diagnosis in various environmental conditions. Therefore, regression analysis was carried out with the pooled data of sampling dates. There were large differences of light intensity among different sampling dates even under overcast days (Table 3, PAR ranging from 145 to 692 μmol m^{-2} s^{-1} in 2011).

Interestingly, regression analysis did not show any evidence that the relationship between SPAD and color index

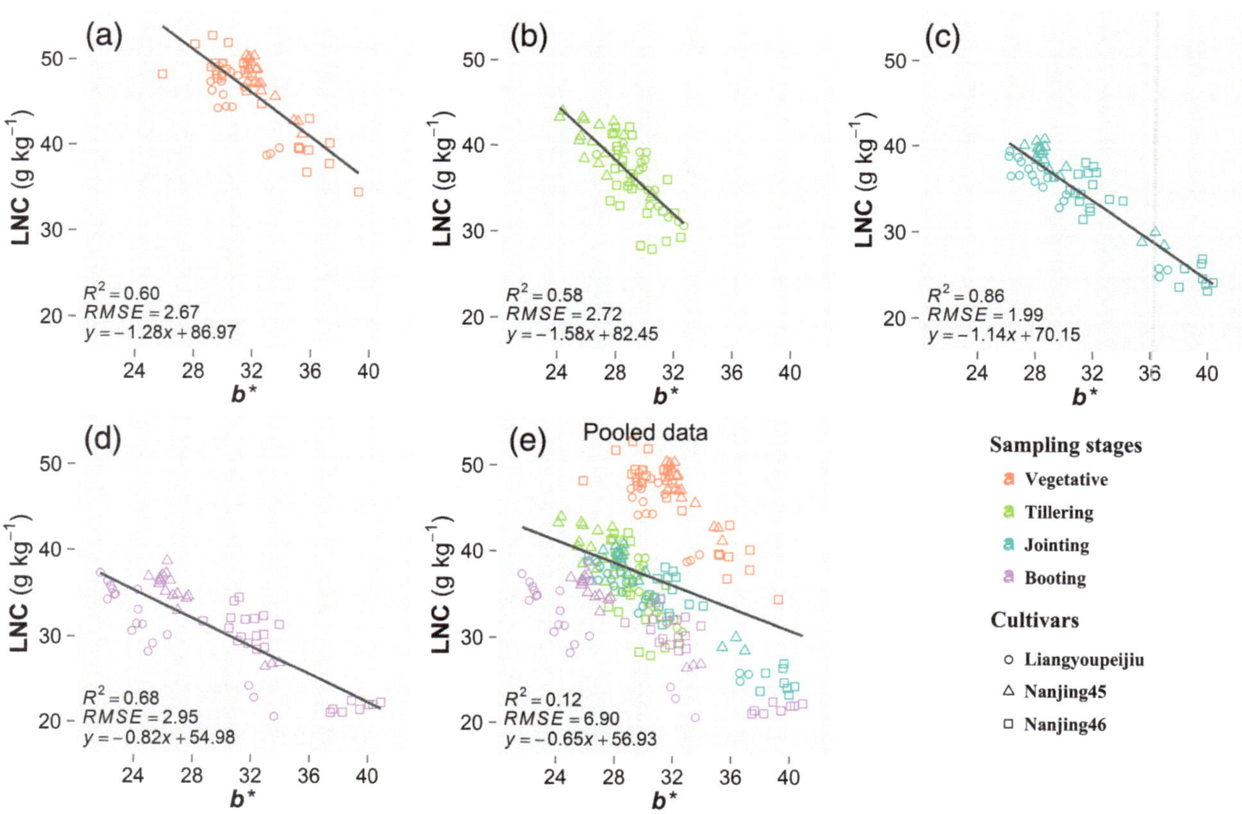

Figure 2 Relationships between leaf nitrogen concentration (LNC, g kg^{-1}) and image color index b^* in vegetative (a), tillering (b), jointing (c), booting (d) stages, and the pooled data of the four stages (e) in 2011.

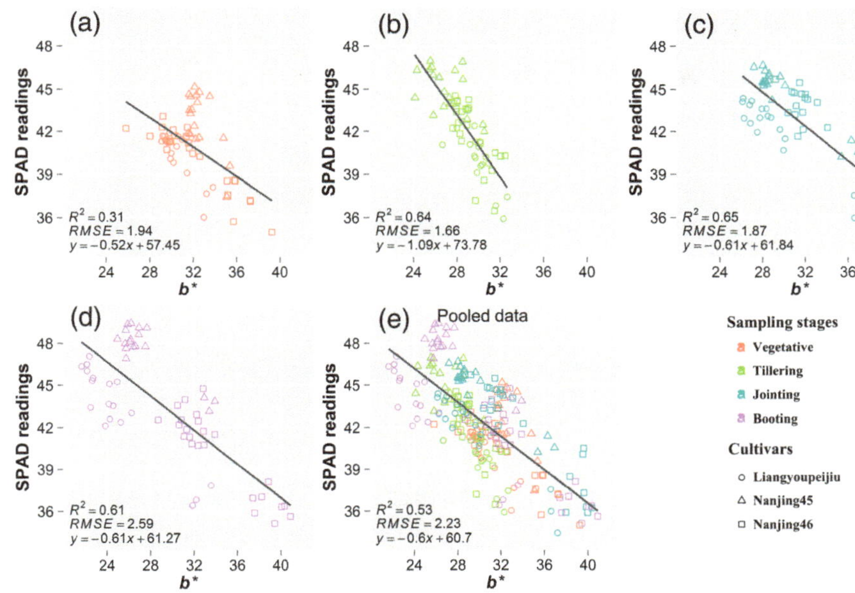

Figure 3 Relationships between SPAD readings and image color index b^* in vegetative (a), tillering (b), jointing (c), booting (d) stages, and the pooled data of the four stages (e) in 2011.

b^* was affected by the varying light intensity (Figure 3e). This might be attributed to the auto exposure controlled by the digital camera which adjusted the exposure time to make compensation for the amount of light reaching the image sensor. However, there were no significant trends between LNC and SPAD readings, or between LNC and color index b^* using data pooled across different sampling dates (Figures 1e and 2e). Previous studies revealed that, for rice and corn, the relationship between LNC and SPAD readings could be improved simply by dividing the readings with specific leaf weight (SLW = dry leaf weight/leaf area) of the sampled leaves or introducing SLW as a second independent variable in the multiple regression [44-46]. The reason is that SPAD readings vary with leaf thickness which can be different in cultivars, developmental stages and environmental conditions [46,47], while the LNC has a relatively consistent value. The uncertain relationship between LNC and color index b^* (Figure 2e) may also be caused by the difference of leaf thickness, because the color index b^* and SPAD readings both reveal the spectral information of leaves, and their relationship keeps consistent with the pooled data of different sampling dates (Table 2 and Figure 3e). Nowadays, destructive sampling or hyperspectral-reflectance [48] is required for the measurement of SLW, however, including this defeats the purpose of using a cheap and simple digital camera.

Table 2 Statistics of the linear regression analysis between color index b^* and SPAD readings in different development stages and cultivars in 2011

		Vegetative	Tillering	Jointing	Booting	All stages
Liangyoupeijiu	R^2	0.74**	0.62**	0.92**	0.86**	0.80**
	RMSE[a]	0.87	1.18	0.86	1.23	1.24
Nanjing45	R^2	0.42**	0.28**	0.90**	0.79**	0.60**
	RMSE	1.61	1.41	0.64	0.91	1.60
Nanjing46	R^2	0.83**	0.52**	0.82**	0.83**	0.59**
	RMSE	0.94	1.62	1.47	1.26	1.85
All cultivars	R^2	0.31**	0.64**	0.65**	0.61**	0.53**
	RMSE	1.94	1.66	1.87	2.59	2.23

Dependent variable: SPAD.
**indicate the significance at $P < 0.01$.
[a]Root mean square error.

Table 3 Rice cultivars, sampling dates (indicated as days after transplanting, DAT), photosynthetic active radiation (PAR, $\mu mol\ m^{-2}\ s^{-1}$) and the number of samples in the two experiments

Experiment	Year	Cultivar	Vegetative	Tillering	Jointing	Booting	Number of samples
I	2010	Wuyunjing24	25[a] (937)[b]	36 (1536)	50 (1369)	64 (1532)	18
	2011	Nanjing46	18 (145)	35 (692)	55 (203)	75 (296)	24
II	2010	Nanjing44	22 (1215)	35 (1058)	49 (1759)	63 (1477)	18
		Yangjing48	22 (1449)	35 (1058)	49 (1759)	63 (1477)	18
	2011	Nanjing45	16 (427)	29 (654)	51[c] (589)	57 (289)	18
		Liangyoupeijiu	16 (427)	29 (654)	51 (589)	65 (621)	18

[a]represents sampling dates (the days after transplanting, DAT).
[b]represents the average PAR during the period of image acquisition.
[c]Because of the continuous sunny days in the jointing stage for Nanjing45, sampling dates was delayed about a week.

Model calibration and validation

Since developmental stages in rice affected the response of N parameters to color indices, multiple linear regression analysis was carried out with the consideration of days after transplanting (DAT) to estimate LNC and SPAD (Table 4). The multiple linear models highly improved the R^2 of SPAD readings and b^* with LNC (Table 4, Figure 4a and b). The relationships between SPAD readings and b^* were less affected by the process of development for all the cultivars (Table 2 and Figure 3e). Therefore, the consideration of DAT in the multiple linear regression did not improve the R^2 effectively (Figure 4c). The slope of the regression lines in Figure 4 was all less than 1, which indicated that predicted LNC or SPAD were generally smaller at the high value area and bigger at the low value area, than the observed ones. It was noteworthy that most of the predicted LNC in the vegetative stage (red symbols) were underestimated in Figure 4a and b. This is mainly because that the plants at this stage has higher concentration of N but lower concentration of chlorophyll. In this case, the lower concentration of chlorophyll would cause an underestimation of the LNC in the vegetative stage.

Validations were performed on the four models in Table 4 with all data obtained in 2010 under sunny days. The images used in model calibration and validation were taken under different weather conditions, the objective of this combination was to evaluate whether the model was robust under different light conditions. In general, good performances on the predicted models were observed for the estimation of LNC and SPAD (Figure 5). Model 1 showed the best performance on the prediction of LNC with a normalized mean bias of 1.19% (Table 4 and Figure 5a). Model 2, 3 and 4 showed relatively lower R^2 and smaller negative bias. The model for the prediction of LNC with color index b^* (Figure 5b) was not severely affected by the different light conditions in 2010 and 2011, compared to the prediction of LNC with SPAD (Figure 5a). As with the calibration results, most of the data in the vegetative stage were below the 1:1 line in models 1 and 2. In addition, the data in tillering stage deviated from the 1:1 line in models 3 and 4, which resulted in the low R^2 (Figure 5). The similar R^2

Table 4 Statistics of the calibration and validation results for estimating leaf nitrogen concentration (LNC, $g\ kg^{-1}$) and SPAD with color index b^*, SPAD and days after transplanting (DAT)

Models	α	β	γ	RMSE[a]	R^2	NMB[b]
Calibration						
Model 1 $LNC = \alpha SPAD + \beta DAT + \gamma$	1.02 ± 0.07	−0.31 ± 0.01	7.01 ± 2.95	3.07 $g\ kg^{-1}$	0.78	
Model 2 $LNC = \alpha b^* + \beta DAT + \gamma$	−0.67 ± 0.06	−0.29 ± 0.01	69.27 ± 2.10	3.36 $g\ kg^{-1}$	0.71	
Model 3 $SPAD = \alpha b^* + \beta DAT + \gamma$	−0.60 ± 0.04	0.024 ± 0.007	59.66 ± 1.14	1.62	0.55	
Model 4 $SPAD = \alpha b^* + \gamma$	−0.60 ± 0.04		60.70 ± 1.12	1.62	0.53	
Validation						
Model 1				2.43 $g\ kg^{-1}$	0.75	1.19%
Model 2				2.59 $g\ kg^{-1}$	0.62	−1.32%
Model 3				2.01	0.46	−1.94%
Model 4				1.89	0.47	−2.00%

The dataset in 2011 was used for model calibration and the dataset in 2010 used for model validation.
[a]Root mean square error.
[b]Normalized mean bias.

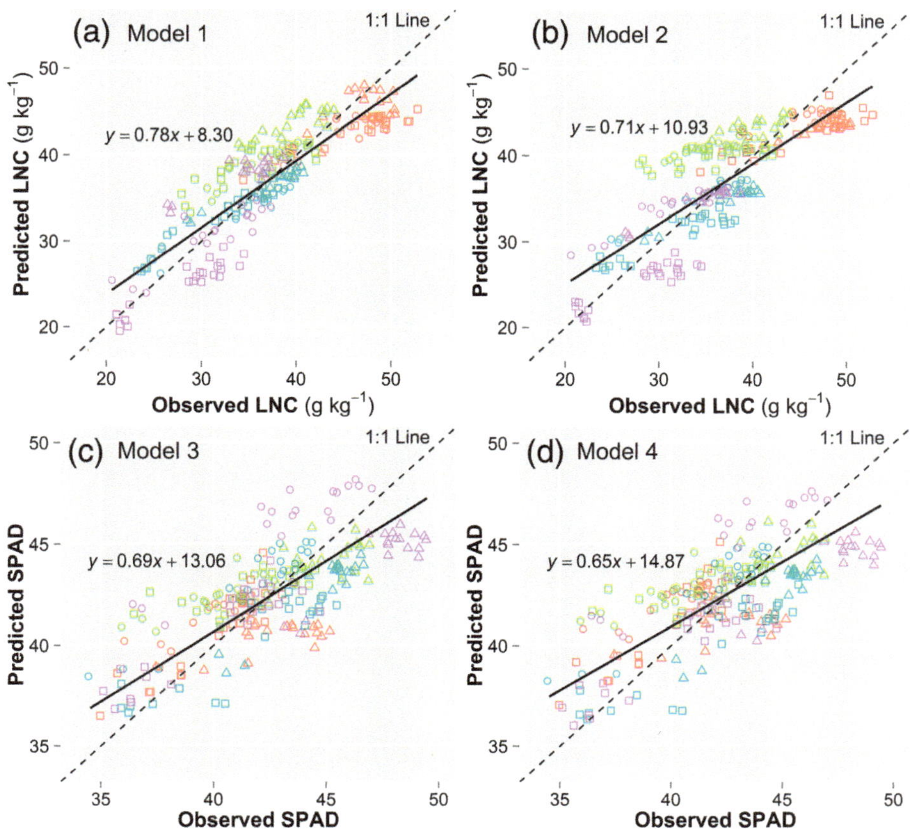

Figure 4 Calibration of the four models in Table 4 for the estimation of leaf nitrogen concentration (LNC, g kg^{-1}) and SPAD. Model 1 **(a)**, $LNC = \alpha SPAD + \beta DAT + \gamma$, model 2 **(b)**, $LNC = \alpha b^* + \beta DAT + \gamma$, model 3 **(c)**, $SPAD = \alpha b^* + \beta DAT + \gamma$, model 4 **(d)**, $SPAD = \alpha b^* + \gamma$. Different colors denote different developmental stages (red: vegetative, green: tillering, blue: jointing, purple: booting). Different symbols denote different cultivars used for model calibration (○ Liangyoupeijiu, Δ Nanjing45, □ Nanjing46).

and RMSE in Figure 5c and d indicated that the relationship between SPAD and color index b^* was not affected by the developmental stage.

In our study, the light conditions during image acquisition were different between 2010 and 2011. Image acquisition in 2011 was under overcast days with low but stable diffused light. In contrast, images were taken under sunny days in 2010 with strong and variable (PAR ranging from 937 to 1759 μmol m^{-2} s^{-1}) light (Table 3). We can remove the effect of illumination change on images when the change is over the whole image by normalizing the image [49] or using a calibration panel [36]. However, there were many other differences caused by the different light conditions. The strong illumination in 2010 caused many white spots and shadows which affected the image color (Additional file 1: Figure S1) while this phenomenon was hardly seen in the images taken in 2011. With the rice growth, the light status within rice canopy (transmittance, reflectance, absorption) becoming more and more complex, the change of illumination will cause different degrees of influence on canopies that in different height (Additional file 2: Figure S2). In this case, it is difficult to calibrate the image color accurately. Sakamoto *et al.* [15] calibrated image indices with the introduction of exposure value (EV) and obtained reliable camera retrieved vegetation indices (VIs). However, this method cannot apply to our study for the calibration of canopy color because of the different influence in one image.

Besides the light conditions, the prediction of crop N status with leaf color may be affected by many other environmental factors, such as developmental stage, diseases and drought stress [50]. These factors may be detectable from high-resolution canopy images [22,23]. Color indices analysis associated with these factors and other image characteristics (e.g. canopy cover, plant shape, leaf texture or even soil status), will provide more reliable results to N diagnosis. Further studies will be devoted to the exploration of image characteristics, leaf color correction and the calibration of the established model with physiological parameters (e.g. SLW) in the evaluation of N status under different environmental conditions.

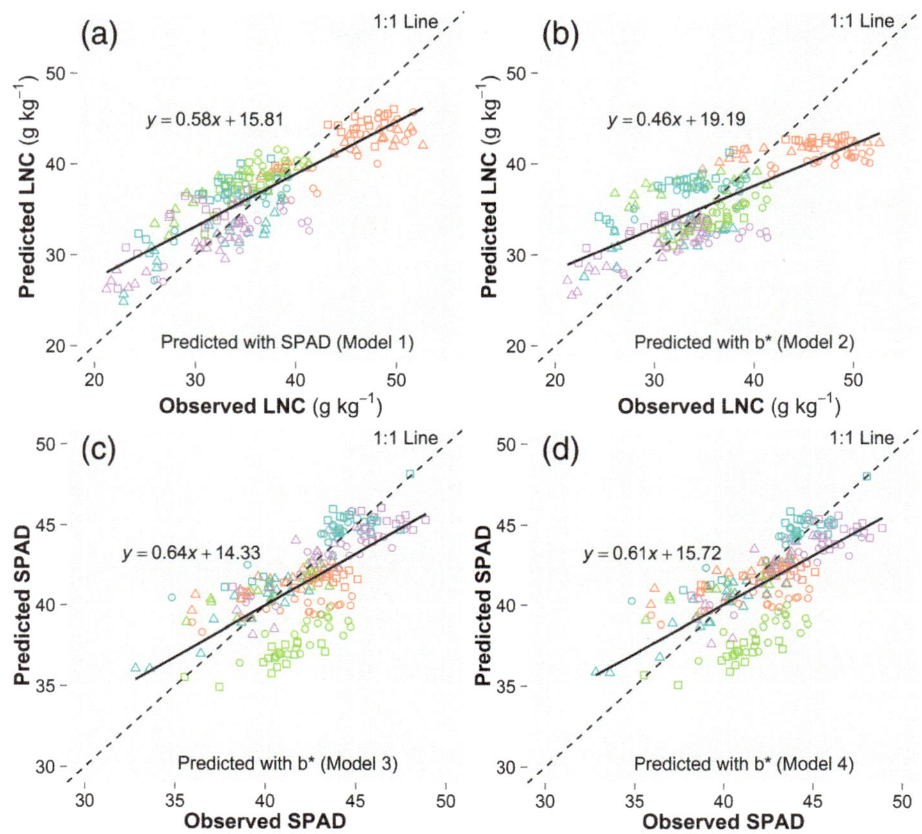

Figure 5 Validation of the four models using the fitted parameters in Table 4 for the estimation of leaf nitrogen concentration (LNC, g kg^{-1}) and SPAD. Model 1 **(a)**, $LNC = \alpha SPAD + \beta DAT + \gamma$, model 2 **(b)**, $LNC = \alpha b^* + \beta DAT + \gamma$, model 3 **(c)**, $SPAD = \alpha b^* + \beta DAT + \gamma$, model 4 **(d)**, $SPAD = \alpha b^* + \gamma$. Different colors denote different developmental stages (red: vegetative, green: tillering, blue: jointing, purple: booting). Different symbols denote different cultivars used for model validation (○ Nanjing44, Δ Wuyunjing24, □ Yangjing48).

Conclusions

Image color indices calculated from RGB, HSV and $L^*a^*b^*$ color models have significant correlations with SPAD readings and leaf N concentration (LNC) of rice leaves. Among these color indices, the index b^*, which represents the visual perception of yellow-blue chroma, had the highest correlation coefficients with SPAD readings and LNC. Regression analysis showed significant linear relationships between index b^* and N parameters. However, the relationship between LNC and SPAD reading, LNC and index b^* were affected by the rice developmental stage. This is mainly caused by the leaf thickness which can be different in cultivars, developmental stages and environmental conditions. In this case, linear regression models were established between color index b^*, LNC and SPAD readings by considering the developmental process in rice. The multiple linear models improved the R^2 of SPAD readings and b^* with LNC, yet most of the predicted LNC in the vegetative stage were underestimated because of the inconsistent relationship between chlorophyll and N concentration. Validations on the models showed good performance and acceptable predicted precision with different cultivars and sampling dates under different natural light conditions. These results indicated that digital color image analysis could be a simple method for assessing rice N status under natural light conditions.

Materials and methods

General information of the experimental site

The experiment was laid out at Changshu Agricultural Ecology Experiment Station, Changshu, Jiangsu, China (31°33′N, 120°42′E). Located in the humid subtropical climate zone, the station receives average annual solar radiation of 4930 MJ m^{-2}, sunshine of 1800 hours, precipitation of 1200 mm and cumulative temperature above 10°C of 4933 degree-days (°C·d). The soil type for the field experimental site is a gleyed paddy soil of the Taihu Lake region, which contains total nitrogen (N) of 1.79 g kg^{-1}, total phosphorus (P) of 0.93 g kg^{-1}, total potassium (K) of 18.7 g kg^{-1}, organic matter of 30.8 g kg^{-1}, alkali-extractable N of 123 mg kg^{-1}, Olsen-P of 13.1 mg kg^{-1}, plant available K of 121 mg kg^{-1} and pH of 7.4 (soil: water, 1:2) in the 0–15 cm soil layer.

Experimental design

Two independent experiments with different N fertilization gradients were implemented in our study. Experiment I was a long-term site-specific rice-wheat rotation experiment that started in 1997. The trial comprised six fertilizer treatments represented as CK, N0, N1, N2, N3, and N4 for N application of 0, 0, 180, 225, 270 and 315 kg N ha^{-1} in rice season, respectively. Each treatment had four replicates which were arranged in a randomized block design. The data used in this paper were from the period May to November in 2010 and 2011 with cultivars Wuyunjing24 and Nanjing46, respectively. Experiment II was carried out in paddy fields with a rice-wheat rotation in 2010 and 2011. Six N application rates with three replicates were designed in this trial, which were represented as N0, N1, N2, N3, N4 and N5 with N application of 0, 120, 180, 240, 270 and 300 kg N ha^{-1}, respectively. The cultivars were Nanjing44 and Yangjing48 in 2010, and Nanjing45 and Liangyoupeijiu in 2011. For both experiments, the N was split into three applications, 40% as basal, 20% at tillering and 40% at booting. In addition, each plot received 90 kg K ha^{-1} and 20 kg P ha^{-1} except the CK treatment in experiment I. The applied K was split into 50% as basal and 50% at booting, and all the P was applied as basal fertilizer. Other crop managements were same as the local traditional practices.

Sample collection and digital image acquisition

For measuring rice growth and nutrition parameters, the above-ground part of rice plant was sampled about every two weeks after transplanting until the booting stage. A total of 4 sets of samples were collected in 2010 and 2011 (Table 3). The plant samples were separated into leaves and stems (including sheaths), and dried at 105°C for half an hour and then at 70°C until constant weight. After that, the samples were weighed for dry weight and analyzed for leaf N concentration (LNC) by the Kjeldahl method [51]. Along with the plant sampling, a chlorophyll meter (SPAD-502, Minolta Camera Co., Osaka, Japan) was used to obtain SPAD values on the four youngest fully expanded leaves. Each blade was measured at three points: on the upper, middle and lower thirds on either side of the midrib. Then, average SPAD readings were calculated for each plot.

On the same day or following day of plant sampling, images of the rice canopy were captured using a digital still color camera (EOS 50D, Canon Inc.) with a resolution of 15 mega pixels. The camera was mounted on a tripod at the nadir position with a constant height of 1 m above the top of the rice canopy. Aperture priority mode was selected, and the camera was set at aperture of f/5.6, ISO of 100, white balance of 4,900 K, auto exposure and auto-focus with the flash turned off. In 2010, the pictures were taken at local time 12:00 – 13:00 in sunny days, while in 2011, the pictures were taken at the same time period but on overcast days. In the days of picture taken (July and August), the deviation between local time and solar noon was within 4 minutes. All the pictures from the experiments were stored in CR2 (Canon raw image file) format. The photosynthetic active radiation (PAR) and illuminance were recorded by a portable light meter (GLZ-C, Top Instrument Co., Zhejiang, P. R. China) during the period of image acquisition. Average PAR was calculated with each set of pictures, and observed 937–1759 μmol photons m^{-2} s^{-1} and 145–692 μmol photons m^{-2} s^{-1} in 2010 and 2011, respectively (Table 3).

Image segmentation and color indices calculation

A raw image file contains minimally processed data from the image sensor of a digital camera. This file saves settings of white balance, color saturation, contrast and sharpness in it, but defers the processing. Therefore, all the modification made on a raw image file is non-destructive.

The canopy images in CR2 format were adjusted for white balance using the 18% gray card (R-27, Kodak) pictures which were taken simultaneously with the canopy images. Then, lens distortion correction was applied, and exposure was set to +1 for all images. After that, images were saved as joint photographic experts group (JPEG) files for further processing. All the procedures above were processed with Adobe Camera Raw (Adobe Systems Inc.).

Since the images contained the rice canopy and some non-canopy elements, such as soil, water and plant residues, images were segmented into canopy portion and non-canopy portion. A computer program was developed based on the G-R thresholding method [25,29] using MatLab® (MathWorks Inc.) to extract the canopy portion of the image. The G-R thresholding method was proposed according to the difference of reflectance spectrum between green vegetation and non-canopy elements in the visible band. There is a reflection peak for green vegetation in the green band, whereas no apparent change for soil or water albedo in the whole visible band. Therefore, the value of green channel minus that of red channel expands the difference between canopy and non-canopy portion.

After the image segmentation, 13 color indices derived from 3 color models were calculated. RGB model is the most common color model for the representation of digital images. A color in the RGB model is described by indicating how much of each of the red, green, and blue is included. The color is expressed as an RGB triplet (R, G, B), with the representation for black of (0, 0, 0) and for the brightest representable white of (255, 255, 255) in an 8-bit image [52]. R, G and B are the mean values of the red, green and blue channels, and r, g and b are the normalized RGB values, respectively. Intensity (INT) is the average of R, G and B. VI_{Green} is a widely used vegetation index [13]. These indices were calculated as follows [27,29]:

$$r = R/(R+G+B) \quad (1)$$

$$g = G/(R+G+B) \quad (2)$$

$$b = B/(R+G+B) \quad (3)$$

$$INT = (R+G+B)/3 \quad (4)$$

$$VI_{Green} = (G-R)/(G+R) \quad (5)$$

In addition, the CIE $L^*a^*b^*$ and HSV color spaces were also tested in this study. The L^* coordinate in CIE $L^*a^*b^*$ [53] closely matches human perception of lightness, a^* and b^* dimensions represent the visual perception of red-green and yellow-blue chroma, respectively. Both a^* and b^* are independent with image lightness (L^*), and take on both negative and positive values ($+a^*$ reds, $-a^*$ greens, $+b^*$ yellows, $-b^*$ blues). The three coordinates of $L^*a^*b^*$ are computed from the tristimulus values X, Y and Z as following equations [32,54]:

$$L^* = 116f(Y/Y_n)-16 \quad (6)$$

$$a^* = 500[f(X/X_n)-f(Y/Y_n)] \quad (7)$$

$$b^* = 200[f(Y/Y_n)-f(Z/Z_n)] \quad (8)$$

$$f(\omega) = \begin{cases} (\omega^{1/3}) & |\omega < 0.008856 \\ 7.787(\omega) + 16/116 & |\omega \geq 0.008856 \end{cases} \quad (9)$$

where X_n, Y_n and Z_n describe a specified white object-color stimulus.

The HSV color space is represented as a cylindrical-coordinate in which the angle around the central vertical axis corresponds to hue (H). The calculation of H was listed below [27,29]:

$$H = \begin{cases} 60 * \{(G-B)/[\max(RGB)-\min(RGB)]\}, \max(RGB) = R \\ 60 * \{2 + \{(B-R)/[\max(RGB)-\min(RGB)]\}\}, \max(RGB) = G \\ 60 * \{4 + \{(R-G)/[\max(RGB)-\min(RGB)]\}\}, \max(RGB) = B \end{cases} \quad (10)$$

Pearson correlation and regression analyses were used to detect the relationship between color indices and crop N status. The significance of linear regressions was evaluated using Student's *t*-test at 95% confidence levels. Significance of ANOVAs was evaluated with the least significant difference test (LSD) at 0.05 probability level. Data analysis and figure production were done using the R v3.0.3 software [55].

Correlation analysis, linear regression analysis and model establishment between color indices and crop N status were based on the data in 2011, and the data in 2010 were used for model validation.

Additional files

Additional file 1: Figure S1. Examples of the "white spots", which are over-exposed areas where the reflected light came into the camera directly.

Additional file 2: Figure S2. Canopy images of Nanjing46 in different developmental stages (a, vegetative; b, tillering; c, jointing; d, booting).

Abbreviations

N: Nitrogen; LNC: Leaf nitrogen concentration; H: Hue from HSV color space; R, G, B: Digital number for the red, green and blue channel of an RGB image; r, g, b: Normalized RGB values; INT: The average of R, G and B.

Competing interests

The authors declare that they have no competing interests.

Authors' contributions

YW and DW designed the study. YW and PS implemented the experiments. YW, PS and KO performed the statistical analysis. YW, DW, PS and KO drafted the manuscript. All authors read and approved the final manuscript.

Acknowledgements

This research was supported by the National Key Technologies R & D Program of China (2012BAJ24B06). The authors acknowledge John Angus of CSIRO Agriculture for his constructive comments and revisions of the manuscript. We wish to thank the editor and two anonymous reviewers for their very helpful comments on earlier drafts.

Author details

[1]Institute of Soil Science, Chinese Academy of Sciences, 71 East Beijing Road, Nanjing 210008, PR China. [2]National Engineering and Technology Center for Information Agriculture, Nanjing Agricultural University, 1 Weigang Road, Nanjing 210095, PR China. [3]Graduate School of Agricultural and Life Sciences, The University of Tokyo, Yayoi 1-1-1, Bunkyo-ku, Tokyo 113-8657, Japan.

References

1. Fageria NK, Baligar VC, Jones CA: *Growth and Mineral Nutrition of Field Crops.* 3rd edition. Boca Raton: CRC Press; 2011.
2. Adhikari C, Bronson KF, Panuallah GM, Regmi AP, Saha PK, Dobermann A, Olk DC, Hobbs PR, Pasuquin E: **On-farm soil N supply and N nutrition in the rice–wheat system of Nepal and Bangladesh.** *Field Crops Res* 1999, **64:**273–286.
3. Prasertsak A, Fukai S: **Nitrogen availability and water stress interaction on rice growth and yield.** *Field Crops Res* 1997, **52:**249–260.
4. Kaushal SS, Groffman PM, Band LE, Elliott EM, Shields CA, Kendall C: **Tracking nonpoint source nitrogen pollution in human-impacted watersheds.** *Environ Sci Technol* 2011, **45:**8225–8232.
5. Zhang F, Cui Z, Fan M, Zhang W, Chen X, Jiang R: **Integrated soil–crop system management: reducing environmental risk while increasing crop productivity and improving nutrient use efficiency in China.** *J Environ Qual* 2011, **40:**1051–1057.
6. Miao YX, Stewart BA, Zhang FS: **Long-term experiments for sustainable nutrient management in China: a review.** *Agron Sustain Dev* 2011, **31:**397–414.
7. Li F, Mistele B, Hu Y, Chen X, Schmidhalter U: **Reflectance estimation of canopy nitrogen content in winter wheat using optimised hyperspectral spectral indices and partial least squares regression.** *Eur J Agron* 2014, **52:**198–209.
8. Chen P, Haboudane D, Tremblay N, Wang J, Vigneault P, Li B: **New spectral indicator assessing the efficiency of crop nitrogen treatment in corn and wheat.** *Remote Sens Environ* 2010, **114:**1987–1997.
9. Peng S, Buresh RJ, Huang J, Zhong X, Zou Y, Yang J, Wang G, Liu Y, Hu R, Tang Q, Cui K, Zhang F, Dobermann A: **Improving nitrogen fertilization in**

rice by site-specific N management: a review. *Agron Sustain Dev* 2010, **30**:649–656.
10. Lin FF, Qiu LF, Deng JS, Shi YY, Chen LS, Wang K: **Investigation of SPAD meter-based indices for estimating rice nitrogen status.** *Comput Electron Agric* 2010, **71**(Supplement 1):S60–S65.
11. Cabangon RJ, Castillo EG, Tuong TP: **Chlorophyll meter-based nitrogen management of rice grown under alternate wetting and drying irrigation.** *Field Crops Res* 2011, **121**:136–146.
12. Reyniers M, Walvoort DJJ, De Baardemaaker J: **A linear model to predict with a multi-spectral radiometer the amount of nitrogen in winter wheat.** *Int J Remote Sens* 2006, **27**:4159–4179.
13. Li Y, Chen D, Walker CN, Angus JF: **Estimating the nitrogen status of crops using a digital camera.** *Field Crops Res* 2010, **118**:221–227.
14. Sakamoto T, Shibayama M, Kimura A, Takada E: **Assessment of digital camera-derived vegetation indices in quantitative monitoring of seasonal rice growth.** *ISPRS J Photogramm Remote Sens* 2011, **66**:872–882.
15. Sakamoto T, Gitelson AA, Nguy-Robertson AL, Arkebauer TJ, Wardlow BD, Suyker AE, Verma SB, Shibayama M: **An alternative method using digital cameras for continuous monitoring of crop status.** *Agric For Meteorol* 2012, **154–155**:113–126.
16. Blackmer TM, Schepers JS: **Use of a chlorophyll meter to monitor nitrogen status and schedule fertigation for corn.** *J Prod Agric* 1995, **8**:56–60.
17. Rorie RL, Purcell LC, Mozaffari M, Karcher DE, King CA, Marsh MC, Longer DE: **Association of "greenness" in corn with yield and leaf nitrogen concentration.** *Agron J* 2011, **103**:529–535.
18. Zhang J, Blackmer AM, Ellsworth JW, Koehler KJ: **Sensitivity of chlorophyll meters for diagnosing nitrogen deficiencies of corn in production agriculture.** *Agron J* 2008, **100**:543–550.
19. Zhang C, Kovacs JM: **The application of small unmanned aerial systems for precision agriculture: a review.** *Precis Agric* 2012, **13**:693–712.
20. Samborski SM, Tremblay N, Fallon E: **Strategies to make use of plant sensors-based diagnostic information for nitrogen recommendations.** *Agron J* 2009, **101**:800–816.
21. Scharf PC, Lory JA: **Calibrating corn color from aerial photographs to predict sidedress nitrogen need.** *Agron J* 2002, **94**:397–404.
22. Berger B, Parent B, Tester M: **High-throughput shoot imaging to study drought responses.** *J Exp Bot* 2010, **61**:3519–3528.
23. Fanourakis D, Briese C, Max JF, Kleinen S, Putz A, Fiorani F, Ulbrich A, Schurr U: **Rapid determination of leaf area and plant height by using light curtain arrays in four species with contrasting shoot architecture.** *Plant Methods* 2014, **10**:9.
24. Golzarian MR, Frick RA, Rajendran K, Berger B, Roy S, Tester M, Lun DS: **Accurate inference of shoot biomass from high-throughput images of cereal plants.** *Plant Methods* 2011, **7**:1–11.
25. Wang Y, Wang D, Zhang G, Wang C: **Digital camera-based image segmentation of rice canopy and diagnosis of nitrogen nutrition.** *Trans Chin Soc Agric Eng* 2012, **28**:131–136.
26. Abramoff MD, Magalhães PJ, Ram SJ: **Image processing with Image J.** *Biophotonics Int* 2004, **11**:36–42.
27. Lee K-J, Lee B-W: **Estimation of rice growth and nitrogen nutrition status using color digital camera image analysis.** *Eur J Agron* 2013, **48**:57–65.
28. Liu J, Pattey E: **Retrieval of leaf area index from top-of-canopy digital photography over agricultural crops.** *Agric For Meteorol* 2010, **150**:1485–1490.
29. Wang Y, Wang D, Zhang G, Wang J: **Estimating nitrogen status of rice using the image segmentation of G-R thresholding method.** *Field Crops Res* 2013, **149**:33–39.
30. Hunt ER Jr, Doraiswamy PC, McMurtrey JE, Daughtry CS, Perry EM, Akhmedov B: **A visible band index for remote sensing leaf chlorophyll content at the canopy scale.** *Int J Appl Earth Obs Geoinf* 2013, **21**:103–112.
31. Pagola M, Ortiz R, Irigoyen I, Bustince H, Barrenechea E, Aparicio-Tejo P, Lamsfus C, Lasa B: **New method to assess barley nitrogen nutrition status based on image colour analysis: comparison with SPAD-502.** *Comput Electron Agric* 2009, **65**:213–218.
32. Wiwart M, Fordoński G, Żuk-Gołaszewska K, Suchowilska E: **Early diagnostics of macronutrient deficiencies in three legume species by color image analysis.** *Comput Electron Agric* 2009, **65**:125–132.
33. Vollmann J, Walter H, Sato T, Schweiger P: **Digital image analysis and chlorophyll metering for phenotyping the effects of nodulation in soybean.** *Comput Electron Agric* 2011, **75**:190–195.
34. Golzarian MR, Frick RA: **Classification of images of wheat, ryegrass and brome grass species at early growth stages using principal component analysis.** *Plant Methods* 2011, **7**:1–11.
35. McCarthy CL, Hancock NH, Raine SR: **Applied machine vision of plants: a review with implications for field deployment in automated farming operations.** *Intell Serv Rob* 2010, **3**:209–217.
36. Noh H, Zhang Q, Han S, Shin B, Reum D: **Dynamic calibration and image segmentation methods for multispectral imaging crop nitrogen deficiency sensors.** *Trans ASAE* 2005, **48**:393–401.
37. Afshari-Jouybari H, Farahnaky A: **Evaluation of Photoshop software potential for food colorimetry.** *J Food Eng* 2011, **106**:170–175.
38. Graeff S, Pfenning J, Claupein W, Liebig HP: **Evaluation of image analysis to determine the N-fertilizer demand of broccoli plants (Brassica oleracea convar. botrytis var. italica).** *Adv Opt Technol* 2008, **2008**:1–8.
39. Xue L, Cao W, Luo W, Dai T, Zhu Y: **Monitoring leaf nitrogen status in rice with canopy spectral reflectance.** *Agron J* 2004, **96**:135–142.
40. Huang J, He F, Cui K, Buresh RJ, Xu B, Gong W, Peng S: **Determination of optimal nitrogen rate for rice varieties using a chlorophyll meter.** *Field Crops Res* 2008, **105**:70–80.
41. Zong S, Lu C, Zhao L, Wang C, Dai Q, Zou J: **Physiological basis of high yield of an inter-subspecific hybrid rice, Liangyoupeijiu.** *J Nanjing Agric Tech Coll* 2000, **16**:8–12.
42. Shibayama M, Watanabe Y: **Estimating the mean leaf inclination angle of wheat canopies using reflected polarized light.** *Plant Prod Sci* 2007, **10**:329–342.
43. Homem Antunes MA, Walter-Shea EA, Mesarch MA: **Test of an extended mathematical approach to calculate maize leaf area index and leaf angle distribution.** *Agric For Meteorol* 2001, **108**:45–53.
44. Chapman SC, Barreto HJ: **Using a chlorophyll meter to estimate specific leaf nitrogen of tropical maize during vegetative growth.** *Agron J* 1997, **89**:557–562.
45. Yang WH, Peng S, Huang J, Sanico AL, Buresh RJ, Witt C: **Using leaf color charts to estimate leaf nitrogen status of rice.** *Agron J* 2003, **95**:212–217.
46. Peng S, García FV, Laza RC, Cassman KG: **Adjustment for specific leaf weight improves chlorophyll meter's estimate of rice leaf nitrogen concentration.** *Agron J* 1993, **85**:987–990.
47. Shukla AK, Ladha JK, Singh V, Dwivedi B, Balasubramanian V, Gupta RK, Sharma S, Singh Y, Pathak H, Pandey P: **Calibrating the leaf color chart for nitrogen management in different genotypes of rice and wheat in a systems perspective.** *Agron J* 2004, **96**:1606–1621.
48. Baret F, Fourty T: **Estimation of leaf water content and specific leaf weight from reflectance and transmittance measurements.** *Agronomie* 1997, **17**:455–464.
49. Drew MS, Wei J, Li ZN: **Illumination–invariant image retrieval and video segmentation.** *Pattern Recognit* 1999, **32**:1369–1388.
50. Pydipati R, Burks TF, Lee WS: **Identification of citrus disease using color texture features and discriminant analysis.** *Comput Electron Agric* 2006, **52**:49–59.
51. Lu RK: *Soil Chemical Nutrient Analysis Techniques.* Beijing: China Agriculture and Technology Press; 2000.
52. Solomon C, Breckon T: *Fundamentals of Digital Image Processing: a Practical Approach with Examples in Matlab.* Chichester: John Wiley & Sons; 2011.
53. Robertson AR: **The CIE 1976 color-difference formulae.** *Color Res Appl* 1977, **2**:7–11.
54. Fairchild MD: *Color Appearance Models.* 2nd edition. Chichester: John Wiley & Sons; 2005.
55. R Core Team: **R: A language and environment for statistical computing.** In Vienna, Austria: R Foundation for Statistical Computing; 2014. http://www.R-project.org/.

14

Use of diffusion magnetic resonance imaging to correlate the developmental changes in grape berry tissue structure with water diffusion patterns

Ryan J Dean[1], Timothy Stait-Gardner[1], Simon J Clarke[2], Suzy Y Rogiers[2,3], Gabriele Bobek[4] and William S Price[1,2,4*]

Abstract

Background: Over the course of grape berry development, the tissues of the berry undergo numerous morphological transformations in response to processes such as water and solute accumulation and cell division, growth and senescence. These transformations are expected to produce changes to the diffusion of water through these tissues detectable using diffusion magnetic resonance imaging (MRI). To assess this non-invasive technique diffusion was examined over the course of grape berry development, and in plant tissues with contrasting oil content.

Results: In this study, the fruit of *Vitis vinfera* L. cv. Semillon at seven different stages of berry development, from four weeks post-anthesis to over-ripe, were imaged using diffusion tensor and transverse relaxation MRI acquisition protocols. Variations in diffusive motion between these stages of development were then linked to known events in the morphological development of the grape berry. Within the inner mesocarp of the berry, preferential directions of diffusion became increasingly apparent as immature berries increased in size and then declined as berries progressed through the ripening and senescence phases. Transverse relaxation images showed radial striation patterns throughout the sub-tissue, initiating at the septum and vascular systems located at the centre of the berry, and terminating at the boundary between the inner and outer mesocarp. This study confirms that these radial patterns are due to bands of cells of alternating width that extend across the inner mesocarp. Preferential directions of diffusion were also noted in young grape seed nucelli prior to their dehydration. These observations point towards a strong association between patterns of diffusion within grape berries and the underlying tissue structures across berry development. A diffusion tensor image of a post-harvest olive demonstrated that the technique is applicable to tissues with high oil content.

Conclusion: This study demonstrates that diffusion MRI is a powerful and information rich technique for probing the internal microstructure of plant tissues. It was shown that macroscopic diffusion anisotropy patterns correlate with the microstructure of the major pericarp tissues of cv. Semillon grape berries, and that changes in grape berry tissue structure during berry development can be observed.

Keywords: Development, Diffusion anisotropy, Diffusion tensor imaging, Grape berry, Nucellus, Nuclear magnetic resonance imaging, Olive, Seeds, Striation patterns, *Vitis vinifera*

* Correspondence: w.price@uws.edu.au
[1]Nanoscale Organisation and Dynamics Group, University of Western Sydney, Penrith NSW 2751, Australia
[2]National Wine & Grape Industry Centre, Charles Sturt University, Locked Bag 588, Wagga Wagga, New South Wales 2678, Australia
Full list of author information is available at the end of the article

Background

Magnetic resonance imaging (MRI) is a powerful tool for investigating biological systems; it is non-invasive, does not use ionising radiation and can draw on a large variety of image contrasts [1,2]. For example, in the field of viticulture, MRI has demonstrated its ability to produce informative proton (^1H) density and transverse magnetic relaxation images of the internal structure of the grape berry [3,4]. Despite the good spatial resolution of the images from the aforementioned studies, it is possible to extract further microstructural (i.e. beyond the resolution limit of an MRI spectrometer) information from a grape berry. This can be achieved using diffusion MRI techniques to observe the distribution of diffusing molecules as they probe the edges of grape berry tissue microstructures. Diffusion MRI has been used to study a variety of different botanical species; for example, maize stems [5,6], barley seeds [7], carrot roots [8], celery stems [9,10] and asparagus stems [11,12].

A recent study has produced diffusion tensor (DT) images (i.e. three-dimensional diffusion images) of the grape berry [13]. While the primary focus of this study was the grape berry vascular system, the orientation of anisotropic diffusion of the berry pericarp was also visible. However, the anisotropy of these observed diffusion patterns did not appear to be related to the known pericarp tissue structure. This is contrary to the commonly shared expectation that anisotropic diffusion should correlate with the orientation of organised tissue structures [14,15].

The objective of the current study was to verify whether diffusion anisotropy features can reflect the relative size, orientation and organisation of the cells that constitute the different tissues of the grape berry pericarp. As previous MRI studies of grape berry tissue structure have focussed on grape berries at single time-points, the second objective of the study was to examine structural changes across grape berry development via diffusion and relaxation MRI. Additional experiments using a postharvest olive as a test sample demonstrated that the diffusion and relaxation MRI techniques are applicable to the study tissue structure of fruits with high oil content.

Since the interpretation of the results of the current investigation require some understanding of relaxation and diffusion MRI, the salient points of both imaging techniques are presented in the following section. More detailed coverage of these topics is available elsewhere [14,16].

Transverse relaxation and diffusion MRI theory

MRI image contrast depends on the pulse sequence used to weight the measured magnetic resonance (MR) signal. The two forms of MR image contrast used in the current study were the rate of transverse magnetic relaxation and translational self-diffusion.

The transverse relaxation (i.e., spin-spin relaxation) of magnetisation relates to the rate at which the transverse component of the nuclear spin magnetisation vanishes after the system has been excited by the application of electromagnetic radio frequency pulses [17]. The relaxation rate is mainly related to the reorientational motions of the molecule (or that part of the molecule containing the nuclear spin) with slowly moving nuclei (i.e. being in a solid or more viscous environment) relaxing more quickly. Unbound paramagnetic ions (e.g. iron and manganese ions), can also increase the rate at which spins relax. Transverse relaxation weighted images are commonly measured using a Carr-Purcell-Meiboom-Gill (CPMG) MRI pulse sequence [18,19].

Translational self-diffusion, on the other hand, is the random, thermal motion of freely moving molecules such as water. This is distinct from mutual diffusion (i.e. molecular motion due to the presence of a chemical potential). For molecules in bulk solution, translational self-diffusion (henceforth referred to as 'diffusion') can be described using the scalar diffusion coefficient (D). For water molecules in bulk solution at standard room temperature, $D \approx 2.29 \times 10^{-9}$ m^2 s^{-1} [20]. The root mean square displacement (RMSD) of a species diffusing in one dimension is given by

$$\sqrt{2D\Delta}, \tag{1}$$

where Δ is the timescale of the measurement [21]. Water molecules in bulk solution at room temperature diffusing over a period of 50 ms have a RMSD of ~15 μm (i.e. the length scale of eukaryotic cell compartments [22]). The pulsed gradient spin-echo (PGSE) diffusion MRI sequence measures the displacement of the diffusing molecules along the direction of the applied magnetic field gradient used in the sequence [14,23,24]. The intensity of the diffusion weighted MR signal is given by;

$$\ln(S/S_0) = \ln(E) = -\gamma^2\delta^2 g^2(\Delta - \delta/3)D \tag{2}$$

where S is the diffusion weighted MR signal intensity, S_0 is the diffusion unweighted MR signal intensity, E is the MR signal attenuation, γ is the gyromagnetic ratio of the nuclei of interest, g is the diffusion magnetic gradient strength, δ is the length of the applied magnetic field gradient pulse, and Δ is the diffusion period. The diffusion weighting of the pulse sequence can be summarised as a scalar factor, b [14,23], where

$$b = \gamma^2\delta^2 g^2(\Delta - \delta/3). \tag{3}$$

When diffusing molecules encounter physical obstructions (e.g. cellular boundaries) their molecular displacement is restricted [14,25]. Consequently, if D is computed from a measured RMSD using Eq. (1) an 'apparent' diffusion coefficient (ADC), which is equal to or smaller than the bulk D, will be determined. The ADC of water measured across a span of cells is dependent on various morphological features, such as cell size and orientation, and the permeability of the cellular membrane [26]. As such, the measured ADC will depend upon the direction of measurement. If molecular mobility is equal in all directions, the apparent diffusion is isotropic and a single ADC is sufficient to characterise diffusion. Conversely, if mobility is not equal in all directions due to the arrangement of microscopic tissue structures, the apparent diffusion is anisotropic (Figure 1). The direction of greatest diffusion mobility (i.e. the direction of least diffusive restriction) can be indicated with a diffusion vector. In the case of anisotropic diffusion, the directional dependence of diffusion can be included in Eq. (2) by replacing D with a diffusion tensor, $\mathbf{D}$ [24,27]. In order to reconstruct $\mathbf{D}$, PGSE measurements along six or more non-collinear directions are performed. This form of MRI is termed diffusion tensor imaging (DTI).

Methods

Grape berry growing conditions

Vitis vinifera L. cv. Semillon (clone DA16162, own roots) berries were obtained from 18 grapevines grown in a glasshouse at the National Wine & Grape Industry Centre, Wagga Wagga, NSW, Australia. The plants were maintained in 35 L pots containing sandy loam. In late winter each plant was pruned to a pair of two-bud spurs and brought into the glasshouse to commence a fifth season of vegetative growth. Average daily maximum/minimum temperature and relative humidity in the glasshouse across the growing season were 30/17°C and 75/40%, respectively. Each plant was watered beyond field capacity four times per day via a pair of drip emitters. Diluted fertiliser (Megamix Plus, Rutec Pty Ltd, Tamworth, Australia, 20 mL per plant) was applied fortnightly to the soil and wettable sulphur/tribasic copper sulphate was sprayed on the shoots. The shoots were trained vertically and pruned to one inflorescence each approximately six weeks after bud burst. The date of bud burst was assessed on each bud [28] and the date of flowering (approximately 100% capfall) was assessed on each inflorescence.

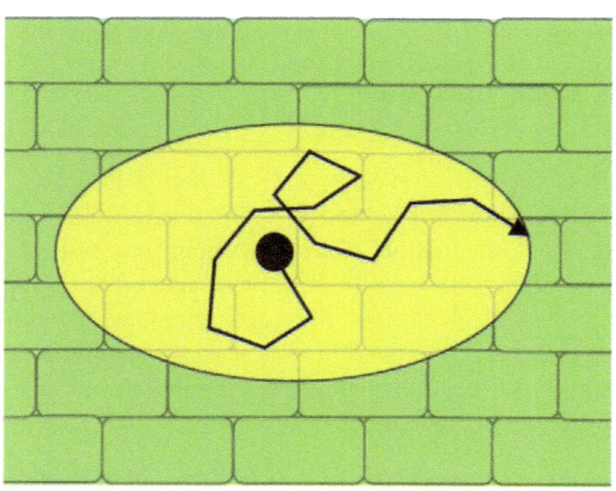

Figure 1 Anisotropic water diffusion due to diffusive restriction (two dimensional representation). Here water molecules (•) encounter cellular boundaries as they diffuse. As there are more cellular boundaries on the longitudinal axis than the lateral axis, water displacement along the longitudinal axis is reduced relative to the lateral axis (represented by the yellow ellipse).

MRI hardware and software

All ^{1}H MRI was performed on a 500 MHz (11.7 T) wide-bore nuclear magnetic resonance spectrometer (AVANCE II; Bruker Biospin Co., Ltd., Germany), equipped with triple axis gradients capable of generating 1.5 T m^{-1} magnetic field gradients. Each imaging experiment employed a 30 mm birdcage radio frequency coil insert. Standard Bruker sequences were used (see Experimental procedures). Data acquisition, post-processing and imaging was controlled from a computer terminal running a Linux operating system, using ParaVision (5.1; Bruker Biospin Co., Ltd.). Image analysis and additional post-processing was performed on a separate computer terminal running Microsoft Windows 7. To analyse the images, the raw MRI data were imported into MATLAB (8.0.0.783, the MathWorks, USA) for analysis using in-house programs.

Experimental procedures

Berries were sampled fortnightly for imaging, beginning four weeks after flowering and ending three months later. To minimise developmental variability at each sampling event, grape bunches were assigned to classes according to flowering date (28, 41, 55, 70, 84, 95 and 109 days after flowering (DAF) respectively). Seven bunches were randomly chosen per sampling event. Each sample consisted of the distal portion of the bunch (approximately 10 berries), which were obtained by cutting the rachis while it was momentarily submersed in tap water. The detached, distal portion of the bunch was wrapped in a moist paper towel, placed in a zip-lock bag and shipped overnight to the Biomedical Magnetic Resonance Facility located at the University of Western Sydney, Campbelltown, NSW, Australia.

Three berries were examined per age class (for a total of twenty-one grape berries) to monitor changes in the diffusion pattern across the berry over the course of its development. Berries of an average size (with respect to the bunch) were cut 3 to 4 mm above the pedicel. All MRI protocols were completed within 11 hours of berry detachment. After imaging, the berry was dried at 70°C and then weighed. All grape bunches were stored at 4°C while not in use. Once selected for imaging, a grape berry had approximately 30 minutes to equilibrate to room temperature, maintained at 22 ± 0.1°C, during the spectrometer setup and calibration stages. The interior temperature of the spectrometer was also maintained at 22 ± 0.1°C. The total soluble solids (°Brix) was estimated for each grape berry imaged, based on the average of three refractormeter (PAL-1, Atago Co., Ltd., Tokyo, Japan) readings obtained from berries of the same bunch as the berry imaged.

Each berry was imaged using the MRI sequences described below. Pulse sequence parameters include the slice thickness (THK); the matrix size (MTX), which indicates the number of voxels (i.e. volume elements) which comprise the image; the repetition time (TR) and the echo time (TE), which can be independently adjusted to alter the magnetisation relaxation weighting of the image; number of echoes (if applicable) and the number of times the signal was averaged (NA). For most MRI experiments, two slices of the berry were imaged consecutively; one slice with a transverse orientation (through the centre of the berry, perpendicular to the central vascular bundles) and a second with a longitudinal orientation (through the centre of the berry, parallel to the central vascular bundles). The field of view (FOV) required adjustment for each berry imaged due to the variation in berry size. As a result, the voxels of each image differ in volume (as indicated in the figure captions), although all voxels had a volume less than 0.02 mm^3. This is in contrast to clinical MRI scanners which have voxel volumes of 1 – 8 mm^3.

A standard multi-shot multi-echo imaging (MSME) sequence (a CPMG based imaging sequence) was employed to produce two-dimensional images weighted by transverse relaxation. The sequence parameters used included THK 1 mm, MTX 256 × 256, a train of 16 echoes spaced 10 ms apart, TR 5 s and NA 2 (total acquisition time ~32 min).

A standard PGSE echo planar DTI sequence [29] was used to produce three-dimensional images weighted by diffusion. The sequence parameters used included THK 1 mm, MTX 128 × 128, δ 1 ms, Δ 25 – 50 ms, TE 35 – 65 ms, TR 6500 – 15000 ms and NA 1 (total acquisition time ~4.5 h). The Δ, TE and TR for the DTI sequences were adjusted for each age class of grape berry due to changes in the transverse and longitudinal relaxation properties of the berries across their development (Table 1). The TE was set in order to maximise the length of Δ, and minimise the loss of MR signal due to transverse relaxation. The TR was adjusted to be five times the length of the maximum longitudinal relaxation time ($5 \times T_1$) of each grape berry in order to minimise unintended longitudinal relaxation weighting. Two diffusion measurements (b weighting of 250 and 500 s mm^{-2}) were performed along forty-two directions (the vertices of a pentakis icosidodecahedron). One S_0 image (which had a negligible diffusion weighting) was also acquired in order to normalise the diffusion weighted images.

Table 1 Summary of the Δ, TE and TR values used for the DTI sequences

Berry age (DAF)	Δ (ms)	TE (ms)	TR (ms)
28	25	40	12500
41	50	65	15000
55	50	60	15000
70	50	60	10000
85	25	35	8000
95	25	35	7000
109	25	35	7000

These values were adjusted for each age class of grape berry due to changes in the transverse and longitudinal relaxation properties of the berries across their development.

For the analysis of the DT and transverse relaxation images, each grape berry was divided into four pericarp tissue groups (Figure 2); the exocarp, outer mesocarp, inner mesocarp and septum [30,31]. The exocarp was defined as the outer epidermis and outer hypodermis tissue of the grape berry. The outer mesocarp was considered to be the tissue between the outer hypodermis and the tissue exterior to the peripheral vascular bundles, while the inner mesocarp was considered to be the tissue inwards from the peripheral vascular bundles. The septum was defined as the irregular tissue found at the centre of the grape berry adjacent to the berry seed(s) and locule(s). The endocarp was not considered in the analysis because it was beyond the resolution of the relaxation images. The seed interior was considered separately from the listed pericarp tissues. Each region in the images was analysed independently from all other regions by the application of image masks. These image masks were based on the tissue groups visible in the

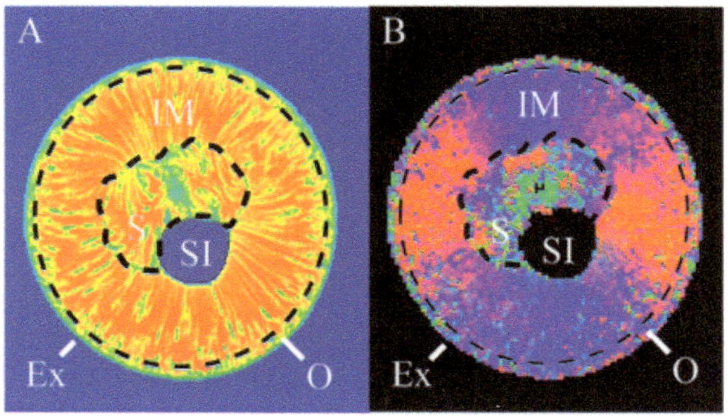

Figure 2 Tissue regions of the grape berry (transverse plane). Here the five tissue regions of the grape berry are provided with reference to a transverse relaxation image **(A)** and a diffusion tensor image **(B)**. Ex: exocarp, OM: outer mesocarp, IM: inner mesocarp, S: septum, SI: seed interior. The outer, black dashed curve indicates the border between the outer mesocarp and the inner mesocarp while the inner, black dashed curve indicates the border between the inner mesocarp and the septum.

transverse relaxation images (in the manner explained previously).

For the analysis of the relaxation images, a mean T_2 (i.e., spin-spin relaxation time) was determined for each of the pericarp tissues, from each grape berry in the same age group. A standard error was also calculated for each tissue based on the mean T_2 values across each respective age group. The acquired DT image data, on the other hand, was used to create three images of each grape berry; a diffusivity map (a map of the average ADC for each voxel), a diffusion vector field map and a diffusion colour map (both of which indicate the direction of least restricted diffusion). The DT colour maps used a symmetrical additive red/blue/green colour scheme to represent that the direction of least restricted diffusion in each voxel was lateral/longitudinal/perpendicular with respect to the FOV. Mean diffusivity values and accompanying standard errors were calculated from the diffusivity maps on a per tissue basis in the same manner as the transverse relaxation maps.

A high spatial resolution DT image of a grape berry 55 DAF was used to confirm whether the radial striation patterns visible in the relaxation images were linked to variations in cellular size across the inner mesocarp tissue. This was done by using the S_0 image (which has a small T_2 weighting, and had visible striation patterns in the inner mesocarp region) to create an MR signal intensity threshold mask. The threshold value was qualitatively chosen in order to divide the inner mesocarp into radial bands of voxels with 'relatively high' and 'relatively low' signal intensity. The image mask was then applied to the DT image data to create two populations of voxels. The three eigenvalues of the DT (i.e. three orthogonal ADCs, the largest of which corresponds to the direction of least diffusion restriction) for each voxel were averaged over the two populations. This resulted in two groups of primary, secondary and tertiary eigenvalues. The mean primary eigenvalue was proportional to the average length of the cells in the voxel, while the mean secondary and tertiary eigenvalues corresponded to either the average width or depth of the cells. The Tukey-Kramer test (P =0.05) was then performed to determine whether differences between corresponding mean eigenvalues were statistically significant. A statistically significant difference between the values would indicate that there is a significant difference in average cell size across the bands of the inner mesocarp striation patterns.

A DT and transverse relaxation image of a postharvest olive (*Olea europaea* L., cv. Correggiolla) was also acquired in order to determine the applicability of diffusion MRI to fruits with high oil contents. The olive was stored and imaged under the same conditions as the grape berries. The transverse relaxation images were obtained using the MSME pulse sequence and parameters outlined above. The DT images, on the other hand, were acquired using a pulsed gradient stimulated echo (PGSTE) pulse sequence. The sequence parameters used included THK 1 mm, MTX 64 × 64, δ 6 ms, Δ 80 ms, TE 20 ms, TR 3500 ms and NA 4 (total acquisition time ~10 h). Two diffusion measurements (b weighting of 80000 and 160000 s mm^{-2}) were performed along twenty directions (the vertices of a dodecahedron). One S_0 image was also acquired in order to normalise the diffusion weighted images. For the analysis of the transverse relaxation and DT images of the olive, the pericarp was treated as a single tissue. This was due to the larger voxel size of the images and lack of distinctive tissue groups.

Results

The grape berries increased in size and weight and then declined as the berries progressed through the ripening and senescence phases (Figure 3). The post-harvest olive had a fresh weight of 3.30 g and dry weight of 1.63 g. The concentration of soluble solids in the grape berries increased sigmoidally with respect to time (adjusted R^2 = 0.99). Véraison occurred at approximately 60 – 65 DAF, and full ripeness was placed at 95 DAF (based on the mean concentration of soluble solids of the sampled berries which plateaued at 26 °Brix).

In the transverse relaxation images, the exocarp was difficult to distinguish from the outer mesocarp (Figures 4 and 5). The exocarp was consistently associated with the shortest mean T_2 at all sampled stages

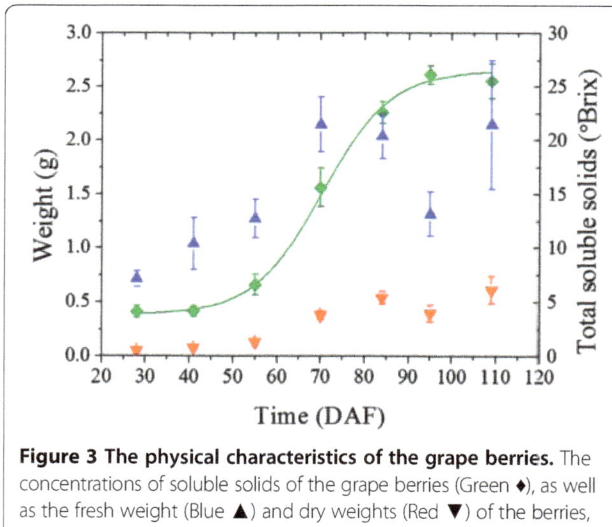

Figure 3 The physical characteristics of the grape berries. The concentrations of soluble solids of the grape berries (Green ♦), as well as the fresh weight (Blue ▲) and dry weights (Red ▼) of the berries, are presented with respect to the number of days after flowering. A sigmoidal function (solid green curve) of the form $a_1 + (a_2 - a_1)/(1 + \exp(-(DAF - x_0)/w))$ was fitted to the soluble solids values by nonlinear regression (adjusted R^2 = 0.99), where a_1 = 26.1 (the approximate maximum soluble solids value), a_2 = 3.9 (the approximate minimum soluble solids value), x_0 = 69.7 (the inflection point) and w =7.4 (the change in DAF which yielded the greatest change in the soluble solids value). The error bars are given by the standard deviation of soluble solids values at each time point.

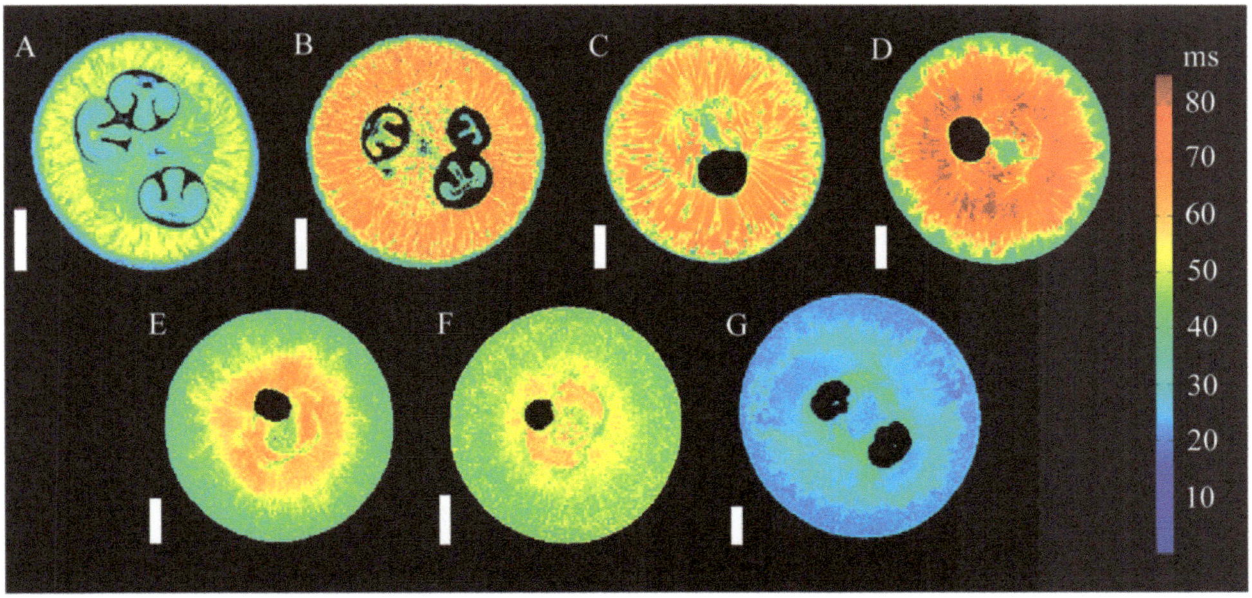

Figure 4 Transverse relaxation images of grape berries at seven different stages of berry development (transverse plane). The images include three pre-véraison grapes, at 28 DAF (**A**, voxel size 59 × 59 × 1000 μm), 41 DAF (**B**, voxel size 78 × 78 × 1000 μm) and 55 DAF (**C**, voxel size 78 × 78 × 1000 μm), a grape undergoing véraison at 70 DAF (**D**, voxel size 82 × 82 × 1000 μm), a ripening grape at 85 DAF (**E**, voxel size 74 × 74 × 1000 μm), a grape which is at oenological maturity at 95 DAF (**F**, voxel size 63 × 63× 1000 μm) and a post-maturity berry at 109 DAF (**G**, voxel size 86 × 86 × 1000 μm). The transverse relaxation values are indicated by the colour bar to the right of the figure. Scale bar: 3 mm.

of grape berry development, relative to the other tissues examined (Figure 6). The diffusion vectors of the exocarp were prominently aligned tangential to the berry surface (Figures 7, 8 and 9). This pattern was consistent for the exocarp of all grape berries imaged, regardless of berry age. The exocarp was also found to have short mean diffusivity values relative to the other tissues (Figure 10).

The transverse relaxation images of the outer mesocarp generally lacked consistent or well defined features (Figures 4 and 5). The mean T_2 of the outer-mesocarp

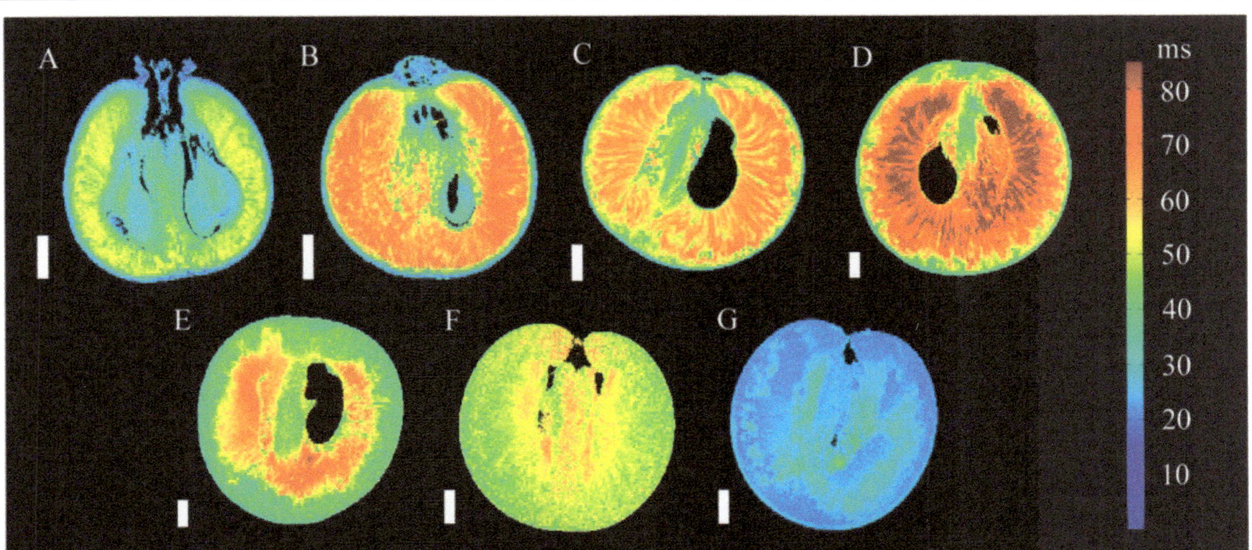

Figure 5 Transverse relaxation images of grape berries at seven different stages of berry development (longitudinal plane). The images include three pre-véraison grapes, at 28 DAF (**A**, voxel size 59 × 59 × 1000 μm), 41 DAF (**B**, voxel size 78 × 78 × 1000 μm) and 55 DAF (**C**, voxel size 78 × 78 × 1000 μm), a grape undergoing véraison at 70 DAF (**D**, voxel size 133 × 133 × 1000 μm), a ripening grape at 85 DAF (**E**, voxel size 82 × 82 × 1000 μm), a grape which is at oenological maturity at 95 DAF (**F**, voxel size 63 × 63 × 1000 μm) and a post-maturity berry at 109 DAF (**G**, voxel size 86 × 86 × 1000 μm). The transverse relaxation values are indicated by the colour bar to the right of the figure. Scale bar: 3 mm.

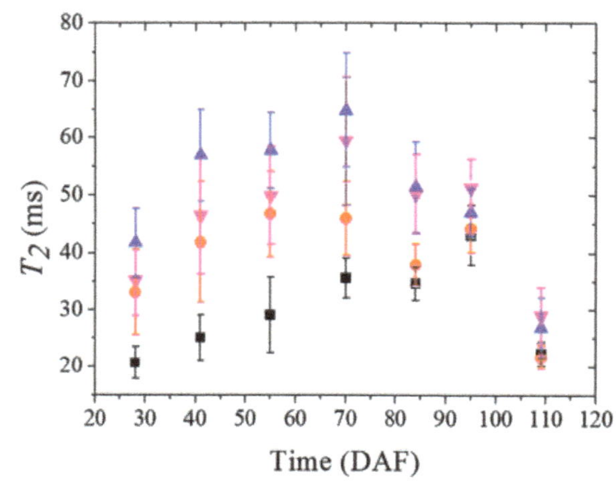

Figure 6 Transverse (T_2) relaxation of the major tissue groups of the grape berry with respect to time (transverse plane). The transverse relaxation values for the exocarp (■), outer mesocarp (Red ●), inner mesocarp (Blue ▲) and septum (Pink ▼). The error bars are given by the standard deviation of the transverse relaxation values at each time point.

tended to be longer than those of the exocarp, but shorter than other tissues (Figure 6). The diffusion vectors had a rotational dependency (Figures 7, 8 and 9) and were aligned so as to radiate from the centre of the fruit. In pre-véraison berries, the diffusivity values of the outer mesocarp were amongst the highest observed. For post-véraison grape berries, however, the mean diffusivity values declined further than other tissues (Figure 10).

A notable feature in the transverse relaxation images of the inner mesocarp was the clear radial striation patterns with distinctly different (Tukey-Kramer test, P =0.05) T_2 values. These patterns were observed in both transverse and longitudinal image orientations for the inner mesocarp (Figures 4 and 5) of grape berries between 28 DAF and 109 DAF. The striation pattern radiated throughout the entire sub-tissue, starting close to the septum and vascular systems at the centre of the berry (i.e., at the ovular and axial vascular network), and terminating at the interface between the inner and outer mesocarp. In addition, upon analysing the high resolution DT image of a grape berry 55 DAF (Figure 11), a statistically significant difference (Tukey-Kramer test, P =0.05) was noted between the mean secondary and tertiary eigenvalues across the striation bands (Table 2). Prior to 95 DAF, the mean T_2 of the inner mesocarp was consistently higher than other tissues (Figure 6). After the concentration of soluble solids plateaued, the mean T_2 of this tissue declined, accompanied by a partial loss of its radial striation pattern. The diffusion vectors of the inner mesocarp had a rotational dependency similar to that of the outer mesocarp. Furthermore, the diffusion vectors were predominantly parallel to the radial striation bands (Figures 7, 8 and 9), except for grape berries 28 DAF or past 95 DAF (Figure 12). After véraison, the

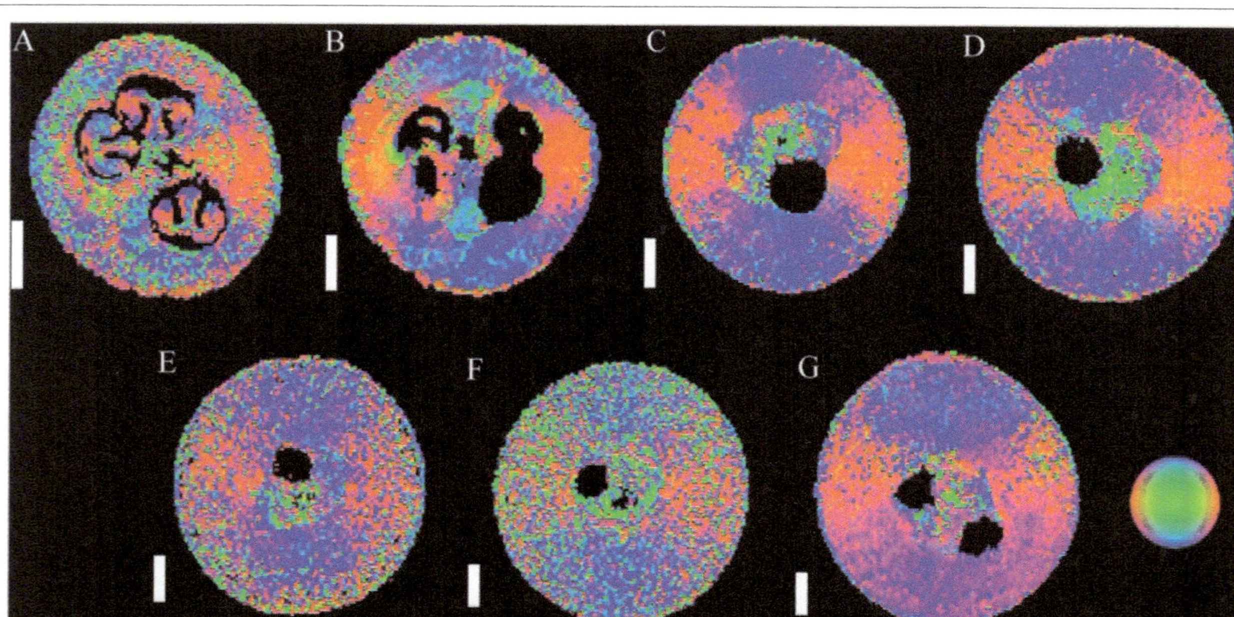

Figure 7 DT images of grape berries at seven different stages of berry development (transverse plane). The images include three pre-véraison grapes, at 28 DAF (**A**, voxel size 117 × 117 × 1000 μm), 41 DAF (**B**, voxel size 156 × 156 × 1000 μm) and 55 DAF (**C**, voxel size 156 × 156 × 1000 μm), a grape undergoing véraison at 70 DAF (**D**, voxel size 164 × 164 × 1000 μm), a ripening grape at 85 DAF (**E**, voxel size 148 × 148 × 1000 μm), a grape at oenological maturity at 95 DAF (**F**, voxel size 125 × 125 × 1000 μm) and a post-maturity berry, at 109 DAF (**G**, voxel size 171 × 171 × 1000 μm). The colours in the figure indicate the direction of least restricted diffusion, as indicated by the image at the bottom right of the figure. Scale bar: 3 mm.

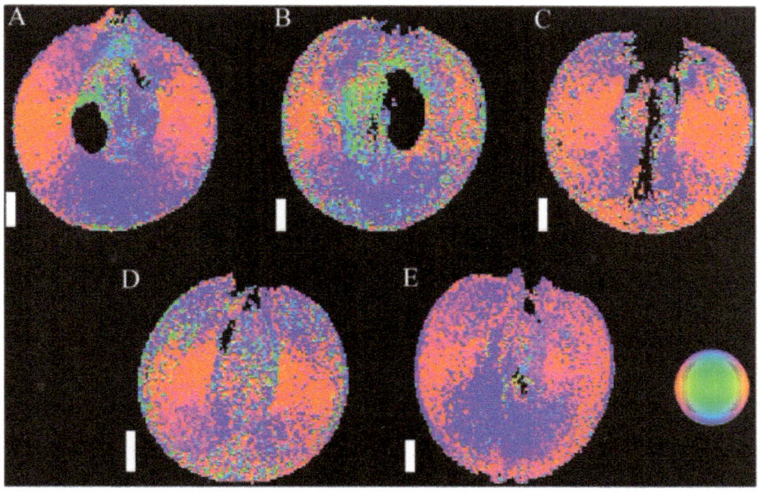

Figure 8 DT images of grape berries at five different stages of berry development (longitudinal plane). The images include a pre-véraison grape at 55 DAF (**A**, voxel size 156 × 156 × 1000 μm), a grape undergoing véraison at 70 DAF (**B**, voxel size 164 × 164 × 1000 μm), a ripening grape at 85 DAF (**C**, voxel size 172 × 172 × 1000 μm), a grape which is at oenological maturity at 95 DAF (**D**, voxel size 125 × 125 × 1000 μm) and a post-maturity berry at 109 DAF (**E**, voxel size 172 × 172 × 1000 μm). No images are available for 28 and 41 DAF. The colours in the figure indicate the direction of least restricted diffusion, as indicated by the image in the bottom right side of the figure. Images are not available for 28 and 41 DAF. Scale bar: 3 mm.

inner mesocarp was also consistently associated with the largest mean diffusivity values, relative to the other tissues examined (Figure 10).

The septum was readily distinguishable from the surrounding mesocarp tissue in the transverse relaxation images. This was due to the presence of curved striation patterns which were perpendicular to the striation patterns of the inner mesocarp (Figures 4 and 5). These striation patterns were curled about the central vascular bundles and extended through the septum to the seeds and locules. Prior to 95 DAF, the mean T_2 of the septum was consistently higher than the exocarp, but lower than the inner mesocarp. After 95 DAF, the septum had the highest mean T_2 (Figure 6). Unlike the inner mesocarp, the diffusion vectors of the septum were not consistently parallel to the striation patterns of the tissue (Figures 7, 8 and 9). The anisotropic diffusion pattern of the septum differed on a per berry basis. The septal-mesocarp boundary could be discerned by abrupt changes in the local orientation of diffusion vectors. The mean diffusivity value of the septum was the lowest of all the examined tissues pre-véraison, but had a mean diffusivity greater than that of the exocarp and outer mesocarp post-véraison (Figure 10).

In the grape seed interior, the nucellus (located towards the bulbous distal end of the seed) was identified in grape berries aged 28 and 41 DAF. It could be readily

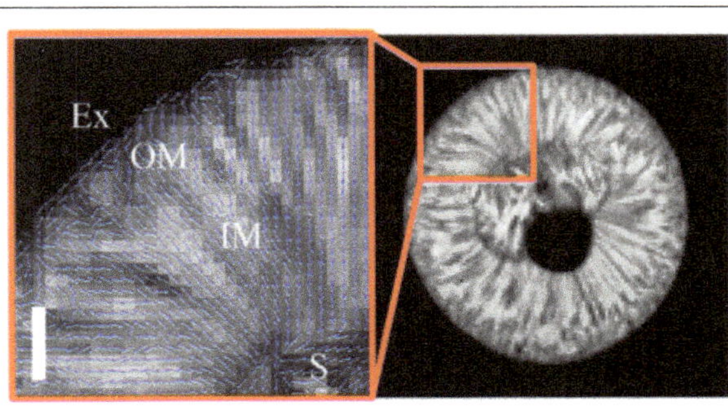

Figure 9 Diffusion vector field map overlaying the S_0 image of a grape berry 41 DAF (transverse plane). The diffusion vectors (blue arrows) indicate the direction of least restricted diffusion in each voxel. Voxel size: 156 × 156 × 1000 μm, bar length: 1000 μm.

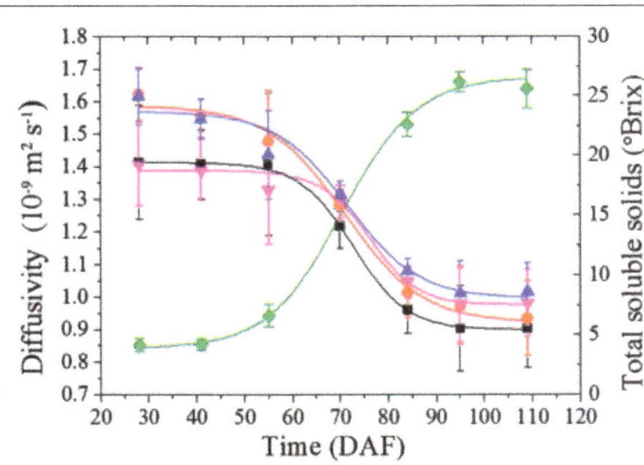

Figure 10 Mean ADC of the major tissue groups of the grape berry with respect to time and total soluble solids. The mean ADC for the exocarp (■), outer mesocarp (Red ●), inner mesocarp (Blue ▲) and septum (Pink ▼) decrease sigmoidally (adjusted $R^2 = 0.99$) with respect to sigmoidally increasing (adjusted $R^2 = 0.99$) dissolved solids content (Green ♦). The error bars are given by the standard deviation of the mean ADC at each time point.

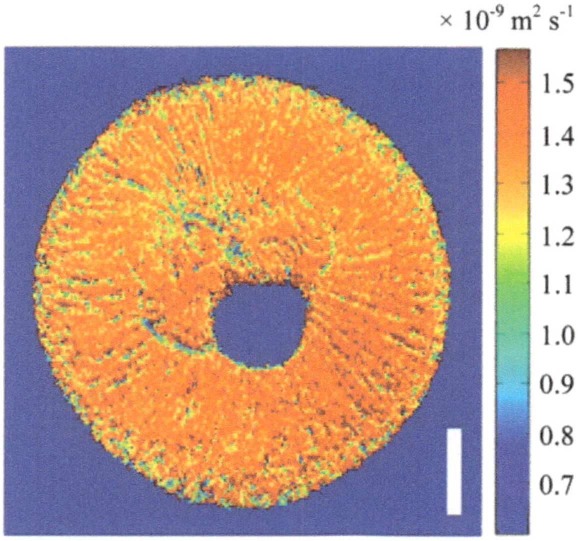

Figure 11 Mean ADC map of a grape berry 55 DAF (transverse plane). Voxel size 78 × 78 × 1000 μm, bar length: 3000 μm.

identified in berries of this age, as it occupies a sizeable portion of the seed and has a stylised '3' shape when imaged through the transverse plane [4]. The nucellus demonstrated a highly characteristic anisotropic diffusion pattern which was rotationally dependent, similar to the anisotropic diffusion pattern of the inner mesocarp tissue (Figures 7, 8 and 13). After 41 DAF, the MR signal from this region decayed too rapidly to be visible in the DT and transverse relaxation images.

The transverse relaxation images of the postharvest olive pericarp lacked well-defined macroscopic features, such as the radial striation patterns noted in the mesocarp of the grape berries. However, the anisotropic diffusion patterns (Figure 14A) and diffusion vectors (Figure 14B) associated with the olive pericarp exhibited a rotational dependency and were aligned so as to radiate from the centre of the fruit. The mean T_2 of the olive pericarp was 45 ± 3 ms while the mean diffusivity value of the olive pericarp was $6.07 \pm 0.46 \times 10^{-12}$ m^2 s^{-1}.

Discussion

The anisotropic diffusion patterns observed in the grape berry pericarp were due to the restricting effects of cell membranes on diffusion. For example, in the mesocarp of grape berries 28 DAF, the diffusion anisotropy exhibited low coherence. This was because the parenchyma cells of the mesocarp were not fully elongated [32], as demonstrated by confocal microscopy (Figure 15). Between 41 DAF and 95 DAF, however, the anisotropic diffusion pattern was clearly radially dependent, thus reflecting the radial orientations of the elongated inner mesocarp cells [3,30,32], as shown by confocal microscopy (Figure 16). The orientations of the diffusion vectors in the inner mesocarp were also preferentially parallel to the radial striation patterns visible in the transverse relaxation images, a relationship that will be discussed further below.

Upon approaching full ripeness, regions of the berry inner mesocarp exhibited a loss of coherence in orientation of diffusion anisotropy (Figure 12). This was accompanied by a decline in T_2 values and partial loss of the radial striation pattern in the transverse relaxation images of these same regions. The decline of T_2 values would begin at the interface between the outer and inner mesocarp, and would steadily shift toward the centre of berry as the grape approached the fully ripened stage. Between 85 and 109 DAF, the T_2 values for all berry tissues noticeably decreased. An increase in cellular fluid viscosity could reduce the rotational velocity of spin bearing molecules in the berry; however, there is no straightforward relationship between the observed transverse relaxation rate and the measured concentration of soluble solids in the final stages of berry ripening. It is also doubtful that the increased transverse relaxation rate could be caused by a sudden influx of free paramagnetic ions. Due to the small width of the voxels constituting the T_2 images (i.e. 60 – 80 μm) the read gradients of the T_2 MR imaging sequence may have caused unaccounted diffusion-weighted signal relaxation [33]. However, this seems unlikely, as T_2 MR images have previously been taken of a geranium leaf petiole with a voxel width of 39 μm [34]. Despite the small voxel width, these images exhibited a strong contrast between the primary tissues of the petiole. The maximum observed T_2 value of the petiole was also less than 85 ms, similar to the maximum observed T_2 values of the grape berries. It is possible that the observed loss of diffusion vector coherence and the decrease in T_2 values is linked to widespread cell death occurring throughout the berry

Table 2 Comparing the mean eigenvalues of diffusion tensors associated with different relaxation striation bands

	Mean primary eigenvalue ($\times 10^{-9}$ m^2 s^{-1})	Mean secondary eigenvalue ($\times 10^{-9}$ m^2 s^{-1})	Mean tertiary eigenvalue ($\times 10^{-9}$ m^2 s^{-1})
Striation bands with high signal intensity	1.53*	1.44	1.36
Striation bands with low signal intensity	1.54*	1.41	1.30

The standard error was not included as it was insignificant (<1% of the values listed). * : values between which there was no statistically significant difference (Tukey-Kramer test, P =0.05).

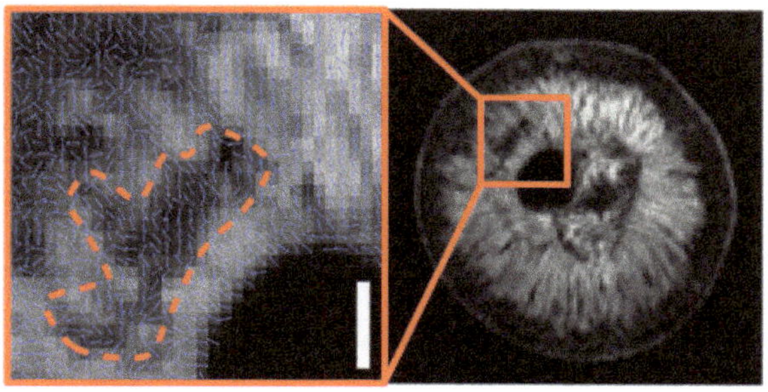

Figure 12 Diffusion vector field map overlaying the S_0 image of a grape berry 109 DAF (transverse plane). The orientation of diffusion vectors are indicated by the blue arrows. There was a loss of diffusion-weighted signal in the region denoted by the red dashed line. Voxel size: 133 × 133 × 1000 μm, bar length: 1000 μm.

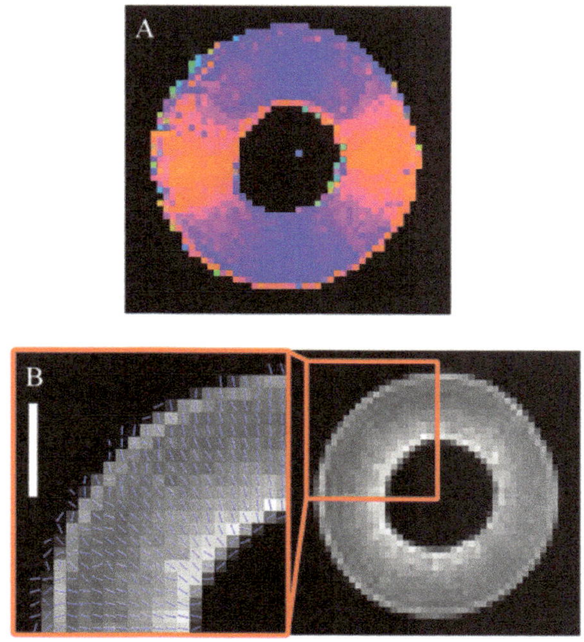

Figure 14 DT and transverse relaxation images of a postharvest olive (transverse plane). The DT image (**A**, voxel size 390 × 390 × 1000 μm) and diffusion vector map (**B**, voxel size 390 × 390 × 1000 μm) of an olive pericarp. Bar length 1000 μm.

mesocarp. This phenomenon is known to occur across the mesocarp of certain wine grape berries during the later stages of ripening [35,36]. Although Semillon grape berries were not included in these studies, for the varieties tested cellular vitality began to decrease at a point between 85 and 109 DAF.

The radial striation patterns noted in the grape berry mesocarp were also noted in a previous investigation [3]. Bright field light micrographs of the berry inner mesocarp indicated that these patterns arise in the MR images due to the arrangement of alternately sized radially elongated parenchyma cells in the grape berry mesocarp tissue. The results of the current study agree with this theory. There was no statistically significant difference between mean primary eigenvalues from striation bands with 'relatively high' and 'relatively low' MR signal intensity, thus indicating that the cells belonging to both populations have the same mean length. However, there was a statistically significant difference between the mean secondary and tertiary eigenvalues of the two populations. The 'relatively low' MR signal intensity population was associated with the lower eigenvalues, indicating that the cells belonging to this population were less wide than their counterparts in the 'relatively high' MR signal intensity population. As the parenchyma cells are radially elongated in grape berries older than 28 DAF, the transverse relaxation striation patterns thus aligned with the anisotropic diffusion patterns in the mesocarp of these berries. While the transverse relaxation images of the septum demonstrated striation patterns similar to those found in the inner mesocarp, the orientation of anisotropic diffusion within the septum was largely not aligned to these septal striation bands. The lack of correlation between the septal striation patterns and diffusion anisotropy is

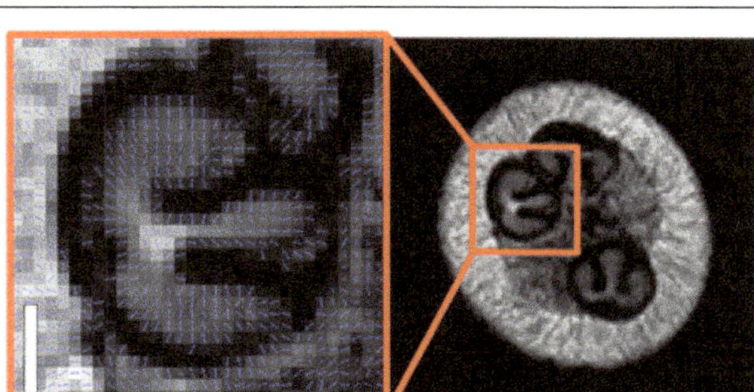

Figure 13 Diffusion vector field map overlaying the S_0 image of a grape berry seed interior 28 DAF (transverse plane). The diffusion vectors (blue arrows) indicate the direction of least restricted diffusion. Voxel size: 117 × 117 × 1000 μm, bar length: 1000 μm.

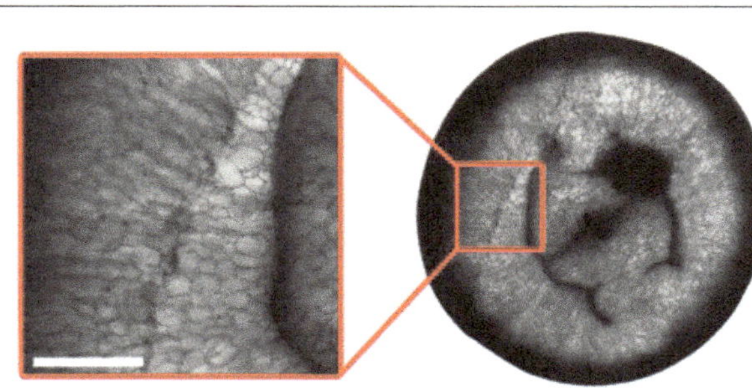

Figure 15 Confocal micrograph of the pericarp of a grape berry prior to véraison 41 DAF (transverse plane). The image was acquired using a confocal microscope (LSM5 Pascal; Zeiss, Germany) which employed a 488 nm Argon laser and a 10 × objective Plan-Apochromatic lens. Bar length 1000 μm.

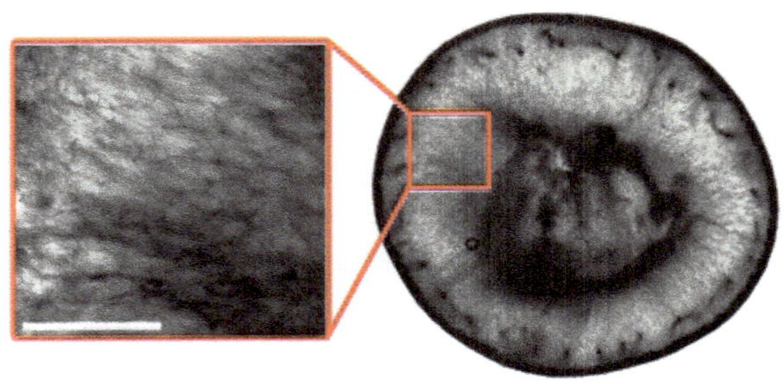

Figure 16 Confocal micrograph of the pericarp of a grape berry undergoing véraison 55 DAF (transverse plane). The image was acquired using a confocal microscope (LSM5 Pascal; Zeiss, Germany) which employed a 488 nm Argon laser and a 10 × objective Plan-Apochromatic lens. Bar length 1000 μm.

likely due to the irregular size and shape of the septal cells [30].

The rotationally dependent anisotropic diffusion patterns noted in the outer edge of young seed nucelli are due to the diffusion of water through the seed integument. These cells are known to be tabular in shape and are radially arranged [37]. However, after 41 DAF, there was a decline in the measurable MR signal from the integument tissue due to tissue dehydration and mechanical hardening. Concurrently, the seed nucellus deteriorated and was replaced by the liquid endosperm [38,39]. The liquid endosperm has a transverse relaxation significantly less than 30 ms (~10 ms). As a result, after 41 DAF the MR signal from the seed interior decayed too rapidly for the imaging protocols used in the current study.

Comparing the results of the study to the literature

The correlation observed in the present study between anisotropic diffusion patterns and the known structure of grape berry pericarp tissues was in contrast to the recent DTI study of grape berries by Gruwel et al., [13]. While the experimental procedures employed in both studies were largely similar, they differed in two important respects. First, the TE of the PGSE MRI pulse sequence used in the previous study was shorter than those used in the current study. TE limits the maximum length of Δ, so although the length of Δ was not specified in the study by Gruwel et al., it can be inferred from the TE used (26 ms) that it was less than the shortest Δ employed in the current study. This difference is noteworthy because as Δ increases, the water molecules have more time to displace further from their origin and interact with different physical structures. The cells of the berry vasculature are far smaller than the cells that make up the mesocarp, thus a short Δ relative to that used in the current study remains appropriate for probing the cellular membranes of the xylem and phloem. However, in regions where the average cell size is much larger, (e.g. the inner mesocarp) the diffusion anisotropy patterns are more likely to be observed due to interactions with cellular components rather than the cellular membrane. As a result, the observations from both the current study and that of Gruwel et al. could be considered accurate; however, they are observations of the effects of different restricting structures (i.e., on a shorter length scale).

The second difference between the present study and that of Gruwel et al. was the diffusion gradient schemes employed. Gruwel et al. performed diffusion measurements along six unique diffusion gradient vectors. While this is the minimum required for DTI, it is possible that gradient cross-terms (from the imaging and diffusion gradients) have affected the diffusion weighting of the images. The current study used forty-two different diffusion gradient vectors (i.e., twenty-one unique directions and their corresponding opposite). The inclusion of gradient vectors and their corresponding opposite gradient vectors reduces the effects of potential gradient cross-terms [27,40]. Using a large number of unique diffusion gradient vectors also reduces potential directional bias when reconstructing **D**.

The applicability of diffusion MRI to plant tissues differing in oil and water content

The question remains as to the applicability of diffusion MRI in the examination of other plant species and organs. This will chiefly depend upon the amount of unbound water the organ contains. Tissues with low content, such as the solid endosperm of seeds, are difficult to examine. In addition, tissues with smaller cells are easier to examine than tissues with larger cells. However, providing that Δ is long enough to ensure that a large population of water molecules are able to interact with the surface of the cellular boundaries (e.g. $\Delta \sim d^2/D$ where d is the distance between cellular boundaries), diffusion anisotropy will be evident.

The applicability of diffusion MRI to the study of plant species with high oil contents is of great interest; oil is hydrophobic and thus diffusing water molecules are likely to treat any droplets of oil encountered in cell cytoplasm [41] as diffusion restricting obstacles. Depending on the concentration of oil in the cytoplasm, water diffusivity could be noticeably affected. Olives are an ideal candidate for such a study as they are comparable in size to grape berries and also have a similar internal tissue structure and cellular dimensions. These features would aid comparisons to the results of the current study. To determine the practicality of such an investigation, a postharvest olive was imaged using both DT and transverse relaxation MRI.

Due a noticeable loss of radio-frequency magnetic pulse power, fast diffusion MRI sequences could not be used, increasing the length of the total image acquisition. This was a consequence of the high salt content of the fruit. The low mean diffusivity value of olive pericarp also necessitated the use of very high diffusion magnetic gradients to perform the diffusion measurements. A lengthy diffusion period was also required to ensure that the slowly diffusing molecules had sufficient time to probe the restricting tissue microstructure. To prevent prohibitive MR signal loss arising from transverse relaxation, a PGSTE MRI sequence was employed. While the two pulse sequences measure diffusion in the same way, the PGSTE sequence halves the intensity of the acquired MR signal. The number of image averages and the size of the image voxel were thus increased to compensate for this loss of signal.

The resultant diffusion MRI images had voxel volume less than 1 mm^3 and a good signal-to-noise ratio. The radial anisotropic diffusion pattern observed in the olive pericarp was similar to the grape berries close to or past véraison, indicating that the cells of the mesocarp are likewise radially elongated. This result agrees with what is known of the olive mesocarp structure [42]. The mean T_2 value of the olive was comparable to those of the grape berry pericarp tissues, as was expected for fruits with similar cellular dimensions. The large difference between the mean ADC of the olive pericarp and the grape berry pericarp tissues is likely due to the presence of oil in the cell cytoplasm. However a more in-depth investigation is required to speculate further.

Conclusion

The results of the current study have demonstrated that diffusion anisotropy patterns are correlated with the microstructure of the major pericarp tissues of cv. Semillon grape berries, including the exocarp, outer and inner mesocarp, and seed interior. Changes in grape berry tissue structure during berry development, which were observable in the DT and transverse relaxation images, have also been described. These changes in tissue structure included the transition from non-elongated to radially elongated mesocarp cells between 28 and 41 DAF, which was associated with an increase in diffusion vector coherence. Conversely, from the mid-ripe phase and beyond, the loss of diffusion vector coherence was possibly the result of the widespread cell death in the mesocarp that often accompanies this stage of development in wine grapes. Rotationally dependent anisotropic diffusion patterns were also noted in the seed integument of young grape berries. These diffusion patterns could only be observed early in the development of the grape berry, due to the dehydration of the seed integument and the replacement of the nucellus by the liquid endosperm.

The results of the current study agree that the inner mesocarp striation patterns previously noted in the transverse relaxation images of previous studies arise due to variations in cell width across the striation bands. These bands were evident in the images of the grape berries at 28 DAF and largely persisted through all subsequent phases of berry development.

Abbreviations

ADC: Apparent diffusion coefficient; CPMG: Carr-Purcell-Meiboom-Gill; DAF: Days after flowering; DT: Diffusion tensor; DTI: Diffusion tensor imaging; FOV: Field of view; MTX: Matrix size; MR: Magnetic resonance; MRI: Magnetic resonance imaging; NA: Number of averages; PGSE: Pulsed gradient spin-echo; RMSD: Root mean square displacement; TE: Echo time; THK: Slice thickness; TR: Repetition time; T_1: Longitudinal relaxation time; T_2: Transverse relaxation time; δ: Duration of the applied magnetic field gradient pulse; Δ: Timescale of the diffusion measurement.

Competing interest

The authors declare that they have no competing interest.

Authors' contributions

RJD carried out the MR imaging of the grape berries, performed the image reconstruction and the statistical analysis of the results, and drafted the manuscript. SJC participated in the design of the study, tended to the grape vines that were the source of samples for this study and carried out berry sampling as well as providing expertise in plant biology. SYR also contributed important intellectual content regarding plant biology. TSG assisted in troubleshooting the MR imaging pulse sequences used and in drafting the manuscript. GB performed the confocal micrographs of the grape berries. WSP critically revised the manuscript for important intellectual content. All authors read and approved the final manuscript.

Acknowledgements

The research is supported by a NSW BioFirst Award from the NSW Ministry for Science & Medical Research (WSP), as well as the grape growers and winemakers of Australia through their investment body, the Australian Grape and Wine Authority, with matching funds from the Federal Government (SYR). RJD gratefully acknowledges the support of the Professor Kai Yip Cho Memorial Scholarship. The authors acknowledge the facilities, and the scientific and technical assistance of the National Imaging Facility, University of Western Sydney Node.

Author details

[1]Nanoscale Organisation and Dynamics Group, University of Western Sydney, Penrith NSW 2751, Australia. [2]National Wine & Grape Industry Centre, Charles Sturt University, Locked Bag 588, Wagga Wagga, New South Wales 2678, Australia. [3]New South Wales Department of Primary Industries, Locked Bag 588, Wagga Wagga, New South Wales 2678, Australia. [4]School of Medicine, University of Western Sydney, Penrith, NSW 2751, Australia.

References

1. Bernstein M, King K, Zhou X, Fong W: **Handbook of MRI pulse sequences.** *Med Phys* 2005, **32**:1452.
2. Lauterbur P: **Image formation by induced local interactions: examples employing nuclear magnetic resonance.** *Nature* 1973, **242**:190–191.
3. Glidewell SM, Williamson B, Goodman BA, Chudek JA, Hunter G: **An NMR microscopic study of grape (Vitis vinifera L.).** *Protoplasma* 1997, **198**:27–35.
4. Pope J, Jonas D, Walker R: **Applications of NMR micro-imaging to the study of water, lipid, and carbohydrate distribution in grape berries.** *Protoplasma* 1993, **173**:177–186.
5. Van Der Weerd L, Claessens MMAE, Efdé C, Van As H: **Nuclear magnetic resonance imaging of membrane permeability changes in plants during osmotic stress.** *Plant Cell Environ* 2002, **25**:1539–1549.
6. Kuchenbrod E, Haase A, Benkert R, Schneider H, Zimmermann U: **Quantitative NMR microscopy on intact plants.** *Magn Reson Imaging* 1995, **13**:447–455.
7. Ishida N, Ogawa H, Kano H: **Diffusion of cell-associated water in ripening barley seeds.** *Magn Reson Imaging* 1995, **13**:745–751.
8. Sibgatullin T, Vergeldt F, Gerkema E, Van As H: **Quantitative permeability imaging of plant tissues.** *Eur Biophys J* 2009, **39**:699–710.

9. Trudeau JD, Dixon WT, Hawkins J: **The effect of inhomogeneous sample susceptibility on measured diffusion anisotropy using NMR imaging.** *J Magn Reson B* 1995, **108**:22–30.
10. Sotak CH: **Nuclear magnetic resonance (NMR) measurement of the apparent diffusion coefficient (ADC) of tissue water and its relationship to cell volume changes in pathological states.** *Neurochem Int* 2004, **45**:569–582.
11. Boujraf S, Luypaert R, Eisendrath H, Osteaux M: **Echo planar magnetic resonance imaging of anisotropic diffusion in asparagus stems.** *MAGMA* 2001, **13**:82–90.
12. Lätt J, Nilsson M, Rydhög A, Wirestam R, Ståhlberg F, Brockstedt S: **Effects of restricted diffusion in a biological phantom: a q-space diffusion MRI study of asparagus stems at a 3 T clinical scanner.** *MAGMA* 2007, **20**:213–222.
13. Gruwel MLH, Latta P, Sboto-Frankenstein U, Gervai P: **Visualization of water transport pathways in plants using diffusion tensor imaging.** *Prog Electromagn Res C* 2013, **35**:73–82.
14. Price WS: *NMR studies of translational motion.* Cambridge: Cambridge University Press; 2009.
15. Le Bihan D, Mangin J, Poupon C, Clark C, Pappata S, Molko N, Chabriat H: **Diffusion tensor imaging: concepts and applications.** *J Magn Reson Imaging* 2001, **13**:534–546.
16. Levitt M: *Spin dynamics: Basics of nuclear magnetic resonance.* 2nd edition. New York: Wiley; 2001.
17. Hahn EL: **Spin Echoes.** *Phys Rev* 1950, **80**:580–594.
18. Meiboom S, Gill D: **Modified spin-echo method for measuring nuclear relaxation times.** *Rev Sci Instrum* 1958, **29**:688–691.
19. Carr HY, Purcell EM: **Effects of diffusion on free precession in nuclear magnetic resonance experiments.** *Phys Rev* 1954, **94**:630–638.
20. Weingärtner H: **Self diffusion in liquid water. A reassessment.** *Z Phys Chem* 1982, **132**:129–149.
21. Einstein A: *Investigations on the theory of the Brownian movement.* New York: Dover Publications; 1956.
22. Mano S, Nakamura T, Kondo M, Miwa T, Nishikawa S-i, Mimura T, Nagatani A, Nishimura M: **The plant organelles database 3 (PODB3) update 2014: Integrating electron micrographs and new options for plant organelle research.** *Plant Cell Physiol* 2014, **55**:e1.
23. Stejskal EO, Tanner JE: **Spin diffusion measurements: spin echoes in the presence of a time-dependent field gradient.** *J Chem Phys* 1965, **42**:288–292.
24. Basser PJ, Mattiello J, LeBihan D: **Estimation of the effective self-diffusion tensor from the NMR spin echo.** *J Magn Reson B* 1994, **103**:247–254.
25. Crank J: *The mathematics of diffusion.* U.S.A.: Oxford University Press; 1979.
26. Marrink S-J, Berendsen HJC: **Simulation of water transport through a lipid membrane.** *J Phys Chem* 1994, **98**:4155–4168.
27. Kingsley PB: **Introduction to diffusion tensor imaging mathematics: part II. Anisotropy, diffusion-weighting factors, and gradient encoding schemes.** *Concepts Magn Reson A* 2006, **28A**:123–154.
28. Coombe BG: **Growth stages of the grapevine: adoption of a system for identifying grapevine growth stages.** *Aust J Grape Wine R* 1995, **1**:104–110.
29. Mansfield P: **Multi-planar image formation using NMR spin echoes.** *J Phys C: Solid State Phys* 1977, **10**:L55.
30. Pratt C: **Reproductive anatomy in cultivated grapes - A review.** *Am J Enol Viticult* 1971, **22**:92–109.
31. Hardie WJ, O'Brien TP, Jaudzems VG: **Morphology, anatomy and development of the pericarp after anthesis in grape, *Vitis vinifera* L.** *Aust J Grape Wine R* 1996, **2**:97–142.
32. Harris JMK PE, Possingham JV: **Anatomical aspects of grape berry development.** *Vitis* 1968, **7**:106–119.
33. Edzes HT, van Dusschoten D, Van As H: **Quantitative T_2 imaging of plant tissues by means of multi-echo MRI microscopy.** *Magn Reson Imaging* 1998, **16**:185–196.
34. Van As H, Scheenen T, Vergeldt FJ: **MRI of intact plants.** *Photosynthesis Res* 2009, **102**:213–222.
35. Tilbrook J, Tyerman S: **Cell death in grape berries: varietal differences linked to xylem pressure and berry weight loss.** *Funct Plant Biol* 2008, **35**:173.
36. Krasnow M, Matthews M, Shackel K: **Evidence for substantial maintenance of membrane integrity and cell viability in normally developing grape (Vitis vinifera L.) berries throughout development.** *J Exp Bot* 2008, **59**:849–859.
37. Cadot Y, Miñana-Castelló MT, Chevalier M: **Anatomical, histological, and histochemical changes in grape seeds from *Vitis vinifera* L. cv Cabernet franc during fruit development.** *J Agric Food Chem* 2006, **54** 9206–9215.
38. Mullins M, Bouquet A, Williams L: *Biology of the grapevine.* Cambridge: Cambridge University Press; 1992.
39. Ebadi A, Sedgley M, May P, Coombe BG: **Seed development and abortion in *Vitis vinifera* L., cv. Chardonnay.** *Int J Plant Sci* 1996, **157**:703–712.
40. Neeman M, Freyer JP, Sillerud LO: **A simple method for obtaining cross-term-free images for diffusion anisotropy studies in NMR microimaging.** *Magn Reson Med* 1991, **21**:138–143.
41. Rangel B, Platt KA, Thomson WW: **Ultrastructural aspects of the cytoplasmic origin and accumulation of oil in olive fruit (*Olea europaea*).** *Physiol Plant* 1997, **101**:109–114.
42. Hammami SB, Rapoport HF: **Quantitative analysis of cell organization in the external region of the olive fruit.** *Int J Plant Sci* 2012, **173**:993–1004.

15

HyperART: non-invasive quantification of leaf traits using hyperspectral absorption-reflectance-transmittance imaging

Sergej Bergsträsser[1*], Dimitrios Fanourakis[1,2,3], Simone Schmittgen[1], Maria Pilar Cendrero-Mateo[1], Marcus Jansen[1,4], Hanno Scharr[1] and Uwe Rascher[1]

Abstract

Background: Combined assessment of leaf reflectance and transmittance is currently limited to spot (point) measurements. This study introduces a tailor-made hyperspectral absorption-reflectance-transmittance imaging (HyperART) system, yielding a non-invasive determination of both reflectance and transmittance of the whole leaf. We addressed its applicability for analysing plant traits, i.e. assessing *Cercospora beticola* disease severity or leaf chlorophyll content. To test the accuracy of the obtained data, these were compared with reflectance and transmittance measurements of selected leaves acquired by the point spectroradiometer ASD FieldSpec, equipped with the FluoWat device.

Results: The working principle of the HyperART system relies on the upward redirection of transmitted and reflected light (range of 400 to 2500 nm) of a plant sample towards two line scanners. By using both the reflectance and transmittance image, an image of leaf absorption can be calculated. The comparison with the dynamically high-resolution ASD FieldSpec data showed good correlation, underlying the accuracy of the HyperART system. Our experiments showed that variation in both leaf chlorophyll content of four different crop species, due to different fertilization regimes during growth, and fungal symptoms on sugar beet leaves could be accurately estimated and monitored. The use of leaf reflectance and transmittance, as well as their sum (by which the non-absorbed radiation is calculated) obtained by the HyperART system gave considerably improved results in classification of *Cercospora* leaf spot disease and determination of chlorophyll content.

Conclusions: The HyperART system offers the possibility for non-invasive and accurate mapping of leaf transmittance and absorption, significantly expanding the applicability of reflectance, based on mapping spectroscopy, in plant sciences. Therefore, the HyperART system may be readily employed for non-invasive determination of the spatio-temporal dynamics of various plant properties.

Keywords: Absorption, *Cercospora beticola*, Chlorophyll content, FieldSpec, FluoWat, Hyperspectral imaging, Non-invasive phenotyping, Reflectance, Transmittance, Imaging spectroscopy

* Correspondence: sergej.bergstraesser@julumni.fz-juelich.de
[1]Institute for Bio- and Geosciences, IBG-2: Plant Sciences, Forschungszentrum Jülich GmbH, 52425 Jülich, Germany
Full list of author information is available at the end of the article

Background

Breeding of stress resistant crops for sustainable agricultural purposes is coupled with the identification of targeted traits and their genetic background [1-4]. In this content, the development and application of non-invasive systems becomes of outmost importance for plant phenotyping applications [4,5]. For this purpose, imaging techniques hold a prominent position in estimating various plant characteristics, ranging from pigment content to infection from diseases.

Light arriving on the leaf surface can be reflected, transmitted, or absorbed by the leaf. Quantifying the spectral characteristics of these components, by employing hyperspectral cameras can be very informative for the physiological status of the plants under research. Such spectroscopic retrieval of plant traits is recently gaining attention in the phenotyping community [6,7], not only due to their explicit connection with the leaf function and composition, but also due to the prospect of applying the method as imaging spectroscopy. This methodology offers the acquisition of information for the plants under study, without making physical contact with them. For instance, by using reflectance data, chlorophyll content of *Arabidopsis thaliana* [8] and disease symptoms' incidence of *Beta vulgaris* [9-13] were recently monitored. Previous studies, however, are mostly limited to find optimal reflectance wavelengths (combination of those in form of indices) or methods that use reflectance signatures related to leaf characteristics such as chlorophyll content, nutritional status, or infection by diseases. This may be explained by the limited possibilities to measure transmittance in a rapid fashion. Despite diverse attempts in this field the technical possibilities still stay unexhausted. Measurements based only on reflectance or only on transmittance obviously neglect a part of the existing spectral information. Currently there are no known imaging approaches that can measure reflectance and transmittance simultaneously and derive absorption of plant leaves. We hypothesize that plant properties will be better estimated when both reflectance and transmittance are taken into account. Their simultaneous quantification will be highly advantageous for academic and industry-oriented research purposes as demonstrated in our results presented below.

Currently available instruments to quantify hyperspectral plant reflectance and transmittance (e.g., integrating sphere and leaf clip) operate in the spot scale (i.e., a few cm^2) [14-16]. Therefore, several spots per leaf need to be assessed, in order to have an average value that is representative for the whole leaf. In this way, scaling plant reflectance and transmittance measurements to the entire leaf level will enable a more comprehensive assessment of a larger number of plants. More importantly, whole-leaf measurements offer spatial resolution allowing non-invasive observation of changes in leaf tissue owing to stressors.

Here we present a novel tailor-made instrument, named HyperART (Hyper = hyperspectral and ART Absorption, Reflectance, Transmittance), which quantifies simultaneously reflectance and transmittance at the whole-leaf scale. Firstly, we compare the data, obtained by the newly-introduced equipment, with a well-established high-performance spot-scale device (ASD- Field-Spec equipped with FluoWat; [17,18]). The FluoWat leaf-clip provides the possibility to record both reflection and transmission spectra of a single-point (i.e., limited in size) on a leaf. Secondly, we evaluated whether *Cercospora beticola* disease progression, a wide-spread pathogen, is more accurately derived by using the *Cercospora* Leaf Spot Index (CLSI) calculated from both reflectance and transmittance, as compared to the estimation that is solely based on reflectance. The CLSI index is computed by using wavelengths that were identified to be sensitive to the *C. beticola* disease symptoms [13]. Further, we tested the additional advantage of the HyperART in the estimation of leaf chlorophyll content of four species (i.e., maize, barley, rapeseed, tomato), based on twenty two commonly-used optical indices.

This paper introduces a novel tailor-made device, called HyperART. It can measure simultaneously spatially- and spectrally- resolved reflection and transmission, allowing the calculation of the spectral absorption image. This additional layer of information is expected to improve the quantification of plant traits. The above-mentioned measurements were performed with a comparable accuracy as with clip-on devices. The HyperART system also provides spatial information that can be potentially used to quantify spatial or temporal dynamics of leaf tissue changes. This is important as almost all processes or changes occurring in a leaf do not take place simultaneously throughout the whole tissue, but have initial points, from where they start spreading. The two employed case studies highlight the possibilities and advantages of the instrument. These are mainly the improved accuracy of results and the upscaling from the spot to the leaf level, encouraging its employment in future phenotyping studies.

HyperART a novel measurement system

The HyperART system is a novel imaging system developed for non-invasive evaluation of leaf properties, based on recordings of reflectance, transmittance and absorption. In contrast to commercially-available instruments that allow point measurements (i.e., limited to only few cm^2), this new equipment offers mapping of the whole leaf. For highest precisions, a geometrically precise system and flat samples deserve careful consideration, as described below.

Currently, two prototypes of the HyperART system are available (Figure 1a, b). The first prototype (Figure 1a) was used for the first case study (sugar beet), while the second prototype (Figure 1b) was employed for the measurements of the second case study (rapeseed, tomato,

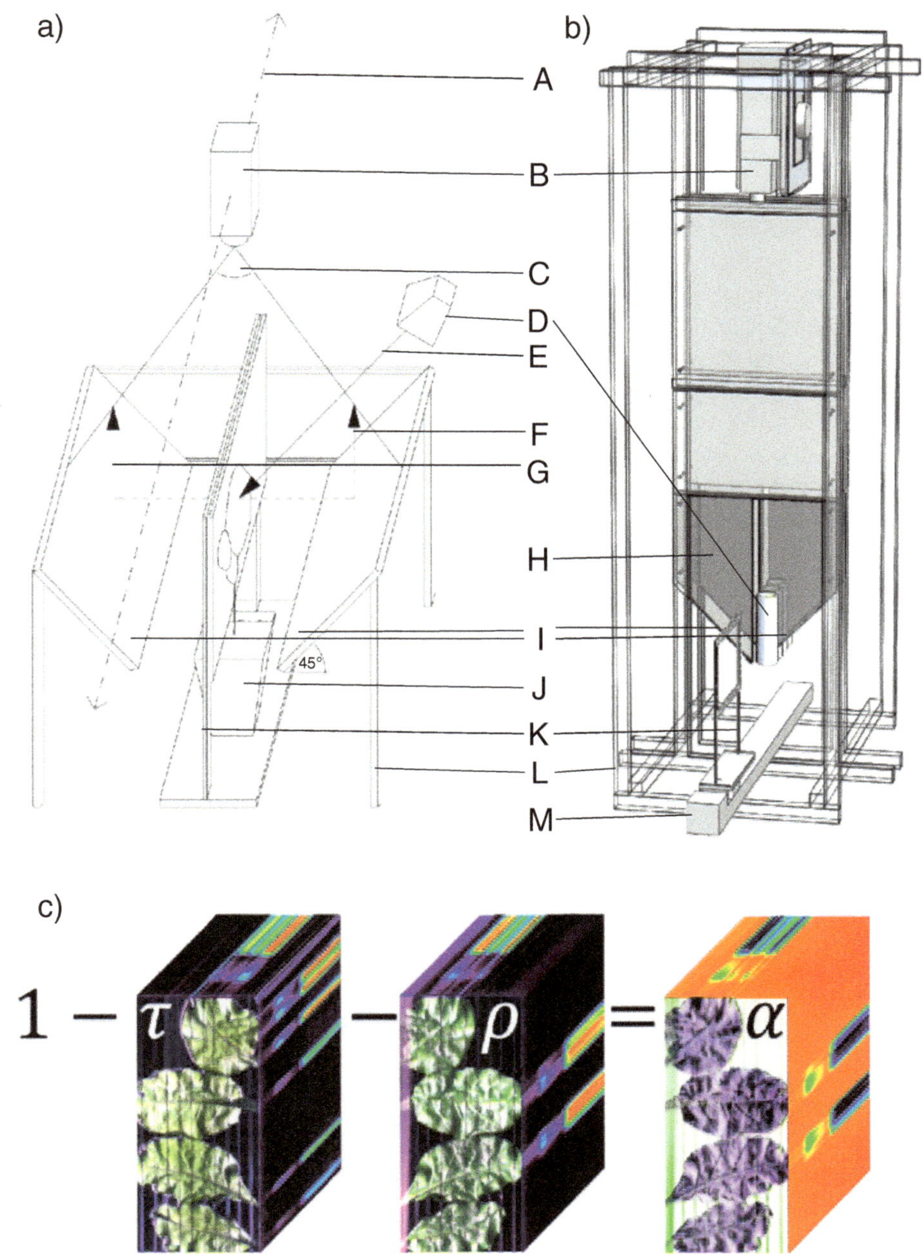

Figure 1 Two prototypes of the HyperART system with an image example. (a, b) Schematics of the hyperspectral absorption reflectance transmittance imaging (HyperART) system, employed for simultaneous recording of both reflectance (ρ) and transmittance (τ). In the first prototype **(a)** the camera is moving, whereas in the second one **(b)** the samples are moved. A, scan direction; B, line scanner; C, field of view; D, illumination source; E, light beam; F, reflected light; G, transmitted light; H, black painted metal sheets (to avoid direct illumination of the mirror from the light source and to reduce light scattering in the scanning process); I, mirrors; J, plant, of which leaves are fixed in the clip; K, clip, where leaf sample is placed; L, framework; M, slide bar. **(c)** Calculation of absorption (α) in the hyperspectral cube (acquired by the HyperART system), based on ρ and τ.

maize and barley). The main components of both prototypes include: i) a custom-made rectangular clip to hold and flatten the leaves (Figure 1K), ii) two mirrors which redirect both the reflected and transmitted light (Figure 1I), iii) an illumination system (Figure 1D), iv) an imaging system, composed by two hyperspectral cameras (together offering a spectral range of 400–2500 nm; Figure 1B), and v) a slide bar which moves the imaging system (first prototype; Figure 1a A) or the custom-made rectangular clip holding the sample (second prototype; Figure 1b M). The main difference between the two prototypes relies on the position of the light source (relative to the mirror) and the scanning procedure (i.e., movement of the imaging system or the clip). The different components of the HyperART system will be described firstly, followed by its working principle.

Custom-made rectangular clip

It consisted of two frames that were connected at the top (Figure 1K). The frames had dimensions of length × width × depth equal to 32 × 41 × 0.5 cm (inner dimensions 28 × 20 cm), and can host samples as large as 560 cm^2. The clip keeps the leaf samples as flat as possible. In the first prototype, the frames were made from metal. Moreover, a metal wire, at a distance of 1.5 cm in the horizontal direction of the frame, was used to position the leaves. The frames and the wire were painted with black mat colour (Dupli Color, MOTIP DUPLI GmbH, Haßmersheim, Germany). In the second prototype, the frames were manufactured from cast polymethylmethacrylat sheet (Plexiglas®GS, Evonik Industries AG, Essen, Germany), and were also painted with black mat colour (described above). Leaf positioning was performed by tight black fishing lines at a distance of 2 cm in both horizontal and vertical (frame) directions.

Mirror

A flat surface mirror (gertenbach.info e. K., Wölltein, Germany) was employed, having a minimum of 85% spectral reflectance performance (in the range of 400–700 nm) at 45° angle relative to illumination. The first prototype contains two (rectangular) mirrors with dimensions of length × width × depth equal to 28 × 20 × 0.5 cm. The second prototype employs four (rectangular) mirrors with the following dimensions: length × width × depth equal to 25.5 × 0.5 × 0.3 cm. In both prototypes, the mirrors were placed exactly at 45° relative to both the leaf clip (thus the leaf sample) and camera planes.

Illumination system

In the first prototype, sample illumination was offered by two double-ended halogen lamps (OMNILUX® 230 V/4000 W R7s 118 mm, OMNILUX, Waldbüttelbrunn, Germany), placed in a halogen spotlight reflector. In the second prototype, similar lamps (OMNILUX® 230 V/120 W R7s 117 mm, OMNILUX, Waldbüttelbrunn, Germany) were placed into a custom made tube reflector from a polished-sheet metal. The lamps were covered with frosted quartz glass, which diffuses the light in order to avoid shadows and enhance the quality of the acquired hyperspectral data. A custom-made converter was employed to transform (rectify) alternating current to direct current (residual ripple ≤ 5%). This was a three-phase power supply unit, manufactured by a company (J.Schneider Elektrotechnik GmbH, Offenburg, Germany). Direct current was preferable for obtaining hyperspectral data by using a line scanner, since it prevents the fluctuation of the halogen lamp emission, which would take place under alternating current. In both prototypes, the lamps were placed in order to illuminate one side of the leaves.

Hyperspectral imaging system

The HyperART system hosted two hyperspectral cameras employed on sequence, and together offering a spectral range between 400 and 2500 nm. Both devices work as push broom line scan cameras. The first hyperspectral camera (PS V10E, Spectral Imaging Ltd., Oulu, Finland) employs a high speed interlaced CCD detector. It is sensitive in the 400 to 1000 nm range, and offers a spatio-spectral resolution of 1392 × 1040 pixels. The nominal spectral resolution is 2.8 nm full width half maximum (FWHM). The spectral sampling interval depends on the spectral binning and varies between 0.63 and 5.06 nm. In our experiments with the first prototype, a binning of 4 (spectral resolution) and 1 (spatial resolution) was employed. In the experiments with the second prototype, a binning of two (spectral resolution) by two (spatial resolution) was employed. The second hyperspectral camera (SWIR, Spectral Imaging Ltd., Oulu, Finland) uses a cooled, temperature-stabilized MCT detector, which is sensitive to the spectral range between 1000 and 2500 nm. The sensor can acquire 320 spatial pixels' and 256 spectral bands' images. The spectral sampling rate is 6.3 nm, combined with a nominal spectral resolution of 10 nm (FWHM).

Scanning procedure

At the first prototype, the two hyperspectral cameras were fixed on a slide bar (BiSlide, Velmex Inc., Bloomfield, NY, USA) that moves along an 1.5 m path at a constant speed (≈15 mm sec^{-1}). The cameras were moved in parallel to the middle line and between the opposite standing mirrors. As the hyperspectral cameras were moved along the slide bar, they recorded the reflected and transmitted light of the whole leaf sample, which was placed in the clip, as explained above. In the second prototype, the hyperspectral cameras were fixed on a framework over the middle point and between the opposite standing mirrors (Figure 1b L). In this case, the clip (holding the leaf sample) was moved along the middle line, between the mirrors that were fixed on the slide bar (Figure 1b M).

The working principle of the HyperART system was identical for both prototypes, and enabled the upward redirection of transmitted and reflected light from the leaf sample towards the hyperspectral camera system through the mirrors. Thus, the unique advantage of the HyperART system is that a reflectance and transmittance image of a leaf can be recorded simultaneously by a single hyperspectral camera. For measurements, leaves must be positioned into the clip (Figure 1K), which can be done without excising them from the plant, since the pot (≤17 cm height) can be placed below the clip frame (Figure 1a J). Placing the leaves between the frames directs the leaf surface into a flat plane. Inside the clip, the leaves were illuminated from one side, while the leaf was positioned between the two mirrors (Figure 1I). Each mirror faced a different side of the clip (thus the sample). In the first prototype, the light source was placed behind one mirror, and opposite to the sample at an angle of 45° (Figure 1a D). In the second prototype, the light source was placed from the sides of mirrors (Figure 1b D) that allowed nearly direct illumination of the sample. The mirrors posed both the clip window with the sample, and the camera, at a 45° angle. One mirror redirected the reflected light (Figure 1a F) to the two line scanners (i.e., upwards; Figure 1B), while the other mirror did the same for the transmitted light (Figure 1a G). The distance of the upper mirror edge to the two line scanners was set to 0.85 m for both prototypes.

To determine the incoming electromagnetic radiation by the first prototype, a 99% reflective white standard (Spectralon, Labsphere Inc., North Dutton, NH, USA) was placed instead of the clip. This was done before and following the measurements. In the second prototype, the transmitted light was determined by placing a white diffuser lambertian transmission sheet (Zenith Polymer® ≈ 50% transmission, SphereOptics GmbH, Uhldingen, Germany) between the mirrors.

Image processing

Hyperspectral images were processed by using ENVI 4.8 (Exelis Visual Information Solutions, Inc., Boulder, CO, USA) and MatLab R2012b (MathWorks®, Natick, Massachusetts, USA) Software. Different protocols were used for image pre-processing obtained from the first and second prototype.

First prototype

Firstly, a dark frame image was subtracted from both the leaf and white reference images (Fastspec extension of ENVI). Secondly, images were cut to separate the reflection and transmission images. The transmission image was flipped to overlay the reflection image, and co-registered by the corresponding points in order to correct slight distortions, produced by image acquisition (VLFeat toolbox in MatLab; [19]). In the next step, each image pixel was divided by the corresponding pixel within the white reference image, and multiplied by a correction factor for white reference material (provided by Labsphere Inc.), aiming at both correcting for illumination differences and achieving relative values (Spectral Math tool of ENVI).

Second prototype

In this prototype images needed modified processing steps since the scan procedure was different as compared to the first prototype. Additionally to the white standard, we here measured a white diffuser lambertian transmission sheet, before scanning the plant samples. By mosaicking both white references we got a white reference image with the same numbers of lines as the sample image. The sample image was normalized by this white reference image for reflection and transmission sides line by line using a normalisation tool from SPECIM (Spectral Imaging Ltd., Oulu, Finland) implemented in ENVI. The transmission image was then flipped and co-registered. With the objective to obtain correct relative values, a correction factor was applied to the reflection and transmission images (provided by Labsphere Inc. and SphereOptics GmbH, respectively).

Absorption mapping

Based on the assumption that the sum of absorption (α), reflectance (ρ) and (flipped) transmittance (τ) equals to 1 (Figure 2), α was calculated by subtracting ρ and τ from the unit (i.e., $\alpha = 1 - \rho + \tau$) for images acquired by using both prototypes. The α contains the information about leaf components and is not sensitive to light scattering effects, caused by leaf surface properties [20]. However, it is not practical to apply different vegetation indices on α, because these indices were developed for approaches based on ρ measurements. To highlight the advantage of the newly-introduced HyperART system and make the data comparison clear, a ρ comparable spectral signature ought to be employed. Therefore, the non-absorbed radiation $\rho + \tau$ images (that are equal to inverted α because $\rho + \tau = 1 - \alpha$) were used in analysis.

Results and discussion

Validation of HyperART spectral data

Reflectance and transmittance spectral signatures of sugar beet leaves were recorded by using either the HyperART system or the FieldSpec spectoradiometer, equipped with a FluoWat device (Figure 2a). In order to validate the HyperART-obtained measurements, reflectance (ρ), transmittance (τ), and the sum of reflectance and transmittance spectra (non-absorbed radiation, $\rho + \tau$) were compared to the FluoWat spectra (Figure 2).

A similar ρ and τ spectrum acquired by either device was found (Figure 2a). Over the full ρ spectral range (except 1400–1500 nm), relative values acquired by the HyperART system were 1–5% lower than those of FluoWat. For the τ spectra, the HyperART system data yielded either higher (>800 nm) or lower (550 nm) values than those of the FluoWat data. However, the difference in the τ spectra was minor (≈1%), as compared to the difference in the ρ spectrum. In the FluoWat device, there was an overlap between the ρ and τ spectra at the near infrared region (NIR, 736–1100 nm) (Figure 2a). HyperART ρ and τ spectra were overlapping in the 1012–1031 nm region. Because of this overlap, the normalized α spectrum is not presented. The noted differences between the values, obtained by the two devices, might be due to the fact that the leaf regions measured by the FluoWat device, are not exactly identical as the selected (for comparison) areas in

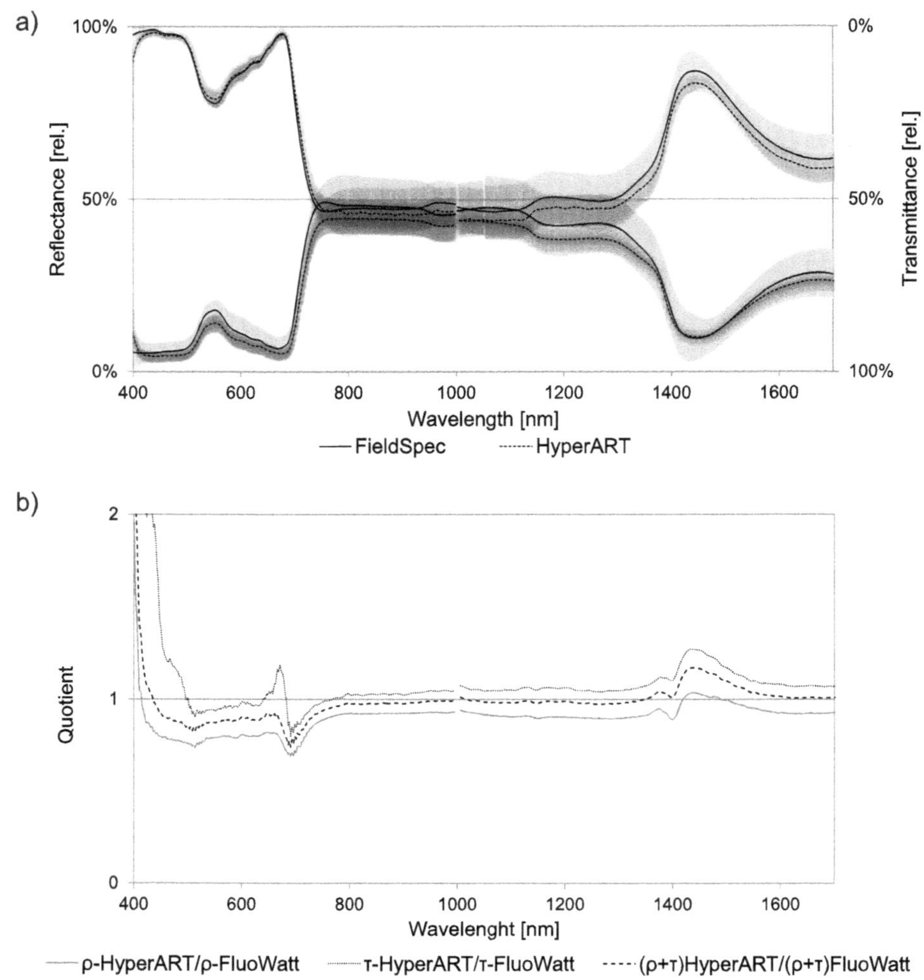

Figure 2 Comparison of spectral signatures obtained by FluoWat and HyperART devices. (a) Reflectance (ρ) and transmittance (τ) spectra of sugar beet leaves acquired by using FluoWat or the hyperspectral absorption reflectance transmittance imaging (HyperART) system. Data are expressed as relative values. Dashed areas indicate SEM ($n = 5$). **(b)** Quotients of ρ, τ and sum of ρ with τ (i.e., ρ + τ) of spectral data acquired by using FluoWat and the HyperART system.

the images obtained by using the HyperART system. The gap of the values observed at 1000 nm at both ρ and τ spectra, is due to the configuration of the FieldSpec and the hyperspectral imaging system. Both devices combine two different sensors, where the first one covers the region up to 1000 nm, while the second one includes longer wavelengths (i.e., ≥ 1000 nm).

For the ρ, τ, and ρ + τ HyperArt normalized spectra (Figure 2b) the highest deviations were observed in the photosynthetic active radiation region (PAR, 420–730 nm) and around the water absorption band at 1450 nm. For the whole spectrum (420–1700 nm), standard deviations of 0.07 (ρ), 0.19 (τ) and 0.07 (ρ + τ) were found. These low standard deviation values indicate that the data obtained with the HyperART system are comparable to those taken by using FluoWat.

The coupling of two hyperspectral cameras (i.e., VNIR and SWIR) might be important for future work, where additional traits, from those studied here, are of interest. For instance, the spectral region, offered by the SWIR camera (i.e., 1000–2500 nm), is often employed to detect senescence and water stress symptoms [21,22]. We were able to show that the HyperART system performs well with the SWIR camera, though its potential in detecting plants characteristics is not dealt in the current study. The SWIR images and the VNIR images can be matched together using corresponding points. This matching of VNIR and SWIR images may assist extended spectral signature for every pixel in only one image. However the spatial resolution in the VNIR image needs to be adjusted to the SWIR image or vice versa.

Cercospora beticola symptoms

Following the data comparison between the HyperART system and FluoWat device we tested if the HyperART is suitable for the detection of *Cercospora* leaf spot disease. Therefore normalized histograms (Figure 3) of the CLSI images with the leaves of one infected and one non-infected sugar beet plant were analysed and compared among each other. Additionally the precision of the unsupervised classification on the ρ, τ and ρ + τ CLSI images of the infected leaves were verified (Figure 4).

Three leaves of the infected sugar beet plant (shown in Figure 4) can be described by visual inspection as: i) symptom-free leaf (lower leaf in Figure 4), ii) leaf

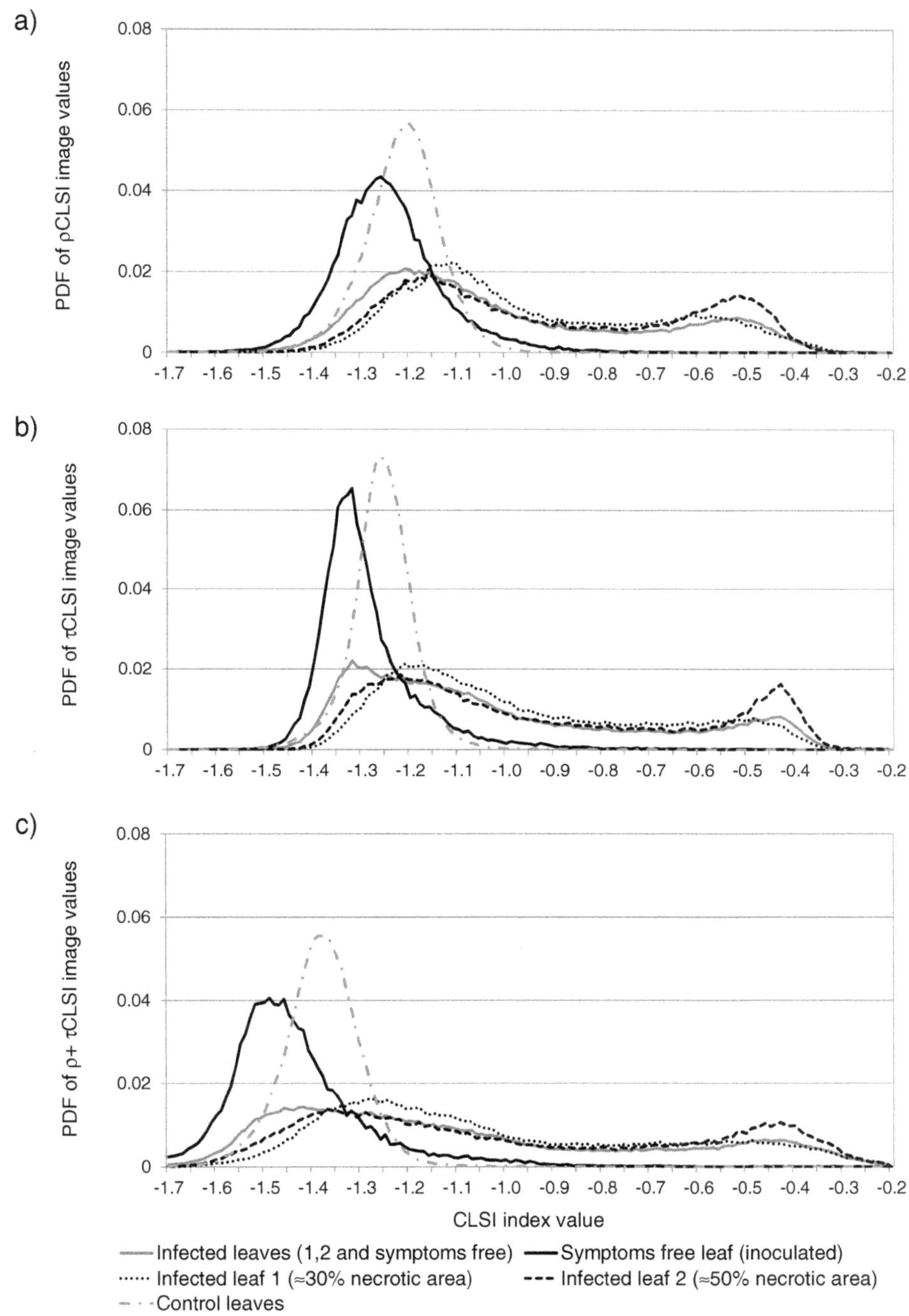

Figure 3 Values difference in normalized histograms of Cercospora Leaf Spot Index (CLSI) images. Normalized histograms (estimated probability functions PDF) of the calculated CLSI values based on **(a)** reflectance (ρ), **(b)** transmittance (τ) or **(c)** combination of ρ with τ (i.e., ρ + τ). Spectra were acquired by using the hyperspectral absorption, reflectance, transmittance (HyperART) imaging system. Measurements were conducted on three leaves of an infected sugar beet plant, and on three leaves of another non-infected (control) sugar beet plant. The two leaves of the former plant showed visual symptoms of infection, whereas the third one was symptom-free.

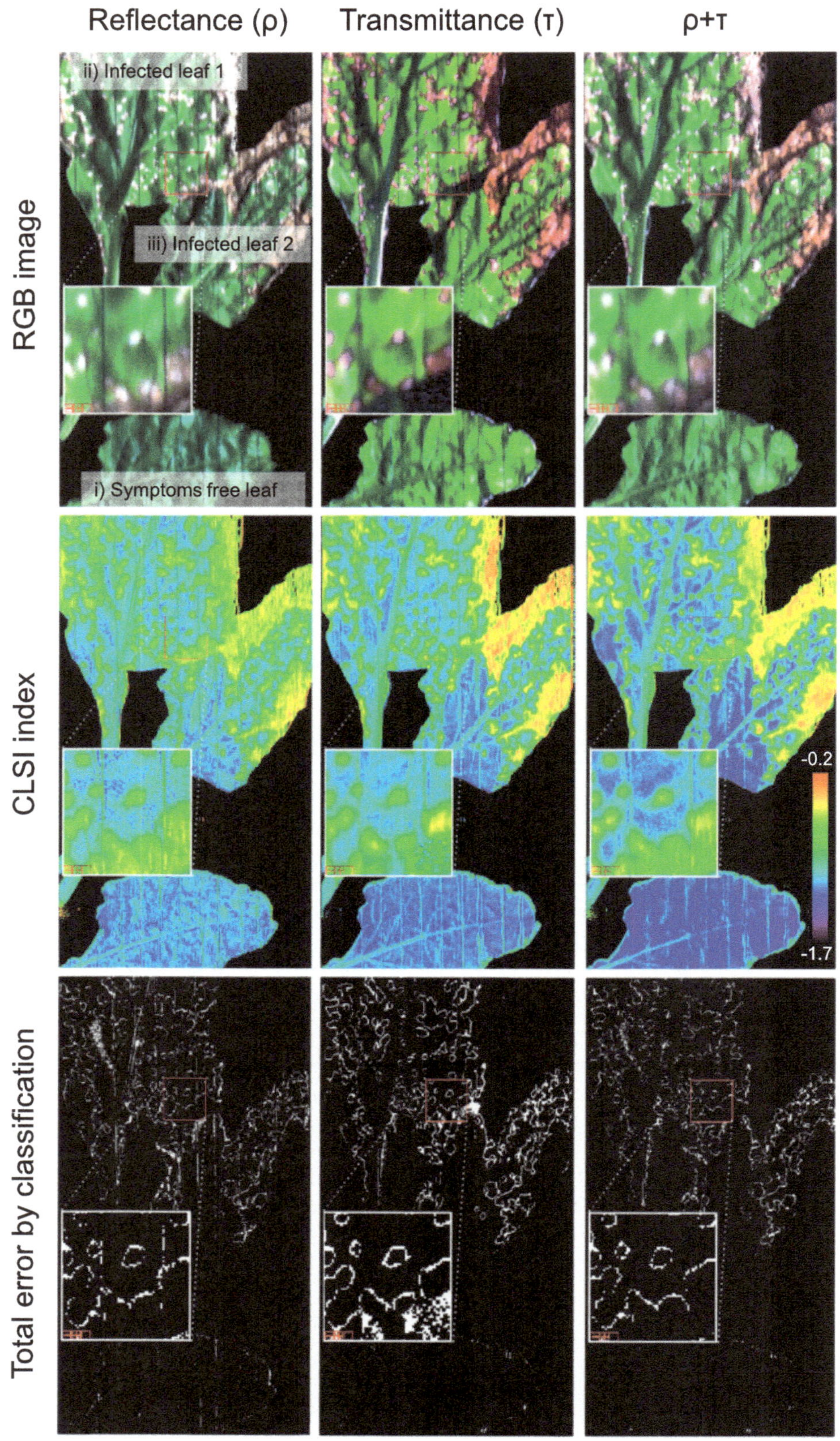

Figure 4 (See legend on next page.)

(See figure on previous page.)

Figure 4 RGB, colour coded *Cercospora* Leaf Spot Index (CLSI) and total classification error images. Images were calculated from reflectance (ρ), transmittance (τ) and combination of ρ with τ (i.e., ρ + τ) of sugar beet leaves infected by *Cercospora beticola*. The classification was performed by supervised SVM (support vector machine) classification on hyperspectral images, and unsupervised K-Mean (on CLSI images) using two classes (visible symptoms and plat tissue without visible symptoms). Images were acquired by using the hyperspectral absorption reflectance transmittance imaging (HyperART) system. The rectangular panels in the images show the enlargement of red bordered infected leave surface.

with ≈ 30% infected area (upper leaf in Figure 4), and iii) leaf with ≈ 50% necrotic area (middle leaf in Figure 4). In order to assess the CLSI value changes induced by pathogen infection, the CLSI histogram from three leaves of the non-infected control plant was also computed.

The CLSI shows similar bimodal distribution when calculated from ρ (ρCLSI), τ (τCLSI), or ρ + τ (ρ + τCLSI) images of leaves with infection symptoms. The normalized histogram of the leaves attached on the control plant, or the symptom-free leaf, attached on the infected plant, were clearly different (Figure 3). Furthermore there is a difference between the normalized histogram of inoculated but still symptom-free leaf and the normalized histogram of control leaves. They both show a single maximum only, however, the distribution of the symptom-free leaves show a heavy right tail, whereas the control leaves are normally distributed. As a right tail is an index shift towards infection symptoms, this indicates that pre-symptomatic fungal development changed leaf spectral properties. Therefore the CLSI offers a perspective for using the method in early detection of fungal infection stages.

The normalized histograms of the symptomatic leaves showed a shift from the right peak that is related to the degree of leaf infestation. While the normalized ρ and ρ + τ histograms are Gaussian-like, the normalized τ histograms were sharper, indicating a lower noise influence. Peak locations varied for the different measurement modalities. The distance between the peak locations were larger in τCLSI and largest in ρ + τCLSI compared to ρCLSI. This might enable a better assignment of the pixels into the categories infected or non-infected. The bimodal amplitude (A_b) was equal to 0.45, 0.48 and 0.43 for ρCLSI, τCLSI and ρ + τCLSI, respectively, being in accordance with our observation of sharper peaks in the normalized τCLSI-histogram. Considering all CLSI histograms, it can be stated that over 90% of the pixels in the index images were distributed between −1.7 and −0.2. Higher values indicate increasing disease severity and vice-versa.

The computed CLSI value range in this study differs due to variation in plant properties and measurement method from the range (−0.45 to 0.3) reported in the literature [13,23]. For instance, leaf pigment content, as well as leaf internal and external structure have been shown to affect plant reflectance properties [24]. In addition, illumination conditions (e.g., shadow or illumination angle) during measurement or measurement methodology (e.g., point measurements, leaf scale or canopy scale) can also raise differences in the CLSI values' range. The differences in CLSI value may be explained by the use of 698 nm and 734 nm wavelengths in CLSI equation. These wavelengths are located in the red edge spectral region, and are sensitive to chlorophyll content [25-28]. The sensitivity at 698 nm is translated to decreased reflectivity (thus lower CLSI values in the normalized part of equation), as chlorophyll content increases. When we subtract lower reflectance value (due to chlorophyll absorption at 734 nm) from the normalised part of equation then this might produce slightly higher CLSI values of the leaves attached on the control plant, as compared to the symptom-free leaf attached to the infected plant.

To investigate the classification performance, the tissue having CLSI values lower than the lower peak center was categorized as 'non-infected' (≈27%, 14%, and 21% of the pixels for ρCLSI, τCLSI, and ρ + τCLSI, respectively), while tissue giving higher values than the higher peak center is termed as 'infected' (≈14%, 19%, and 36% of the pixels for ρCLSI, τCLSI, and ρ + τCLSI, respectively). The remaining pixels (≈59%, 67%, and 43% for ρCLSI, τCLSI, and ρ + τCLSI, respectively) between the peaks allow the presence of classification errors. This observation leads to the conclusion that the separation of infected from non-infected leaf tissue is better done with the ρ + τCLSI image, since most of the tissue area is unambiguously classified.

In the RGB τ image, the reddish colour may be an indication of anthocyanin production in the (sugar beet) leaves, due to infection (Figure 4). The CLSI images have a rainbow colour code, ranging from −1.7 (blue) to −0.2 (red) (Figure 4), which is similar to the histograms in Figure 3. The ρCLSI image seems to have a slightly higher noise level, as compared to the other images. The τCLSI image shows a better contrast between extremes, while the ρ + τCLSI image differentiates the initial symptom and its growing borders. The unsupervised classification yields 93.4%, 92.3% and 94.7% overall accuracy (automatically calculated in ENVI using contingency matrix which did compare supervised classification with unsupervised classification) for the whole ρCLSI, τCLSI and ρ + τCLSI images respectively, being well in accordance with our previous observation. In total, the ρCLSI index classification quantifies 33.9% of the tissue as infected (i.e., 66.1% non-infected), while the τCLSI index classification led to an estimation of 29.3% infected tissue area (i.e., 70.7% non-infected). The $\rho + \tau$CLSI image classification results in 31.8% infected leaf area, and gave the best overall accuracy, as explained below.

The total classification error images (Figure 4) suggest that the noise in the ρCLSI image leads to an overestimation of infected area. This overestimation may be due to the leaf veins or shadow parts, as a result of leaf waviness. Another error source is an overlay of leaves by the black wire taut in the leaf clip. The opposite is noted for the τCLSI image classification, where the non-infected area seems to be overestimated. This overestimation of the non-infected area might be due to both the overlapping region between the two leaves, which were together placed in the clip, and an underestimation of the symptom border. The τCLSI was found to feature comparably low noise, being also in accordance with the sharper peaks in the respective histogram (Figure 3). The $\rho + \tau$CLSI image gave the best class separation, supressed noise and counted overlapping regions that led to the most precise estimation of disease severity. In the $\rho + \tau$CLSI image, the unsupervised classification gave a 99.4% of non-infected tissue area in the symptom-free leaf. The infected leaves were classified to have 38.7% and 44.6% infected areas, respectively.

In general, our visual estimation had an error between 5.4% and 8.7%, as compared to the image classification results. Therefore, the machine-based classification method (in our case the HyperART system) was found to be more precise and reliable in quantifying disease severity, as compared to assessments based on visual inspection [29].

Leaf chlorophyll content estimation

Leaves of four species (maize, rapeseed, barley and tomato), grown under two different nitrogen regimes, were analysed for chlorophyll content. The relationships between 22 optical indices (summarized in Table 1), calculated from 3 source HyperART images (ρ, τ, and $\rho + \tau$) and destructively-assessed chlorophyll content were analysed (i.e., 66 indices = 22 optical indices × 3 HyperART source images). Both logarithmic (Table 2) and linear (Table 3) regression models were employed for the analysis.

Out of 66 indices, a significant and positive correlation ($P < 0.001$; $R^2 \geq 0.5$) between calculated chlorophyll content and destructively measured content was found in 25 and 18 indices for logarithmic and linear regression, respectively (bold highlighted in Tables 2 and 3). Importantly, 23 (out of the 25) and 17 (out of the 18) indices for logarithmic and linear regressions, respectively, were computed from τ and $\rho + \tau$ source images (bold italic highlighted in Tables 2 and 3). These results indicate that the additional τ measurements *per se* or in combination with ρ (i.e., $\rho + \tau$) result in higher correlation coefficients between measured and estimated chlorophyll content.

Logarithmic regressions always gave higher correlation coefficients, as compared to linear regression models (Tables 2 and 3). Previous work also reports that indices are non-linearly related to chlorophyll content [31]. The highest correlation coefficients (i.e., $0.63 \leq R^2 \leq 0.72$) were noted in the logarithmic regression models of the following five indices: $\rho + \tau$EVI, τmND, $\rho + \tau$mND, τVOG, and ρBGI2 (Table 2, and Additional file 1: Figure S1). Importantly, the good correlation of these five indices with the destructively-measured values was driven by differences in chlorophyll concentration, owing to both the growth nitrogen regime and species.

Indices developed for carotenoids' pigment changes (i.e., SIPI, PSRI) showed a poor relationship with chlorophyll content (i.e., $R^2 \leq 0.52$; Tables 2 and 3). Indices, traditionally used for vegetation monitoring (i.e., NDVI, SR, mSR), did not obtain as good results as red edge and combined indices (Tables 2 and 3). For NDVI specifically, its low performance might be explained by the fact that it changes only when chlorophyll concentration do not exceeds the value of 10 $\mu g\ cm^{-2}$ [31,46]. In this study, chlorophyll concentration was eight- to nine-fold higher than this value. Indices developed for maximal sensitivity to leaf area index (i.e., MACRI, TCARI, TCARI/OSAVI) presented low correlation coefficients (i.e., $R^2 \leq 0.44$; Tables 2 and 3). This is because the aforementioned indices are not sensitive to changes in chlorophyll content at low leaf area index values [43], as those of this study where a set of a few leaves were assessed.

Interestingly, indices based on the off-chlorophyll absorption centre wavebands (690–730 nm; e.g., mND, VOG) seemed to perform better than indices based on the main chlorophyll and carotenoids' absorption regions (450–550 and 670–680 nm, respectively; e.g., NDVI, MCARI) [47,48].

Besides the optical indices, employed here, there are also other approaches to estimate chlorophyll content by using spectral signature [49-51]. In this paper, we look into the first methodology, due to the simplicity of computation and implementation in studying plant properties. However, the performance of the HyperART, when using the aforementioned more elaborate approach, stays unexhausted and necessitates further work.

Conclusion

The HyperART system was designed to provide data with high spectral and spatial resolution, which can be readily employed for phenotypic measurements, such as disease detection and quantification of chlorophyll content. The newly-introduced HyperART system scans entire leaves, attached to the plants, whereas currently-used instruments (e.g., FluoWat) are limited to spot (point) measurements. The system measures both reflectance (ρ) and transmittance (τ) data, by which the absorption (α) spectrum can be calculated, with one sensor. In this study we

Table 1 Formulas and abbreviations of the employed vegetation indices

Index	Equation	Indicator (Scale)	Reference
Area under continuum-removed curve	$ANCB_{650-720}$: Continuum removal based method	Chlorophyll a & b content (Canopy)	[30]
Blue/Green index	$BGI_2 = \frac{R450}{R550}$	Chlorophyll content (Canopy)	[31]
Chlorophyll green index	$Chlgreen = \frac{R790}{R550} - 1$	Chlorophyll (Leaf)	[32]
Chlorophyll red edge index	$Chlred\ edge = \frac{R790}{R705} - 1$	Chlorophyll (Leaf)	[32]
Enhanced vegetation index	$EVI = 2.5 * \frac{R800-R670}{R800+6 * R670-7.5 * R400+1}$	Chlorophyll (Canopy)	[33]
Modified chlorophyll absorption reflectance index	$MCARI = ((R701 - R670) - 0.2 * (R701 - R550)) * \frac{R701}{R670}$	Chlorophyll (Leaf, Canopy)	[34]
Modified normalized difference index	$mND = \frac{R750-R705}{R750+R705-2 * R445}$	Chlorophyll (Leaf)	[35]
Modified simple ratio	$mSR = \frac{R750 - R445}{R705 - R445}$	Green biomass Chlorophyll (Leaf)	[35]
Normalized difference index	$ND = \frac{R750 - R705}{R750 + R705}$	Chlorophyll (Leaf)	[36]
Normalized difference vegetation index	$NDVI = \frac{R800-R670}{R800+R670}$	Biomass, leaf area (Canopy)	[37]
Structure insensitive pigment index	$SIPI = \frac{R800-R455}{R800+R680}$	Carotinoid/chlorophyll a ratio (Leaf)	[38]
Simple ratio 1	$SR_{750/710} = \frac{R750}{R710}$	Chlorophyll (Canopy)	[39]
Simple ratio 2	$SR = \frac{R800}{R670}$	Green biomass (Canopy)	[40]
Pigment specific normalized difference a	$PSNDa = \frac{R800-R680}{R800+R680}$	Chlorophyll a (Leaf)	[41]
Pigment specific normalized difference b	$PSNDb = \frac{R800 - R635}{R800 + R635}$	Chlorophyll b (Leaf)	[41]
Plant senescence index	$PSRI = \frac{R680 - R500}{R750}$	Plant senescence (Leaf)	[42]
Pigment specific simple ratio a	$PSSRa = \frac{R800}{R680}$	Chlorophyll a (Leaf)	[41]
Pigment specific simple ratio b	$PSSRb = \frac{R800}{R635}$	Chlorophyll b (Leaf)	[41]
Transformed chlorophyll absorption in reflectance Index	$TCARI = 3 * ((R700-R670)-0.2 * (R700-R550) * (\frac{R700}{R670}))$	Chlorophyll (Canopy)	[43]
Transformed chlorophyll absorption in reflectance Index/Optimized soil-Adjusted vegetation index	$\frac{TCARI}{OSAVI} = 3 * \frac{(R700-R670)-0.2 * (R700-R550) * (\frac{R700}{R670})}{(1+0.16) * \frac{R800-R670}{R800+R670+0.16}}$	Chlorophyll (Canopy)	[43]
Triangular vegetation index	$TVI = 0.5 * (120 * (R750 - R550) - 200 * (R670 - R550))$	Leaf area and chlorophyll content (Canopy)	[44]
Vogelmann	$VOG = \frac{R740}{R720}$	Chlorophyll (Leaf)	[45]

The scale (leaf or canopy), at which these are commonly used, is also presented.

demostrated that HyperART measurements were consistent with data acquired with the well-established FluoWat device. We were able to show that the estimation of chlorophyll content of entire leaves, sampled from four different crops, was reasonably accurate. Moreover, it was shown that the system leads to an improved detection and quantification of disease symptoms (model plant-pathogen interaction *Cercospora beticola* and *Beta vulgaris* spp. *vulgaris*). Finally, it was noted that the non-absorbed radiation signal (i.e., $\rho + \tau$) reduced data noise in disease detection and significantly improved the estimation of leaf chlorophyll content across species with different leaf structure. Based on our results, we suggest that the HyperART system holds a high potential for studying plant responses to both biotic and abiotic stressors at the entire leaf level.

Methods

Plant material and growth conditions

Two experiments were conducted, each case study focused on another set of species. Both experiments took place in a greenhouse located in the western part of Germany (Jülich, 50°N; 6°E).

Experiment I: Cercospora beticola infection

Two single plants of a sugar beet line (*Beta vulgaris* ssp. *vulgaris*; identification number 8RF5006 (HS) KWS SAAT AG, Einbeck, Germany) were cultivated. During growth, air temperature and relative air humidity were 20.9 ± 6.9°C (range: 8.5–47.6) and 59.6 ± 21.7% (range: 8.2–96.9), respectively. Supplementary light was provided by high-pressure sodium lamps (SON-T Agro, 250 W, Philips,

Table 2 Vegetation indices performance based on a logarithmic regression model

Index	Rank	Source	COR	R^2	F	P	Index	Rank	Source	COR	R^2	F	P
ANCB	8	**ρ + τ**	0.80	0.64	66.94	0.00	PSNDb	7	**τ**	0.80	0.64	67.86	0.00
	13	**τ**	0.77	0.60	56.98	0.00		20	**ρ + τ**	0.75	0.56	48.48	0.00
	26	**ρ**	0.71	0.50	37.60	0.00		43	ρ	0.55	0.30	16.15	0.00
BGI2	9	**ρ**	0.80	0.63	65.65	0.00	PSRI	53	τ	0.37	0.14	6.02	0.02
	27	τ	−0.69	0.48	34.56	0.00		54	ρ + τ	0.35	0.12	5.41	0.03
	48	ρ + τ	0.46	0.22	10.41	0.00		61	ρ	0.15	0.02	0.82	0.37
Chlg	38	ρ + τ	0.60	0.36	21.10	0.00	PSSRa	23	**τ**	0.74	0.55	45.88	0.00
	41	ρ	0.56	0.31	17.32	0.00		49	ρ + τ	0.45	0.20	9.49	0.00
	57	τ	0.25	0.06	2.58	0.12		65	ρ	0.08	0.01	0.23	0.64
Chlre	15	**τ**	0.77	0.59	54.90	0.00	PSSRb	12	**τ**	0.78	0.61	58.42	0.00
	22	**ρ + τ**	0.75	0.56	47.42	0.00		21	**ρ + τ**	0.75	0.56	48.36	0.00
	33	ρ	0.63	0.40	25.60	0.00		51	ρ	0.41	0.17	7.64	0.01
EVI	1	**ρ + τ**	0.85	0.72	99.16	0.00	SR1	10	**τ**	0.80	0.63	65.28	0.00
	34	τ	0.62	0.38	23.21	0.00		17	**ρ + τ**	0.76	0.58	53.06	0.00
	44	ρ	0.54	0.29	15.88	0.00		31	ρ	0.66	0.44	30.07	0.00
MCARI	36	ρ	−0.61	0.37	22.07	0.00	SIPI	24	**τ**	0.72	0.52	41.39	0.00
	46	τ	0.51	0.26	13.02	0.00		52	ρ + τ	0.38	0.14	6.38	0.02
	47	ρ + τ	−0.49	0.24	12.24	0.00		66	ρ	−0.03	0.00	0.03	0.86
mND	3	**τ**	0.83	0.69	85.43	0.00	SR2	18	**τ**	0.76	0.58	51.62	0.00
	4	**ρ + τ**	0.82	0.67	76.13	0.00		45	ρ + τ	0.53	0.29	15.16	0.00
	28	ρ	0.69	0.47	34.36	0.00		63	ρ	0.10	0.01	0.42	0.52
mSR	14	τ	0.77	0.60	56.46	0.00	TCARI	37	ρ	−0.60	0.36	21.16	0.00
	40	ρ + τ	−0.57	0.32	18.27	0.00		42	ρ + τ	−0.56	0.31	17.09	0.00
	56	ρ	−0.29	0.08	3.46	0.07		64	τ	0.09	0.01	0.32	0.57
ND	2	**τ**	0.84	0.70	90.52	0.00	TCARI/OSAVI	30	ρ	−0.67	0.44	30.36	0.00
	6	**ρ + τ**	0.80	0.65	69.52	0.00		32	ρ + τ	−0.66	0.43	29.11	0.00
	29	ρ	0.69	0.47	34.36	0.00		59	τ	−0.21	0.04	1.71	0.20
NDVI	16	**τ**	0.77	0.59	54.01	0.00	TVI	50	ρ + τ	0.43	0.19	8.68	0.01
	35	ρ + τ	0.62	0.38	23.20	0.00		55	τ	0.29	0.09	3.57	0.07
	58	ρ	0.23	0.05	2.16	0.15		62	ρ	0.14	0.02	0.81	0.37
PSNDa	19	**τ**	0.76	0.57	50.71	0.00	VOG	5	**τ**	0.82	0.66	75.39	0.00
	39	ρ + τ	0.57	0.33	18.40	0.00		11	**ρ + τ**	0.78	0.61	60.69	0.00
	60	ρ	0.18	0.03	1.22	0.28		25	ρ	0.71	0.50	37.91	0.00

Determination (R^2) and correlation (COR) coefficients, as well as significance level of the relation between the 22 vegetation indices (abbreviations in Table 1) and destructively-measured chlorophyll content. Vegetation indices were calculated based on reflectance (ρ), transmittance (τ) or combination of ρ with τ (i.e., ρ + +τ) source images. Indices were ranked based on the R^2 value (all three source images were included in the ranking). The entire leaf surface was averaged (n = 40). All four species (maize, rapeseed, barley and tomato) were plotted together (examples are shown in Additional file 1: Figure S1). Plants were grown under control or deficient nitrogen levels. Bold text indicates R^2 values greater or equal to 0.5.

Eindhoven, The Netherlands), at 200 μmol m^{-2} s^{-1} photosynthetic photon flux density for 16 h per day. Plants were grown in 1 L pots, filled with soil mixture (1:1, v/v) of Hawita Dachstaudensubstrat (HAWITA GRUPPE GmbH, Vechta, Germany) and Pikiererde (Einheitserde Typ P Pikier, Balster Einheitserdewerk GmbH, Frödenberg, Germany). Plants were kept well-watered, and were weekly provided with nutrient solution (0.2%; Hakaphos Blau, COMPO GmbH & Co. KG, Münster, Germany) starting from the second week following emergence. The amount of nutrient solution was 60 mL per plant (up to 4 weeks following emergence), and was subsequently increased to 80 mL. Plants were grown for six weeks (starting from 17th of July, 2012) up to the developmental stage 18 of the BBCH scale [52].

Before inoculation, the pathogen *Cercospora beticola* was grown on agar plates (1.5%; Sigma-Aldrich Chemie GmbH, Taufkirchen, Germany), containing 50% vegetable

Table 3 Vegetation indices performance based on a linear regression model

Index	Rank	Source	COR	R^2	F	P	Index	Rank	Source	COR	R^2	F	P
ANCB	12	**τ**	0.75	0.56	49.14	0.00	PSNDb	17	**τ**	0.73	0.53	42.17	0.00
	15	**ρ + τ**	0.74	0.55	45.75	0.00		23	ρ + τ	0.65	0.43	28.49	0.00
	28	ρ	0.62	0.39	24.20	0.00		49	ρ	0.45	0.20	9.53	0.00
BGI2	5	**ρ**	0.78	0.61	59.78	0.00	PSRI	52	τ	0.34	0.12	4.97	0.03
	31	τ	−0.60	0.36	21.79	0.00		56	ρ + τ	0.27	0.07	3.05	0.09
	42	ρ + τ	0.50	0.25	12.68	0.00		64	ρ	0.05	0.00	0.09	0.77
Chlg	37	ρ + τ	0.55	0.30	16.15	0.00	PSSRa	18	**τ**	0.71	0.51	39.52	0.00
	43	ρ	0.49	0.24	11.86	0.00		50	ρ + τ	0.37	0.14	6.16	0.02
	57	τ	0.23	0.05	2.12	0.15		66	ρ	0.02	0.00	0.02	0.88
Chlre	13	**τ**	0.75	0.56	47.68	0.00	PSSRb	9	**τ**	0.76	0.58	52.50	0.00
	19	**ρ + τ**	0.71	0.50	38.59	0.00		20	ρ + τ	0.70	0.48	35.76	0.00
	34	ρ	0.59	0.34	19.80	0.00		53	ρ	0.33	0.11	4.67	0.04
EVI	1	**ρ + τ**	0.83	0.70	87.37	0.00	SR1	6	**τ**	0.78	0.60	57.58	0.00
	24	τ	0.65	0.42	27.30	0.00		16	**ρ + τ**	0.73	0.54	43.81	0.00
	40	ρ	0.52	0.27	13.72	0.00		29	ρ	0.62	0.39	23.97	0.00
MCARI	35	ρ	−0.58	0.34	19.23	0.00	SIPI	30	τ	0.61	0.37	22.38	0.00
	39	ρ + τ	−0.52	0.27	14.32	0.00		55	ρ + τ	0.29	0.08	3.39	0.07
	48	τ	0.46	0.21	10.01	0.00		63	ρ	−0.08	0.01	0.27	0.60
mND	4	**τ**	0.80	0.64	67.83	0.00	SR2	14	**τ**	0.74	0.55	45.79	0.00
	7	**ρ + τ**	0.77	0.59	55.58	0.00		47	ρ + τ	0.46	0.21	10.05	0.00
	26	ρ	0.63	0.39	24.65	0.00		65	ρ	0.04	0.00	0.06	0.81
mSR	10	**τ**	0.75	0.57	49.85	0.00	TCARI	38	ρ	−0.54	0.29	15.61	0.00
	33	ρ + τ	−0.59	0.35	20.20	0.00		45	ρ + τ	−0.48	0.23	11.14	0.00
	54	ρ	−0.31	0.10	4.15	0.05		60	τ	0.13	0.02	0.67	0.42
ND	3	**τ**	0.80	0.64	68.15	0.00	TCARI/OSAVI	32	ρ	−0.59	0.35	20.27	0.00
	11	**ρ + τ**	0.75	0.56	49.16	0.00		36	ρ + τ	−0.56	0.31	17.12	0.00
	27	ρ	0.63	0.39	24.65	0.00		59	τ	−0.13	0.02	0.68	0.41
NDVI	22	τ	0.66	0.43	28.78	0.00	TVI	44	ρ + τ	0.48	0.23	11.46	0.00
	41	ρ + τ	0.51	0.26	13.30	0.00		51	τ	0.35	0.12	5.35	0.03
	58	ρ	0.16	0.03	1.00	0.32		61	ρ	0.13	0.02	0.67	0.42
PSNDa	25	τ	0.64	0.41	26.88	0.00	VOG	2	**τ**	0.80	0.64	68.85	0.00
	46	ρ + τ	0.47	0.22	10.87	0.00		8	**ρ + τ**	0.76	0.58	52.70	0.00
	62	ρ	0.12	0.01	0.56	0.46		21	ρ	0.68	0.46	32.31	0.00

Determination (R^2) and correlation (COR) coefficients, as well as significance level of the relation between the 22 vegetation indices (abbreviations in Table 1) and destructively-measured chlorophyll content. Vegetation indices were calculated based on reflectance (ρ), transmittance (τ) or combination of ρ with τ (i.e., ρ + +τ) source images. Indices were ranked based on the R^2 value (all three source images were included in the ranking). The entire leaf surface was averaged ($n = 40$). All four species (maize, rapeseed, barley and tomato) were plotted together. Plants were grown under control or deficient nitrogen levels. Bold text indicates R^2 values greater or equal to 0.5.

juice (Gemüsemix, Eckes-Granini Deutschland GmbH, Nieder-Olm, Germany). Pathogen cultivation took place in a growth chamber, set at 60% relative air humidity and 26°C air temperature. The pathogen was kept on the agar plates for 3 weeks. To stimulate conidia production, plates were placed under ultraviolet light (340–400 nm) for 3 days before conidia collection. Subsequently, conidia were scraped with a slide, and were suspended to an aqueous solution containing Tween20 (0.1%; Sigma-Aldrich Chemie GmbH, Taufkirchen, Germany). The solution concentration was set to 3×10^4 conidia mL^{-1} using a hemocytometer (Thoma chamber, Carl Roth GmbH + Co. KG, Karlsruhe, Germany). Plant inoculation was conducted, as previously described by Schmidt *et al.* (2008) [53]. Following inoculation, plants were kept at 80–100% relative air humidity for one week. During that time, air temperature was 24 ± 0.3°C (range: 15–35). Light intensity was set to 300 μmol m^{-2} s^{-1} one day after inoculation.

The control plants were inoculated with an aqueous solution containing only Tween20 (0.1%). Measurements were conducted about three weeks after inoculation, when symptoms of *Cercospora beticola* infection were visible.

Experiment II: Nitrogen deficit

Two dicotyledonous crops [rapeseed (*Brassica rapa* cv. Campino) and tomato (*Lycopersicon esculentum* cv. Harzfeuer)] and two monocotyledonous crops [maize (*Zea mays* cv. Gelber) and barley (*Hordeum vulgare* cv. Barke)] were grown under controlled conditions in a greenhouse. During growth period, air temperature and relative air humidity were 21.1 ± 0.3°C (range: 19.7–23.9) and 51 ± 1% (range: 43–60), respectively. Supplementary light was provided by high-pressure sodium lamps, as described above, at 50 µmol m^{-2} s^{-1} photosynthetic photon flux density for 16 h per day (from 0600 to 2200 hours). The average daily light integral, including natural light, was 9.6 ± 0.5 mol m^{-2} day^{-1}. Seeds were sown and germinated in small pots until plants were grown up to the cotyledons' stage (i.e. fully open cotyledons and before the appearance of the first leaf for dicotyledonous species and the second leaf stage for monocotyledonous species). Subsequently, seedlings were transplanted into 17.4 L rectangular boxes containing a ready-made soil mixture ('Nullerde' Archut Erzeugnisse GmbH, Vechta, Germany) with low nitrogen content (50 mg L^{-1}). A planting density of five plants per box was used. Boxes were randomly distributed on a glasshouse table.

After planting, plants were exposed to different levels of nitrogen fertilisation. For this, eight boxes, containing five plants each, were separated into two different treatments: control (10 mM nitrogen concentration) and nitrogen deficit (15% of the control). Once a week, each box received 500 mL of nutrient solution [54]. The experiment lasted for 23 days, following planting (starting from 25th of April, 2013).

Point measurements

The spectral signature of selected leaves was also obtained by using a high spectral resolution point spectroradiometer (ASD FieldSpec® 3, Analytical Spectral Devices, Boulder, CO, USA), operating at a spectral range between 350 and 2500 nm. The nominal spectral resolution is 3 and 10 nm FWHM in the 350–1050 and 1050–2500 nm regions, respectively. The sampling interval is 1.4 and 2 nm in the 350–1050 and 1050–2500 nm regions, respectively. The FieldSpec fibre optic was connected to a FluoWat leaf clip [17,18]. The FluoWat leaf clip offers the possibility to measure both reflectance and transmittance of either sample side. Illumination was provided by a cold light source (Schott KL 1500 LCD 150 W, Lighting and Imaging SCHOTT AG, Mainz, Germany). Due to the spectral output of the illumination source, spectral readings were limited to the 400 to 1700 nm range. The intensity of the electromagnetic radiation emission was determined by using a 99.9% reflective white standard (Spectralon, Labsphere Inc., North Dutton, NH, USA). The spectra of the samples were divided by the white standard spectra and multiplied by a correction factor (provided by Labsphere Inc.) for calculation of the relative values. In all measurements, leaf reflectance was assessed on the adaxial (upper) surface.

Spectra comparison

To assess the validity of the spectral data acquired using the HyperART system (first prototype), simultaneous measurements using both this and the FluoWat device were performed on three disease-free sugar beet leaves from the control plant. Five FluoWat point measurements were averaged, and these were tested against five comparable points that were selected in the hypespectral image obtained by using the HyperART system. In this study we combine the spectra of two cameras by selecting nearly the same regions in both hyperspectral images and averaging them. Since the investigated plant tissue was homogeneous, and thus signal changes spatially smooth, we assume that the spectral differences do not change much when the regions of interest do not fit perfectly.

The sensors of these two compared instruments have similar spectral resolution, whereas spectral sampling differs. For comparison of the spectra obtained by the two instruments, the data were converted to a similar spectral sampling. This was done by reducing the FieldSpec spectral sampling to the one of the line scanners, by matching the wavelength.

Quantification of *Cercospora beticola* symptoms

The potential of the HyperART system in detecting and quantifying the symptoms of *Cercospora beticola* infection, on sugar beet leaves was evaluated. Three leaves of an infected plant and three leaves of an infection-free plant, were placed in the leaf clip between the mirrors of the HyperART system (first prototype). Two leaves of the infected plant had necrotic areas, whereas the third one did not have any disease symptoms. Disease severity was also estimated visually (expressed in percentage) for each leaf.

The background of the hyperspectral image was masked out using a threshold in the EVI image. After image processing, a disease specific index (*Cercospora leaf spot index* $(CLSI) = \frac{\rho698-\rho750}{\rho698+\rho750} - \rho734$; [13]) was calculated from the ρ, τ and ρ + τ data, respectively. Low values indicate low disease severity, and vice-versa. From the CLSI images, histograms with a binning of 0.01 were computed and divided by overall number of pixels. By this the normalized histograms represent estimated probability density function of CLSI values. As an indicator of how well the two modes can be distinguished, we calculated the so-called bimodal amplitude (referred as A_b). The bimodal amplitude equation ($A_b = \frac{A_m - A_v}{A_m}$; [55]) was used to analyse the CLSI histograms. For calculating A_b, the smaller peak amplitude (referred as A_m) and the amplitude of the minimum between the two peaks (referred as A_v) were computed. Peak heights (normalized frequencies) and locations were 0.021 at ρCLSI −1.21 value, 0.022 at τCLSI −1.32 value and 0.014 at ρ + τCLSI −1.42 value. The minima normalized frequencies were 0.005 at −0.75 value, 0.004 at −0.75 value, and 0.004 at −0.75 value for ρCLSI, τCLSI and ρ + τCLSI, respectively. A_b values close to one indicate a better double peak distinction.

In order to evaluate the separability performance in the index images of ρCLSI, τCLSI and ρ + τCLSI, these were firstly classified by using the K-Means unsupervised classification (ENVI 4.8; [56]). The automatic algorithm was set to separate the data in the following two classes: infected and non-infected leaf tissue. The same classes were used for a supervised classification performed by support vector machine on the ρ, τ and ρ + τ hyperspectral images. Therefore, symptomatic and symptom-free regions were selected manually in the hyperspectral images. After all images were classified, the validity of unsupervised classification performance was evaluated using post classification. Unsupervised classification results for the index images of ρCLSI, τCLSI and ρ + τCLSI were compared by confusion matrix with the respective supervised classification (ENVI 4.8). The supervised classification results were used as ground true images.

Vegetation indices

Twenty two commonly-used indices for chlorophyll content estimation at both leaf and canopy levels (Table 1) were applied to ρ, τ, and ρ + τ pre-processed hyperspectral images. Throughout the paper, the letters ρ, τ, and ρ + τ before an index indicate the source of the hyperspectral image, which was used for the calculation. Sixty six outcomes (i.e., 22 indices × 3 sources) were obtained. Both a simple linear regression and a logarithmic model were applied to the relationships between predicted (by the indices) and destructively-measured (described below) chlorophyll content.

Assessment of chlorophyll content

Sampling for biochemical assessment of chlorophyll content was conducted 23 days after planting. The third and fourth (counting from the apex) fully-expanded leaves were collected for rapeseed, barley and tomato (two leaves per plant; n = 10). The middle portion of the third fully-expanded was sampled for maize (one leaf per plant; n = 5). Chlorophyll concentration was assessed in the same leaves, where hyperspectral measurements were done. Non-destructive (hyperspectral imaging) and destructive (chlorophyll content) measurements were performed within 24 h.

Collected plant material was immediately frozen in liquid nitrogen, and stored at −80°C for further analysis. Plant material was homogenized by grinding the tissue in liquid nitrogen. About 50 mg of the homogenized sample were ground in 6 ml acetone (100% pure solvent), by using a mortar and pestle. Since chlorophyll is light sensitive, extraction took place in a dark room. Liquid nitrogen was continuously used to cool down the sample. The resulting homogenate was centrifuged in test tubes for 6 min for precipitation of the cell debris. Absorption spectra of the supernatants were recorded. Chlorophyll content was calculated by equations described by Lichtenthaler [57].

Additional file

Additional file 1: Figure S1. Logarithmic relationships between calculated indices and destructively-measured chlorophyll content. The leaves were sampled from four different crop species. The green colour indicates control plants, while the red colour refers to the nitrogen-deficient plants. Indices were calculated based on reflectance (ρ), transmittance (τ) or combination of ρ with τ (i.e., ρ + +τ). Spectra were acquired by using the hyperspectral absorption reflectance transmittance imaging (HyperART) system. Abbreviations are explained in Table 1.

Abbreviations

α: Absorption; CLSI: *Cercospora* leaf spot index; A_b: Bimodal amplitude; HyperART: Hyperspectral absorption-reflectance-transmittance imaging; NIR: Near infrared; PAR: Photosynthetic active radiation; ρ: Reflectance; τ: Transmittance.

Competing interests

The authors declare that they have no competing interests.

Authors' contributions

SB developed the HyperART system, acquired the spectral data, carried out the data analysis and interpretation, and wrote the manuscript. DF carried out the second case study, and contributed to the writing of the paper. SS was responsible for the biotic stress handling (first case study) and supported the respective measurements. MPC assisted the data analysis and interpretation. MJ, HS and UR designed experiments, participated in data interpretation and supervised the study. All authors contributed in writing, editing, proof-reading and approving the final manuscript.

Acknowledgements

This work has been conducted in frame of a research program funded by the BMBF with the project number 315309/CROP.SENSe and by the European Regional Development Fund [ERDF (En), EFRE (De)]. It was partly funded by the European Plant Phenotyping Network (EPPN -European Union FP7 Capacities Programme) through a post-doctoral grant to DF (grant agreement 284443). We thank the company KWS SAAT AG for providing the plant and pathogen material of the first case study. We are grateful to Hans Jung and the FZJ IBG-workshop/ZEA-1 personnel for assembling the prototypes. Thanks also to Dr. Mark Müller-Linow and Dr. Zbyněk Malenovský for providing the MatLab script. The authors are thankful to Anna Krapp for assistance in the preparation of figures and tables. Insightful discussions with Dr. Fabio Fiorani are greatly appreciated. Finally, we thank the editor and two anonymous reviewers for their valuable comments on the manuscript.

Author details

[1]Institute for Bio- and Geosciences, IBG-2: Plant Sciences, Forschungszentrum Jülich GmbH, 52425 Jülich, Germany. [2]Present address: Department of Crop Science, Technological Educational Institute of Crete, GR 71004 Heraklio,

Greece. [3]Present address: Institute of Viticulture, Floriculture and Vegetable Crops, Hellenic Agricultural Organization 'Demeter' (NAGREF), P.O. Box 2228, GR 71003 Heraklio, Greece. [4]Present address: LemnaTec GmbH, Pascalstraße 59, 52076 Aachen, Germany.

References

1. Jansen M, Pinto F, Nagel KA, Dusschoten D, Fiorani F, Rascher U, et al. Non-invasive phenotyping methodologies enable the accurate characterization of growth and performance of shoots and roots. In: Tuberosa R, Graner A, Frison E, editors. Genomics of plant genetic resources. Netherlands: Springer; 2014. p. 173–206.
2. Fiorani F, Schurr U. Future scenarios for plant phenotyping. Annu Rev Plant Biol. 2013;64:267–91.
3. Dhondt S, Wuyts N, Inzé D. Cell to whole-plant phenotyping: the best is yet to come. Trends Plant Sci. 2013;18:428–39.
4. Furbank RT, Tester M. Phenomics – technologies to relieve the phenotyping bottleneck. Trends Plant Sci. 2011;16:635–44.
5. Fanourakis D, Briese C, Max J, Kleinen S, Putz A, Fiorani F, et al. Rapid determination of leaf area and plant height by using light curtain arrays in four species with contrasting shoot architecture. Plant Methods. 2014;10:9.
6. Fiorani F, Rascher U, Jahnke S, Schurr U. Imaging plants dynamics in heterogenic environments. Curr Opin Biotechnol. 2012;23:227–35.
7. Rascher U, Blossfeld S, Fiorani F, Jahnke S, Jansen M, Kuhn AJ, et al. Non-invasive approaches for phenotyping of enhanced performance traits in bean. Funct Plant Biol. 2011;38:968–83.
8. Matsuda O, Tanaka A, Fujita T, Iba K. Hyperspectral imaging techniques for rapid identification of arabidopsis mutants with altered leaf pigment status. Plant Cell Physiol. 2012;53:1154–70.
9. Rumpf T, Romer C, Plumer L, Mahlein AK, Ieee. Optimal wavelengths for an early identification of cercospora beticola with support vector machines based on hyperspectral reflection data. New York: Ieee; 2010.
10. Rumpf T, Mahlein AK, Steiner U, Oerke EC, Dehne HW, Plümer L. Early detection and classification of plant diseases with support vector machines based on hyperspectral reflectance. Comput Electron Agr. 2010;74:91–9.
11. Mahlein AK, Steiner U, Dehne HW, Oerke EC. Spectral signatures of sugar beet leaves for the detection and differentiation of diseases. Precis Agr. 2010;11:413–31.
12. Mahlein AK, Steiner U, Hillnhütter C, Dehne HW, Oerke EC. Hyperspectral imaging for small-scale analysis of symptoms caused by different sugar beet diseases. Plant Methods. 2012;8:3.
13. Mahlein AK, Rumpf T, Welke P, Dehne HW, Plümer L, Steiner U, et al. Development of spectral indices for detecting and identifying plant diseases. Remote Sens Environ. 2013;128:21–30.
14. Van Wittenberghe S, Verrelst J, Rivera JP, Alonso L, Moreno J, Samson R. Gaussian processes retrieval of leaf parameters from a multi-species reflectance, absorbance and fluorescence dataset. J Photochem Photobiol B Biol. 2014;134:37–48.
15. Gitelson AA, Buschmann C, Lichtenthaler HK. Leaf chlorophyll fluorescence corrected for re-absorption by means of absorption and reflectance measurements. J Plant Physiol. 1998;152:283–96.
16. Lukes P, Stenberg P, Rautiainen M, Mottus M, Vanhatalo K, Lukeš P, et al. Optical properties of leaves and needles for boreal tree species in Europe. Rem Sens Lett. 2013;4:667–76.
17. Alonso L, Gomez-Chova L, Vila-Frances J, Amoros-Lopez J, Guanter L, Calpe J, et al. Sensitivity analysis of the fraunhofer line discrimination method for the measurement of chlorophyll fluorescence using a field spectroradiometer. 2007. 7–9 February edition. Florence, Italy: Proceedings of the 3th International Workshop on Remote Sensing of Vegetation Fluorescence.
18. Van Wittenberghe S, Alonso L, Verrelst J, Hermans I, Delegido J, Veroustraete F, et al. Upward and downward solar-induced chlorophyll fluorescence yield indices of four tree species as indicators of traffic pollution in Valencia. Environ Pollut. 2013;173:29–37.
19. Vedaldi A, Fulkerson B. VLFeat: an open and portable library of computer vision algorithms. In: Proceedings of The International Conference on Multimedia. Firenze, Italy: ACM; 2010. p. 1469–72.
20. Eller BM. Leaf pubescence - significance of lower surface hairs for spectral properties of upper surface. J Exp Bot. 1977;28:1054–9.
21. Kokaly RF, Asner GP, Ollinger SV, Martin ME, Wessman CA. Characterizing canopy biochemistry from imaging spectroscopy and its application to ecosystem studies. Remote Sens Environ. 2009;113(Supplement 1):S78–91.
22. Delegido J, Verrelst J, Rivera JP, Ruiz-Verdú A, Moreno J. Brown and green LAI mapping through spectral indices. Int J Appl Earth Observation Geoinform. 2015;35:350–8.
23. Jansen M, Bergsträsser S, Schmittgen S, Müller-Linow M, Rascher U. Non-invasive spectral phenotyping methods can improve and accelerate cercospora disease scoring in sugar beet breeding. Agriculture. 2014;4:147–58.
24. Buschmann C, Lenk S, Lichtenthaler HK. Reflectance spectra and images of green leaves with different tissue structure and chlorophyll content. Israel J Plant Sci. 2012;60:49–64.
25. Horler DNH, Dockray M, Barber J. The red edge of plant leaf reflectance. Int J Rem Sens. 1983;4:273–88.
26. Gitelson AA, Gritz Y, Merzlyak MN. Relationships between leaf chlorophyll content and spectral reflectance and algorithms for non-destructive chlorophyll assessment in higher plant leaves. J Plant Physiol. 2003;160:271–82.
27. Filella I, Penuelas J. The red edge position and shape as indicators of plant chlorophyll content, biomass and hydric status. Int J Rem Sens. 1994;15:1459–70.
28. Thomas JR, Gausman HW. Leaf reflectance vs. leaf chlorophyll and carotenoid concentrations for eight crops. Agron J. 1977;69:799–802.
29. Bock CH, Poole GH, Parker PE, Gottwald TR. Plant disease severity estimated visually, by digital photography and image analysis, and by hyperspectral imaging. Crit Rev Plant Sci. 2010;29:59–107.
30. Malenovský Z, Homolová L, Zurita-Milla R, Lukeš P, Kaplan V, Hanuš J, et al. Retrieval of spruce leaf chlorophyll content from airborne image data using continuum removal and radiative transfer. Remote Sens Environ. 2013;131:85–102.
31. Zarco-Tejada PJ, Berjón A, López-Lozano R, Miller JR, Martín P, Cachorro V, et al. Assessing vineyard condition with hyperspectral indices: leaf and canopy reflectance simulation in a row-structured discontinuous canopy. Remote Sens Environ. 2005;99:271–87.
32. Gitelson AA, Keydan GP, Merzlyak MN. Three-band model for noninvasive estimation of chlorophyll, carotenoids, and anthocyanin contents in higher plant leaves. Geophys Res Lett. 2006;33:L11402.
33. Huete AR, Liu HQ, Batchily K, van Leeuwen W. A comparison of vegetation indices over a global set of TM images for EOS-MODIS. Remote Sens Environ. 1997;59:440–51.
34. Daughtry CST, Walthall CL, Kim MS, de Colstoun EB, McMurtrey III JE. Estimating corn leaf chlorophyll concentration from leaf and canopy reflectance. Remote Sens Environ. 2000;74:229–39.
35. Sims DA, Gamon JA. Relationships between leaf pigment content and spectral reflectance across a wide range of species, leaf structures and developmental stages. Remote Sens Environ. 2002;81:337–54.
36. Gitelson A, Merzlyak MN. Quantitative estimation of chlorophyll-a using reflectance spectra: Experiments with autumn chestnut and maple leaves. J Photochem Photobiol B Biol. 1994;22:247–52.
37. Rouse JW, Haas RH, Schell JA, Deering DW. Monitoring Vegetation Systems in the Great Plains with ERTS, Third ERTS Symposium, NASA SP-351. 1973. p 309–17.
38. Penuelas J, Baret F, Filella I. Semiempirical indexes to assess carotenoids chlorophyll-a ration from leaf spectral reflectance. Photosynthetica. 1995;31:221–30.
39. Zarco-Tejada PJ, Miller JR, Noland TL, Mohammed GH, Sampson PH. Scaling-up and model inversion methods with narrowband optical indices for chlorophyll content estimation in closed forest canopies with hyperspectral data. Geosci Rem Sens IEEE Trans On. 2001;39:1491–507.
40. Birth GS, McVey GR. Measuring color of growing turf with a reflectance spectrophotometer. Agron J. 1968;60:640–9.
41. Blackburn GA. Spectral indices for estimating photosynthetic pigment concentrations: a test using senescent tree leaves. Int J Rem Sens. 1998;19:657–75.
42. Merzlyak MN, Gitelson AA, Chivkunova OB, Rakitin VYU. Non-destructive optical detection of pigment changes during leaf senescence and fruit ripening. Physiol Plant. 1999;106:135–41.
43. Haboudane D, Miller JR, Tremblay N, Zarco-Tejada PJ, Dextraze L. Integrated narrow-band vegetation indices for prediction of crop chlorophyll content for application to precision agriculture. Remote Sens Environ. 2002;81:416–26.
44. Broge NH, Leblanc E. Comparing prediction power and stability of broadband and hyperspectral vegetation indices for estimation of green leaf area index and canopy chlorophyll density. Remote Sens Environ. 2001;76:156–72.

45. Vogelmann JE, Rock BN, Moss DM. Red edge spectral measurements from sugar maple leaves. Int J Rem Sens. 1993;14:1563–75.
46. Buschmann C, Nagel E. In vivo spectroscopy and internal optics of leaves as basis for remote sensing of vegetation. Int J Rem Sens. 1993;14:711–22.
47. Main R, Cho MA, Mathieu R, O'Kennedy MM, Ramoelo A, Koch S. An investigation into robust spectral indices for leaf chlorophyll estimation. ISPRS J Photogram Rem Sens. 2011;66:751–61.
48. le Maire G, François C, Dufrêne E. Towards universal broad leaf chlorophyll indices using PROSPECT simulated database and hyperspectral reflectance measurements. Remote Sens Environ. 2004;89:1–28.
49. Asner GP, Martin RE. Spectral and chemical analysis of tropical forests: scaling from leaf to canopy levels. Remote Sens Environ. 2008;112:3958–70.
50. Delegido J, Alonso L, González G, Moreno J. Estimating chlorophyll content of crops from hyperspectral data using a normalized area over reflectance curve (NAOC). Int J Appl Earth Observation Geoinform. 2010;12:165–74.
51. Ji-Yong S, Xiao-Bo Z, Jie-Wen Z, Kai-Liang W, Zheng-Wei C, Xiao-Wei H, et al. Nondestructive diagnostics of nitrogen deficiency by cucumber leaf chlorophyll distribution map based on near infrared hyperspectral imaging. Sci Hortic. 2012;138:190–7.
52. Hess M, Barralis G, Bleiholder H, Buhr L, Eggers TH, Hack H, et al. Use of the extended BBCH scale—general for the descriptions of the growth stages of mono; and dicotyledonous weed species. Weed Res. 1997;37:433–41.
53. Schmidt K, Pflugmacher M, Klages S, Maser A, Mock A, Stahl DJ. Accumulation of the hormone abscisic acid (ABA) at the infection site of the fungus Cercospora beticola supports the role of ABA as a repressor of plant defence in sugar beet. Mol Plant Pathol. 2008;9:661–73.
54. Hoagland DR, Arnon DI. Growing Plants without Soil by The Water-Culture Method. 1938
55. Zhang C, Mapes BE, Soden BJ. Bimodality in tropical water vapour. Q J Roy Meteorol Soc. 2003;129:2847–66.
56. MacQueen J. Some methods for classification and analysis of multivariate observations. In: Proceedings of the Fifth Berkeley Symposium on Mathematical Statistics and Probability, Volume 1: Statistics; 1967. Berkeley, Calif: University of California Press; 1967. p. 281–97.
57. Lichtenthaler HK. Chlorophylls and carotenoids: Pigments of photosynthetic biomembranes. Meth Enzymol. 1987;148:350–82.

EMS mutagenesis in mature seed-derived rice *calli* as a new method for rapidly obtaining TILLING mutant populations

Xavier Serrat[1,2*], Roger Esteban[1], Nathalie Guibourt[1], Luisa Moysset[2], Salvador Nogués[2] and Eric Lalanne[1]

Abstract

Background: TILLING (Targeting Induced Local Lesions IN Genomes) is a reverse genetic method that combines chemical mutagenesis with high-throughput genome-wide screening for point mutation detection in genes of interest. However, this mutation discovery approach faces a particular problem which is how to obtain a mutant population with a sufficiently high mutation density. Furthermore, plant mutagenesis protocols require two successive generations (M1, M2) for mutation fixation to occur before the analysis of the genotype can begin.

Results: Here, we describe a new TILLING approach for rice based on ethyl methanesulfonate (EMS) mutagenesis of mature seed-derived *calli* and direct screening of *in vitro* regenerated plants. A high mutagenesis rate was obtained (i.e. one mutation in every 451 Kb) when plants were screened for two senescence-related genes. Screening was carried out in 2400 individuals from a mutant population of 6912. Seven sense change mutations out of 15 point mutations were identified.

Conclusions: This new strategy represents a significant advantage in terms of time-savings (i.e. more than eight months), greenhouse space and work during the generation of mutant plant populations. Furthermore, this effective chemical mutagenesis protocol ensures high mutagenesis rates thereby saving in waste removal costs and the total amount of mutagen needed thanks to the mutagenesis volume reduction.

Keywords: *Oryza sativa*, *Scutellum*, Mature seed, *Calli*, Mutagenesis, EMS, TILLING, Reverse genetics

Background

Rice (*Oryza sativa*) is one of the most important food crops in the world. It is also a model cereal plant [1] for molecular biology and genetics due to its small genome size relative to other cereals, the availability of the entire genome sequence [2], it's ease of transformation and regeneration, and the availability of a variety of mutants. Since sequencing of the rice genome was completed in December 2004 [2], functional genomics has been used to determine the function of all of the approximately 50,000 annotated genes [3,4]. This objective has already been reached through the development of a variety of gene knockout strategies [5-7].

There are 3 ways in which to induce mutations, by either using: 1) biological agents such as transposons and T-DNA, 2) physical agents such as fast neutron, UV and x-ray radiation, or 3) chemical agents such as N-methyl-N-nitrosourea (MNU), 1,2:3,4-diepoxybutane (DEB) or ethyl methanesulfonate (EMS). Among these compounds, EMS has become one of the most effective, reliable, powerful and frequently used chemical mutagens in plants [8]. EMS mainly induces C–to-T substitutions resulting in C/G to T/A transitions [9,10] and at a low frequency, EMS generates G/C to C/G or G/C to T/A transversions through 7-ethylguanine hydrolysis or A/T to G/C transitions through 3-ethyladenine pairing errors [9-12].

Irradiation and chemical mutagenesis have long been used to produce mutant plants for breeding proposes [13,14]. Molecular screening of mutations was developed much later after the efficiency of chemical mutagens producing small deletions and point mutations had been improved [8,9,15,16] and DNA sequencing allowed for the identification of such point mutations. Nevertheless, direct sequencing in large populations is a slow and

* Correspondence: xserrat1@hotmail.com
[1]Oryzon Genomics, S.A., Cornellà de Llobregat, Spain
[2]Departament de Biologia Vegetal, Universitat de Barcelona, Barcelona, Spain

expensive process. Therefore, several mutation detection techniques based on physical properties were developed [17-24] before enzymatic mismatch detection methods provided the key to efficient mutant population screening. The endonucleases obtained from *Aspergillus* [25], *Vigna radiata* [26] or *Penicillium* [27] were the first enzymes used to detect DNA mismatches. Oleykowski et al. [28] improved this technique by using the CEL I endonuclease obtained from *Apium graveolens* combined with an electrophoresis step, thereby paving the way for high throughput screening technology. Since then, the enzymatic detection methods in combination with high throughput genotyping have been improved for the efficient detection of genetic polymorphisms [29-32]. Consequently, there has been growing interest in using irradiation and chemical mutagenesis in model organisms for use in functional genomics research [33,34].

Chemically induced mutant populations have been generated in different plant species [11] and efficiently screened following Targeting Induced Local Lesions IN Genomes (TILLING) high-throughput screening protocols [35-37]. These combine random chemical mutagenesis with polymerase chain reaction (PCR) amplification of target genes, heteroduplex formation and identification of a range of allele changes [38] by using enzymatic mismatch cleavage and electrophoresis. Achieving a genome-wide saturated mutant population in plant species with large genomes is challenging. Small genome species such as rice are more suitable for TILLING [30,39,40]. As a result, many rice mutant populations have been efficiently screened using this technique [36,41-43].

In TILLING chemical mutagenesis protocols, germinating seeds are incubated in a mutagenic solution. The first generation (*M1*) that is produced directly from the mutagenic treatment cannot be screened because the majority of generated mutations are somatic and are not transmitted to the progeny [30]. To solve this problem, the *M1* mutant population has to be grown and then self-fertilized. The mutations in *M1* sexual structures can produce whole mutant *M2* descendants, thereby avoiding any ambiguities caused by mosaicism. The resulting *M2* progeny can be screened for mutations.

Tissue culture methods and mutagenesis techniques currently available could significantly shorten the breeding process and overcome some substantial agronomic and environmental problems. In most cases, the embryo-derived rice *callus* regeneration is only generated from a few cells. Thus, the regenerated *M1* plantlets from mutant *calli* could be screened directly without waiting for a self-pollinated *M2* population. Few attempts at mutagenesis for breeding purposes in rice using immature embryos, *calli* derived from mature seeds or single zygotic cells in recently fertilized spikelets have been reported [44-47]. Recently, mutagenesis of suspension-cultured rice cells for phenotypic detection of mutants has been reported [48]. However, to date no studies of chemical mutagenesis in mature seed-derived rice *calli* in order to obtain mutant populations for TILLING have been reported.

The aim of this work is to carry out a new TILLING strategy based on the production of a plant mutant population from EMS mutagenised embryo-derived *calli* followed by a mutational screening on the regenerated plants. This mutational screening focusses on two genes related to senescence since this developmental process in annual cereal crop plants overlaps with the reproductive phase and may reduce crop yield when it is induced prematurely under adverse environmental conditions.

Results

Mutagenised population

In order to till rice we followed a new approach that differs from the traditional TILLING procedure (described in the introduction section, see above, Figure 1) in two aspects: i) mutagenesis was applied to embryo-derived *calli*, and ii) mutational screening was carried out on regenerated plantlets after acclimatization.

Rice seeds were cultured in *callus* induction *media* (OryCIM) for three weeks before *Scutellum*-derived *callus* masses were picked and mutagenised avoiding those *calli* obtained from the radicle (Figure 2a and b). No apparent differences were detected between partially disaggregated mutagenised and non-mutagenised *calli* except that the first showed a certain degree of browning. Both *calli* grew normally when cultured in OryCIM *media* for four weeks. Plantlets from both mutagenised and control *calli* started regenerating just three weeks after the growing *callus* masses were transferred to regeneration (MSM) media (Figure 2d). Mutant plantlets did not show any apparent phenotypic differences with respect to control plantlets. The control material was discarded after observing that the regeneration rate was satisfactory and similar in both treatments.

The *callus* regeneration process yielded 6912 individual plantlets obtained from 395 different mutagenised *callus* masses. From these, 2400 plantlets were sampled and their DNA was extracted, pooled fourfold and organized into a 96-well format for TILLING screening.

Molecular screening

Two target genes of agronomic interest were selected: *OsACS1* (Os03g0727600) and *OsSGR* (Os09g0532000). Among the six ACS isozymes identified in rice, *Os*ACS1 is the most closely related (86% identity) to ACS6 of *Zea mays* (Swiss-prot: Q3ZTU2), a protein encoded by the *ZmACS6* gene whose expression is largely responsible for directing natural, dark-induced and drought-induced senescence in maize [49,50]. *OsSGR* exists as a single

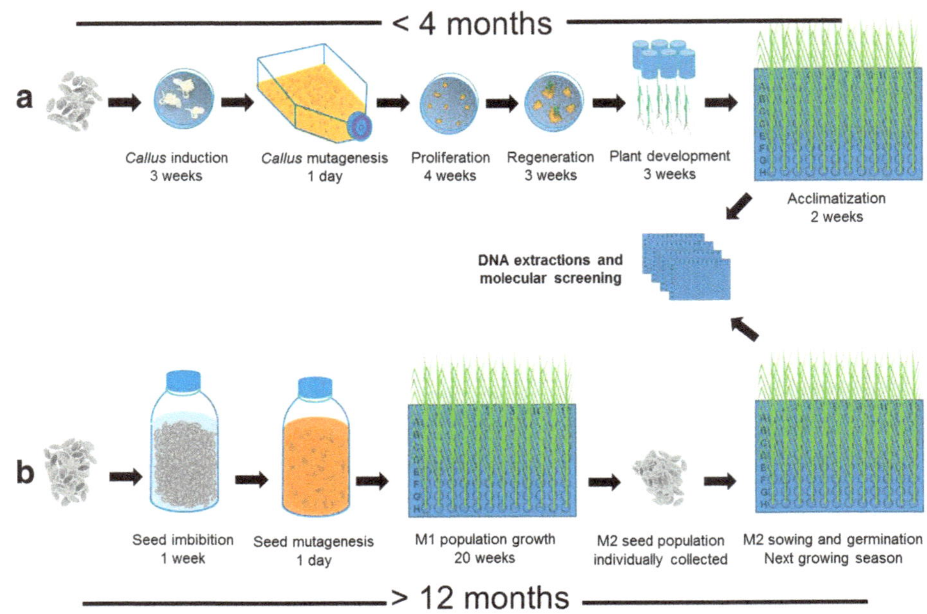

Figure 1 Diagram comparing callus mutagenesis and seed mutagenesis protocols in rice TILLING. (a) In the proposed rice TILLING protocol through *callus* mutagenesis, *calli* are induced, mutagenised, and the regenerated plants provide DNA for molecular screening of mutations. **(b)** In the basic TILLING method, seeds are mutagenised, the resulting M1 plants are self-fertilized and the M2 generation of individuals is used to prepare DNA samples for mutational screening while their seeds are inventoried. The steps represented cover from *callus* induction **(a)** or seed imbibition **(b)** to DNA extractions for molecular screening of mutations. Duration of each step is indicated.

Figure 2 Plantlets regeneration of *Oryza sativa* var. Hispagram from EMS mutagenised *calli*. (a) Oryza sativa cv. Hispagran dehusked seeds forming callus masses after 18 days culture in darkness using N6 medium supplemented with 0.5 mg L^{-1} casaminoacids, 1 g L^{-1} L-proline, 2 mg L^{-1} 2,4 dichlorophenoxyacetic acid and 0.5 g L^{-1} 2-(N-morpholino) ethane sulphonic acid. **(b)** development of a scutellum-derived callus mass (black arrow) and a callus mass growing from radicle (black circle), **(c)** and **(d)** plantlet regeneration from mature seed-derived callus (0.2% EMS mutagenised) in MS medium supplemented with 1 g L^{-1} casein hydrolisate, 3 mg L^{-1} kinetin, 0.5 mg L^{-1} 6-benzylaminopurin, 0.5 mg L^{-1} 1-naphtalenacetic acid and 0.5 g L^{-1} 2-(N-morpholino) ethane sulphonic acid under 18/6 h light/dark cycles (details in Methods). Scale bar 1 cm.

copy in the rice genome and is mapped onto the long arm of chromosome 9 [51].

OsACS1 is a four exon rice gene that codes for a 487 amino acid protein [Swiss-prot: Q10DK7]. *OsACS1* was screened in two fragments called *OsACS1* 1–3 and *OsACS1* 4. *OsACS1* 1–3 is a 1014 base pair (bp) fragment harboring the first three exons, while *OsACS1* 4 is a 1480 bp fragment spanning exon 4 (Figure 3).

Figure 4 shows the detection of three *OsACS1* mutants (*acs1* 152 s3, *acs1* 228 s1 and *acs1* 576 s1) based on different heteroduplex banding patterns (Figure 4a) and the identification of nucleotides changes by sequencing (Figure 4b). *Acs1* 152 s3 resolve into four bands of approximately 250, 450, 550 and 750 bp and has two missense mutations, T→G and G→A transitions at gene nucleotide positions 58 and 174, in the target fragment *OsACS1* 1–3 (Table 1). The G→A transition generates a modification (GT→AT) in the splicing donor site of the first intron. The T→G change resulted in the amino acid substitution C→G at position 20 in exon 1 (C20G, Table 1). *Acs1 228 s2, acs1* 398 s4 and *acs1* 576 s1 are missense mutations in the target fragment *OsACS1* 4 which result in the amino acid substitutions S314N, A246P and L354P at gene positions 1266, 1220 and 1351 bp respectively (Table 1). In addition, one mutation in intron 3 (*acs1* 43 s3, T→A at gene position 707 bp), one silent mutation in exon 1 (*acs1* 418 s2, C→T at gene position 84 bp) and one silent mutation in exon 3 (*acs1* 558 s2, G→A at gene position 535 bp) were detected in the *OsACS1* 1–3 fragment. Furthermore, two silent mutations in exon 4 coded *acs1* 83 s2 (C→T at gene position 1177 bp) and *acs1* 364 s2 (G→A at gene position 1669 bp), and one +73 bp C→T downstream mutation were also detected in the *OsACS1* 4 fragment (Table 1). A total of 11 nucleotide changes in the *OsACS1* gene were detected after screening 2400 individuals. This is a one in every 457 Kb, 2.19 e^{-6} mutation frequency.

The analysis of the deduced mutant amino acid sequences using the GOR (Garnier-Osguthorpe-Robson) Secondary Structure Prediction application [52] is summarized in Table 1. According to the Eukaryotic Linear Motif resource for Functional sites in Proteins (ELM) [53], the conserved functional motif (50–432 amino acids) of the wild type rice protein ACS1 is divided into two sub-domains 50–148 amino acids and 275–404 amino acids.

The C20G (*acs1* 152 s3 mutant) change is located in the first functional domain and generates a weak α-helix domain modification (+1) (Table 1). The mutations L354P (576 s1 mutant) and S314N (228 s1 mutant) are located in the second functional sub-domain resulting in a loss and a weak modification of an α-helix motif respectively (Table 1). The change A246P (*acs1* 398 s4 mutant) does not affect any functional domain; it produces a complete α-helix domain lost mutation. Finally, the *acs1* 152 s3 mutant which has a GT→AT change at the intron 1 splicing acceptor site may affect splicing.

OsSGR contains three exons that code for a 274 amino acid protein [Swiss-prot:Q652K1]. A single 926 bp fragment harbouring the first two exons was screened in this study (Figure 3b). Following the same strategy described above, one mutation in exon 1 (*sgr* 24 s1 G→A at gene position 67 bp), one mutation in exon 2 (*sgr* 389 s2 G→A at gene position 484), one silent mutation in exon 1 (*sgr* 855 s2 C→T at gene position 78 bp) and one mutation in intron 1 (*sgr* 854 s1 G→A gene position

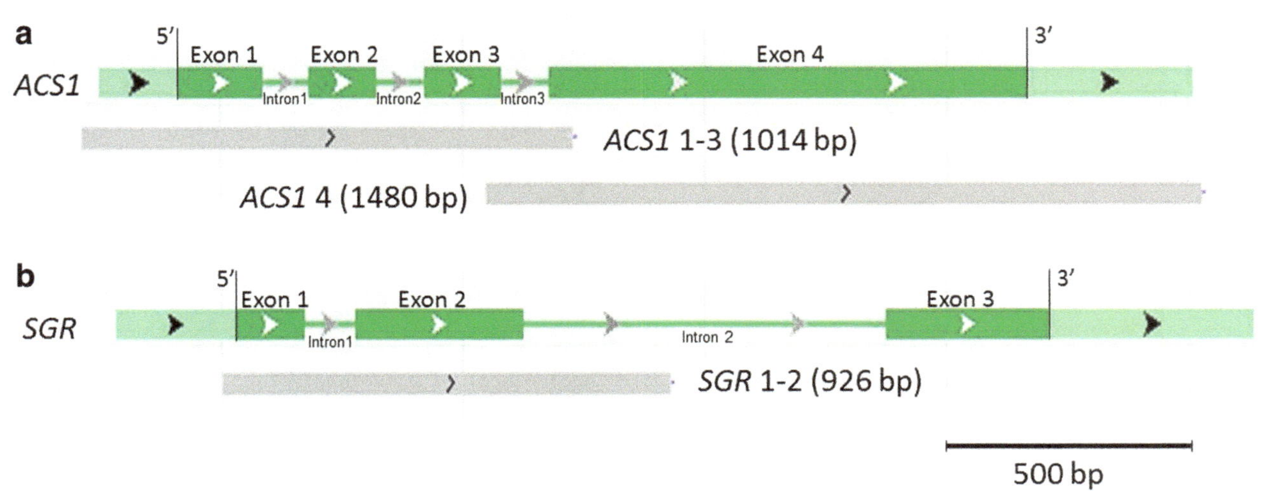

Figure 3 Targeted genes diagrams and PCR amplicons. Diagrammatic representation of the OsACS1 (Os03g0727600) **(a)** and OsSGR (Os09g0532000) **(b)** genes obtained by blast searches at NCBI. Exons are represented by dark green boxes, introns by dark red lines and 5′-UTR and 3′-UTR by pale green boxes. Exons are designated Exon 1-Exon 4 (OsACS1) or Exon 1-Exon 3 (OsSGR). Intron numbering follows exon numbering. Targeted screened gene segments (grey boxes) are designated ACS1 1–3, ACS1 4 and SGR 1–2.

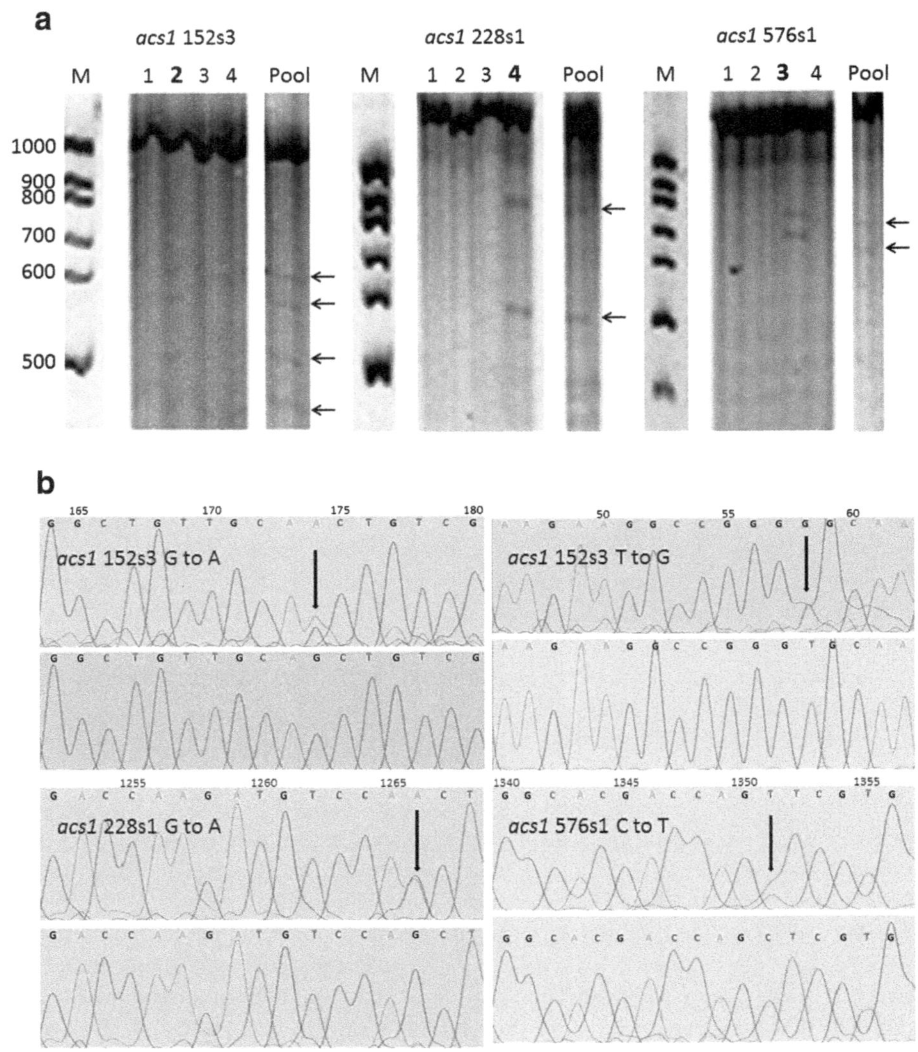

Figure 4 Identification of *OsACS1* mutants. (a) Heteroduplex mobility assays identifying three of the OsACS1 mutants (acs1 152 s3, acs1 228 s1 and acs1 576 s1). Denaturating polyacrylamide gel electrophoresis showing different heteroduplex DNA band patterns formed after PCR amplification and mismatch cleavage by Fennel Crude Extract (FCE) incubation. The presence of two bands of about 600 and 850 bp in the case of acs1 228 s1 sample and 700 and 750 bp in the case of acs1 576 s1 sample, indicate heteroduplex digestions in the 1480 bp amplicon (ACS1 4). The presence of two mutations in the same 1014 bp amplicon (ACS1 1–3) of the same individual (acs1 152 s3) generates four bands of about 250, 450, 550 and 750. M: molecular weight marker (100 bp marker, Thermo Fisher Scientific Inc.). Pool: positive mismatch pool formed by mixing four individual DNA samples. Arrows indicate mismatch digested bands. Lanes 2–5: individuals DNA samples of the positive mismatch pool. Mutant samples are indicated by bold number lane. **(b)** Identification of nucleotide changes by sequencing. Heterozygous acs1 mutants (upper panel) and wt (lower panel) sequences are represented. Black arrows indicate the base substitution. Direct nucleotide sequencing of the acs1 152 s3 mutant revealed the heterozyogus G to A and T to G transitions at gene nucleotides position 174 and 58 respectively which resulted in substitution of GT/AG at intron 1 5'UTR and the amino acid C to G change at position 20 (C20G). Direct nucleotide sequencing of acs1 228 s1 and acs1 576 s1 mutants revealed the heterozigous G to A and C to G transitions at gene nucleotides position 1266 and 1351 which resu ted in substitution of amino acids S to N and L to P at amino acid positions 314 (S314N) and 354 (L354P) respectively.

238 bp) were detected after screening 2400 individuals. This is a one in every 436 Kb, 2.30 e^{-6} mutation frequency. The GOR application predicted that amino acid substitutions A23T and V127M induced modification in α-helix domains in exon 1 (*sgr* 24 s1) and exon 2 (*sgr* 389 s2) respectively (Table 1).

Individual mosaicism was discarded after re-analysing 8 new independent leaf samples from all of the 14 detected mutant individuals, while the incidence of multiple clone regeneration was studied by revising mismatch cleavage results of all plants that had originated from the same *callus* which had regenerated the 14 detected mutant plants. In total, 112 leaf samples were subjected to mismatch cleavage detection and sequencing, resulting in no individual mosaicism (Additional file 1: Table S1). All the mismatch cleavage results from samples of plants

Table 1 Mutations discovered in the *Oryza sativa* cv. Hispagran population obtained from 0.2% EMS-mutagenised *callus*

Mutant gene	*Mutant code*	*Target fragment*	*Nucleotide change*	*Gene bp position*	*Effect*	*Predicted protein structure modification*
acs1	152 s3	*ACS1* 1-3	T- > G	58	C20G	Exon 1 α-helix domain modification (weak effect)
acs1	418 s2	*ACS1* 1-3	C- > T	84	S28S	Exon 1 silent mutation
acs1	152 s3	*ACS1* 1-3	G- > A	174	GT- > AT	Intron 1 5'UTR splicing donor site GT modification that may affect splicing
acs1	558 s2	*ACS1* 1-3	G- > A	535	Q114Q	Exon 3 silent mutation
acs1	43 s3	*ACS1* 1-3	T- > A	707		Intron 3 mutation
acs1	83 s2	*ACS1* 4	C- > T	1177	F231F	Exon 4 silent mutation
acs1	398 s4	*ACS1* 4	G- > C	1220	A246P	Exon 4 α-helix domain lost
acs1	228 s1	*ACS1* 4	G- > A	1266	S314N	Exon 4 α-helix domain modification (+1)
acs1	576 s1	*ACS1* 4	C- > T	1351	L354P	Exon 4 α-helix domain modification affecting functional domain
acs1	364 s2	*ACS1* 4	G- > A	1669	L395L	Exon 4 silent mutation
acs1	408 s3	*ACS1* 4	C- > T	+73		Downstream mutation
sgr	24 s1	*SGR9* 1-2	G- > A	67	A23T	Exon 1 α-helix domain modification (weak effect)
sgr	855 s2	*SGR9* 1-2	C- > T	78	L26L	Exon 1 silent mutation
sgr	854 s1	*SGR9* 1-2	G- > A	238		Intron 1 mutation
sgr	389 s2	*SGR9* 1-2	G- > A	484	V127M	Exon 2 α-helix domain modification (+2)

Confirmed *acs1* and *sgr* mutants detected by TILLING screening and its predicted effect using GOR (Garnier-Osguthorpe-Robson) Secondary Structure Prediction application.

that had originated from the same *callus* which had generated the 14 mutant plants resulted negative (Additional file 1: Table S1).

In total 15 mutations were obtained in both target genes after screening 2400 individuals. This is a one in every 451 Kb, a 2.22 e^{-06} mutation frequency which is useful for reverse genetic studies and breeding proposes [54].

Homozygous mutant lines

Mutants were self-pollinated in order to obtain homozygous mutants. Selection was completed in two steps. First, in order to detect and discard heterozygous mutants, individual DNA was subjected to PCR amplification, denaturation and re-annealing to perform heteroduplexes, and FCE incubation and electrophoretic analysis to detect FCE-cut products. Then, with the aim of detecting and selecting homozygous mutants, individual samples of potential homozygous mutants selected in step 1 were mixed with wild type DNA (1:1 w/w). From this point onwards, screening was identical including PCR amplification, FCE digestion and electrophoresis. An example of this screening using 16 descendants of a self-fertilized *sgr* 389 s2 mutant is presented in Figure 5. In the case of heterozygous descendants, mismatch cleaved bands of 380 and 506 bp were observed, while no bands were detected in homozygous descendants (either wild type or mutants) (Figure 5a). In the second step, mismatch cleaved bands only appear in samples where wild type DNA is mixed with DNA from homozygous mutants (Figure 5b).

Four wild type, 7 heterozygous and 5 homozygous mutant lines were obtained from the *sgr* 389 s2 descendants, X^2 statistic for "goodness of fit" with the expected Mendelian segregations was $X^2 = 0.375$ confirming the null hypothesis (2 degrees of freedom, $p = 0.05$) (Additional file 1: Table S1). Preliminary results indicated that they were completely fertile and showed a delayed senescence phenotype. In contrast, strong effect mutations of *OsACS1* such as *acs1* 228 s1, 398 s4 and 576 s1 (Additional file 2: Table S2) were partially sterile, no mutant homozygotes were obtained and about 25% of seeds were unable to germinate. Segregations did not fit Mendelian segregation, but when scoring non-germinated seeds as lethal homozygous mutants, X^2 test for goodness-of-fit with expected segregation resulted in X^2 = 1.500, $X^2 = 1.444$ and $X^2 = 0{,}200$ respectively, being less than 5.991 when considering 2 degrees of freedom and $p = 0.05$ (Additional file 2: Table S2).

Discussion

A new rice *callus* mutagenesis protocol using EMS was established. Mutant plants were efficiently obtained by mutagenizing *scutellum*-derived *callus* masses containing primary, embryogenic and non-embryogenic *calli* [55] (Figure 2). Root *callus* was the only *callus* mass to be discarded as it is unable to regenerate plants and it is

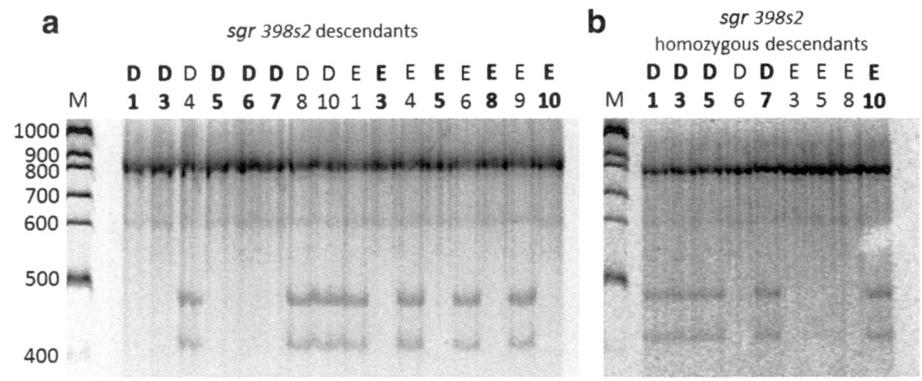

Figure 5 Identification of homozygous mutants among *OsSGR* 389 s2 mutant descendants. (a) Electrophoretic analysis to discard heterozygous mutant plants obtained by self-fertilization of sgr 389 s2. Individual DNA samples were subjected to PCR amplification, denaturation, re-annealing, Fennel Crude Extract (FCE) heteroduplex mismatch cleavage and denaturing polyacrylamide gel analysis. The 926 bp band corresponding to the amplified fragment is present in all samples, while cleaved bands (380 and 506 bp) were detected in heterozygous mutant descendants (D4, D8, D10, E1, E4, E6 and E9). No cleavage bands were observed either in wild type or mutant homozygous descendants (D1, D3, D5, D6, D7, E3, E5, E8 and E10) highlighted in bold. **(b)** Electrophoretic analysis to discard homozygous wild type descendants. Homozygous descendants (D1, D3, D5, D6, D7, E3, E5, E8 and E10) individual DNA was mixed with wild type DNA (1:1 w/w) and subjected to PCR amplification, denaturation, re-annealing, FCE mismatch cleavage incubation and polyacrylamide gel analysis. Cleaved bands (380 and 506 bp) were detected in homozygous mutant descendants (D1, D3, D5, D7 and E10 in bold). Homozygous wild type individuals DNA could not form mismatches, and consequently no mismatch cleaved bands were visualized (D6, E3, E5 and E8). M: Molecular weight marker (100 bp marker, Thermo Fisher Scientific Inc.).

easy to identity as it grows separately from the radicle. The expensive and time consuming embryogenic *callus* selection process which is commonly used in transformation protocols proved to be unnecessary for successful *callus* mutagenesis, since embryogenic *calli* are contained in *scutellum*-derived *callus* masses [55] (Figure 2b).

In order to ensure sufficient callus availability 1,200 Hispagran seeds were sown, although only less than 20% of total *calli* was used for mutagenesis. It would have been possible to mutagenise the same amounts of *calli* using less than 200 *callus* forming seeds, although the capability of obtaining and regenerating embryogenic *callus* depends on the cultivar and culture media used [55]. Nevertheless, it is highly recommendable to use mutagenesis flask replicates in order to ensure that fungal and/or bacterial contamination free *in vitro* material is obtained.

In general, in seed-propagated plants, the chemical mutagenesis protocols use a seeds under germination process, so that the mutagen has to be absorbed by the germinating embryo and reach the meristematic region where the germ cells are contained. Other alternative plant material has been mutagenised, such as pollen, microspores, single zygotic cells in recently fertilized eggs and suspension cultured cells. Although pollen mutagenesis has been performed in maize, no pollen mutagenesis attempts have been reported to date in rice as pollen lifetime is too short and manual pollination is much more complicated than in maize [56-58]. Iftikhar and Mumtaz [59] mutagenised microspores using EMS for a practical mutation breeding programme in the genetic improvement of oilseed brassicas. On the other hand, Suzuki et al. [46] found high mutagenesis rates when treating single zygotic cells in recently fertilized rice spikelets by using N-methyl-N-nitrosourea. However, a low cell survival was obtained and the resulting seeds (*M1*) had to be grown, fertilized and harvested until the *M2* population was ready for screening. Recently, suspension-cultured cell mutagenesis using EMS has been reported [48] where regenerated mutant plants were self-pollinated in order to obtain 302 *M2* lines for phenotypic analysis in field conditions which subsequently achieved high mutagenesis rates.

In this context, *callus* mutagenesis is more effective when compared to the traditional mutagenesis technique in seeds since this technique allows the mutagen agent to easily reach the target uncoated embryogenic cells rather than complex fully-formed embryos. Considering that somatic embryogenesis is the main regeneration method in the culturing of rice *in vitro*, and that somatic embryos arise from single cells, each mutated single cell can develop into a somatic embryo and regenerate a mutant plant. [48]. Furthermore, all mutants were unique; no clones were detected in the mutant population after screening the whole population and upon revision of all the results involving any plant that had shared the same *callus* of origin.

The intensity of mutagenesis applied is an important component in a TILLING project, and it is necessary to find a compromise between mutagen toxicity, genome mutation saturation and possible accumulation of undesirable phenotypes. In our study, 0.2% EMS for 2 hours was effective to generate a whole rice mutant population with a sufficiently high mutation density. This mutagen dose applied in order to mutagenize rice seeds is in the range of values (0.2-2%), however, the duration of treatment is lower, 2 hours versus 6 hours reported by Chakravarti et al. [60] (0.2% EMS), and 12 hours reported by Wu et al. [43] (0.4% to 1% EMS) and Talebi et al. [61] (0.25 to 2% EMS). Regarding rice cell culture, Chen et al. [48] reported that a treatment of 0.4% EMS for 18–22 hours is optimal in order to induce mutagenesis. Consequently, the effectiveness of our low doses and low volume of mutagen presents an advantage since it implies a considerable financial saving and reduces the amount of residues generated. The mutagenized *callus* regeneration rate was similar to that of non-mutagenized control *calli*, which means that this treatment neither causes apparent lethality, which is one of the main problems associated with chemical mutagenesis [61], nor affects the plantlet regeneration rate.

Characterization of the two mutant populations obtained through screening with *ACS* and *SGR* target genes revealed mutation densities of 1/457 kb and 1/436 Kb, respectively, which are satisfactory and suitable for high throughput TILLING [42]. These mutations densities are in the range of those found by Wu et al. (1 per 1 MB for cv. IR64) [43], Till et al. (1 per 300 Kb for cv. Nipponbare) [42] and Suzuki et al. (1 per 135 Kb for cv. Taichung 65) [46] using rice seed mutagenesis protocols.

In twelve out of 15 mutants the most common EMS induced mutation was present, which is the C/G to T/A ($C \rightarrow T$ or $G \rightarrow A$) substitution [9,12], while one case of G/C to C/G ($C \rightarrow G$) transition was detected as expected in *Arabidopsis* EMS mutagenesis experiments [10]. Although $T \rightarrow A$ transversions are the second largest expected in *Drosophila* EMS mutantagenesis experiments [62], $T \rightarrow A$ and $T \rightarrow G$ detected transversions were unexpected in EMS rice mutagenesis. The origin of these mutations is unknown. We used 1200 seeds obtaining 395 *callus* masses which subsequently regenerated 6912 plantlets, and from these, 2400 individual mutants were screened. Polymorphisms in the Hispagran cv. seeds batch used to generate the mutant population could not explain these mutations since at least more than one other identical mutant should have been found, however all mutants were unique and furthermore, no other identical mutants were observed within the mutant population, even in those sharing the same *callus* of origin.

In this work, mutations were detected using a heteroduplex digestion assay with crude fennel juice extract (FCE) instead of crude celery juice extract (CJE) [63] or CEL I nuclease from celery, the most common enzyme used in TILLING projects [42,64]. The disadvantage of using juice extracts is that they contain multiple mismatch-cleaving enzymes which collaborate in the digestion of heteroduplex DNA substrates. However, Till et al. [65] concluded that juice extracts and highly purified preparations in optimal conditions yielded similar mismatch detection results. Crude FCE exhibits lower mismatch-cleavage activity than that of CJE [66] however, it cleaves A/C and T/G mismatches preferentially, matching them with the most likely substitutions induced by EMS treatment, while the commercial CEL1 purified enzyme cleaves C/C mismatches preferentially [28]. In addition, it is inexpensive when compared with the substantially higher cost of the CEL I purified enzyme. Nevertheless, FCE nucleases mismatch cleavage was under study, and its 3′-5′ exonuclease activity and other aspects implied the need for specific incubation conditions [67].

Functional analysis of the *acs* and *sgr* rice mutant lines would be needed in order to shed some light on the segregation of these genes and the roles that they play in the improvement of rice culture (i.e. delayed senescence and increase in rice yield), through performing field trials. To date, we are not able to predict the effect of *OsACS1* mutations on senescence since the amino acidic changes detected do not affect critical amino acids for catalysis, interaction and correct orientation of pyridoxal 5′-phosphate and substrate recognition [68] or any of the seven strongly conserved regions described by Wong et al. [69]. With respect to *sgr* mutants, the A23T and V127M changes do not correspond to amino acids critical for the correct functioning SGR proteins of rice [51,70], pepper and tomato [71]. As far as we currently know, all rice *sgr* recessive mutants obtained to date [51,70] belong to non-functional type C–*sgr* mutants in which chlorophylls are retained in senescent leaves as their photosynthesis efficiency decreases [72]. Given that SGR is a highly conserved protein in plants and does not show a large degree of similarity to any other proteins, we are not able to predict the effect of the two mutations which were obtained. Homozygous lines obtained from both mutants are completely fertile, and they remain greener longer than in Hispagran plants. Therefore, these results suggest that they could be functional stay-green rice mutants. If this is the case, these mutants could be the first functional *sgr* mutants in rice to have been found and they could potentially be useful for rice improvement. On the other hand, if these mutants produce a non-functional SGR phenotype, they would be useful for the study of the chlorophyll degradation pathway. Further work is now underway to understand the effect of these mutations on rice.

Conclusions

In conclusion, our results showed that combined EMS mutagenesis in *callus* with FCE heteroduplex digestion assay is a powerful tool for the identification and genetic characterization of rice mutants. In our study, we were able to identify 15 nucleotide changes. The estimated mutation density is in the range of that previously reported for rice.

Our mutagenesis protocol avoids the problem of the inhibitory effect of the mutagen on seed germination and can be adapted to any *callus* induction/regeneration media with two modifications which are: the removal of agar and the addition of antioxidants during mutagenesis and rinsing. In addition, *callus* mutagenesis makes it possible to rapidly obtain a mutant population with a time saving of more than eight months when compared to classical seed mutagenesis. Furthermore, it saves greenhouse resources and work, the amount of mutagen needed to produce a mutant population and, consequently, the amount of residues generated.

This methodological approach could be easily adapted to any rice variety or even other plant species (e.g. cereals) and also cell suspensions.

Methods

Plant material

Hispagran temperate japonica rice (*Oryza sativa*) cultivar is a high yielding variety grown in Extremadura and Seville (Spain). Certified Hispagran seeds were supplied by the *Instituto Hispánico del Arroz, S.A.* (Hisparroz). The possibility of polymorphisms in targeted gene fragments of Hispagran was studied in advance by sequencing the targeted fragments in 80 wild type Hispagran individuals. The nucleotidic sequences were compared with the GenBank database of Nipponbare japonica rice cultivar (GenBank: AP008209.2).

Callus induction

To ensure enough *calli* were obtained, 1200 Hispagran certified seeds were dehusked and surface disinfected after soaking and stirring for 1 minute in 70% ethanol followed by 30 minutes in bleach sterilization solution; 40% commercial bleach supplemented with 8 drops of Tween 20 (Sigma-Aldrich, Madrid, Spain) per litre. After five rinses (5 minutes per rinse) using sterilized water, seeds were sown in solid *Callus* Induction Media (OryCIM), based on rice *Callus* Induction Media [73] that was optimized to Mediterranean japonica rice varieties; Chu N6 [74] standard salts and vitamins were supplemented with 0.5 g L^{-1} casaminoacids, 1 g L^{-1} L-proline, 2 mg L^{-1} 2,4 dichlorophenoxyacetic acid (2,4 D), 0.5 g L^{-1} 2-(N-morpholino) ethane sulphonic acid (MES), and 30 g L^{-1} sucrose. The pH 5.7 was adjusted using 1 M KOH solutions and 2.5 g L^{-1} Gelrite™ was added before autoclaving. Sterilin 90 mm petri dishes (Sterilin LTD, Cambridge) were filled with 25 mL media after autoclaving. Fourteen sterilized seeds were sown in each OryCIM petri dish, and plates were sealed and incubated for three weeks in complete darkness, at 28°C (Figure 6). Contaminated seeds were discarded and *scutellum*-derived *callus* masses growing close to the embryo were selected avoiding smaller root producing *calli* growing from the radicle [55]. All media components were supplied by Duchefa (Duchefa Biochemie BV, The Netherlands) with the exception of casaminoacids (Becton, Dickinson and Company).

Chemical mutagenesis

Callus masses were transferred into three 375 mL NUNC EasyFlasks™ with filter caps for continuous venting (Thermo Fisher Scientific Inc.) containing 96 mL OryCIM liquid media without 2,4-D, but supplemented with 125 mg L^{-1} L-ascorbic acid and 125 mg L^{-1} citric acid (Liquid OryCIM) to prevent *callus* oxidation in further *callus* mutagen treatments. *Callus* masses were transferred until the final volume reached 110 mL in each flask. Next, two flasks were immediately mutagenised by adding EMS (0.2% v/v) (Sigma-Aldrich, Madrid, Spain) following the safety instructions provided by the manufacturer's Material Safety Data Sheet (MSDS). No EMS was added to the control flask. Flasks were placed on an orbital shaker (150 rpm) and covered with aluminium foil to avoid light for two hours. After incubation, every *callus* batch was rinsed ten times using 200 mL liquid OryCIM media and removing it after 3 to 4 minutes. After rinsing was complete, *calli* were incubated on modified liquid OryCIM media which was supplemented with 2 mg L^{-1} 2–4 D and shaken (120 rpm) for two additional days (28°C, darkness).

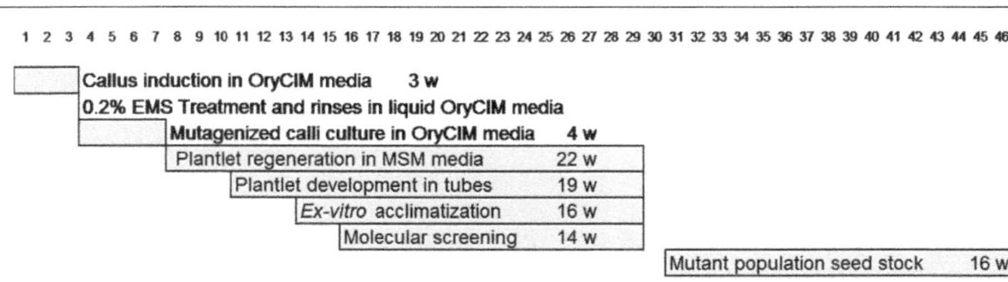

Figure 6 Schedule of the experiment. Schematic representation of TILLING schedule based on callus mutagenesis. The duration of each step (weeks, w) is indicated.

Plantlet regeneration

Calli were partially dried in a laminar flow cabinet on sterile cellulose paper for about 30 minutes before sowing them in solid OryCIM fortified with 4.5 g L^{-1} Gelrite™ (Duchefa) in 90 mm diameter dish. Fourteen *callus* masses per dish were transferred and cultured for four weeks in darkness at 28°C (Figure 6).

Calli were transferred to regeneration media (MSM) petri dishes (90 mm diameter) and cultured under a 18/6 hours light/dark photoperiod under 70 μmol m^{-2} s^{-1} fluorescent light, and at 28°C until plantlets were fully-formed 30 days later. This MSM media was based on MS [75] standard salts and vitamins, fortified with 1 g L^{-1} casein hydrolisate, 3 mg L^{-1} kinetin, 0.5 mg L^{-1} 6-benzylaminopurine, 0.5 mg L^{-1} 1-naphthaleneacetic acid, 0.5 g L^{-1} MES and 30 g L^{-1} sucrose. The pH was adjusted (5.8) by using 0.5 M HCl solution and 4.5 g L^{-1} Gelrite™ was added before autoclaving. *Calli* were transferred every three/four weeks to fresh MSM media until the end of the experiment ca. 22 weeks (Figure 6).

Regenerated plantlets were individually numbered and its *callus* mass origin was recorded as soon as they were sub-cultured into culture tubes (17 cm high/2 cm width) filled with 10 mL of root media (RM) [73]. Plants were cultured for three more weeks to enhance rooting and leaf development under a 18/6 hours light/dark photoperiod 70 μmol m^{-2} s^{-1} fluorescent light, at 28°C (Figure 6).

Mutant population seed stock

Rooted plantlets were *ex vitro* acclimatized (Figure 6) as follows. Media was carefully removed from roots using tap water and plants were transplanted into 96-well multi-pots, each pot filled with 35 cm^3 of specially designed Floratorf™ peat moss (Floragard Vertriebs, Oldenburg, Germany) - vermiculite (2:1 v/v) substrate that was supplemented with Osmocote™ (The Scotts Company LLC, Ohio, USA) controlled release fertilizer mix [Osmocote Exact™ (15 + 9 + 9 + 3 MgO + micronutrients) 6-month release and Osmocote™ high K (11 + 5 + 15 + 1.2 MgO + micronutrients) 9-month release (1:1), 1gr L^{-1} of substrate]. Then 1 gr of $CaCO_3$ per peat litre was added to adjust the substrate to pH 6. Multi-pots were placed in 54 × 31 × 4.6 cm plastic trays with holes ensuring a maximum 1 cm flooding irrigation. Supplementary fertilization was supplied during panicle formation period by adding high ammonium (50% total N) soluble fertilizer (NPK 19 + 6 + 6) supplemented with 4% w/w micronutrients and 4% w/w iron chelate diluted in osmotized water, final electro-conductivity was adjusted to 1000–1200 micro Siemens.

The mutants identified by the TILLING screening were transplanted from multi-pot wells to 4 litre pots to obtain as many seeds as possible, while the rest of plantlets growing in multi-pots where discarded as soon as a minimum 20 seeds per plant was obtained. Seeds were collected in individually labelled envelopes, and stored as if they were an *M3* mutant population generated by seed mutagenesis (Figure 6).

DNA extraction and sample pooling

About 0.5 cm^2 of newly formed leafs were sampled from each plant using 96-well sample boxes containing 50–100 μg quartz (Merck KGaA, Darmstadt, Germany) and two 4 mm glass balls (Merck) placed on ice in polystyrene boxes to avoid DNA degradation. Samples were stored at −80°C and the completely frozen material was disrupted for 1 minute using a Mixer Mill (Retsch). Genomic DNA phenol-chloroform extraction: 200 μl buffer 1 (100 mM Tris–HCl pH 8, 50 mM EDTA pH 8, 500 mM NaCl and SDS 20%) was added to each sample and boxes were incubated at 65°C for 30 minutes. Two hundred μl of phenol:chloroform:isoamyl alcohol (24:25:1) was added to every extraction tube and the entire box was vortexed before being centrifuged at 3,000 rpm for 30 minutes at room temperature. Two hundred and twenty μl from the upper phase was transferred to a clean 96-well plate and kept for further analysis. DNA quality and quantity were determined with a combination of electrophoresis gel analysis and gel determination software and the same amount of DNA from 4 distinct plants was pooled together and organized into 96-well format, this being the starting material for TILLING mutant screening.

Target genes selection and primer design

The *ZmACS6* gene is involved in senescence regulation and drought tolerance in maize [49,50]. Several *ACS* genes were found in rice and maize after a BLAST was performed using the *ZmACS6* protein sequence. *OsACS1* was selected as the best candidate after homology studies using CLUSTAL W2 [76]. *OsSGR* has been related to Chlorophyll catabolism [51]. These two genes were selected as target genes.

Primer design was carried out using the Primer 3 program application [77]. A set of primers (*OsSGR 1–2*) was designed to amplify the first two exons *of OsSGR*. Two sets of primers were designed to amplify 2 segments of the *OsACS1* gene: The first segment (*OsACS1 1–3*) was designed to amplify from the first to the third exon, and the second (*OsACS1 4*) was designed to amplify the fourth exon (Figure 3) (Table 2). Eighty plants were sequenced for *OsACS1* and *OsSGR* fragments and compared with Nipponbare japonica cultivar sequences available at NCBI.

Target gene amplification and heteroduplex DNA formation

Fragments of target genes were amplified by polymerase chain reaction (PCR) using 1–4 ng DNA, 200 μM dNTPs

Table 2 Gene specific primers used for the amplification of targeted genes *OsACS1* and *OsSGR* design for target genes amplification

Target gene	*Primer name*	*Nucleotide sequences 5'-3'*	*Amplicon size (bp)*	*Amplicon name*
OsACS1	*ACS1* 1–3 F	TAAGCAGCTCGTCCAACCTT	1014	*ACS1* 1-3
	ACS1 1–3 R	CAGTGCACGGGTACGATCT		
	ACS1 4 F	CTCATCCCCACCCCATACTA	1480	*ACS1* 4
	ACS1 4 R	CCCAAATGTGGGAGTGGTAG		
OsSGR	*SGR* 1–2 F	TAAGAGATCCGAGGGAGCAG	926	*SGR*
	SGR 1–2 F	ACAGATGGATGGATGCCAAT		

Primers were designed using Primer3 application and used for the *OsACS1* (Os03g0727600) and *OsSGR* (Os09g0532000) molecular screening of mutations. F: forward. R: reverse. Position of amplicons are shown in Figure 2.

and 0.4 μM Hispagran genome specific primer sets (Table 2), and 5U Pfu DNA polymerase in 20 μL 1X Pfu DNA polymerase reaction buffer including 1X final concentration of 2 mM $MgCl_2$ (Fermentas). The number of cycles, the times and the annealing temperatures were optimized for each specific primer pairs. DNA denaturation and slow re-annealing step (from 95°C to 4°C) processes were added to the end of the PCR amplification thermal cycle programs in order to induce heteroduplex DNA formation within the different amplification products obtained from the pools.

An aliquot of each PCR product (ca. 5 μL) was checked on a 1% w/v agarose gel and submitted to electrophoresis to verify the efficiency of the PCR amplification and the size of the amplified product.

Fennel crude extract heteroduplex mismatch cleavage detection

The amplified DNA from each pool was subsequently incubated with FCE [67] for heteroduplex mismatch cleavage: 80 ng of heteroduplex DNA, 1 μl FCE, 20 mM Tris–HCl pH 7.5, 25 mM KCl and 20 mM $MgCl_2$ (final volume 10 μl). After 30 minutes incubation at 45°C, the reaction was stopped by adding 2 μl of stop buffer (20 mM Tris–HCl pH 8, 10 mM EDTA, 12.5% v/v Glycerol, 50% v/v Sybr gold-dimethyl sulfoxide and 0.05% w/v Bromophenol Blue). Digested samples were size-fractionated by polyacrylamide gel electrophoresis (4% acrylamide:bisacrilamide 19:1 in Tris/Borate/EDTA supplemented with 0.008% v/v tetramethylethylenediamine and 0.002% w/v ammonium persulphate) at 300 V and 15–20 mA for 1.5 hours [67]. Results were analysed using a Typhoon 8600 scan (GE Healthcare) with a fluorescein filter (526 nm) for Sybr-Gold stain detection.

Mutant individual identification

Individual plates containing the four individual DNA samples from each positive pool were similarly analysed after mixing (1:1, w/w) with control wild type DNA. PCR product from validated individuals were purified using Microcon™ filters purification kit (Millipore) and 5–40 ng (depending on DNA fragment length) were sequenced on a ABI Prism 3700 (Applied Biosystems) sequencer using Terminator polymerase (Applied Biosystems). In order to detect mosaicism, 8 DNA samples from eight different leafs from all 15 identified mutants (120 samples) were extracted and subjected to targeted fragment amplification, mismatch formation, enzymatic digestion gel analysis and sequencing. Furthermore, wherever a mutant plant was detected, plants that had originated from the same *callus* were tracked and the results of the analysis were carefully revised.

Data analysis

The nucleotide sequences were analysed using Chromas Lite 2.1 (Technelysium Pty Ltd) giving the nature and the exact localisation of the mutation within the gene. Mutant amino acid sequences were submitted to the GOR (Garnier-Osguthorpe-Robson) Secondary Structure Prediction application [52] and listed in Table 1.

Mutation frequencies

The mutation rate was calculated by dividing the total number of observed mutations by the total surveyed DNA length. The total surveyed DNA length was calculated by the sum of the screened individuals multiplied by the base pair (bp) sum of amplified fragments. We subtracted 200 bp from each TILLED fragment to account for the difficulty in detecting digested DNA products that migrate in the top and bottom range of the gel like other authors [42].

Additional files

Additional file 1: Table S1. Individual mosaicism and multiple clone regeneration detection. All the independent leaf samples from each mutant individual yield identical results, consequently no individual mosaicism was detected. The incidence of clones regenerated from the same callus is represented as the number of TILLED plants originating from the same *callus* related to the number of additional mutant clones detected, being only 1 in all cases.

Additional file 2: Table S2. Total number of wild type (WT), heterozygous mutant (het) and homozygous mutant descendants obtained from predicted strong effect mutants. Sgr 389 s2 segregation X^2 test for goodness-of-fit ($X^2 = 0.375$) confirms the null hypothesis being less than 0.5991 (2 degrees of freedom, p = 0.05). In the case of acs1 398 s4, 228 s1 and 576 s1 mutants, non-germinated seeds have to be considered homozygous mutants in order to predict recessive lethality inheritance ($X^2 = 1.500$, $X^2 = 1.444$ and $X^2 = 0.200$ respectively). For the rest of mutations, the progeny was not studied.

Competing interests
The authors declare that they have no competing interests.

Authors' contributions
XS adapted *in vitro* culture media, mutagenesis and regeneration protocols. XS and EL conceived the study and carried out the *callus* mutagenesis. XS directed the In Vitro process. NG and RE designed the molecular screening and developed the FCE endonuclease mismatch cleavage detection. XS drafted the manuscript. The final script main contributors were XS, SN, LM and RE. All authors read and approved the final script.

Acknowledgements
This work has been carried out as part of the HISPA Project for the INSTITUTO HISPANICO DEL ARROZ SA (HISPARROZ) "Mejora del rendimiento de la variedad Hispagran y reducción del tamaño y mejora del rendimiento de la variedad Bomba de arroz mediante mutagénesis química". S.N. would like to thank the OPTIMA and PALEOISOCROP projects for their financial support.

References
1. Shimamoto K, Kyozuka J: **Rice as a model for comparative genomics of plants.** *Annu Rev Plant Biol* 2002, **53**:399–419.
2. IRGSP: **The map-based sequence of the rice genome.** *Nature* 2005, **436**(7052):793–800.
3. Feng Q, Zhang Y, Hao P, Wang S, Fu G, Huang Y, Li Y, Zhu J, Liu Y, Hu X, *et al*: **Sequence and analysis of rice chromosome 4.** *Nature* 2002, **420**(6913):316–320.
4. Sasaki T, Matsumoto T, Yamamoto K, Sakata K, Baba T, Katayose Y, Wu J, Niimura Y, Cheng Z, Nagamura Y, *et al*: **The genome sequence and structure of rice chromosome 1.** *Nature* 2002, **420**(6913):312–316.
5. Miyao A, Tanaka K, Murata K, Sawaki H, Takeda S, Abe K, Shinozuka Y, Onosato K, Hirochika H: **Target site specificity of the Tos17 retrotransposon shows a preference for insertion within genes and against insertion in retrotransposon-rich regions of the genome.** *Plant Cell* 2003, **15**(8):1771–1780.
6. Jiang SY, Ramachandran S: **Natural and artificial mutants as valuable resources for functional genomics and molecular breeding.** *Int J Biol Sci* 2010, **6**(3):228–251.
7. Bhat RS, Upadhyaya NM, Chaudhury A, Raghavan C, Qiu F, Wang H, Wu J, McNally K, Leung H, Till B, *et al*: **Chemical- and irradiation-induced mutants and TILLING.** In *Rice functional genomics: challenges, progress and prospects.* Edited by Upadhyaya NM. New York: Springer; 2007:149–180.
8. Brockman HE, de Serres FJ, Ong TM, DeMarini DM, Katz AJ, Griffiths AJ, Stafford RS: **Mutation tests in *Neurospora crassa*. A report of the U.S. Environmental Protection Agency Gene-Tox Program.** *Mutat Res* 1984, **133**(2):87–134.
9. Krieg DR: **Ethyl methanesulfonate-induced reversion of bacteriophage T4rII mutants.** *Genetics* 1963, **48**:561–580.
10. Kim Y, Schumaker KS, Zhu JK: **EMS mutagenesis of Arabidopsis.** *Methods Mol Biol* 2006, **323**:101–103.
11. Kodym A, Afza R: **Physical and chemical mutagenesis.** *Methods Mol Biol (Clifton, NJ)* 2003, **236**:189–204.
12. Greene EA, Codomo CA, Taylor NE, Henikoff JG, Till BJ, Reynolds SH, Enns LC, Burtner C, Johnson JE, Odden AR, *et al*: **Spectrum of chemically induced mutations from a large-scale reverse-genetic screen in Arabidopsis.** *Genetics* 2003, **164**(2):731–740.
13. Stadler LJ: **Genetic effects of x rays in maize.** *Proc Natl Acad Sci USA* 1928, **14**:69–75.
14. Oehlkers F: **Die Auslosung von Chromosomenmutationen in der Meiosis durch Einwirkung von Chemikalien.** *Z Vererbungs Lehre* 1943, **81**:313–341.
15. Oeschger NS, Hartman PE: **ICR-induced frameshift mutations in Histidine operon of Salmonella.** *J Bacteriol* 1970, **101**(2):490.
16. Carlson PS: **Induction and isolation of auxotrophic mutants in somatic cell cultures of *Nicotiana tabacum*.** *Science* 1970, **168**(3930):487.
17. Oefner PJ, Underhill PA: **Comparative DNA-sequencing by denaturing high-performance liquid-chromatography (DHPLC).** *Am J Hum Genet* 1995, **57**(4):1547.
18. Ganguly A, Rock MJ, Prockop DJ: **Conformation-sensitive gel electrophoresis for rapid detection of single-base differences in double-stranded PCR products and DNA fragments: evidence for solvent-induced bends in DNA heteroduplexes.** *Proc Natl Acad Sci USA* 1994, **91**(11):5217.
19. Rozycka M, Collins N, Stratton MR, Wooster R: **Rapid detection of DNA sequence variants by conformation-sensitive capillary electrophoresis.** *Genomics* 2000, **70**(1):34–40.
20. Nataraj AJ, Olivos-Glander I, Kusukawa N, Highsmith WE: **Single-strand conformation polymorphism and heteroduplex analysis for gel-based mutation detection.** *Electrophoresis* 1999, **20**(6):1177–1185.
21. Myers RM, Maniatis T, Lerman LS: **Detection and localiztion of single base changes by denaturing gradient del-electrophoresis.** *Methods Enzymol* 1987, **155**:501–527.
22. Khrapko K, Hanekamp JS, Thilly WG, Belenkii A, Foret F, Karger BL: **Constant denaturant capillary electrophoresis (CDCE): a high resolution approach to mutational analysis.** *Nucleic Acids Res* 1994, **22**(3):364–369.
23. Hsia AP, Wen TJ, Chen HD, Liu ZW, Yandeau-Nelson MD, Wei YL, Guo L, Schnable PS: **Temperature gradient capillary electrophoresis (TGCE) - a tool for the high-throughput discovery and mapping of SNPs and IDPs.** *Theor Appl Genet* 2005, **111**(2):218–225.
24. Ainsworth PJ, Surh LC, Coultermackie MB: **Diagnostic Single-Strand Conformational Polymorphism, (SSCP). A simplified nonradiosisotopic method as applied to a Tay-Sachs-B1 variant.** *Nucleic Acids Res* 1991, **19**(2):405–406.
25. Germond JE, Vogt VM, Hirt B: **Activity of single-strand-specific nuclease S1 on double-stranded DNA.** *Experientia* 1973, **29**(6):771.
26. Kole R, Sierakow H, Szemplin H, Shugar D: **Mung bean nuclease: mode of action and specificity vs synthetic esters of 3'-nucleotides.** *Nucleic Acids Res* 1974, **1**(5):699–706.
27. Fujimoto M, Kuninaka A, Yoshino H: **Purification of a nuclease from *Penicillium citrinum*.** *Agric Biol Chem* 1974, **38**(4):777–783.
28. Oleykowski CA, Bronson Mullins CR, Godwin AK, Yeung AT: **Mutation detection using a novel plant endonuclease.** *Nucleic Acids Res* 1998, **26**(20):4597–4602.
29. Borevitz JO, Nordborg M: **The impact of genomics on the study of natural variation in Arabidopsis.** *Plant Physiol* 2003, **132**(2):718–725.
30. Henikoff S, Comai L: **Single-nucleotide mutations for plant functional genomics.** *Annu Rev Plant Biol* 2003, **54**:375–401.
31. Winzeler EA, Shoemaker DD, Astromoff A, Liang H, Anderson K, Andre B, Bangham R, Benito R, Boeke JD, Bussey H, *et al*: **Functional characterization of the *S. cerevisiae* genome by gene deletion and parallel analysis.** *Science* 1999, **285**(5429):901–906.
32. Pattemore J, Henry R: **High-throughput genotyping of cereals.** In *Cereals 2007: proceedings of 57th Australian Cereal Chemistry Conference, Vic, 5–10.* Edited by Panozzo JF, Black CK. Melbourne: Cereal Chemistry Division, Royal Australian Chemical Institute; 2007.
33. Liu LX, Spoerke JM, Mulligan EL, Chen J, Reardon B, Westlund B, Sun L, Abel K, Armstrong B, Hardiman G, *et al*: **High-throughput isolation of *Caenorhabditis elegans* deletion mutants.** *Genome Res* 1999, **9**(9):859–867.
34. Nadeau JH, Frankel WN: **The roads from phenotypic variation to gene discovery: mutagenesis versus QTLs.** *Nat Genet* 2000, **25**(4):381–384.
35. McCallum CM, Comai L, Greene EA, Henikoff S: **Targeted screening for induced mutations.** *Nat Biotechnol* 2000, **18**(4):455–457.
36. Sikora P, Chawade A, Larsson M, Olsson J, Olsson O: **Mutagenesis as a tool in plant genetics, functional genomics, and breeding.** *Int J Plant Genomics* 2011, **2011**:314829.
37. Till BJ, Colbert T, Reynolds SH, Enns L, Codomo C, Johnson JE, Burtner C, Young K, Henikoff JG, Greene EA, *et al*: **High-throughput TILLING for functional genomics.** *Plant Biology (Rockville)* 2002, **2002**:18.
38. Xu Y, McCouch S, Zhang Q: **How can we use genomics to improve cereals with rice as a reference genome?** *Plant Mol Biol* 2005, **59**(1):7–26.

39. Koornneef M, Dellaert LW, van der Veen JH: **EMS- and radiation-induced mutation frequencies at individual *loci* in *Arabidopsis thaliana* (L.) Heynh.** *Mutat Res* 1982, **93**(1):109–123.
40. Hirochika H, Guiderdoni E, An G, Hsing YI, Eun MY, Han CD, Upadhyaya N, Ramachandran S, Zhang Q, Pereira A, *et al*: **Rice mutant resources for gene discovery.** *Plant Mol Biol* 2004, **54**(3):325–334.
41. Krishnan A, Guiderdoni E, An G, Hsing YI, Han CD, Lee MC, Yu SM, Upadhyaya N, Ramachandran S, Zhang Q, *et al*: **Mutant resources in rice for functional genomics of the grasses.** *Plant Physiol* 2009, **149**(1):165–170.
42. Till BJ, Cooper J, Tai TH, Colowit P, Greene EA, Henikoff S, Comai L: **Discovery of chemically induced mutations in rice by TILLING.** *BMC Plant Biol* 2007, **7**:19.
43. Wu JL, Wu C, Lei C, Baraoidan M, Bordeos A, Madamba MR, Ramos-Pamplona M, Mauleon R, Portugal A, Ulat VJ, *et al*: **Chemical- and irradiation-induced mutants of indica rice IR64 for forward and reverse genetics.** *Plant Mol Biol* 2005, **59**(1):85–97.
44. Gao MW, Cai QH, Liang ZQ: ***In Vitro* culture of hybrid indica rice combined with mutagenesis.** *Plant Breed* 1992, **108**(2):104–110.
45. Maluszynski M, Ahloowalia B, Sigurbjörnsson B: **Application of in vivo and in vitro mutation techniques for crop improvement.** *Euphytica* 1995, **85**(1–3):303–315.
46. Suzuki T, Eiguchi M, Kumamaru T, Satoh H, Matsusaka H, Moriguchi K, Nagato Y, Kurata N: **MNU-induced mutant pools and high performance TILLING enable finding of any gene mutation in rice.** *Mol Genet Genomics* 2008, **279**(3):213–223.
47. Wu D, Shu Q, Xia Y: ***In Vitro* mutagenesis induced novel thermo/photoperiod-sensitive genic male sterile indica rice with green-revertible xanthan leaf color marker.** *Euphytica* 2002, **123**(2):195–202.
48. Chen Y-L, Liang H-L, Ma X-L, Lou S-L, Xie Y-Y, Liu Z-L, Chen L-T, Liu Y-G: **An efficient rice mutagenesis system based on suspension-cultured cells.** *J Integr Plant Biol* 2013, **55**(2):122–130.
49. Young TE, Meeley RB, Gallie DR: **ACC synthase expression regulates leaf performance and drought tolerance in maize.** *Plant J* 2004, **40**(5):813–825.
50. Gallie DR, Meeley R, Young T: **Zea mays seeds and plants with reduced expression of the ACS6 gene.** In *Pioneer Hi Bred International Inc The Regents of the University of California*; 2012.
51. Park SY, Yu JW, Park JS, Li J, Yoo SC, Lee NY, Lee SK, Jeong SW, Seo HS, Koh HJ, *et al*: **The senescence-induced staygreen protein regulates chlorophyll degradation.** *Plant Cell* 2007, **19**(5):1649–1664.
52. Kloczkowski A, Ting TL, Jernigan RL, Garnier J: **Combining the GOR V algorithm with evolutionary information for protein secondary structure prediction from amino acid sequence.** *Proteins* 2002, **49**(2):154–166.
53. Dinkel H, Michael S, Weatheritt RJ, Davey NE, Van Roey K, Altenberg B, Toedt G, Uyar B, Seiler M, Budd A, *et al*: **ELM-the database of eukaryotic linear motifs.** *Nucleic Acids Res* 2012, **40**(D1):D242–D251.
54. Kurowska M, Daszkowska-Golec A, Gruszka D, Marzec M, Szurman M, Szarejko I, Maluszynski M: **TILLING - a shortcut in functional genomics.** *J Appl Genetics* 2011, **52**(4):371–390.
55. Pons M, Marfà V, Melé E, Messeguer J: **Regeneration and genetic transformation of Spanish rice cultivars using mature embryos.** *Euphytica* 2000, **114**(2):117–122.
56. Nuffer MG: **Additional evidence on the effect of X-Ray and ultraviolet radiation on mutation in maize.** *Genetics* 1957, **42**(3):273–282.
57. Amano E, Smith HH: **Mutations induced by ethyl methanesulfonate in maize.** *Mutat Res* 1965, **2**(4):344–351.
58. Mottinger JP: **The effects of X rays on the bronze and shrunken loci in maize.** *Genetics* 1970, **64**(2):259–271.
59. Iftikhar A, Mumtaz A: ***In vitro* mutagenesis in oil seed brassica.** *Pakistan J Biotechnol* 2004, **1**(1):1–5.
60. Chakravarti SKR, Kumar H, Lal JP, Vishwakarma MK: **Induced mutation in traditional aromatic rice - frequency and spectrum of viable mutations and characterizations of economic values.** *The Bioscan, an international quarterly jounal of life sciences* 2012, **7**(4):739–742.
61. Talebi R, Rahemi MR, Arefi H, Nourozi M, Bagheri N: **In vitro plant regeneration through anther culture of some Iranian local rice (*Oryza sativa* L.) cultivars.** *Pak J Biol Sci* 2007, **10**(12):2056–2060.
62. Cooper JL, Till BJ, Henikoff S: **Fly-TILL Reverse genetics using a living point mutation resource.** *Fly* 2008, **2**(6):300–302.
63. Uauy C, Paraiso F, Colasuonno P, Tran RK, Tsai H, Berardi S, Comai L, Dubcovsky J: **A modified TILLING approach to detect induced mutations in tetraploid and hexaploid wheat.** *BMC Plant Biol* 2009, **9**(115):1–14.
64. McCallum CM, Comai L, Greene EA, Henikoff S: **Targeting Induced Local Lesions IN Genomes (TILLING) for plant functional genomics.** *Plant Physiol (Rockville)* 2000, **123**(2):439–442.
65. Till BJ, Burtner C, Comai L, Henikoff S: **Mismatch cleavage by single-strand specific nucleases.** *Nucleic Acids Res* 2004, **32**(8):2632–2641.
66. Zolala J, Bahrami AR, Farsi M, Matin MM, Yassaee VR: **Comparison of CEL I gene expression and mismatch-cleavage activity in some Apiaceae plants.** *Mol Breed* 2009, **24**(1):17–24.
67. Esteban R: *Generació i detecció de mutacions en organismes d'interès comercial.* Barcelona: Universitat de Barcelona; 2010.
68. Jakubowicz M: **Structure, catalytic activity and evolutionary relationships of 1-aminocyclopropane-1-carboxylate synthase, the key enzyme of ethylene synthesis in higher plants.** *Acta Biochim Pol* 2002, **49**(3):757–774.
69. Wong WS, Ning W, Xu PL, Kung SD, Yang SF, Li N: **Identification of two chilling-regulated 1-aminocyclopropane-1-carboxylate synthase genes from citrus (Citrus sinensis Osbeck) fruit.** *Plant Mol Biol* 1999, **41**(5):587–600.
70. Jiang HW, Li MR, Liang NB, Yan HB, Wei YL, Xu X, Liu JF, Xu Z, Chen F, Wu GJ: **Molecular cloning and function analysis of the stay green gene in rice.** *Plant J* 2007, **52**(2):197–209.
71. Barry CS, McQuinn RP, Chung M-Y, Besuden A, Giovannoni JJ: **Amino acid substitutions in homologs of the STAY-GREEN protein are responsible for the green-flesh and chlorophyll retainer mutations of tomato and pepper.** *Plant Physiol* 2008, **147**(1):179–187.
72. Thomas H, Howarth CJ: **Five ways to stay green.** *J Exp Bot* 2000, **51**:329–337.
73. Toki S: **Rapid and efficient Agrobacterium-mediated transformation in rice.** *Plant Mol Biol Report* 1997, **15**(1):16–21.
74. Chu CC, Wang CC, Sun CS, Chen H, Yin KC, Chu CY, Bi FY: **Establishment of an efficient medium for another culture of rice through comparative experiments on the nitrogen sources.** *Sci Sinica* 1975, **18**(5):659–668.
75. Murashige T, Skoog F: **A revised medium for rapid growth and bio assays with tobacco tissue cultures.** *Physiol Plant* 1968, **15**:473–497.
76. Larkin MA, Blackshields G, Brown NP, Chenna R, McGettigan PA, McWilliam H, Valentin F, Wallace IM, Wilm A, Lopez R, *et al*: **Clustal W and clustal X version 2.0.** *Bioinformatics* 2007, **23**(21):2947–2948.
77. Ye J, Coulouris G, Zaretskaya I, Cutcutache I, Rozen S, Madden TL: **Primer-BLAST: a tool to design target-specific primers for polymerase chain reaction.** *BMC Bioinformatics* 2012, **13**:134.

17

WallProtDB, a database resource for plant cell wall proteomics

Hélène San Clemente[1,2] and Elisabeth Jamet[1,2*]

Abstract

Background: During the last fifteen years, cell wall proteomics has become a major research field with the publication of more than 50 articles describing plant cell wall proteomes. The *WallProtDB* database has been designed as a tool to facilitate the inventory, the interpretation of cell wall proteomics data and the comparisons between cell wall proteomes.

Results: *WallProtDB* (http://www.polebio.lrsv.ups-tlse.fr/WallProtDB/) presently contains 2170 proteins and ESTs identified experimentally in 36 cell wall proteomics studies performed on 11 different plant species. Two criteria have to be met for entering *WallProtDB*. First one is related to the identification of proteins. Only proteins identified in plant with available genomic or ESTs data are considered to ensure unambiguous identification. Second criterion is related to the difficulty to obtain clean cell wall fractions. Indeed, since cell walls constitute an open compartment difficult to isolate, numerous proteins predicted to be intracellular and/or having functions inside the cell have been identified in cell wall extracts. Then, except proteins predicted to be plasma membrane proteins, only proteins having a predicted signal peptide and no known intracellular retention signal are included in the database. In addition, *WallProtDB* contains information about the strategies used to obtain cell wall protein extracts and to identify proteins by mass spectrometry and bioinformatics. Mass spectrometry data are included when available. All the proteins of *WallProtDB* are linked to *ProtAnnDB*, another database, which contains structural and functional bioinformatics annotations of proteins as well as links to other databases (Aramemnon, CAZy, Planet, Phytozome). A list of references in the cell wall proteomics field is also provided.

Conclusions: *WallProtDB* aims at becoming a cell wall proteome reference database. It can be updated at any time on request and provide a support for sharing cell wall proteomics data and literature references with researchers interested in plant cell wall biology.

Keywords: Cell wall, Mass spectrometry, Plant, *ProtAnnDB*, Proteomics

Background

The plant cell wall is an external matrix containing polysaccharides and proteins. The interest in plant cell wall proteomes has been increasing during the last years with the discovery that plant cell walls are dynamic compartments constantly modified during development and in response to environmental cues [1,2]. The physiology of plant cell walls is strongly linked to its enzyme and structural protein content. The full description of the proteins present in various cell walls at precise stages of development or in response to biotic and abiotic stresses is now a main goal for many laboratories [3-5]. Besides, the search for procedures efficiently deconstructing cell walls to produce bioethanol has renewed the interest in cell wall physiology and especially in proteins playing roles in the remodeling of cell wall polysaccharides which are the major constituents of biomass [6-9].

Recent progresses in mass spectrometry (MS) technologies have led to the identification of cell wall proteins (CWPs) allowing the description of many cell wall proteomes. The next challenge is to gain biological messages out of these data. The first problem is the validation of the proteins identified as *bona fide* CWPs. This point is critical in plant cell wall proteomics. Indeed, it is difficult (i) to extract proteins by non-destructive methods avoiding the leakage of plasma membranes and the release of

* Correspondence: jamet@lrsv.ups-tlse.fr
[1]Université de Toulouse; UPS; UMR 5546, Laboratoire de Recherche en Sciences Végétales, BP 42617, F-31326 Castanet-Tolosan, France
[2]CNRS; UMR 5546, BP 42617, F-31326 Castanet-Tolosan, France

intracellular proteins and (ii) to purify cells walls because they form an open compartment which is not delimited by membranes [3,10]. Two kinds of methods have been employed: non-destructive methods consist in the analysis of extracellular fluids collected by vacuum infiltration of different types of solutions or of culture medium; destructive methods comprise several steps starting with the grinding of plant material followed by the purification of cell walls and the extraction of proteins with salt solutions [3]. The type of identified proteins and the ratio between identified proteins predicted to be secreted and identified leaderless proteins depends on the type of method used and on the type of plant material [10,11]. The issue of the non-canonical CWPs, *i.e.* proteins having no predicted signal peptide, has been a matter of debate since the first cell wall proteomics studies [12-14]. The second problem is the quality of functional annotations of proteins in databases. They are often not sufficiently reliable to allow an appropriate biological interpretation of proteomics data because they are mostly based on sequence comparisons [15,16]. The third problem occurs with plants for which sequence data are not available. In this case, the proteins cannot be unambiguously identified. This is a major problem in plants since most cell wall proteins belong to multigene families [17]. All these difficulties make the comparison between different cell wall proteomes a challenging task.

In order to answer such questions, *WallProtDB* (http://www.polebio.lrsv.ups-tlse.fr/WallProtDB/) was built in 2008 as a tool (i) to collect cell wall proteomics data, (ii) to facilitate their biological interpretation, and (iii) to allow comparisons between cell wall proteomes of different plant species. A new version of *WallProtDB* has been recently launched with new tools allowing the comparison between cell wall proteomes from different organs of the same plants or from different plants. *WallProtDB* contains experimental published data which are manually curated and is restricted to plants for which sequence data, genomic or ESTs, are available. Protein accession numbers are linked to another database, *ProtAnnDB* (http://www.polebio.lrsv.ups-tlse.fr/ProtAnnDB/), which provides bioinformatics predictions of sub-cellular localization and functional domains of diverse plant proteins using programs available online [15].

Construction and content

Construction and updating

The construction and the updating of *WallProtDB* are supported by the following steps performed to ensure a reliable database (Figure 1):

(1) Literature survey of plant cell wall proteomics papers. Selection of papers describing cell wall proteomes of plants with available sequenced genomes. Gathering of experimental data.
(2) Bioinformatic prediction of sub-cellular localization of proteins. This annotation is done using *ProtAnnDB* which is regularly enriched with new proteins. Depending on the plant of interest, protein sequences are from different databases (Table 1).
(3) Selection of proteins having predicted signal peptide, but no intracellular retention signal such as an ER canonical retention signal (IPR011679, http://www.ebi.ac.uk/interpro/entry/IPR011679; PS00014,

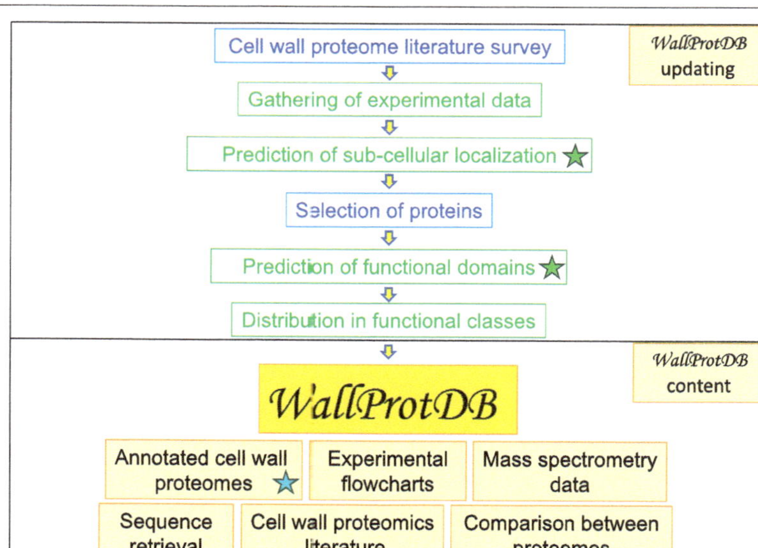

Figure 1 The *WallProtDB* flowchart (upper part) and content (lower part). *ProtAnnDB* is used at two different steps of the annotation procedure: prediction of sub-cellular localization and of functional domains (indicated by green stars). The accession number of identified protein is linked to *ProtAnnDB* (blue star).

Table 1 Proteomes included in *WallProtDB*

Plant species	Common name	Sequence databases	Proteome size[a]
Arabidopsis thaliana	Thale cress	TAIR10 http://www.ncbi.nlm.nih.gov/	495
Brachypodium distachyon	False brome	http://www.brachypodium.org/database	314
Brassica oleracea	cabbage	COMPBIO http://compbio.dfci.harvard.edu/index.html	162 (ESTs)
Gossypium hirsutum	cotton	COTTONGEN http://www.cottongen.org/search/search_by_gene	116
Linum usitatissimum	flax	Genolin flax unigenes https://urgi.versailles.inra.fr/Species/Flax/Download-sequences	106
		Phytozome http://www.phytozome.net/	
Medicago sativa	Alfalfa	JCVI: *Medicago truncatula* Genome Project http://medicago.jcvi.org/medicago/index.php	199
Oryza sativa	Rice	RAPD http://rapdb.dna.affrc.go.jp	270
Populus spp	Poplar	DOE Joint Genome Institute http://genome.jgi-psf.org/Poptr1_1/Poptr1_1.home.html	142
		Phytozome http://www.phytozome.net/	
Saccharum officinarum	Sugarcane	SUCEST http://www.sucest-fun.org/index.php/overview/sucest-fun-project	69 (ESTs)
Solanum lycopersicum	Tomato	sol genomics network http://solgenomics.net/	161
Solanum tuberosum	Potato	COMPBIO http://compbio.dfci.harvard.edu/index.html	136

[a]ESTs mean that protein identification was done using translated ESTs because of the lack of genomic data. Note that the distribution of proteins into functional classes is indicated in Additional file 1 for the cell wall proteome of all the plant species listed in this table.

http://prosite.expasy.org/PS00014) and no more than one trans-membrane domain as predicted by TMHMM (http://www.cbs.dtu.dk/services/TMHMM-2.0/).

(4) Bioinformatics prediction of functional domains. This annotation is done using *ProtAnnDB* (see below).

(5) Definition of a dictionary for the functional annotation of proteins, based on Pfam (http://pfam.xfam.org) [18] or InterPro (http://www.ebi.ac.uk/interpro/) [19] domain repertoires. This step ensures that the same annotation is used for all the proteins sharing the same predicted functional domains.

(6) Classification of proteins into 8 functional classes on the basis of the presence of predicted functional domains [10]: proteins acting on cell wall polysaccharides, oxido-reductases, proteases, proteins related to lipid metabolism, proteins with interaction domains (with proteins or polysaccharides), proteins possibly involved in signaling, structural proteins, proteins with yet unknown function. All the other proteins are included in a ninth class named "miscellaneous proteins" (Additional file 1).

(7) Design of a flowchart form allowing the description of most of the possible strategies usable to isolate CWPs and to identify them by MS and bioinformatics [3]. Customize the form for each set of experimental data (for an example, see Analysis of the cell wall proteome of *Brachypodium distachyon* young leaves: http://www.polebio.lrsv.ups-tlse.fr/WallProtDB_data/biblio/biblio26.html).

(8) Addition of MS data when available using either the X! Tandem software [20] or links to excel sheets found as supplementary data in articles of interest. When the data are in the X!Tandem format, it is possible to visualize the sequenced peptides on the protein sequence and their MS/MS fragmentation data.

(9) Search for homologous proteins in closely related genomes when only ESTs are available. The identification of homologous genes allows completing the bioinformatics prediction of signal peptide and/or of functional domains when EST sequences are not full-length. This is the case for *Saccharum officinarum* and *Brassica oleracea* for which homologous genes have been searched for in *Sorghum bicolor* and *Arabidopsis thaliana* respectively.

(10) Addition of cell wall proteomics literature. Direct links to articles or to their abstract in PubMed-NCBI (http://www.ncbi.nlm.nih.gov/pubmed) are available.

Data are stored in a mySQL database. *WallProtDB* is queried through a web interface constructed in the PHP code (http://www.php.net/).

Bioinformatics annotation of proteins using *ProtAnnDB*

ProtAnnDB is an annotation tool used for (i) selecting proteins to be included in *WallProtDB* and (ii) providing annotation of selected proteins. *ProtAnnDB* collects the results of bioinformatics predictions of sub-cellular localization and functional domains using available programs [17]. The following programs or databases have been used for prediction of sub-cellular localization: SignalP (http://www.cbs.

dtu.dk/services/SignalP/) [21], TargetP (http://www.cbs.dtu.dk/services/TargetP/) [22], Predotar (http://urgi.versailles.inra.fr/predotar/predotar.html) [23], Aramemnon (http://aramemnon.botanik.uni-koeln.de/) [24], TMHMM (http://www.cbs.dtu.dk/services/TMHMM-2.0/), GPIsom (http://gpi.unibe.ch/) [25] and PredGPI (http://gpcr.biocomp.unibo.it/predgpi/pred.htm) [26]. The databases used for the prediction of functional domains are Pfam [18], InterPro [19] and PROSITE (http://prosite.expasy.org/) [27]. *ProtAnnDB* also offers links to other databases providing genomic or gene regulation data such as Phytozome which collects genomic data (http://www.phytozome.net) and PlaNet which provides co-expression networks (http://aranet.mpimp-golm.mpg.de/) [28]. PlaNet has been chosen because it gives information on all the *A. thaliana* genes as well as on other plant species. *ProtAnnDB* has also links to Aramemnon which presently contains membrane protein data for nine plant species (http://aramemnon.botanik.uni-koeln.de/index.ep) and to two databases which collect expert annotation on cell wall protein families: (i) the PeroxiBase which is dedicated to peroxidases (http://peroxibase.toulouse.inra.fr/) [29] and (ii) CAZy which provides annotation of carbohydrate active enzymes (http://www.cazy.org/, http://csbl.bmb.uga.edu/dbCAN/) [30,31].

Tools for browsing *WallProtDB*

They are three ways to query *WallProtDB*: "Detailed search", "Summarized search" or "Blast search".

(1) The "Detailed search" interface offers several criteria: (1) protein accession number; (2) plant species; (3) plant material; (4) protein functional class; (5) protein family; (6) keyword. These criteria can be combined to refine comparisons. The result of the query is a customizable table that can be exported in different formats such as a tab delimited text, an excel sheet or a pdf file (Figure 2). Alternatively, they can be directly printed. Hyperlinks lead to *ProtAnnDB* bioinformatics annotation, experimental flowcharts and MS data (Figure 2). Protein sequences can be retrieved in FASTA format.

(2) The "Summarized search" interface provides tools for overall proteome comparisons. The result of the query is a table in which the numbers of proteins in each (i) protein functional class, (ii) protein family or (iii) protein (putative) function are indicated (Figure 3). As mentioned above, different formats are available for export of query results. It is also possible to draw a Venn diagram to visualize proteome comparisons within a plant species (Figure 4). All the figures are clickable, thus enabling retrieval of lists of the corresponding proteins.

(3) The "Blast search" [32] permits finding sequences homologous to a given nucleic or protein sequence in *WallProtDB*. A list of hits is proposed together with the possibility to visualize sequence comparisons and to collect the protein sequences in the FASTA format. It allows clustering newly identified CWPs with proteins present in the database. Then, it is easier to link the presence of some protein clusters to different physiological conditions and/or to cell wall types.

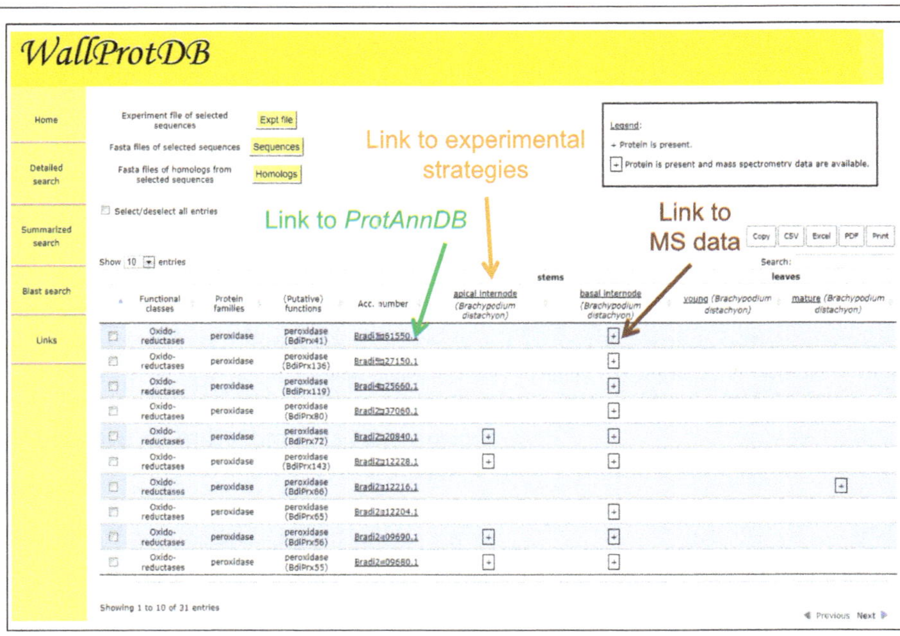

Figure 2 An example of results of a Detailed search: *B. distachyon* as a plant; peroxidases as a protein family; leaves and stems as plant organs. Note that only a part of the result sheet is shown.

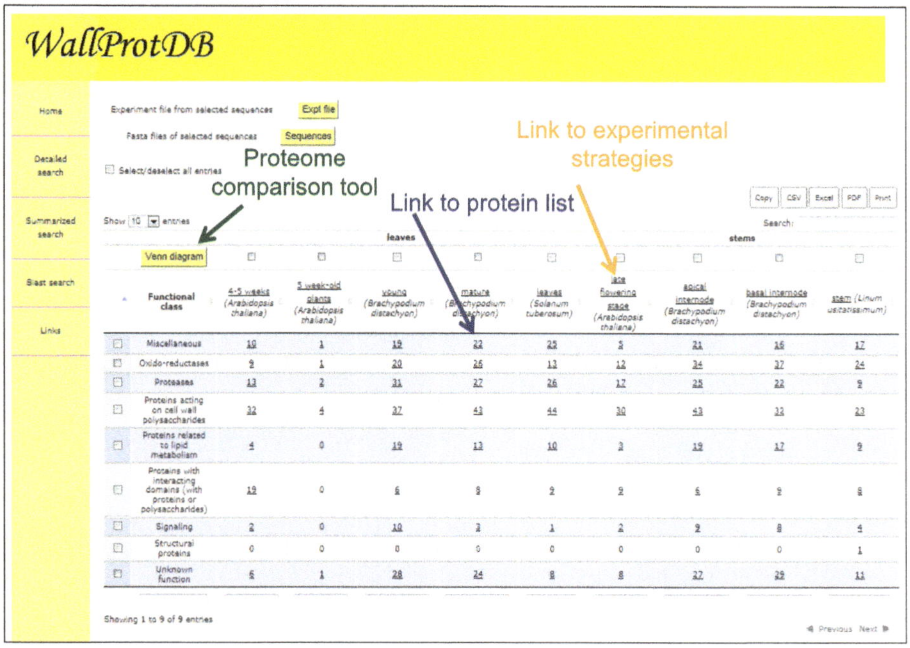

Figure 3 **An example of results of a Summarized search performed using the following criteria: group by functional class; *A. thaliana*, *B. distachyon*, *L. usitatissimum* and *S. tuberosum* as plants; leaves and stems as plant organs.**

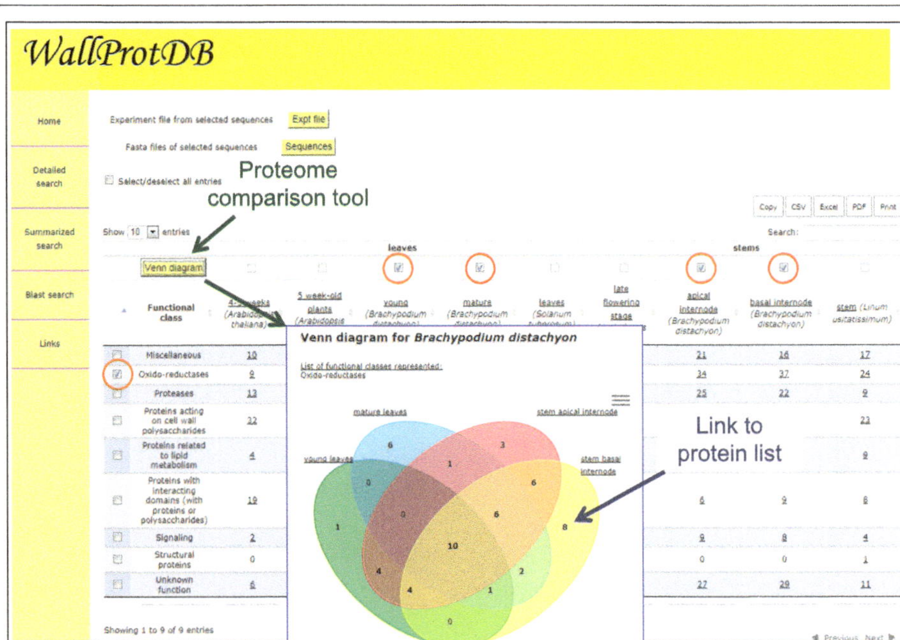

Figure 4 **An example of comparison between different proteomes within a plant species using the Summarized search.** The same criteria as in Figure 3 were used to obtain a table of results. Then, the Venn diagram was obtained after selection of *B. distachyon* as a plant, oxido-reductases as a functional class, and the four following experiments: young leaves (green), mature leaves (blue), apical internodes (pink), basal internodes (yellow).

Utility and discussion

At present, *WallProtDB* contains 2170 proteins and expressed sequence tags (ESTs) identified in 36 cell wall proteomics studies performed on 11 different plant species (8 dicots and 3 monocots) (Table 1, Additional file 1). It also offers tools for comparisons between proteomes. *WallProtDB* is regularly updated with newly published experimental data which are manually curated to obtain a homogeneous annotation (prediction of sub-cellular localization and functional domains of proteins). Only proteins having a signal peptide to address proteins to the secretion pathway and no known intracellular retention signal are included in the database. Proteins predicted to be plasma membrane proteins have been introduced in the database such as cellulose synthase, callose synthase or receptor kinases. They have been identified through peptides located in their extracellular domain. They are not true CWPs, but since they are involved in cell wall biogenesis or in signal transduction, they might be of interest for people working in the plant biology field. In addition, *WallProtDB* contains information about the protocols used to obtain cell wall protein extracts and about the strategies to identify proteins by MS and bioinformatics, as well as MS data when available. Furthermore, *WallProtDB* provides a list of references in the cell wall proteomics field which is regularly updated and comprises experimental articles as well as reviews.

WallProtDB complements other databases such as SUBcellular location database for Arabidopsis proteins (SUBA3, http://suba.plantenergy.uwa.edu.au/) [33], The Plant Proteome Database (PPDB, http://ppdb.tc.cornell.edu/) [34] and the cellwallgenomics database (http://cellwall.genomics.purdue.edu/intro/index.html). In each of these databases, the gathering of information is done in a different way. In SUBA3, only *A. thaliana* proteins are listed and all the proteins identified in published proteomes are included. This can be misleading because proteins known to be intracellular can be found in proteomes called "cell wall proteomes". For example, the At5g38410 *A. thaliana* small subunit of RUBISCO is mentioned as an extracellular or a plasma membrane protein although it is a well described chloroplastic protein. However, since all the proteins identified in cell wall proteomes are listed, SUBA3 is useful to get access to leaderless proteins identified in cell wall proteomes. PPDB is devoted to *A. thaliana*, *Oryza sativa* and *Zea mays*. It contains experimental MS data on proteins identified in different organs or sub-cellular compartments including the cell wall. Finally, the cellwallgenomics database provides repertoires of genes involved in cell wall biogenesis in *A. thaliana*, *O. sativa*, *S. bicolor* and *Z. mays* including intracellular proteins such as glycosyl transferases involved in the biosynthesis of cell wall polysaccharides. It also gives information on some mutants and on techniques useful to study cell wall biology, but no cell wall proteomics data.

Conclusions

To date, *WallProtDB* describes the content of cell wall proteomes and proposes tools for their analysis. It contains proteins with a high probability of being *bona fide* CWPs with regard to our present knowledge of the secretion pathway and of cell wall physiology. However, in the future, it could also include proteins with no predicted signal peptides, but experimentally proven to be located in cell walls by alternative methods such as localization of proteins tagged with fluorescent proteins or immunolabeling [3]. So far, there are only a few examples of such proteins in plants. The symplastic mannitol deshydrogenase has been shown to be secreted upon pathogen infection, and the secretion can occur in the presence of brefeldin A. However, the mechanism of secretion has not been described [35]. The exocyst-positive organelle (EXPO) could mediate the exocytosis from cytosol to cell wall of learderless proteins such as SAMS2 (S-adenosylmethionine synthetase 2) [36]. Since all the proteins included in *WallProtDB* are annotated in the same way, it allows fine comparisons between cell wall proteomes of different species and of various plant materials. In addition, it allows clustering proteins on a sequence homology basis. New proteomes can be introduced in *WallProtDB* on request, providing EST or genomic sequences of the plant of interest are available. The distribution of proteins into functional classes will certainly evolve when the functions of the proteins are experimentally determined. New functional classes can be easily created. Finally, the possibility to introduce wall proteomes of plant pathogens or symbionts will be considered since they share common protein families with plant cell wall proteomes. Altogether, *WallProtDB* aims at becoming a cell wall proteome reference database.

Availability and requirements

WallProtDB is freely available at the following address: http://www.polebio.lrsv.ups-tlse.fr/WallProtDB/. It is compatible with all major web browsers.

Additional file

Additional file 1: The *WallProtDB* content.

Competing interests

The authors declare that they have no competing interests.

Authors' contributions

EJ has supervised the project and written the manuscript. HSC has created the *WallProtDB* database and is updating it regularly. Both authors have read and approved the manuscript.

Acknowledgements
The authors are thankful to Université Paul Sabatier (Toulouse, France) and Centre National de la Recherche Scientifique (CNRS) for funding their research. They are thankful to Drs Cécile Albenne, Hervé Canut, Christophe Dunand, Valérie Pacquit, and late Pr Rafael Pont-Lezica for stimulating discussions. They also wish to thank Pr Klaas J van Wijk at Cornell University for linking *WallProtDB* to the Plant Protein Database (PPDB). The GenoToul Bioinformatics platform is acknowledged for providing the JQuery jvenn plug-in and facilities (http://www.genotoul.fr/index.php?id=12).

References

1. Fry SC. Primary cell wall metabolism: tracking the careers of wall polymers in living plant cells. New Phytol. 2004;161:641–75.
2. Passardi F, Penel C, Dunand C. Performing the paradoxical: how plant peroxidases modify the cell wall. Trends Plant Sci. 2004;9:534–40.
3. Albenne C, Canut H, Jamet E. Plant cell wall proteomics : the leadership of *Arabidopsis thaliana*. Front Plant Sci. 2013;4:111.
4. Lee SJ, Saravanan RS, Damasceno CM, Yamane H, Kim BD, Rose JK. Digging deeper into the plant cell wall proteome. Plant Physiol Biochem. 2004;42(12):979–88.
5. Jung Y-H, Agrawal G, Rakwal R, Jwa N-S. Secretome: toward deciphering the secretory pathways and beyond. In: Agrawal G, Rakwal R, editors. Plant proteomics: technologies, strategies and applications. Hoboken, NJ: John Wiley & Sons Inc; 2008. p. 764.
6. Gomez L, Steele-King C, McQueenMason S. Sustainable liquid biofuels from biomass: the writing of the walls. New Phytol. 2008;178:473–85.
7. Jordan D, Bowman M, Braker J, Dien B, Hector R, Lee C, et al. Plant cell walls to ethanol. Biochem J. 2012;442:241–52.
8. Ragauskas A, Williams C, Davison B, Britovsek G, Cairney J, Eckert C, et al. The path forward for biofuels and biomaterials. Science. 2006;311:484–9.
9. Carpita N. Progress in the biological synthesis of the plant cell wall: new ideas for improving biomass for bioenergy. Curr Opin Biotechnol. 2012;23:330–7.
10. Jamet E, Albenne C, Boudart G, Irshad M, Canut H, Pont-Lezica R. Recent advances in plant cell wall proteomics. Proteomics. 2008;8:893–908.
11. Albenne C, Canut H, Hoffmann L, Jamet E. Plant cell wall proteins: a large body of data, but what about runaways? Proteomes. 2014;2:224–42.
12. Borderies G, Jamet E, Lafitte C, Rossignol M, Jauneau A, Boudart G, et al. Proteomics of loosely bound cell wall proteins of *Arabidopsis thaliana* cell suspension cultures: a critical analysis. Electrophoresis. 2003;24:3421–32.
13. Chivasa S, Ndimba BK, Simon WJ, Robertson D, Yu X-L, Knox JP, et al. Proteomic analysis of the *Arabidopsis thaliana* cell wall. Electrophoresis. 2002;23:1754–65.
14. Rose JKC, Lee S-J. Straying off the highway: trafficking of secreted plant proteins and complexity in the plant cell wall proteome. Plant Physiol Biochem. 2010;153:433–6.
15. San Clemente H, Pont-Lezica R, Jamet E. Bioinformatics as a tool for assessing the quality of sub-cellular proteomic strategies and inferring functions of proteins: plant cell wall proteomics as a test case. Bioinform Biol Insights. 2009;3:15–28.
16. Jamet E. Bioinformatics as a critical prerequisite to transcriptome and proteome studies. J Exp Bot. 2004;55:1977–9.
17. Yokoyama R, Nishitani K. Genomic basis for cell-wall diversity in plants. A comparative approach to gene families in rice and Arabidopsis. Plant Cell Physiol. 2004;45:1111–21.
18. Finn R, Tate J, Mistry J, Coggill P, Sammut J, Hotz H, et al. The PFAM protein families dabase. Nucleic Acids Res. 2008;36:D281–8.
19. Quevillon E, Silventoinen V, Pillai S, Harte N, Mulder N, Apweiler R, et al. InterProScan: protein domains identifier. Nucleic Acids Res. 2005;33(Web Server issue):W116–20.
20. Fenyö D, Eriksson J, Beavis R. Mass spectrometric protein identification using the global proteome machine. Methods Mol Biol. 2010;673:189–202.
21. Bendtsen J, Nielsen H, von Heijne G, Brunak S. Improved prediction of signal peptides: SignalP 3.0. J Mol Biol. 2004;340:783–95.
22. Emanuelsson O, Nielsen H, Brunak S, von Heijne G. Predicting subcellular localization of proteins based on their N-terminal amino acid sequence. J Mol Biol. 2000;300:1005–16.
23. Small I, Peeters N, Legeai F, Lurin C. Predotar: a tool for rapidly screen proteomes for N-terminal targeting sequences. Proteomics. 2004;4:1581–90.
24. Schwacke R, Schneider A, van der Graaff E, Fischer K, Catoni E, Desimone M, et al. ARAMEMNON, a novel database for Arabidopsis integral membrane proteins. Plant Physiol. 2003;131(1):16–26.
25. Fankhauser N, Mäser P. Identification of GPI anchor attachment signals by a Kohonen self-organizing map. Bioinformatics. 2005;21:1846–52.
26. Pierleoni A, Martelli P, Casadio R. PredGPI: a GPI-anchor predictor. BMC Bioinformatics. 2008;9:392.
27. Hulo N, Bairoch A, Bulliard V, Cerutti L, Cuche B, De Castro E, et al. The 20 years of PROSITE. Nucleic Acids Res. 2008;36:D245–9.
28. Mutwil M, Klie S, Tohge T, Giorgi F, Wilkins O, Campbell M, et al. PlaNet: combined sequence and expression comparisons across plant networks derived from seven species. Plant Cell. 2011;23:895–910.
29. Fawal N, Li Q, Savelli B, Brette M, Passaia G, Fabre M, et al. PeroxiBase: a database for large-scale evolutionary analysis of peroxidases. Nucleic Acids Res. 2013;41(Database issue):441–4.
30. Lombard V, Golaconda Ramulu H, Drula E, Coutinho P, Henrissat B. The carbohydrate-active enzymes database (CAZy) in 2013. Nucleic Acids Res. 2013;42(Database issue):D490–5.
31. Yin Y, Mao X, Yang J, Chen X, Mao F, Xu Y. dbCAN: a web resource for automated carbohydrate-active enzyme annotation. Nucleic Acids Res. 2012;40(Web Server issue):W445–51.
32. Altschul SF, Gish W, Miller W, Myers EW, Lipman D. Basic local alignement search tool. J Mol Biol. 1990;215:403–10.
33. Tanz S, Castleden I, Hooper C, Vacher M, Small I, Millar H. SUBA3: a database for integrating experimentation and prediction to define the SUBcellular location of proteins in Arabidopsis. Nucleic Acids Res. 2013;41(Database issue):D1185–91.
34. Sun Q, Zybailov B, Majeran W, Friso G, Olinares P, van Wijk K. PPDB, the Plant Proteomics Database at Cornell. Nucleic Acids Res. 2008;37(Database issue):D969–74.
35. Cheng F, Zamski E, Guo W, Pharr D, Williamson J. Salicylic acid stimulates secretion of the normally symplastic enzyme mannitol dehydrogenase: a possible defense against mannitol-secreting fungal pathogens. Planta. 2009;230:1093–103.
36. Wang J, Ding Y, Wang J, Hillmer S, Miao Y, Lo S, et al. EXPO, an exocyst-positive organelle distinct from multivesicular endosomes and autophagosomes, mediates cytosol to cell wall exocytosis in Arabidopsis and tobacco cells. Plant Cell. 2010;22:4009–30.

18

Direct comparison of MRI and X-ray CT technologies for 3D imaging of root systems in soil: potential and challenges for root trait quantification

Ralf Metzner[1*†], Anja Eggert[2†], Dagmar van Dusschoten[1], Daniel Pflugfelder[1], Stefan Gerth[2], Ulrich Schurr[1], Norman Uhlmann[2] and Siegfried Jahnke[1]

Abstract

Background: Roots are vital to plants for soil exploration and uptake of water and nutrients. Root performance is critical for growth and yield of plants, in particular when resources are limited. Since roots develop in strong interaction with the soil matrix, tools are required that can visualize and quantify root growth in opaque soil at best in 3D. Two modalities that are suited for such investigations are X-ray Computed Tomography (CT) and Magnetic Resonance Imaging (MRI). Due to the different physical principles they are based on, these modalities have their specific potentials and challenges for root phenotyping. We compared the two methods by imaging the same root systems grown in 3 different pot sizes with inner diameters of 34 mm, 56 mm or 81 mm.

Results: Both methods successfully visualized roots of two weeks old bean plants in all three pot sizes. Similar root images and almost the same root length were obtained for roots grown in the small pot, while more root details showed up in the CT images compared to MRI. For the medium sized pot, MRI showed more roots and higher root lengths whereas at some spots thin roots were only found by CT and the high water content apparently affected CT more than MRI. For the large pot, MRI detected much more roots including some laterals than CT.

Conclusions: Both techniques performed equally well for pots with small diameters which are best suited to monitor root development of seedlings. To investigate specific root details or finely graduated root diameters of thin roots, CT was advantageous as it provided the higher spatial resolution. For larger pot diameters, MRI delivered higher fractions of the root systems than CT, most likely because of the strong root-to-soil contrast achievable by MRI. Since complementary information can be gathered with CT and MRI, a combination of the two modalities could open a whole range of additional possibilities like analysis of root system traits in different soil structures or under varying soil moisture.

Keywords: X-ray Computed Tomography (CT), Magnetic Resonance Imaging (MRI), Root system architecture, Common bean (*Phaseolus vulgaris* L.) 3D imaging, Roots in soil, Non-destructive

* Correspondence: r.metzner@fz-juelich.de
†Equal contributors
[1]Institute of Bio- and Geosciences, IBG-2: Plant Sciences, Forschungszentrum Jülich GmbH, Wilhelm-Jonen- Str., 52425 Jülich, Germany
Full list of author information is available at the end of the article

Background

Roots are vital for higher plants to gather water and nutrients and thus critical for performance and productivity [1]. Considering the impact on crop yields, more knowledge is needed about root system development belowground and interaction with the surrounding soil [2-4]. Approaches to investigate roots and their development reach from excavation and measurement of root system traits with manual or optical methods [5-7], mini-rhizotrons [7] or trenches dug into the ground [1], transparent artificial growth media [8,9] to soil based 2D growth monitoring in rhizoboxes [1,10,11]. All these are valid for answering specific questions concerning root development; however, none of these is able to follow 3D root development in soil where a multitude of biotic and abiotic interaction takes place. Quantitative knowledge of 3D root traits would help to achieve better mechanistic understanding of root architecture as it develops in a soil environment. Non-invasive approaches which have been used already to investigate plant roots in soil are magnetic resonance imaging (MRI) [12-15], neutron computed tomography [16,17] and X-ray computed tomography (CT) [3,18-21]. Neutron tomography requires access to a nuclear reactor or a high energy particle accelerator whereas MRI and CT on the other hand, though still expensive, are becoming available to a steadily growing number of plant biologists.

CT employs an X-ray beam passing through the sample which absorbs part of it thereby reducing the intensity of the beam. This process is called attenuation. The sample is rotated between an X-ray source and a detector, recording series of 2D projections of the object from which a 3D volume dataset can be reconstructed. X-ray attenuation is mainly determined by material properties, in particular electron-density. Thereby the internal structure of the sample becomes visible by contrast according to density and atomic number of the elements [22]. Further information about the basics of CT can be found in various articles [23,24] or text books [25]. Recent applications in plant sciences include lateral root development [26] or root elongation rates [27]. A number of other applications covering a range of plant species and root traits are reviewed in [3]. A major problem is the often very similar attenuation of roots and some structures in the soil such as water filled pores [3]. This leads to low contrast hindering simple and straightforward segmentation of the roots from the soil background. Different approaches were tried to tackle this problem, such as adjusting soil humidity to optimize contrast [28] or sophisticated software tools for data analysis [29]. Together with scanners that can produce high resolution scans in short time [3] these approaches may lead to a wider use of CT in plant sciences.

MRI is based on the magnetic moment of atomic nuclei like ^1H (protons) which are highly abundant in living tissues particularly in water molecules. The magnetic moment can be manipulated using strong magnetic fields and radio frequency fields to produce 3D datasets of samples. The magnetic fields require a substrate with low ferro-magnetic particle content for high quality images. A range of contrast parameters can be exploited, highlighting differences within the sample such as density of the protons or their physical and chemical microenvironment. This can be exploited to produce a strong difference between 'root water signal' and 'soil water signal' which provides a very high contrast between roots and soil background [13,14]. The basic principles of MRI and its use in biomedical sciences are described in detail in several textbooks [30,31] or review articles of plant biology [12,32-34]. Research applications to plant roots range from phytopathology [35], across storage root internal structures [15] to combined studies with positron emission tomography for structure function relations [13]. Also water mobility in roots and soil has been shown to be detectable with MRI [36,37].

Both CT and MRI are able to image root systems in soil. Their fundamental different physical principles give each of them a unique perspective but also different challenges for application in the plant sciences. Therefore, a direct comparison of both techniques using the same samples and analyzing for the same root system may help to decide which method is better suited for a specific research question. A whole range of instrument parameters can affect root imaging with either method but also sample parameters such as soil type [14,38] or moisture [28]. Pot sizes, on the other hand has rarely been considered despite its importance for root system and plant development [39] making it an important parameter of experimental design. Since pot size may also affect root imaging and data analysis depending on the applied method, we focus in our CT-MRI comparison approach on this parameter using common bean (*Phaseolus vulgaris*) as a model plant. We evaluated both techniques for imaging roots in soil, segmentation of roots from the soil background and extracting root length as a basic parameter of root system architecture (RSA). For comparison the roots were finally excavated, washed and scanned using WinRHIZO as a widely used standard technique.

Results and discussion

We compared reconstructed root systems of bean plants in pots of three different sizes imaged by both CT and MRI. The small pots had an inner diameter (I.D.) of 34 mm and a height of 200 mm which could be used to study early growth stage in crop plants. Pots of this size or smaller were also used in other high resolution CT studies [3,19]. The medium pots with an I.D. of 56 mm and a height of 200 mm would allow plant growth for a few weeks or even a whole growth cycle of small plants.

The large pots had I.D. of 81 mm and were 300 mm high, thereby suited to cultivate plants for a longer time and even plants with large storage roots like sugar beet [15]. Here, we focused on the potentials and challenges for image quality and extracting root length as a major trait of RSA [1,11] to compare the two imaging modalities.

CT imaging of roots is based on high spatial resolution and a good segmentation procedure

An example of a CT image with a spatial resolution of 28 μm (voxel size) is shown in Figure 1a which was obtained from a bean plant growing in a small pot. Roots that grew at the inner surface of the pot can be directly recognized by removing the pot material from the image without special segmentation efforts (Figure 1a). Due to their elongated shape they can be distinguished visually from sand grains and darker areas where air or water filled pores and finer soil particles were located. The ability to image both soil structures and plant roots is a very intriguing feature of CT [3,40,41]. However, roots developing as a 3D structure in a soil filled pot have to be detected by signal contrast based on X-ray attenuation which may be low between particular soil structures such as water filled pores and the roots [3,28]. Segmentation of the roots from the soil background is therefore a major step in extracting root structures and traits from CT images. So far many different approaches exist for this step in data analysis, exploiting different algorithms and filters in manual [19,40,42] or automated fashion [43,44]. However, a standard procedure remains yet to be defined. Figure 1b shows the result of a root segmentation procedure applied here to the dataset partly shown in Figure 1a. The segmentation follows an image analysis procedure including threshold filtering for manually selected grey values and for objects of a size and (cylindrical) shape typical for roots [45]. For the root segmentation procedure the spatial resolution of the images had to be reduced by a factor of 2 in each dimension because otherwise it would not have been possible to run the analysis on a standard workstation-sized computer with enhanced memory. The 3D rendered image of the segmented root system (Figure 1b) shows not only the roots at the soil surface but also a range of roots of variable diameter deep in the soil core. The 3D visualization allowed identification of primary, basal and hypocotyl-borne roots [46] by tracing them back to their respective origin (data not shown). In this way, the lowest root tips were identified as belonging to basal roots or the primary root (Figure 1b) while the thinner roots in the upper part of Figure 1b were lateral roots of each of these root types. The high spatial resolution of CT shown here also enables a precise mapping of root diameters.

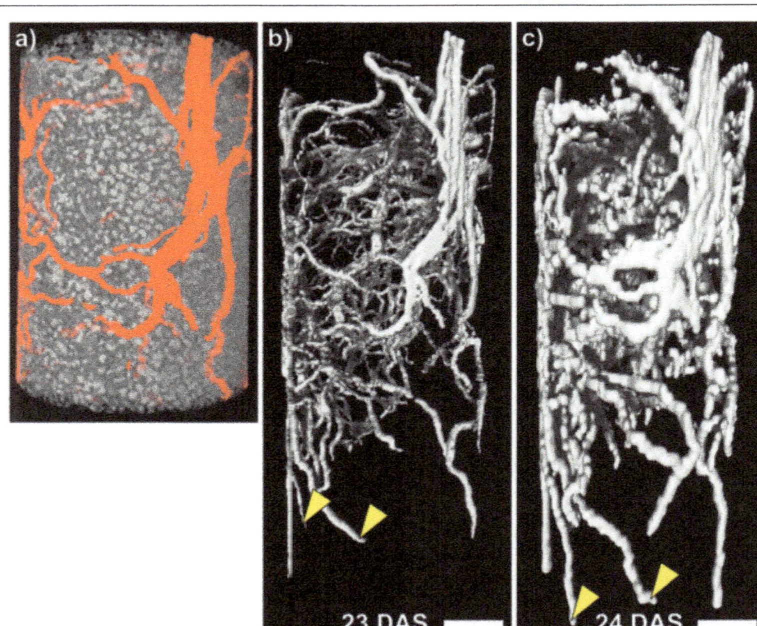

Figure 1 CT and MRI images of soil and bean roots in a small pot. The same plant grown in a soil filled pot with an inner diameter (I.D.) of 34 mm and a height of 200 mm was imaged sequentially with X-ray Computed Tomography (CT) and Magnetic Resonance Imaging (MRI). **(a)** Shows the upper part of the soil column below the pot material imaged with CT (gray-scale) measured with voxel size of 28 × 28 × 28 μm^3 with the roots of a bean plant (*Phaseolus vulgaris* L.) highlighted (red). **(b)** The root system segmented from the same CT image as in **(a)** on a voxel size of 56 × 56 × 56 μm^3. **(c)** Shows the same root system imaged one day later with MRI and a voxel size of 333 × 333 × 1000 μm^3 with the last value representing the vertical dimension. Arrowheads denote the same roots in both images. DAS: days after sowing Scale bar: 10 mm.

The limit for the thinnest observable root is determined mainly by two factors: contrast (between roots and soil background) and spatial resolution (voxel size). Both factors influence the probability that a structure can be recognized as a root and thereby segmented. Indirectly, resolution may also influence contrast as voxels comprised of different materials with attenuation both higher and lower than roots (e.g. air and solid soil particles) may result in similar attenuation values as roots, the so called partial volume effect. Therefore a high spatial resolution may also enhance contrast. Other factors include signal-to-noise ratio and imaging artefacts such as beam hardening [3,41] but these were found to be of less importance in this study. The thinnest roots detected here were about 100 μm in diameter, corresponding to 2 voxel diameters in the segmentation. This relation between voxel resolution and minimal detectable object size fits well with the values found by other authors [41,47]. On the other hand, not all roots of 100–250 μm thick were quantitatively detected: when comparing the CT data with WinRHIZO data only about 70% of the total root length was found by CT segmentation (Table 1). However, for WinRHIZO roots classes thicker than 300 μm 78% and for those thicker than 400 μm 92% were found by CT. These values are in the order of the 86% Flavel et al. [19] found for wheat roots thicker than 250 μm in similarly sized pots considering the differences in plant species, scanner type and segmentation procedure. The imaging resolution used here was two times better than that achieved by Flavel et al. [19]; we therefore expect that, when the segmentation procedure can be further improved, a much larger fraction of a root system can be detected, at least in small pots. Other authors found values above 90% [48] but, according to their images, the segmentation between roots and water filled pores may have been suboptimal and lead to an overestimation of root length. To evaluate the effectiveness of the different segmentation approaches it might be useful to compare them directly on the same datasets as has been exemplarily attempted for soil parameter quantification with CT [49].

MRI imaging of roots is based on high root to soil contrast

The MRI image (Figure 1c) showed a similar 3D structure of the root system as segmented from the CT data (Figure 1b). Due to high contrast between roots and soil background, root segmentation from an MRI image is not necessary which means that Figure 1c shows the raw data with only a noise cut-off and no further processing. While the MRI scan took about as long as the CT scan, it must be noted that for the CT scan it took another two hours to reach the segmented image shown in Figure 1b. Though the root images from CT and MRI appear rather similar, there are some distinct differences due to fundamental differences in the imaging and segmentation principles. Roots which appear to be very thin in CT often look much thicker in MRI. This is caused by the much coarser spatial resolution of MRI (Figure 1c) with a voxel size of 330 × 330 × 1000 μm^3, so any root detected is shown with voxels of this size whereas in the CT image (Figure 1b) the voxel size was 56 × 56 × 56 μm^3. As already stated, in MRI the root detection is based on contrast against soil background which in this study was about two orders of magnitude, rather than on geometrical recognition as in CT. That means that roots even thinner than voxel resolution can be recognized as roots when only a (detectable) small fraction of the actual volume is comprised of root tissue. Therefore parts of the roots displayed in Figure 1c as one voxel thick were in reality thinner due to this sub-voxel resolution. We estimated that the thinnest roots detected with MRI were about 250 μm in diameter. The root length found with MRI was still only 70% of the total root length found with WinRHIZO after harvest (Table 1). Upon taking only the length of roots in the WinRHIZO classes thicker than 400 μm into account this fraction was 96%. Therefore, similar to CT, roots above this diameter appear to be largely detected, and in bean plants the primary, basal and hypocotyl borne roots are typically all above this threshold [46], the major root structure can be visualized in 3D. The high contrast between roots and soil background in MRI results from the fact that the signal decay from soil water is much faster than that from roots, so with the appropriate measurement settings only the latter is recorded. Since the signal decay of the soil water depends on soil properties, the measurement settings need to be adjusted adequately to the specific substrate. For MRI imaging of roots in soil the choice of substrate is important also because, due to the strong magnetic fields required for the measurements, disturbances of these fields have adverse effects on image quality. Ferromagnetic particles are thus

Table 1 Root traits calculated from CT and MRI root images of the volumes shown in Figures 1, 2 and 3

	Small pots (I.D. 34 mm)			Medium pots (I.D. 56)			Large pots (I.D. 81)	
Measurement modality	CT	MRI	Win-RHIZO	CT	MRI	Win-RHIZO	CT	MRI
Total root length [mm]	2775	2910	4012	2787	4602	4855	6632	10496
Percent root length of WinRHIZO [%]	69	73		57	95			

most critical and their presence in the substrate for MRI should be minimized. This can either be done by selecting soils with naturally low content of ferromagnetic particles [14] or remove them from the sieved substrate with a simple arrangement of permanent magnets [15] as was used in this study. Paramagnetic ions such as Mn^{2+} described to be also problematic by some authors [50,51] did not appear to have any strong effect in our studies as demonstrated by the good image quality in the chosen soil (Figure 1c). The removal process for ferromagnetic particles led to a sieved and homogenized soil, and this substrate with good MRI properties was also used for CT. In contrast to pure MRI studies on roots [15,35] a higher fraction of coarse sand was used here, which may cause problems for plant cultivation because of low water retention capacity. However, pretests had shown that this allowed better segmentation results for CT due to higher contrast between roots and substrate, whereas MRI image quality was only marginally affected. Homogenized substrates appears to be advantageous for CT as it provides a rather homogenous background with small pores optimal for segmentation, which may be the reason why this type of substrate was also used in many CT studies targeting finer roots [28,52,53]. The need of homogenized substrates seems to be a common limitation for current MRI or CT applications in plant studies. Compared to natural soil from the field, which would be the ultimate condition to grow e.g. crop plants in, measure and characterize root traits, sieved soil comprises loss of texture and aggregate structure. Nevertheless, both methods are in principle able to investigate roots in soil cores collected in the field, as has been shown for CT [44] and also in a first MRI experiment in our lab (data not shown).

For small pots both modalities recovered a comparable fraction of a root system with more details visible in the CT

When comparing the root images provided by both techniques, it is obvious that CT (Figure1b) with its higher spatial resolution showed many of the thinner roots with a higher level of detail than MRI (Figure 1c). For studies targeting different root classes in plants such as bean [46] or maize [54] this feature may be advantageous for identification and tracking of lateral roots of each class which are difficult to follow with MRI. Also thinner root systems in general may be investigated with CT as shown for *Arabidopsis* roots [55] which can be hardly achieved with MRI.

Visual comparison between root systems in 3D can be quite useful for qualitative description of treatment effects but, in addition to that, a quantitative evaluation of trait development is required. For the root system in the small pots shown in Figure 1, a root length of 2.8 m with CT and 2.9 m with MRI on the following day was found (Table 1) fitting well with the slightly longer roots in the lower part of the pot shown with MRI (Figure 1c) compared to CT (Figure 1b). Considering this, the fraction of the WinRHIZO-detected roots found by CT and MRI (69% and 73% respectively; Table 1) was almost identical, i.e. for such a (small) pot there was no obvious difference between MRI and CT in acquiring root lengths. Another experiment with a similar plant yielded longer total root length for all methods (Additional file 1), but the fraction of the roots found by WinRHIZO of CT (61%) and MRI (71%) was similar to that in Figure 1.

For medium sized pots, CT showed more details but MRI revealed a larger fraction of the root system

A bean plant growing in a medium sized pot (I.D. 56 mm, height 200 mm) was imaged with MRI (Figure 2a) and CT (Figure 2b). A larger pot size is needed for working with larger root systems and later developmental stages but also for bulky belowground structures such as storage organs to minimize growth limitations due to pot size [15,39]. However, the pot size affects also the quality of root imaging and segmentation of both CT and MRI. Either modality produced images of the root system of sufficient quality to see the overall structure of the specimen (Figure 2a,b). In contrast to the smaller pots, distinct differences can be seen in the overlay of the CT and MRI images (Figure 2c). The CT again showed detailed structures, a few of which were not visible in the MRI image such as small roots in the center. While it is possible that these have grown over the two days between the measurements, it is more likely that they were just too thin to be detected with MRI. Considering that the spatial resolution of MRI on the medium sized pots was only slightly lower than for the small ones, we estimated that roots above 280 μm should be largely detectable. Root segmentation from the CT dataset was more difficult for the medium pots compared to the small ones since resolution was limited to a voxel size of 68 μm due to the tradeoff between resolution and sample size apparent for a given CT system [3,24].

On the other hand, many roots detected with MRI also appeared to be successfully segmented with CT (Figure 2c) with two significant exceptions: on the left hand side and the bottom ~20 mm of the pot (Figure 2c) almost no roots could be successfully segmented from CT data, while MRI showed an abundance of roots. The lack of segmentation of roots from the CT image at these different positions can be taken as an example to explain different challenges frequently met when extracting roots from CT measurements. High water content can be an issue for segmentation as attenuation of roots and water-filled pores may overlap [28,42]. Gravity will inevitably lead to higher water content at the bottom, thereby degrading CT root segmentation particularly at the bottom of the pot; as dry soil

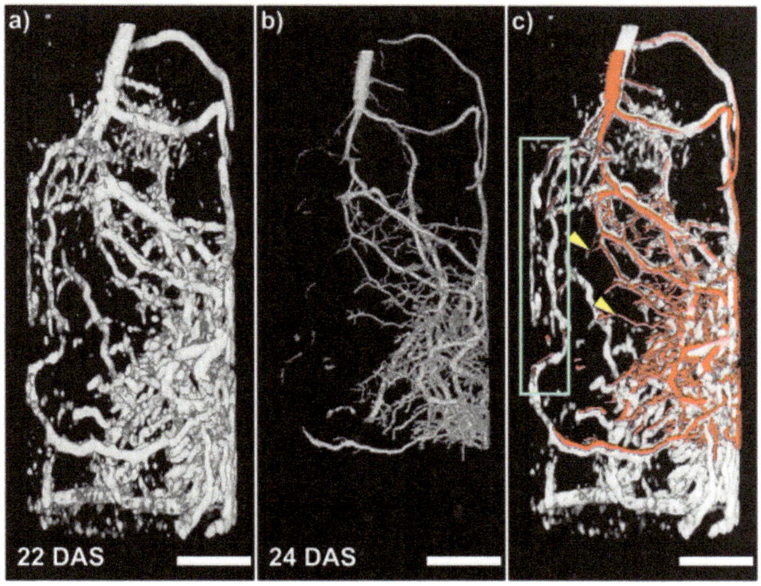

Figure 2 MRI and CT images of a bean root system in a medium soil filled pot. The same plant grown in a soil filled pot with an I.D. of 56 mm and a height of 200 mm was imaged sequentially by MRI and CT. **(a)** MRI image measured with a voxel size 375 × 375 × 1000 μm^3. The CT image **(b)** shows the root system measured of the same plant segmented on a voxel size of 68 × 68 × 68 μm^3. Roots in the lowest part of the pot could not be segmented and are therefore not shown. **(c)** CT-MRI co-registration: the CT image is in red and the MRI image in grey. Arrowheads highlight roots visible in CT but not in MRI. Box highlights area where few roots are visible in CT. Scale bar: 10 mm.

also may increase artefacts from air filled pores in segmentation, a compromise may be to keep the soil in the pot at the soil-specific field capacity [28]. However, especially in higher pots needed to study the development of long roots or in sandy soils, the problem of water gradients in the pot still remains leading to segmentation problems towards the bottom of the pot. Additionally, water uptake by plants would also alter water distribution in a pot. It is therefore advisable to measure these effects for the chosen substrate-pot combination, water contents, CT-scanner and imaging system as a very basic control. Since the lowest part of a pot is the most likely to cause segmentation problems with CT, Figure 2b showed no roots in this area whereas a large number of roots was monitored at the bottom of the pot with MRI (Figure 2a). The MRI image also showed more unconnected voxels in this area with grey values similar to the roots compared to the upper part. We observed such unconnected voxels to be an artefact in MRI in the soil mixture used here when the water content surpassed about 10% to 15% (v/v). Since these voxels are not connected to the roots or to each other they were not counted in the quantification of root length unless they were directly adjacent to the roots. This also points to a high water content at the bottom of the pot as the reason for failed CT root segmentation in this volume. However, we occluded roots found by MRI in this volume from root length quantification in order to compare similar volumes for both CT and MRI measurements.

For the left side of the medium pot there was no hint in the MRI image (Figure 2a) of high water content. The failure in root segmentation by CT in this region might be therefore different from what has been discussed before. Another problem in CT root segmentation can be an imprecise removal of the pot material that accidentally removed root segments from the image during the image processing pipeline. If part of a major root is removed, all the attached roots will also be lost for root length calculation if the resulting gap is too large. For MRI this was not an issue since PVC, similar to most other common plastic materials, gave no detectable signal at the applied measurement settings and therefore did not show up on the images. Nevertheless, the strong difference in root length extracted from the medium pot by CT or MRI (2.8 m and 4.6 m respectively), was unlikely based exclusively on this effect. With reference to the WinRHIZO data, MRI found 95% of the roots which, considering that the lowest part of the pot was not used for the MRI root length calculation, advises root length to be somewhat overestimated here by MRI. Possibly, some signal caused by high water content in the soil adjacent to the roots could not be excluded from root length quantification. Therefore, also for MRI it seems to be advisable to investigate effects of high soil water content, probably close to or above field capacity, for the specific plant-soil system on root detection and root length quantification as a basic control.

For the large pots, MRI showed a larger fraction of the root system and more details compared to CT

The biggest pots used here had an inner diameter of 81 mm and were 300 mm long. Again the roots system of a bean plant was consecutively imaged by MRI (Figure 3a) and CT (Figure 3b). A large number of the roots was found in the CT image with a range of diameters allowing root class identification. However, at this pot size, some roots appeared to have no laterals in CT while, on the same roots, they were visible in the MRI image. The MRI scan was accomplished two days prior to the CT scan, illustrated by one thick root which had elongated clearly (middle arrowhead), so growth of the laterals can be excluded. This indicates that, for larger pots, MRI may have a better imaging capability compared to CT. Also, the detected root length with MRI was about 50% higher than with CT (Table 1). A likely explanation for this is the loss in spatial resolution necessary to cope with larger pots. Segmenting roots with CT mainly relies on the high spatial (voxel) resolution achievable in a particular measurement, allowing extraction of root structures despite the low contrast between the roots and the soil background [3,19]. Due to the tradeoff between sample size and resolution the voxel size had to be adjusted to 99 μm for the large pot, compared to 56 μm for the small pot, where both methods performed equally. The resolution in MRI also had to be adjusted as well, from 333 μm to 521 μm, but since MRI root detection was based on the strong root-soil contrast allowing sub-voxel root detection it was much less affected by a loss in spatial resolution.

Conclusions

Both CT and MRI produced high quality 3D images of root systems *in vivo* as demonstrated here for bean

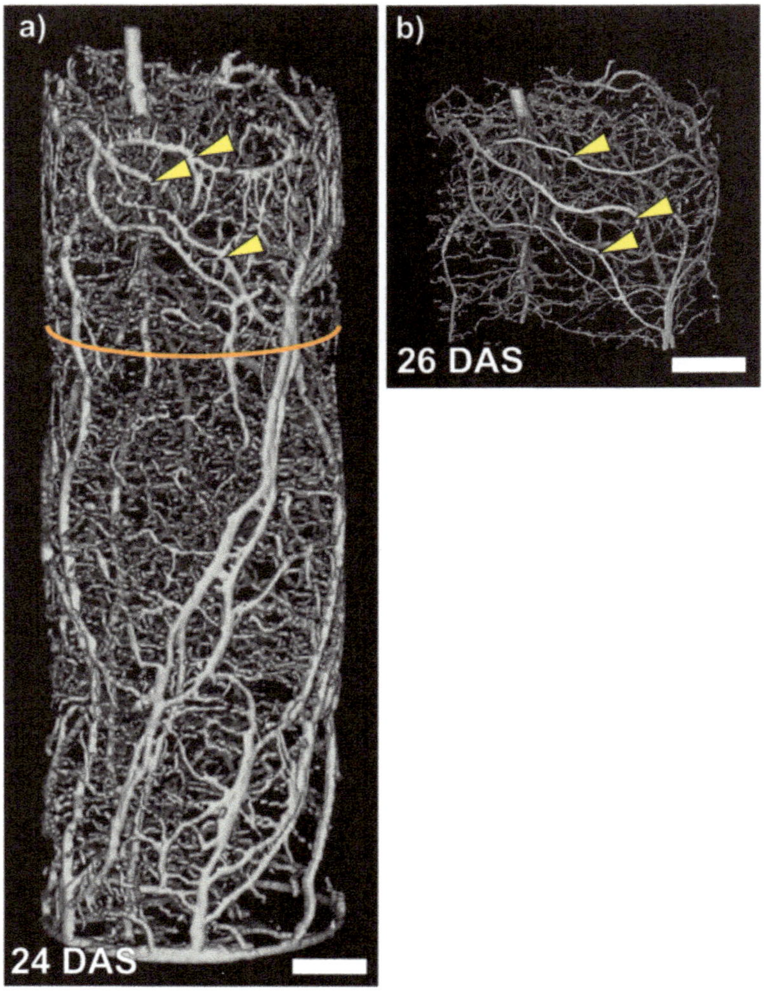

Figure 3 MRI and CT images of a bean root system in a large pot. The same plant grown in a soil filled pot with an I.D. of 81 mm and a height of 300 mm was imaged sequentially with MRI and CT. **(a)** MRI image shows the roots system in the whole pot with a voxel size of 521 × 521 × 1000 μm^3. Root systems of the same plant segmented from CT data **(b)** on a voxel size of 99 × 99 × 99 μm^3. Since the CT dataset was not complete only data of the upper ~75 mm of the pot could be reconstructed and are shown here. The orange colored ring denotes the volume measured also by CT. Arrowheads highlight the same roots in both images. Scale bar: 20 mm.

plants grown in soil-filled pots of three different sizes (34, 56 and 81 mm in diameter). For comparison, root lengths of the same plants were calculated from data obtained in consecutive measurement with both modalities. The obtained length of the investigated root systems and image details were distinctly different between CT and MRI and, in particular dependent on the pot size. Compared to CT, MRI was less affected by a lower spatial resolution associated with larger pot sizes and appeared to be better suited for imaging large root systems, e.g. at later developmental stages. On the other hand, CT gave almost identical results as MRI for small pots with respect to root length but provided more detailed information such as finely graduated root diameter estimation, in particular of small roots, due to higher spatial resolution. The possibility to visualize soil structures is a further intriguing feature of CT while MRI opens the possibility of measuring the water content of roots or soils. Considering the time needed to analyze a full root system, the scan time for both modalities was comparable but segmentation of roots from CT images took much longer than the scans while MRI data could be used for quantification immediately. Segmentation of the CT data, however, may be performed separately while scanning on other plants continues. According to WinRHIZO data, however a considerable portion of the root system present could not be detected by either MRI or CT even in the small pots. Most likely these are roots thinner than 400 μm. Altogether combining the two modalities on the same samples would be intriguing as, with direct root and water content detection by MRI and soil structure imaging by CT, many questions regarding root-soil interactions can be addressed. Interesting sub-volumes identified with MRI in larger pots could then be targeted and investigated in detail with high resolution CT. As demonstrated here, both methods have their specific pros and cons, but are excellent tools to non-destructively investigate properties of roots growing in soil.

Methods

Plant material

Common bean plants (*Phaseolus vulgaris* L. cv. 'Shiny Fardenlosa') were grown in a growth chamber from seed in a mixture of homogenized agricultural topsoil and coarse sand (1:9; v/v). A mix with less sand and the same soil was already shown to be suited for MRI application [35]. The agricultural soil, characterized as a gleyic cambisol, was collected by removing the top 30 cm from an agricultural field (Kaldenkirchen, Germany) and air dried. Subsequently, the soil was powdered and homogenized in a drum hoop mixer (J. Engelsmann, Ludwigshafen, Germany), sieved to 2 mm and freed of stronger ferromagnetic particles by moving it in a thin layer on a conveyor belt through a magnetic field provided by rare earth magnets (NdFeB N42, 1.3 T; Webcraft GmbH, Gottmardingen, Germany). Coarse quartz sand (grain size 0.71 to 1.4 mm; Quartzwerke Witterschlick, Alfter, Germany) was similarly freed of ferromagnetic particles. The ready mixture was filled into PVC tubes of three different sizes with (a) an inner diameter of 34 mm, a height of 200 mm, a volume of 0.182 l, (b) and I.D. of 56 mm, a height of 200 mm, a volume of 0.49 l or (c) an inner diameter (I.D.) of 81 mm, a height of 300 mm and a volume of 1.5 l. All had holes drilled into their bottom caps for drainage and aeration covered with nylon mesh (grid size 200 μm) to prevent loss of substrate and roots growing out.

The pots were watered to above container capacity and, after excess water had drained away, seeds were laid down in holes 2 cm deep and covered with soil. After germination the pots were watered automatically once per day with tap water. The growth chamber was set to 16 h light/ 8 h dark and 20°C/16°C, respectively, while relative humidity was kept constant at 60 ± 3%. Lighting was provided by 5 × 400 W HPI and 5 × 400 W SON-T lamps (both Philips, Hamburg, Germany) which alternated every 2 hours with 5 min overlap giving PAR intensity between 350 and 450 μmol m^{-2} s^{-1} at canopy level.

Computed tomography

For the root detection we used the CT setup at the Fraunhofer Development Center X-ray Technology (EZRT) [23]. The CT setup was used with a FXE 225.99 X-ray tube and a Perkin Elmer XRD 1620 detector, which is a flat panel type and operates at a frame rate of 2 images s^{-1} in 14-bits full frame mode (2048 × 2048 pixels).

The measurement parameters were adapted to the three different pot sizes to ensure an optimal image quality. The tube was operated in 'high power' mode. Measurement parameters for the different pot sizes are listed in Table 2. For the 34 mm and 56 mm pots data set with and without a 1 mm Copper filter between X-ray tube and sample were acquired to investigate the effect of beam hardening on the image quality. Although the filter did remove the beam hardening artefact, this artefact was nevertheless only present on a thin surface layer of the PVC pot (Additional file 2). As the image quality inside the pot was identical for both settings, the used filter was found to be of no importance in this study. Since the pots were 200–300 mm long, multiple scans of the samples were made at different heights. The obtained single part volumes were merged together afterwards to a whole recording of the plant.

The scans were made with continuous sample rotation (fly-by) over 360°. For improved handling and decreased calculation times the root structure was analyzed after a 2 × 2 binning with a voxel sampling of 56 μm for pot size

Table 2 Parameters of CT measurements

	I.D. 34	I.D. 56	I.D. 81
Tube voltage [kV]	220	220	160
Tube current [μA]	180	180	160
Power [W]	39.6	39.6	25.6
Exposure time [ms]	499	499	999
Filter	Cu 1 mm	Cu 1 mm	none
Number of projections	1200	1200	1200
Voxel size [μm]	28	34	49.5
FOV in horizontal plane [mm]	33	58	82
FOV vertical [mm]	74	104	75
Measurement time [min]	40	60	20*

*For the upper 75 mm that could be segmented as shown in Figure 3b. Total scan time for the whole pot ca. 1.5 h.
I.D. pot inner diameter [mm], FOV field of view, Voxel were isotropic with the same size in all three dimensions.

34 mm, 68 μm for pot size 56 mm and 99 μm for pot size 81 mm. The effective dose to the sample during the scan was measured with an additional pot in the center of which a Dose Rate Meter 6150 AD with Scintillator Probe from automess (Automation und Messtechnik GmbH, Ladenburg, Germany) was buried in the same soil and subjected to the same X-ray exposure as the plant samples described above. This measurement gave a value of 274 mGy.

Segmentation of roots from CT images

For the root segmentation we used the Volume Player Plus (VPP) software of the EZRT and the Modular Algorithms for Volume Images (MAVI) software package (Fraunhofer ITWM Kaiserslautern, Germany). In the first step the pot material was removed by manually fitting an ellipse to the inner pot wall in the first slice and removing everything outside this ellipse in all slices of the vertical stack. In cases where thin roots are directly abutting the pot wall or the pot is not perfectly rounded this may cause the loss of root voxels. Afterwards the images were binarised using a range of different gray values. First we determined for every data set which gray value area contains the roots. This area is quite wide due to the different thickness of the roots. Big roots show higher gray values than small ones. According to this we limited the gray value area by defining gray value ranges for small, middle and big roots based on empirical knowledge. In summary we cover the whole gray value area containing roots, but split this area in three parts linked to root thickness by hysteresis binarization. As a result we receive three partial volumes to work with. In the following steps every partial volume is cleaned from remaining non-root parts (described later on). At the end the three volumes containing small, medium and big roots were combined to one volume again and a closing algorithm is used to generate a closed architecture. The closing algorithm is dilation followed by erosion of the objects and was used with a closing radius of three times the respective voxel size. With the MAVI software the remaining non-root parts in the binarised image like stones, soil and air or water filled pores were removed using the labeling function and the object filter. The labeling function assigns to every connected object in the image a unique value. For defining the connectivity of the objects a neighborhood system is used. The information to the different labels is stored in the image history. With this history an object filter can separate them working with object size and sphericity. Small and roundish labels are defined by their sphericity as non-root objects and removed from the image. A sphere would have a value of 1 and a line one of 0. The threshold was set to 0.3. Further information about the image processing can be found in the textbook written by Ohser and Schladitz [45]. The whole procedure for segmenting roots took 1 h for one dataset, in case of the 34 mm pots 2 and for 56 mm pots 3 datasets were acquired. Both pots were of the same height but for the small pots MRI showed that they had not reached the bottom. In case of 81 mm pots 4 datasets were necessary to cover the whole pot, whereas only one could be used for segmentation.

MRI measurement

MRI measurements were performed on a plant dedicated vertical bore 4.7 T magnet equipped with gradient coils providing up to 300 mT m^{-1} (Varian, Palo Alto, USA). For plants grown in 34 mm and 56 mm I.D. pots we used a 63 mm I.D. RF coil (sensitive vertical length 60 mm; Varian, Palo Alto, USA). For plants grown in 81 mm I.D. tubes we used a 100 mm I.D. RF coil (sensitive vertical length 100 mm; Varian, Palo Alto, USA). Experimental control was run on a Varian VNMRS console and a Linux PC using the Varian software VnmrJ. For 3D images a multislice spin echo sequence was used (single echo; provided as part of the instrument package by Varian).

Table 3 Parameters of MRI measurements

	I.D. 34	I.D. 56	I.D. 81
Voxel size in horizontal plane [μm]	333	375	521
Voxel size vertical [μm]	1000	1000	1000
Repetition time T_R [s]	3	3	3
Echo time T_E [ms]	9	9	9
Number of averages	2	2	2
FOV in horizontal plane [mm]	64	72	100
FOV vertical [mm]	156	160	300
Measurement time [min]	40	40	60

I.D. inner pot diameter [mm], Voxel were anisotropic with the same size in the two horizontal dimensions and a larger size in the vertical dimension.

Measurement parameters are listed in Table 3. For the standard MRI data handling, several tools were written in IDL (ITT, Boulder, USA), including data re-ordering, Fourier transformation, filtering and concatenation of blocks of virtual slices that were generated for each pot. Three-dimensional image rendering was performed using MeVisLab (Mevis Medical Solutions AG, Bremen, Germany).

Segmentation of roots from MRI images

Segmentation of roots from MRI images was achieved using a simple threshold on the MRI signal. The threshold was set to four times the noise level of the MRI data. For the medium sized pot, the excessive soil water was visible in the MRI images. For this data set the threshold was manually increased to remove the soil voxels.

Image analysis

Prior to root length calculation, the MRI images were co-registered to the CT data by automated image registration using the program FSL [56] and interpolated onto the same grid. As some parts of the CT images could not be segmented, this co-registration step assures that the images of both modalities contain the same part of the root system. Furthermore, effects due to different voxel sizes are avoided by using the same data grid. Root length calculation was done as described by [57], with some adaptations to be able to process not only MRI images but also segmented root masks obtained from the CT data. First, to be able to process the data with our developed software tools the images were downsampled by an integer factor, resulting in isotropic voxel sizes for the individual plant between 300-400 μm. The root system architecture was then extracted by calculating the cheapest path [58] from each segmented root voxel to the manually selected shoot. In Schulz et al. [57], the cost of a path element was determined by the geometrical length and the MRI image intensity at the respective position. For the CT data, the actual image intensity is not a useful measure to describe the root structure, thus we only worked with the segmented root mask. As the 'image value' for the segmented mask was either 0 or 1, all paths had equal costs and multiple parallel traces in one root segment did occur during the root extraction. These were avoided by additionally weighting the path costs with the distance transform of the root mask, thus favoring paths that run in the center of the segmented mask. The resulting root trees then needed pruning: paths shorter than 3 mm were removed, as well as paths bridging more than 3.5 mm of non-segmented voxels. After visual inspection of the resulting root system architecture, the total length was finally calculated from it.

Additional files

Additional file 1: Root length data for additional 34 mm pot measured with CT, MRI and WinRHIZO. A second bean plant grown under the same conditions as the one shown in Figure 1 was also analyzed with CT, MRI and WinRHIZO. Here the results of the root length measurements for all three methods are shown.

Additional file 2: Comparison of 34 mm pot scanned with CT with and without filter. The same bean plant grown in soil in a 34 mm I.D. pot was scanned with CT a) with a 1 mm thick Copper filter on the X-ray tube at 220 kV tube voltage and 100 μA tube current and without the filter at 100 kV and 120 μA. The same horizontal slice from both measurements was selected and is shown here, depicting roots within the sand/soil matrix. The X-shaped structure is the cross section of a plastic support for the bean pushed into the soil. The white ring at the outer circumference of the pot in b) indicates beam hardening artefacts in this area.

Competing interests

The authors declare that they have no competing interests.

Authors' contributions

NU, SJ, DvD and US designed the experiments. RM, AE and DvD performed the experiments. DP, DvD, AE and SG analyzed the data. AE and RM wrote the manuscript. NU and US supervised the study. All authors contributed in reading, editing and approval of the final manuscript.

Acknowledgements

This study was supported by the German Federal Ministry for Education and Science (BMBF) within the German-Plant-Phenotyping Network (project identification number 031A053). The authors would like to thank Esther Breuer for help with plant cultivation and doing the WinRHIZO analysis, Johannes Kochs for technical support in designing and building hardware for MRI and Vitali Weizel for technical support with the CT instrument.

Author details

[1]Institute of Bio- and Geosciences, IBG-2: Plant Sciences, Forschungszentrum Jülich GmbH, Wilhelm-Jonen- Str., 52425 Jülich, Germany. [2]Development Center X-Ray Technology EZRT, Fraunhofer Institute for Integrated Systems IIS, Flugplatzstraße 75, 90768 Fürth, Germany.

References

1. Gregory PJ. Plant roots. Oxford, UK: Blackwell Publishing Ltd; 2006.
2. Gregory PJ, Bengough AG, Grinev D, Schmidt S, Thomas WTB, Wojciechowski T, et al. Root phenomics of crops: opportunities and challenges. Funct Plant Biol. 2009;36:922–9.
3. Mooney SJ, Pridmore TP, Helliwell J, Bennett MJ. Developing X-ray computed tomography to non-invasively image 3-D root systems architecture in soil. Plant Soil. 2012;352:1–22.
4. Giehl RFH, von Wirén N. Root nutrient foraging. Plant Physiol. 2014;166:509–17.
5. Trachsel S, Kaeppler SM, Brown KM, Lynch JP. Shovelomics: high throughput phenotyping of maize (Zea mays L.) root architecture in the field. Plant Soil. 2011;341:75–87.
6. Lynch JP. Root phenes for enhanced soil exploration and phosphorus acquisition: tools for future crops. Plant Physiol. 2011;156:1041–9.
7. Eshel A. Plant roots: the hidden half. 4th ed. Boca Raton, Florida, USA: CRC Press; 2013.
8. Iyer-Pascuzzi AS, Symonova O, Mileyko Y, Hao YL, Belcher H, Harer J, et al. Imaging and analysis platform for automatic phenotyping and trait ranking of plant root systems. Plant Physiol. 2010;152:1148–57.
9. Downie H, Holden N, Otten W, Spiers AJ, Valentine TA, Dupuy LX. Transparent soil for imaging the rhizosphere. PLoS One. 2012;7:e44276.
10. Nagel KA, Kastenholz B, Jahnke S, Van Dusschoten D, Aach T, Muhlich M, et al. Temperature responses of roots: impact on growth, root system architecture and implications for phenotyping. Funct Plant Biol. 2009;36:947–59.

11. Neumann G, George TS, Plassard C. Strategies and methods for studying the rhizosphere-the plant science toolbox. Plant Soil. 2009;321:431–56.
12. Blümler P, Windt CW, Van Dusschoten D. Magnetic resonance in plants. Nova Acta Leopoldina. 2009;357:17–30.
13. Jahnke S, Menzel MI, van Dusschoten D, Roeb GW, Bühler J, Minwuyelet S, et al. Combined MRI-PET dissects dynamic changes in plant structures and functions. Plant J. 2009;59:634–44.
14. Rogers HH, Bottomley PA. *In situ* Nuclear Magnetic Resonance imaging of roots - Influence of soil type, ferromagnetic particle content and soil water. Agron J. 1987;79:957–65.
15. Metzner R, van Dusschoten D, Bühler J, Schurr U, Jahnke S. Belowground plant development measured with magnetic resonance imaging (MRI): exploiting the potential for non-invasive trait quantification using sugar beet as a proxy. Front Plant Sci. 2014;5:469.
16. Tumlinson LG, Liu HY, Silk WK, Hopmans JW. Thermal neutron computed tomography of soil water and plant roots. Soil Sci Soc Am J. 2008;72:1234–42.
17. Leitner D, Felderer B, Vontobel P, Schnepf A. Recovering root system traits using image analysis exemplified by two-dimensional neutron radiography images of lupine. Plant Physiol. 2014;164:24–35.
18. Jenneson PM, Gilboy WB, Morton EJ, Gregory PJ. An X-ray micro-tomography system optimised for the low-dose study of living organisms. Appl Radiat Isot. 2003;58:177–81.
19. Flavel RJ, Guppy CN, Tighe M, Watt M, McNeill A, Young IM. Non-destructive quantification of cereal roots in soil using high-resolution X-ray tomography. J Exp Bot. 2012;63:2503–11.
20. Zhu JM, Ingram PA, Benfey PN, Elich T. From lab to field, new approaches to phenotyping root system architecture. Curr Opin Plant Biol. 2011;14:310–7.
21. Koebernick N, Weller U, Huber K, Schluter S, Vogel HJ, Jahn R, et al. In situ visualization and quantification of three-dimensional root system architecture and growth using X-Ray computed tomography. Vadose Zone J. 2014;13:8.
22. Plews AG, Atkinson A, McGrane S. Discriminating structural characteristics of starch extrudates through X-ray micro-tomography using a 3-D watershed algorithm. Int J Food Eng. 2009;5:11.
23. Hanke R, Fuchs T, Uhlmann N. X-ray based methods for non-destructive testing and material characterization. Nucl Instrum Meth A. 2008;591:14–8.
24. Salvo L, Cloetens P, Maire E, Zabler S, Blandin JJ, Buffiere JY, et al. X-ray micro-tomography an attractive characterisation technique in materials science. Nucl Instrum Meth B. 2003;200:273–86.
25. Kalender WA. Computed tomography: fundamentals, system technology, image quality, applications. 3 revth ed. Erlangen, Germany: Publicis Publ; 2011.
26. Bao Y, Aggarwal P, Robbins NE, Sturrock CJ, Thompson MC, Tan HQ, et al. Plant roots use a patterning mechanism to position lateral root branches toward available water. Proc Natl Acad Sci U S A. 2014;111:9319–24.
27. Tracy SR, Black CR, Roberts JA, Sturrock C, Mairhofer S, Craigon J, et al. Quantifying the impact of soil compaction on root system architecture in tomato (Solanum lycopersicum) by X-ray micro-computed tomography. Ann Bot (London). 2012;110:511–9.
28. Zappala S, Mairhofer S, Tracy S, Sturrock CJ, Bennett M, Pridmore T, et al. Quantifying the effect of soil moisture content on segmenting root system architecture in X-ray computed tomography images. Plant Soil. 2013;370:35–45.
29. Mairhofer S, Zappala S, Tracy S, Sturrock C, Bennett MJ, Mooney SJ, et al. Recovering complete plant root system architectures from soil via X-ray mu-Computed Tomography. Plant Methods. 2013;9:1–7.
30. Haacke EM, Brown RW, Thompson MR, Venkatesan R. Magnetic Resonance Imaging. New York, USA: John Wiley and Sons; 1999.
31. Callaghan PT. Principles of nuclear magnetic resonance microscopy. New York, USA: Oxford University Press; 1993.
32. Köckenberger W, De Panfilis C, Santoro D, Dahiya P, Rawsthorne S. High resolution NMR microscopy of plants and fungi. J Microsc. 2004;214:182–9.
33. Van As H, Scheenen T, Vergeldt FJ. MRI of intact plants. Photosynth Res. 2009;102:213–22.
34. Borisjuk L, Rolletschek H, Neuberger T. Surveying the plant's world by magnetic resonance imaging. Plant J. 2012;70:129–46.
35. Hillnhütter C, Sikora RA, Oerke EC, van Dusschoten D. Nuclear magnetic resonance: a tool for imaging belowground damage caused by Heterodera schachtii and Rhizoctonia solani on sugar beet. J Exp Bot. 2012;63:319–27.
36. Gruwel MLH. In situ magnetic resonance imaging of plant roots. Vadose Zone J. 2014;13:3.
37. MacFall JS, Johnson GA. Plants, seeds, roots, and soils as applications of magnetic resonance microscopy. In: Encyclopedia of magnetic resonance. New York, USA: John Wiley & Sons, Ltd; 2012.
38. Lontoc-Roy M, Dutilleul P, Prasher SO, Han LW, Brouillet T, Smith DL. Advances in the acquisition and analysis of CT scan data to isolate a crop root system from the soil medium and quantify root system complexity in 3-D space. Geoderma. 2006;137:231–41.
39. Poorter H, Bühler J, van Dusschoten D, Climent J, Postma JA. Pot size matters: a meta-analysis of the effects of rooting volume on plant growth. Funct Plant Biol. 2012;39:839–50.
40. Schmidt S, Bengough AG, Gregory PJ, Grinev DV, Otten W. Estimating root-soil contact from 3D X-ray microtomographs. Eur J Soil Sci. 2012;63:776–86.
41. Helliwell JR, Sturrock CJ, Grayling KM, Tracy SR, Flavel RJ, Young IM, et al. Applications of X-ray computed tomography for examining biophysical interactions and structural development in soil systems: a review. Eur J Soil Sci. 2013;64:279–97.
42. Kaestner A, Schneebeli M, Graf F. Visualizing three-dimensional root networks using computed tomography. Geoderma. 2006;136:459–69.
43. Mairhofer S, Zappala S, Tracy SR, Sturrock C, Bennett M, Mooney SJ, et al. RooTrak: automated recovery of three-dimensional plant root architecture in soil from X-ray microcomputed tomography images using visual tracking. Plant Physiol. 2012;158:561–9.
44. Tracy SR, Roberts JA, Black CR, McNeill A, Davidson R, Mooney SJ. The X-factor: visualizing undisturbed root architecture in soils using X-ray computed tomography. J Exp Bot. 2010;61:311–3.
45. Ohser J, Schladitz K. 3D images of materials structures: processing and analysis Weinheim. Germany: Wiley-VCH; 2009.
46. Rubio G, Lynch JP. Compensation among root classes in Phaseolus vulgaris L. Plant Soil. 2007;290:307–21.
47. Wildenschild D, Hopmans JW, Vaz CMP, Rivers ML, Rikard D, Christensen BSB. Using X-ray computed tomography in hydrology: systems, resolutions, and limitations. J Hydrol. 2002;267:285–97.
48. Perret JS, Al-Belushi ME, Deadman M. Non-destructive visualization and quantification of roots using computed tomography. Soil Bio Biochem. 2007;39:391–9.
49. Baveye PC, Laba M, Otten W, Bouckaert L, Dello Sterpaio P, Goswami RR, et al. Observer-dependent variability of the thresholding step in the quantitative analysis of soil images and X-ray microtomography data. Geoderma. 2010;157:51–63.
50. Bottomley PA, Rogers HH, Foster TH. NMR Imaging shows water distribution and transport in plant-root systems *in situ*. Proc Natl Acad Sci U S A. 1986;83:87–9.
51. Heeraman DA, Hopmans JW, Clausnitzer V. Three dimensional imaging of plant roots in situ with x-ray computed tomography. Plant Soil. 1997;189:167–79.
52. Gregory PJ, Hutchison DJ, Read DB, Jenneson PM, Gilboy WB, Morton EJ. Non-invasive imaging of roots with high resolution X-ray micro-tomography. Plant Soil. 2003;255:351–9.
53. Hargreaves CE, Gregory PJ, Bengough AG. Measuring root traits in barley (Hordeum vulgare ssp vulgare and ssp spontaneum) seedlings using gel chambers, soil sacs and X-ray microtomography. Plant Soil. 2009;316:285–97.
54. Hochholdinger F, Tuberosa R. Genetic and genomic dissection of maize root development and architecture. Curr Opin Plant Biol. 2009;12:172–7.
55. Lucas M, Swarup R, Paponov IA, Swarup K, Casimiro I, Lake D, et al. SHORT-ROOT Regulates Primary, Lateral, and Adventitious Root Development in Arabidopsis. Plant Physiol. 2011;155:384–98.
56. Jenkinson M, Smith S. A global optimisation method for robust affine registration of brain images. Med Image Anal. 2001;5:143–56.
57. Schulz H, Postma J, Dusschoten D, Scharr H, Behnke S. Plant root system analysis from MRI images. In: Csurka G, Kraus M, Laramee R, Richard P, Braz J, editors. Computer vision, imaging and computer graphics theory and application. Volume 359. Berlin Heidelberg: Springer; 2013. p. 411–25. Communications in Computer and Information Science.
58. Dijkstra EW. A note on two problems in connexion with graphs. Numer Math. 1959;1:269–71.

An optical imaging chamber for viewing living plant cells and tissues at high resolution for extended periods

Grant Calder, Chris Hindle, Jordi Chan and Peter Shaw*

Abstract

Background: Recent developments in both microscopy and fluorescent protein technologies have made live imaging a powerful tool for the study of plant cells. However, the complications of keeping plant material alive during a long duration experiment while maintaining maximum resolution has limited the use of these methods.

Results: Here, we describe an imaging chamber designed to overcome these limitations, which is flexible enough to support a range of sizes of plant materials. We were able use confocal microscopy to follow growth and development of plant cells and tissues over several days. The chamber design is based on a perfusion system, so that the addition of drugs and other experimental treatments are also possible.

Conclusions: In this article we present a design of imaging chamber that makes it possible to image plant material with high resolution for extended periods of time.

Keywords: Arabidopsis, Imaging chamber, Time-lapse microscopy, Perfusion, Live cell imaging, Plant tissue

Background

The progress in optical microscopy technology (faster acquisition, more sensitive cameras, specialised immersion objectives, confocal, multi photon etc.) combined with direct labelling of target proteins using fluorescent protein markers has made live cell imaging a powerful tool for investigating dynamic processes in plants [1-5] This technology enables us to follow dynamic events in live experiments, rather than taking snapshot images of chemically fixed material. This allows many biological processes to be studied from start to finish. For example, the processes involved in the formation of a tissue or organ are regulated by developmental cues or circadian rhythms. To understand these developmentally regulated patterns it is necessary to follow cells in a developing tissue throughout organ formation, which can take from hours to days [6].

Short term imaging of living samples is relatively easy and the sample preparation can be as simple as putting the sample onto a slide in a suitable medium with a cover glass on top. However, successful high resolution imaging for a long period of time is more difficult. Three key factors need to be addressed: a constant environment surrounding the sample; stably holding the sample's position within the imaging volume; and, for larger samples, imaging a large enough volume. To track a sample for long durations it is necessary to make specialised sample holders or chambers to address these issues [7].

In general plant material is very robust to large variations in many environmental conditions. However, plant samples are highly susceptible to decreasing levels of nutrients and oxygen (anoxia) during live cell imaging experiments. Effects like anoxia have been shown to alter the localisation of a range nuclear and nucleolar proteins from a nucleoplasmic distribution to a speckled pattern [8]. Therefore imaging chambers for plants must be designed to maintain the sample's environment, in order to keep the sample healthy over periods of days. In static systems, as a plant grows, the nutrients in the medium are used up, changing the specimen's environment. It is therefore preferable to use a perfusion system to replace depleted media (nutrients and oxygen) around the sample, giving a stable environment throughout the duration of the experiment.

* Correspondence: peter.shaw@jic.ac.uk
Department of Cell & Developmental Biology, John Innes Centre, Norwich Research Park, Norwich NR4 7UH, UK

When imaging at high resolution, optical constraints limit the imaging volume. A typical high magnification immersion objective has a working distance in z (the optical axis) of 100–200 μm and a field of view in x and y (the image plane) of 200-400 μm. It is therefore necessary to position different sized samples within this working distance (close to the cover-glass) without damaging them, and then stably hold their position within this narrow imaging volume throughout an experiment. An additional stability issue can be caused by the laminar flow of fresh media into the chamber when using a perfusion-based system.

The microscope's field of view is a limiting factor in imaging large or fast-growing specimens. Low magnification lenses can be used to increase the viewable area but this decreases the resolution. To collect high resolution images of large specimens, it is necessary to increase the field of view of the microscope by another means. This can be achieved by translating the specimen to create a series of overlapping images, which can be stitched together to form a high resolution tiled image. This technique requires a computer-controlled motorised x-y stage and suitable image processing software to stitch the images together such as, for example, ImageJ/Fiji [9-12]. Most microscope manufactures are now able to equip their systems with motorised stages and image stitching software.

Although there are several good commercial imaging chambers available [13], these chambers have not been specifically designed for the high resolution imaging of plant tissues over extended periods. Instead, commercial systems are primarily designed for imaging animal and microbial cell cultures using inverted microscopes. However, gravitropism in plants causes some complications, since aerial parts such as leaves tend to grow upwards and roots grow downwards. Since it damages the plant to clamp it too tightly, this generally means that the smaller parts – usually those parts of interest that are growing rapidly – are not fully constrained between the front and back of the chamber. Thus, leaves will tend to grow away from the coverglass on an inverted system, but will stay next to it on an upright system. Conversely, roots will tend grow away from the coverglass on an upright system, but stay on the coverglass on an inverted system. This means that it is preferable to have a growth chamber that can be accommodated on either an upright or an inverted microscope. Other things being equal, we prefer the ease of use of an upright microscope.

In plants the majority of long duration time-lapse experiments have either been performed using static systems, which cause the environment issues discussed above over extended periods [7], or using perfusion systems designed specifically for the study of roots on inverted microscopes [14].

Our aim, therefore, was to construct an imaging chamber that allowed the stable imaging of a range of plant tissues over extended periods of time at high resolution. Here we describe the design of a fully enclosed imaging chamber based on a liquid perfusion system that can be attached to an upright microscope, or could be easily used in an inverted orientation adapted for an inverted microscope. This modified perfusion chamber is designed to reduce the negative effects of laminar flow of medium. It has the additional benefit that the environmental medium can be changed or drugs or other soluble components can be added during an imaging experiment, which makes this system a powerful experimental tool.

Results and discussion

The imaging chamber

The system is designed in two parts: an outer containment tray (CT in Figure 1 – L160mm x W110mm x H16mm) that attaches to the microscope stage; and an inner specimen chamber (IC in Figure 1 – L70mm x W48mm x H11mm). The outer tray was designed to fit into a Prior Proscan H101 motorised scanning stage, but is easily attached to different brands of microscope stages with simple adapters. The outer containment tray protects the microscope from possible damage from the liquid medium if the imaging chamber should leak. The imaging chamber is made from stainless steel and glass, which can be sterilised by autoclaving prior to use to minimise any potential contamination problems. An optical cover glass (CG in Figure 1) with a 32 mm diameter and a thickness of 0.17 mm (number 1.5 - Thermo Scientific) is used; this thickness is necessary for high resolution immersion objectives. As the cover glass is very thin it requires a metal holding plate to give added structural strength and rigidity (HP in Figure 1 - outer diameter 40 mm, inner diameter 25 mm, thickness 1 mm). The holding plate has a 25 mm opening but only 20 mm area is accessible due to the physical constraints of immersion objective lenses. The cover glass and holding plate are held in place on the imaging chamber by two sliding locking clamps (LC in Figure 1 - each one L35mm x W55mm x H10mm, with a 35 mm diameter semi-circular cut-out for HP) that have flanges (6 mm × 3 mm) that slide in slots in the inner chamber (LCG in Figure 1). When the two clamps are pushed together the holding plate and coverglass are lightly but firmly held in position by an even force that minimises physical distortion. With this sliding clamp configuration the imaging chamber can be easily opened and resealed to allow the quick addition of biological material, usually under sterile flow hood conditions. To open the chamber, perfusion is stopped and the outlet tube is clamped closed to prevent medium from draining from the

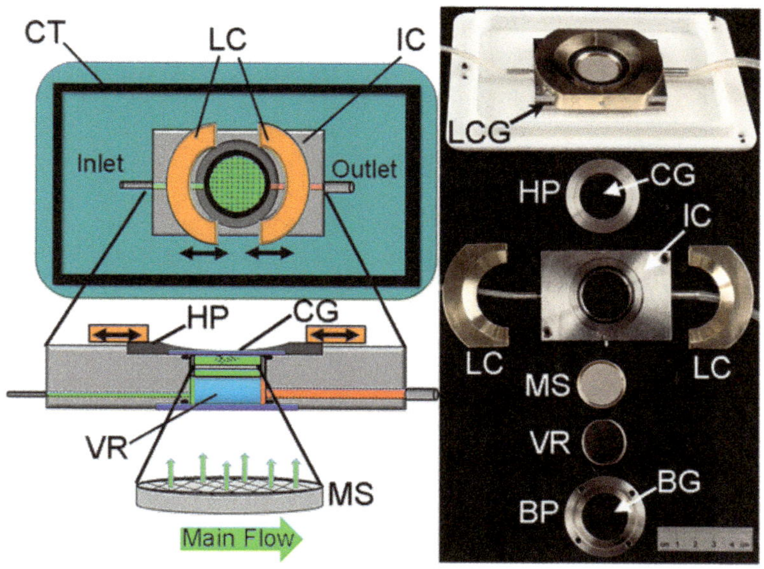

Figure 1 Design of the imaging chamber. The left panel is a schematic diagram of the chamber from top and side views. Note that the flexible mesh support removes the sample from the laminar flow of the incoming liquid protecting it from shear stress. The right panel is a picture of the chamber fully assembled and disassembled. CT - Containment tray, HP - holding plate, IC - imaging chamber, LC - locking clamp, MS - mesh support, VR - volume reducer, BP - base plate, CG - cover glass, BG - base glass, LCG – locking clamp groove.

chamber. The locking clamps are slid apart along the chamber body to expose the holding plate, which can then be removed to give access to the sample. The chamber is reassembled in the reverse order. The base of the imaging chamber is made in two parts - a base glass (BG in Figure 1 - window 35 mm diameter, thickness 2 mm - UQG Optics Ltd), which is held in place with a base plate (BP in Figure 1 - outer diameter 44 mm, inner diameter 25 mm, thickness 2.5 mm), screwed into the body of imaging chamber. The imaging chamber body contains top and bottom O-ring gaskets to create watertight seals between both the cover and base glasses. To lower the cost of an experiment using expensive compounds, the imaging chamber volume can be reduced from 3.5 ml to 2.0 ml by inserting a circular glass volume reducer (VR in Figure 1 - 25 mm diameter, thickness 3 mm) (UQG Optics Ltd).

The specimen is placed on top of a thin stainless steel mesh (Endecotts Ltd) stretched out across a supporting ring to form a mesh support (MS in Figure 1 - 26 mm diameter, thickness 3 mm). The height of the mesh support is adjustable and is usually set to approximately 100-150 μm from the cover glass to form a suitable cavity for Arabidopsis seedlings. The mesh support positions the sample near the cover glass so that it is within the working distance of an immersion objective lens. The mesh is flexible and deforms to the shape of the tissue to some extent, which helps to stabilise the specimen's position in z (the direction of the optical axis). Previous designs of perfusion imaging chambers suffered from sample movement or damage caused by the laminar flow of medium passing through the narrow sample cavity. This chamber design alleviates this problem by using the mesh support to separate the sample cavity from the flowing medium. Medium is exchanged between the two sides by passive diffusion of liquid through the 20 μm pores in the mesh (Figure 1).

A peristaltic pump is used for long duration experiments with large media reservoirs. For adding soluble compounds a syringe pump is more convenient as syringes containing different compounds can be quickly exchanged and the solution can be added manually, allowing for faster application. The flow rate for the peristaltic pump was measured directly. For the syringe pump a calibration table relating the size of syringe and the pump speed was used. The input pressure drives the flow through the chamber, and there is no output pump. The culture medium is pumped into the imaging chamber through an inlet tube of 1 mm inner diameter and exits through a larger 2 mm inner diameter outlet tube, which prevents pressure build-up inside the chamber.

Environmental conditions

A successful imaging chamber needs to maintain a healthy environment over extended periods. This system continuously perfuses fresh media into the chamber to replace depleted nutrients. To assess the system's ability to support plant growth we placed Arabidopsis seeds inside the chamber and then imaged them at daily intervals using a stereo microscope. This experiment showed that seeds could germinate inside the chamber and that the perfusion system provided adequate nutrients for growth into large seedlings, which developed root and leaf systems. An example of germination is shown in Figure 2 (also see movie Additional file 1). The seed germinated after 4 days, the hypocotyl emerging from the seed coat. This was then followed by the expansion of the hypocotyl and cotyledons. New leaves and roots were then seen to emerge from day 9 and continued to grow throughout the duration of the experiment (31 days). These results demonstrate that the system is suitable for tracking the expansion of tissues such as the hypocotyl [15] and cotyledons as well as to follow development of emerging roots and leaves [16].

Here, we imaged for 31 days to demonstrate the system's ability to maintain growth. The growing plant was continuously perfused over the entire period using a peristaltic pump and a 1 litre reservoir of medium which was continuously recycled (the total amount of nutrients used was very small compared to the size of the reservoir, and oxygen was replenished by atmospheric exchange during recycling as the reservoir was not airtight). However, it is important to note that over these extended periods (tens of days) the depth of chamber's sample cavity set at the start of the experiment imposes a physical constraint that can cause tissue squashing or distortion during long periods of growth. To prevent extreme changes in specimen size during experiments, we recommend using specimens at the appropriate developmental stage and adjusting the chamber's internal cavity depth using the mesh support (MS in Figure 1) to match the specimen's size.

Maintaining sample position

Time-lapse experiments using high resolution immersion objective lenses commonly suffer from the part of the specimen of interest drifting out of the small and narrow imaging volume. It is therefore imperative that the imaging chamber stably holds the specimen's position. With a perfusion system used to maintain the sample's health, the flow of fresh medium flowing into the chamber could cause a problem by affecting the sample's position. To prevent this we used a design that separates the sample from the direct flow of medium with a mesh support (MS in Figure 1). We used suspension cells to test the chambers ability to maintain a sample's position, since they exist in small loosely packed groups which are more liable to move out of the imaging volume if strong currents are generated by media flow. In this experiment, Arabidopsis culture cells expressing GFP-tubulin were added to the

Figure 2 Arabidopsis seedling growth inside imaging chamber. Low magnification images showing the germination and growth of an Arabidopsis plant inside the imaging chamber. The moiré patterns in the background are caused by interference between the illumination and the grid of the mesh support. Scale bar = 1 mm.

chamber and allowed to settle for a period of 2 hours. Confocal section stacks were collected every 10 minutes for 14 hours (Figure 3). Despite nearly continuous imaging, the cells underwent growth and division (see white arrows) indicating that they were healthy. Some movement was caused by cell growth but most cells remained within the imaging volume (marked by asterisks in Figure 3, first and last time-points) showing that the flow of fresh medium did not seriously affect the specimen position.

High resolution imaging

Detailed images of sub-cellular structures such as microtubules are only possible if a high resolution immersion objective is used. In turn this will limit how far from the cover glass the sample can be positioned - typically a maximum of about 200 μm. The chamber's design allows the sample to be positioned within this small working distance. Figure 4 shows Arabidopsis suspension cells expressing GFP-tubulin imaged using a confocal microscope equipped with 63/1.4 oil objective lens, which has a working distance of 190 μm. The chamber's mesh support was set to a depth of 100 μm to position the cells within the working distance of the objective. During this experiment, confocal section stacks were collected every 5 minutes for 4 hours. Detailed images of a cell as it progressed through cell division revealed microtubule structures from pro-metaphase through to cytokinesis (see Figure 4 and Additional file 2). This shows that the chamber supports high resolution optics, allowing detailed imaging of subcellular structures.

Imaging large samples

With a high resolution objective the available field of view is limited and plants can quickly grow beyond this area. The chamber was designed to attach to a motorised stage so that image tiling can be used to increase the area imaged. The stage moves the sample to create a series of overlapping images, which are stitched together to form a larger image. To illustrate how image tiling can be used to track the growth of plant tissue at high resolution, we used

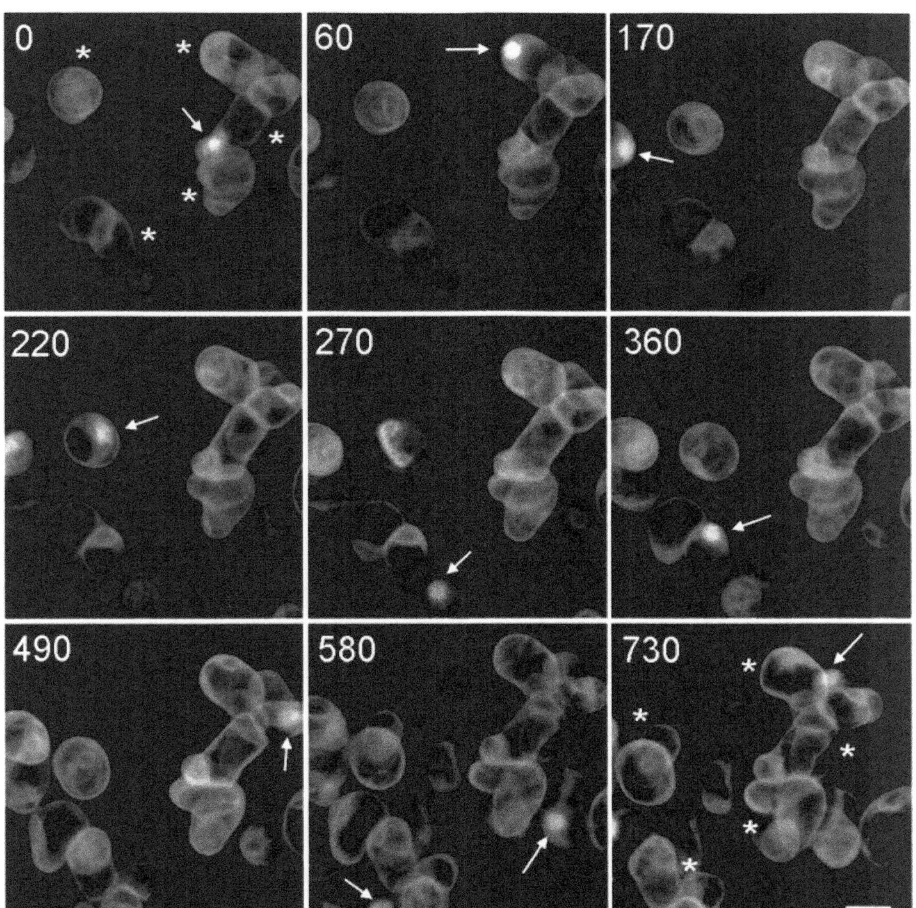

Figure 3 Time-lapse of Arabidopsis suspension cells. Confocal imaging of Arabidopsis suspension cells expressing GFP tubulin. A large field of view containing several cells was imaged every 10 minutes for 780 minutes. Cell growth and division (arrows) were captured throughout experiment with asterisks indicating examples of cells that have maintained stable positions. Scale bar = 20 μm.

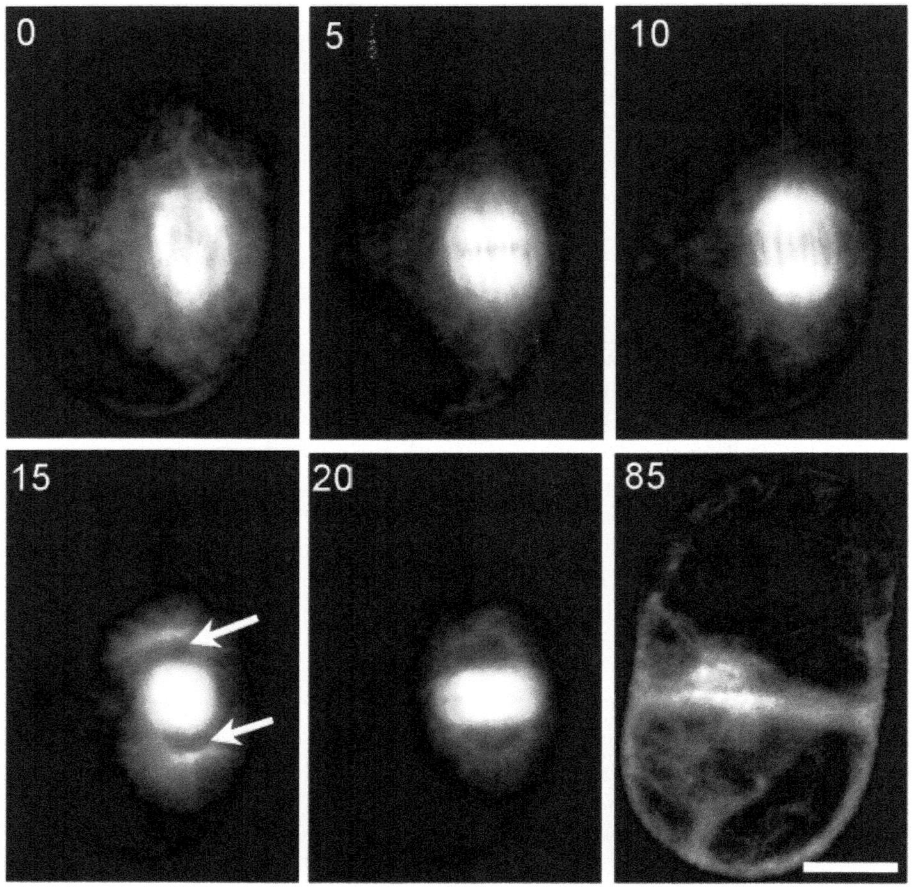

Figure 4 High resolution imaging of a dividing Arabidopsis suspension cell. Confocal microscopy time-lapse images of an Arabidopsis culture cell expressing GFP tubulin. High resolution z stack images were collected using a 63/1.4 oil immersion objective, every 5 minutes throughout division. This illustrates the interchange of different microtubule structures: prometaphase (0 min); metaphase (5 min); anaphase (10 min); telophase (15 min daughter nuclei position indicated by arrows); phragmoplast (20 min); cell division completion (85 min). The numbers indicate the time in minutes from start of time-lapse acquisition and confocal z stacks are shown as summation projections. Scale bar = 10 μm.

a confocal microscope equipped with a motorised stage and a 40x/1.2NA water immersion objective to follow the growth of an Arabidopsis leaf expressing GFP-tubulin. Overlapping confocal section stacks of the growing leaf were collected every hour for 63 hours (Figure 5 and Additional file 3). The available imaging area of this objective lens was limited to 212 × 212 μm (see the dashed box in Figure 5a). At the start of the experiment a 3 × 3 array of overlapping images (595 × 595 μm) was used to collect an image of a complete leaf. By the end of the experiment, the area needed to image the entire leaf had increased, to a 3 ×7 array of overlapping images (1360 × 595 μm). It was possible to follow cell growth and division (arrow at 12 hours shows a mitotic cell) as the leaf developed. Tracking a complete developing organ can reveal how cell division and growth are regulated at the tissue level, for example in hypocotyls [15] and leaf tissue [6,16]. To illustrate cell growth we have extracted from the tile images a smaller sub region of cells near the leaf tip (highlighted in green in Figure 5a) and displayed them in Figure 5b. This sub-region of cells shows the high resolution information that is available with the tiling approach. Some caution is needed when overlapping a large number of areas since, if significant growth occurs during the long acquisition time necessary, the final tiled image will be distorted.

Changing the specimen environment

An advantage of a perfusion system over a static one [7] is that soluble compounds can be quickly added or removed from the imaging chamber and their effects can be tracked without disturbing the specimen. To illustrate this we added and removed propidium iodide (PI) from the chamber containing a growing Arabidopsis seedling expressing eIF4a-III-GFP, which is localised to the nucleus. Propidium iodide fluoresces in the red spectral region, and is excluded from the interior of living plant cells. It is commonly used to label the cell outlines in roots [17]. We used appearance and disappearance of PI

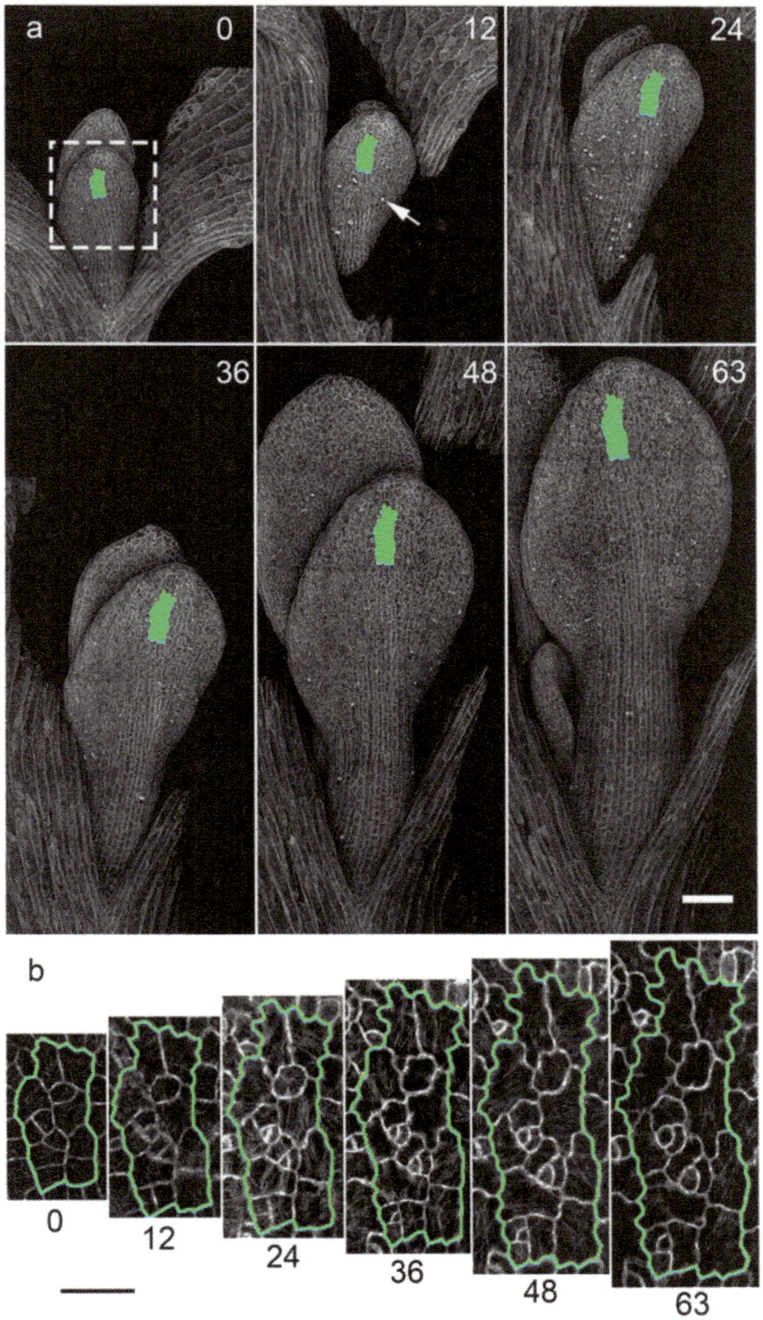

Figure 5 Increasing the field of view using tiled imaging to capture the growth of an Arabidopsis leaf expressing GFP tubulin. (a) The dashed box indicates the limited field of view from 40/1.2 water objective at a single position. Tiled images were captured every hour for 63 hours where cell division (arrow) and growth were captured. Scale bar = 100 µm. **(b)** Displays a sub region from the tiled image (panel (a) highlighted in green) illustrating cellular growth. Maximum projections of the z stacks are shown and the numbers indicate time in hours from start of time-lapse experiment. Scale bar = 30 µm.

fluorescence to monitor the addition and removal of the compound from the plant. Confocal section stacks of growing roots were collected every 5 minutes throughout the experiment (Figure 6 and Additional file 4). At the start of the experiment, before the addition of PI, only the root cell nuclei were visible (green in Figure 6a). PI was added to the medium by manually injecting 6mls of PI solution over 60 secs and then perfusing the chamber. Cell outlines became visible within 5 minutes (red in Figure 6b). To remove the PI, several chamber volumes (10 ml) of fresh MS media were then manually flushed through the system between two time-points

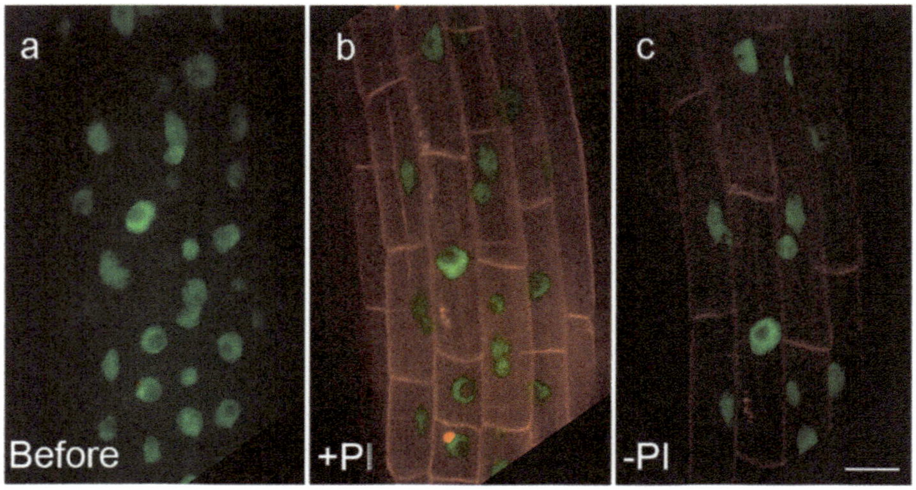

Figure 6 The addition and removal of PI in a growing Arabidopsis root during image acquisition. The root was imaged by confocal microscopy where z stacks were taken every 5 minutes. **(a)** Only green fluorescent nuclei (GFP-eIF4A-III) are visible at the start of the experiment **(b)** Within 5 minutes after the addition of PI cell outlines become clearly visible **(c)** PI was washed out of the chamber leaving only a faint red outline. Images are summation projection of the z stacks. Scale bar = 20 μm.

5 minutes apart, which removed most of the PI, leaving only very faint cell outlines from the small amount of residual PI (Figure 6c). It should be noted that although compounds can be added or removed from the chamber within minutes, the uptake or release rate from the specimen is tissue specific (for example, very fast in roots, but slower in leaf because of the waxy cuticle). The chamber's design allows for uninterrupted imaging during the addition or removal compounds because the specimen is separated from the direct flow of incoming media. The ability to add or remove biologically active compounds makes this system a powerful experimental tool. In a recent example using this chamber gibberellic acid was added to growing Arabidopsis seedlings [15], and the hypocotyls were imaged.

Conclusions

Long duration time-lapse imaging experiments of living plants at sub-cellular resolution present substantial challenges which require a specialised imaging chamber. Here we describe a suitable imaging chamber that can maintain plant material in a healthy state for extended period of time by replacing depleted nutrients with fresh media. Its flexible design enables imaging of a wide variety of plant specimens – cell cultures, hypocotyls, leaves and roots. The system is compatible with high resolution immersion objectives, allowing the samples to be positioned close enough to the cover glass to be within the working distance of high numerical aperture objective lenses. The mesh support helps to maintain the sample's position within the imaging volume but is flexible enough to allow sample growth. The chamber can be attached to microscopes equipped with motorised x,y stages to allow larger samples to be imaged at high resolution by tiling overlapping images. This design of imaging chamber eliminates movement problems caused by the flow of incoming media that can be a problem in perfusion systems. The ability to add or remove soluble biologically active compounds without affecting image acquisition makes this imaging chamber a powerful experiment tool for studying plant development, which has already been used in many published studies.

Methods

Cell culture

Stable cell lines expressing an alpha tubulin 6-GFP fusion protein expressed under the 35S promoter were derived from *Arabidopsis thaliana* Col-0 root explants. Calli were then transferred to Murashige and Skoog liquid culture (MS: Formedium, Hunstanton, UK) containing 3% (w/v) sucrose (Sigma) and grown in the dark at 25°C with 150 rpm agitation and were sub-cultured every 7 to 8 days at a 1/10 dilution as previously described [18]. Cells were transferred to the microscope imaging chamber 2 days after sub-culturing, in a laminar flow hood to prevent contamination. After transfer to the imaging chamber, cells were continuously perfused with MS medium containing 1% (w/v) sucrose at a flow rate of 3 ml/h.

Plant material

Arabidopsis thaliana plants expressing 35S::GFP-eIF4A-III [8] or 35S::alpha tubulin 6-GFP [19] were used. Seeds were sterilised in 5% (w/v) sodium hypochlorite before transfer onto Petri dishes containing MS medium containing 1% (w/v) sucrose (Sigma) and 0.5% (w/v)

Phytagel (Sigma). Seeds were stratified at 4°C for 2 days and grown for 5 days for root experiments or 7 days for young leaf experiments under artificial light at 23°C with a 16 hour photoperiod. 5 or 7-day-old seedlings were transferred to the microscope imaging chamber in a laminar flow hood to prevent contamination and then imaged using a confocal microscope. After transfer the seedlings were continuously perfused with ¼ strength MS liquid medium at a flow rate of 3 ml/h.

Low magnification imaging

Arabidopsis seeds were sterilised and placed inside the imaging chamber and continuously perfused with ¼ strength MS liquid medium at a flow rate of 3 ml/h. The chamber was housed in growth room under artificial light at 23°C with a 16 hour photoperiod. The chamber was removed briefly for low magnification imaging using a Leica M205 FA stereo microscope with both transmitted and reflected illumination. Images were collected over a period of 28 days at intervals of ≥1 day.

Confocal imaging

Arabidopsis cell cultures were imaged with a Zeiss LSM 510 Meta confocal microscope using a 63/1.4 oil immersion objective lens. GFP fluorescence was excited using the 488 nm line from an argon ion laser and the emitted light was filtered through a 505-550 nm band-pass filter. Confocal z section stacks were collected at 1 μm spacing through the depth of the cultured cell, and time-series images were acquired with an interval of either 5 or 10 minutes.

Arabidopsis seedlings were imaged using a Zeiss LSM 780 confocal microscope using a 40/1.2 water immersion objective lens. GFP fluorescence was excited using the 488 nm line from an argon ion laser and the emitted light was capture using a GAsP spectral array 499-579 nm. Confocal z section stacks were collected at 2 μm spacing through the depth of the leaf tissue, and time-series images were acquired with an interval of 1 hour. Image tiling was controlled through Zeiss Zen Black software with 10% overlap between tiles for software alignment and stitching.

The Arabidopsis roots were imaged using a VisiTech QLC100 (Visitech, Sunderland, UK) spinning disc confocal microscope using a 40/1.3 NA oil objective lens. GFP and propidium iodide were both excited using the 488 nm line of a diode laser and emitted light was filtered through band-pass filters (500–550 nm for GFP and 580-654 nm for propidium iodide) and detected using a Hamamatsu Orca ER cooled CCD camera. Confocal z section stacks were collected at a spacing of 1.5 μm through the depth of the root tip and time-series images were acquired with an interval of 5 minutes.

Adding soluble compounds

The red fluorescent dye propidium iodide was chosen to add to Arabidopsis roots expressing 35S::GFP-eIF4A-III as it concentrates in the apoplastic spaces between cells revealing the cell outlines [17]. The propidium iodide was added to the chamber by manually injecting 6 ml (i.e. 3x chamber volume exchanges as the chamber volume was 2 ml) of propidium iodide (10 μg/ml) dissolved in ¼ strength MS medium over approximately 60 secs (i.e. a rate of 360 ml/hour) and was further perfused for 1 hour at 3 ml/h. To remove the propidium iodide 10 ml of ¼ strength MS media was flushed manually through the chamber between two consecutive time-points (5 minute intervals) and was further perfused at a rate of 3 ml/h.

Image analysis

All images collected by confocal microscopy were processed using Fiji [9,12], a version of ImageJ [10]. Time-lapse movies of cell division and root growth were aligned using the StackReg plug-in [20] and the time-lapse movies were aligned manually using "Alignment-Macros". Fiji was used to project z data using either maximum or sum projection. Fiji was also used to rotate, crop, adjust the brightness and contrast, and add scale bars and time stamp to movies [21]. The Zeiss Zen Black software was used to stitch overlapping images (10%) to form a larger tilled image. Adobe Photoshop was used for figure assembly and annotation.

Additional files

Additional file 1: Arabidopsis plant growth inside the imaging chamber at low magnification (QuickTime movie). A low magnification overview of developing Arabidopsis plants growing from seed to seedling as seen in Figure 2. Time interval ≥ 1 day. Scale bar = 1 mm.

Additional file 2: Time-lapse of an Arabidopsis suspension cell going through cell division (QuickTime movie). High resolution confocal microscopy imaging of a single Arabidopsis culture cell expressing GFP tubulin as it progresses through cell division as seen in Figure 3. Time interval = 5 minutes. Scale bar = 10 μm.

Additional file 3: Tiled image time-lapse of a growing Arabidopsis leaf (QuickTime movie). Time-lapse of a developing Arabidopsis leaf expressing tubulin GFP complete tissue captured using image tiling as seen in Figure 4 Time interval = 1 hour. Scale bar = 100 μm.

Additional file 4: The addition and removal of PI from the chamber containing a growing Arabidopsis root (QuickTime movie). The addition and removal of propidium iodide dye to an Arabidopsis root expressing GFP nucleus marker as seen in Figure 5. Time interval = 5 minutes. Scale bar = 20 μm.

Abbreviations

MS: Murashige and Skoog; PI: Propidium Iodide.

Competing interests

The authors declare that they have no competing interests.

Authors' contributions

GC conceived and designed the imaging chamber, performed the time-lapse imaging experimentation, drafted the manuscript. CH helped design and fabricate the imaging chamber. JC helped test proto-type versions of the chamber, design experiments and draft the manuscript. PS coordinated experiments, helped in the conception of the chamber and drafted manuscript. All authors read and approved the final manuscript.

Acknowledgements

We would like to thank John Humble for technical input into early chamber designs, Paul Derbyshire for supplying the Arabidopsis cells and Susana Sauret-Gueto and Samantha Fox with help in with establishing optimum biological conditions for growing seedlings inside the imaging chamber. We would also like to thank Enrico Coen and Jerome Avondo for their encouragement in the development of the imaging chamber. This work was funded by grant BB/J004588/1 from the Biotechnology and Biological Sciences Research Council (BBSRC) and the John Innes Foundation.

References

1. Brandizzi F, Fricker M, Hawes C. A greener world: the revolution in plant bioimaging. Nat Rev Mol Cell Biol. 2002;3(7):520–30. doi:10.1038/nrm861.
2. Chalfie M, Tu Y, Euskirchen G, Ward WW, Prasher DC. Green fluorescent protein as a marker for gene expression. Science. 1994;263(5148):802–5.
3. Shaw SL. Imaging the live plant cell. Plant J. 2006;45(4):573–98. doi:10.1111/j.1365-313X.2006.02653.x.
4. Shaw SL, Ehrhardt DW. Smaller, faster, brighter: advances in optical imaging of living plant cells. Annu Rev Plant Biol. 2013;64:351–75. doi:10.1146/annurev-arplant-042110-103843.
5. Cutler S, Ehrhardt D. Dead cells don't dance: insights from live-cell imaging in plants. Curr Opin Plant Biol. 2000;3(6):532–7.
6. Robinson S, Barbier de Reuille P, Chan J, Bergmann D, Prusinkiewicz P, Coen E. Generation of spatial patterns through cell polarity switching. Science. 2011;333(6048):1436–40. doi:10.1126/science.1202185.
7. Chan J, Calder G, Fox S, Lloyd C. Cortical microtubule arrays undergo rotary movements in Arabidopsis hypocotyl epidermal cells. Nat Cell Biol. 2007;9(2):171–5. doi:10.1038/ncb1533.
8. Koroleva OA, Calder G, Pendle AF, Kim SH, Lewandowska D, Simpson CG, et al. Dynamic behavior of Arabidopsis eIF4A-III, putative core protein of exon junction complex: fast relocation to nucleolus and splicing speckles under hypoxia. Plant Cell. 2009;21(5):1592–606. doi:10.1105/tpc.108.060434.
9. Fiji Is Just ImageJ. [http://fiji.sc/wiki/index.php/Fiji]
10. ImageJ - Image Processing and Analysis in Java. [http://imagej.nih.gov/ij/].
11. Preibisch S, Saalfeld S, Tomancak P. Globally optimal stitching of tiled 3D microscopic image acquisitions. Bioinformatics. 2009;25(11):1463–5. doi:10.1093/bioinformatics/btp184.
12. Schindelin J, Arganda-Carreras I, Frise E, Kaynig V, Longair M, Pietzsch T, et al. Fiji: an open-source platform for biological-image analysis. Nat Methods. 2012;9(7):676–82. doi:10.1038/nmeth.2019.
13. Imaging and Perfusion Chamber Internet Resources. [http://www.microscopyu.com/articles/livecellimaging/chamberresources.html]
14. Grossmann G, Meier M, Cartwright HN, Sosso D, Quake SR, Ehrhardt DW et al. Time-lapse fluorescence imaging of Arabidopsis root growth with rapid manipulation of the root environment using the RootChip. J Vis Exp. 2012(65). doi:10.3791/4290.
15. Sauret-Gueto S, Calder G, Harberd NP. Transient gibberellin application promotes Arabidopsis thaliana hypocotyl cell elongation without maintaining transverse orientation of microtubules on the outer tangential wall of epidermal cells. Plant J. 2012;69(4):628–39. doi:10.1111/j.1365-313X.2011.04817.x.
16. Kuchen EE, Fox S, de Reuille PB, Kennaway R, Bensmihen S, Avondo J, et al. Generation of leaf shape through early patterns of growth and tissue polarity. Science. 2012;335(6072):1092–6. doi:10.1126/science.1214678.
17. Vandenberg C, Willemsen V, Hage W, Weisbeek P, Scheres B. Cell Fate in the Arabidopsis Root-Meristem Determined by Directional Signaling. Nature. 1995;378(6552):62–5. doi:10.1038/378062a0.
18. Pesquet E, Korolev AV, Calder G, Lloyd CW. The microtubule-associated protein AtMAP70-5 regulates secondary wall patterning in Arabidopsis wood cells. Curr Biol. 2010;20(8):744–9. doi:10.1016/j.cub.2010.02.057.
19. Ueda K, Matsuyama T, Hashimoto T. Visualization of microtubules in living cells of transgenic Arabidopsis thaliana. Protoplasma. 1999;206(1–3):201–6. doi:10.1007/Bf01279267.
20. Thévenaz P. StackReg - An ImageJ plugin for the recursive alignment of a stack of images. [http://bigwww.epfl.ch/thevenaz/stackreg/]
21. Rasband W. Time Stamper. [http://rsb.info.nih.gov/ij/plugins/stamper.html]

A portable extensional rheometer for measuring the viscoelasticity of pitcher plant and other sticky liquids in the field

Catherine Collett[1], Alia Ardron[1], Ulrike Bauer[2,3], Gary Chapman[1], Elodie Chaudan[1,4], Bart Hallmark[1], Lee Pratt[1], Maria Dolores Torres-Perez[1,5] and D Ian Wilson[1*]

Abstract

Background: Biological fluids often have interesting and unusual physical properties to adapt them for their specific purpose. Laboratory-based rheometers can be used to characterise the viscoelastic properties of such fluids. This, however, can be challenging as samples often do not retain their natural properties in storage while conventional rheometers are fragile and expensive devices ill-suited for field measurements. We present a portable, low-cost extensional rheometer designed specifically to enable *in situ* studies of biological fluids in the field. The design of the device (named Seymour) is based on a conventional capillary break-up extensional rheometer (the Cambridge Trimaster). It works by rapidly stretching a small fluid sample between two metal pistons. A battery-operated solenoid switch triggers the pistons to move apart rapidly and a compact, robust and inexpensive, USB 3 high speed camera is used to record the thinning and break-up of the fluid filament that forms between the pistons. The complete setup runs independently of mains electricity supply and weighs approximately 1 kg. Post-processing and analysis of the recorded images to extract rheological parameters is performed using open source software.

Results: The device was tested both in the laboratory and in the field, in Brunei Darussalam, using calibration fluids (silicone oil and carboxymethyl cellulose solutions) as well as *Nepenthes* pitcher plant trapping fluids as an example of a viscoelastic biological fluid. The fluid relaxation times ranged from 1 ms to over 1 s. The device gave comparable performance to the Cambridge Trimaster. Differences in fluid viscoelasticity between three species were quantified, as well as the change in viscoelasticity with storage time. This, together with marked differences between *N. rafflesiana* fluids taken from greenhouse and wild plants, confirms the need for a portable device.

Conclusions: Proof of concept of the portable rheometer was demonstrated. Quantitative measurements of pitcher plant fluid viscoelasticity were made in the natural habitat for the first time. The device opens up opportunities for studying a wide range of plant fluids and secretions, under varying experimental conditions, or with changing temperatures and weather conditions.

Keywords: Biological fluids, Filament, Giesekus, *Nepenthes*, Pitcher plants, Polymer solution, Polysaccharide, Rheometry

* Correspondence: diw11@cam.ac.uk
[1]Department of Chemical Engineering and Biotechnology, New Museums Site, Pembroke Street, Cambridge CB2 3RA, UK
Full list of author information is available at the end of the article

Background

Viscoelastic behaviour of biological fluids

Water has long been recognised as the essence of life, and many ubiquitous biological fluids such as cytoplasm, blood and plant sap are based on water. In contrast to pure water, aqueous (and other) biological fluids often exhibit non-Newtonian behaviour such as shear-thinning (e.g. blood [1], bronchial mucus [2], gastropod foot mucus [3] and the adhesive fluids of insects [4]). Soluble long chain polymers give rise to viscoelastic behaviour and the ability to form filaments of liquid that can stretch [5]. This impacts a broad range of biological processes from the locomotion of sperm through cervical mucus [6] to the spinning of spider silk [7] and the trapping of insects by carnivorous plants (genera *Drosera* [8], *Drosophyllum*, *Pinguicula* and *Nepenthes* [9]).

Limitations of current rheometry methods

Accurate measurement of viscoelastic fluid properties, using extensional rheometry, is essential for understanding their contribution to the biological function. As part of living organisms, biological fluids often undergo marked changes over time [10,11], and fluid properties need to be monitored at short intervals in the natural environment in order to investigate these dynamic processes and their effects. Rigorous quantitative measurements are currently not possible in the field (laboratory devices are expensive, fragile and not readily transported), while the viscoelastic properties of many natural liquids change after sampling. Resins and latex are examples which change properties rapidly when exposed to air. Furthermore, the fluid properties depend on environmental factors such as temperature and air humidity, while the size and immobility of traditional extensional rheometers prohibits their use in climate control chambers. There is therefore a need for a portable device to study viscoelastic biological fluids *in situ* or under controlled environmental conditions.

This paper reports the development of such a device, which arose from the desire to study pitcher plant fluids *in situ* in Borneo (it was consequently named Seymour after the owner of a carnivorous plant in the movie 'Little Shop of Horrors'). The device can be used for routine testing as well as field work. It offers the following advantages:

(a) It is lightweight, robust, easy to assemble and has few moving parts;
(b) It is constructed mainly from standard parts which can be replaced readily, and is therefore relatively inexpensive;
(c) It employs small sample volumes (<10 μL), which fits the increasing demand for the miniaturisation of rheometric techniques owing to the limited availability or high cost of samples [12];
(d) It is easy to operate. Data can be analysed directly or remotely;
(e) It is suitable for testing over a broad range of temperatures and humidity levels as it fits readily into a controlled environment chamber: many studies of extensional rheology to date have been limited to standard laboratory conditions [13].

Pitcher plant fluids

We used pitcher plant fluids as an example of a viscoelastic biological fluid in order to test the performance of the Seymour device in the laboratory and in the field. *Nepenthes* pitcher plant fluids are sticky aqueous solutions of polysaccharides [9] held in cup-shaped leaves to trap insects. Prey struggling at the fluid surface quickly cover themselves in sticky threads, much like a piece of bread is covered by molten cheese in a fondue. The plant subsequently digests the drowned insects to release and absorb mineral nutrients. Bauer *et al.* [14] compared different pitcher plant species in the field using a crude measure of extensional viscosity, namely the length of filament that could be formed by stretching the fluid between two fingers. In addition to marked differences in apparent fluid viscoelasticity between species they found strong variation between individual plants. Observations on greenhouse plants of *N. rafflesiana* further suggest that greenhouse cultivation affects fluid viscoelasticity negatively (Bauer, unpublished). It is not clear whether this is a result of suboptimal growth conditions leading to reduced polysaccharide production by the plant, or due to dilution of the pitcher fluid when the plants are watered.

Erni *et al.* [8] studied the extensional behaviour of *Drosera* mucilage using a microscope-based CaBER™ device and reported phenomena such as beads-on-string formation associated with viscoelastic fluids, with a relaxation time of *circa* 0.33 s. Gaume and Forterre [9] measured the filament thinning of *N. rafflesiana* fluid with a simple rod arrangement and reported viscoelastic behaviour with a relaxation time of *circa* 1 s. This relaxation time was longer than the period in which insect prey were observed to flex limbs and was therefore interpreted as an adaptation for prey capture. Gaume and Forterre collected fluid samples in the field [9] but it is not clear if the viscoelastic testing was performed on location. Their method also required a pool of liquid to withdraw the rod from, which is not feasible for smaller sample volumes such as the mucilage droplets of *Drosera*, *Drosophyllum* and *Pinguicula*.

Extensional rheometry

Extensional rheology is the study of the deformation of material under conditions of pure strain; in this paper the exact form of the strain is the Hencky strain. The more commonly studied case is that of deformation under pure shear, where a sample is placed between two surfaces which move past one another, keeping the separating distance constant (or where the fluid is allowed to flow along a pipe so that the fluid shears against itself). Extensional rheometry requires the material to be stretched, and generating a reproducible stretching requires a precise mechanical action. Unlike Newtonian liquids, the extensional behaviour of viscoelastic fluids cannot be estimated reliably from measurements of shear rheology so direct measurements are necessary.

There are several methods for measuring extensional rheology [5]. Filament stretching is now a routine method in the laboratory, using devices such as the FiSER™ (Cambridge Polymer Group, Boston), CaBER™ [15] (Cambridge Polymer Group, Boston and Haake), and the Cambridge Trimaster [16] to measure the necking of an extensionally-strained fluid filament as a function of time. The evolution of the filament diameter can be used to identify the nature of the fluid (*e.g.* Newtonian, viscoelastic *etc.*, see next paragraph) and, given the surface tension of the fluid, to calculate the extensional viscosity. Viscoelastic fluids are characterised by viscous and elastic contributions: an important parameter in the latter is the relaxation time, λ, which is simply speaking the time needed for the polymer molecules to adjust to a change in strain.

A review of filament stretching rheometers can be found in Galindo-Rosales *et al.* [17]. A sample of liquid is placed between two platens (*i.e.* the ends of two rods), and the platens moved apart either (*i*) at controlled separating speed, imparting a known strain rate, or (*ii*) rapidly, causing the formation of a filament which thins under capillary action. The latter mode, shown in Figure 1(a), known as CaBER™ (capillary break-up extensional rheometry [18]) mode, is used here. A high speed camera and image analysis software monitors the filament diameter at its mid-point, D, over time, t. The filament will break at the mid-point under pure extension. Gravity has negligible effect with the devices employed here, as the dimensions give Bond numbers « 1 [9]; this is discussed after Equation (7).

The evolution of D with respect to its initial value, D_0, has been determined for various constitutive fluid models, and those considered here are:

(a) Newtonian liquid, with constant shear viscosity, η_0, and surface tension, α [19]. A correction factor, X, has been included to account for the non-cylindrical nature of the fluid filament: $X = 1$ for ideal, cylindrical filaments; $X = 0.7127$ for non-inertial, smooth, filaments. More information on this correction, and on the systematic differences in Newtonian viscosity calculated by extensional and shear measurements, can be found in McKinley and Tripathi [15] and Liang and Mackley [20].

$$\frac{D}{D_0} = 1 - \frac{(2X-1)\alpha t}{3\eta_0 D_0} \tag{1}$$

(b) The Upper Convected Maxwell (UCM) model, representing the simplest viscoelastic fluid model [19] for negligible viscosity with relaxation time, λ_{UCM};

$$\frac{D}{D_0} = \exp\left(\frac{-t}{3\lambda_{UCM}}\right) \tag{2}$$

(c) The Giesekus model for viscoelastic solutions [21], which includes a relaxation time, λ_G, and a polymer

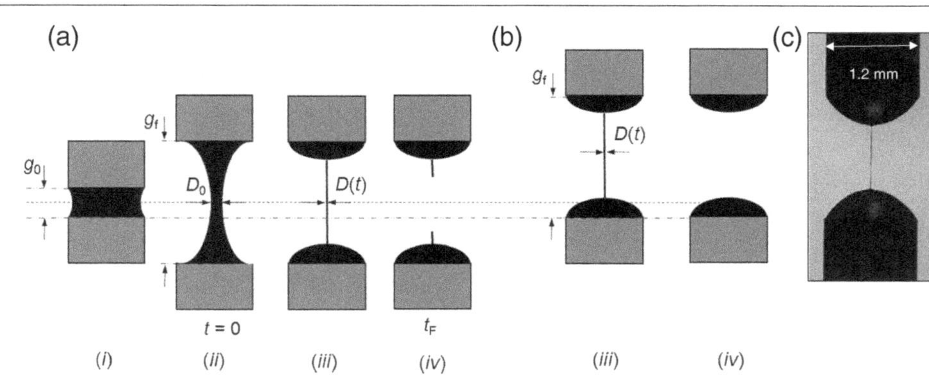

Figure 1 Schematic of CaBER™ operation. (a) standard mode, platens move equal distance apart. Initial gap g_0, final gap g_f. **(b)** Seymour operation, only upper platen moves. **(c)** Example of *N. maxima* filament shortly before breakup, g_0 = 0.4 mm. Time set to zero when platens stop moving, (*ii*); filament width measured (*iii*) until break up at t_F, (*iv*).

interaction term, the mobility parameter, a, gives the following implicit relationship for D/D_0 [22];

$$(4a{-}3)\ln\left(\frac{D/D_0+2\alpha\lambda_G a/\eta_0 D_0}{1+2\alpha\lambda_G a/\eta_0 D_0}\right)-\frac{2D_0\eta_0}{\alpha\lambda_G}(D/D_0{-}1)=\frac{t}{\lambda_G} \quad (3)$$

Numerical fitting techniques are needed to extract the Giesekus model parameters from filament thinning data sets.

As the filament thins it will break due to capillary instability. The time where this occurs will depend on the nature of the filament but an estimate of the filament break-up time, t_F, can be obtained by setting $D/D_0 = 0$. For a Newtonian fluid, Equation (1) gives

$$t_F \approx \frac{3\eta_0 D_0}{\alpha(2X-1)} \quad (4)$$

The Seymour concept

In the Trimaster (and other devices), the platens move apart so that the filament midpoint remains at the same location, greatly reducing the computational effort in analysing images but requiring delicate mechanical action (Figure 1(a)). In contrast, the Seymour only moves one platen using a standard solenoid switch (Figure 1 (b)). This improves the robustness of the device (fewer moving parts) and reduces the cost demonstrably. The major cost component is the digital camera: the device for fieldwork reported here cost approximately €2000 in June 2014. Moving only one platen also makes setting the filament size simpler. Due to the increasing resolution of modern cameras and processing power of laptop computers, the filament midpoint can be easily located by an image analysis code.

Aims and scope of the present study

The portable rheometer (Seymour) and its use are described in detail. We performed a series of experiments in order to answer the following questions: (*i*) Does the Seymour device produce comparable results to the conventional laboratory rheometer Trimaster Mk II, for both standardised synthetic liquids and natural pitcher plant fluids? (*ii*) Does the device yield reliable measurements in the field? (*iii*) Are there quantifiable differences between the fluids of *Nepenthes* pitcher plants sampled in greenhouses and in the field? The last two questions were investigated in order to quantify the advantages of a device that allows for on-site measurements of fluid viscoelasticity.

Results and discussion

Benchmarking studies

CaBER™ testing requires a rapid initial separation of the platens in order to generate a filament, and negligible movement thereafter. Figure 2 compares the gap separations, expressed in terms of the Hencky strain, measured for the Seymour and the Trimaster for identical stretching settings. The separation speed for both devices is similar, at approximately 75 mm s^{-1}. Neither device imparts a constant strain rate during the separation stage, while the Seymour produces less overshoot and faster damping than the belt-driven Trimaster. Filament diameter measurements could therefore be collected after 10 ms on the Seymour (17 ms on the Trimaster). This sets a lower limit on the viscosity of liquids that can be measured as the filament stretching should be measured after the platens have stopped moving. Equation (1) predicts that low viscosity liquids will approach filament breakup (D/D_0 approaching zero) at short times. For example, a Newtonian 20 mPa s silicone oil exhibited filament breakup within the 10 ms initial separation period.

Figure 3 shows the evolution of filament diameter for the Newtonian silicone oil obtained with the Trimaster and Seymour devices; both series exhibit an essentially linear decrease, as predicted by Equation (1). The dashed lines in this Figure represent the best fit of Equation (1) to the Seymour device's experimental data using viscosity as an adjustable parameter: for this fitting, $X = 1$ and $\alpha = 0.0159$ N m^{-1}. The range of reported surface tension values for silicone oil in the literature lie between 0.0159 N m^{-1} and 0.0213 N m^{-1} [23]. Viscosity values of 2.37 Pa s, 2.71 Pa s and 3.1 Pa s were

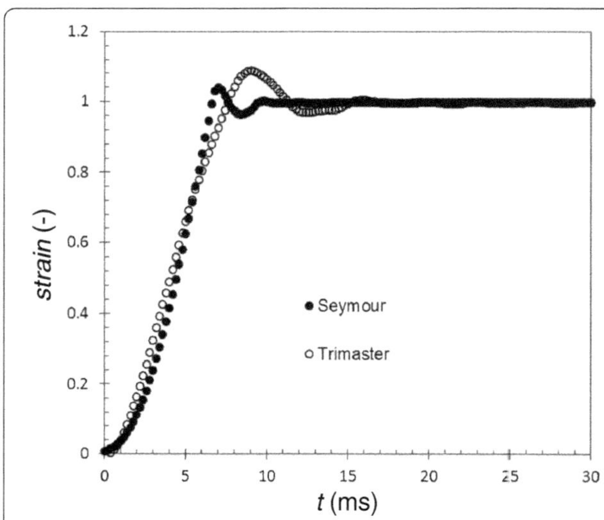

Figure 2 Comparison of linear (Hencky) strain between portable (Seymour) and laboratory (Trimaster) devices.
The initial gap, g_0, was 0.7 mm and reaches a final gap of 1.9 mm. Both devices produced similar separation speed. Seymour produced less overshoot and higher damping than the Trimaster.

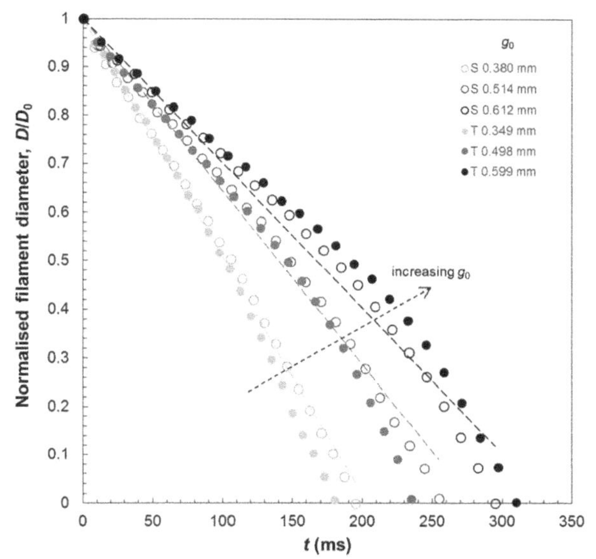

Figure 3 Comparison of filament thinning behaviour for silicone oil. Filament diameters decreased linearly as expected for a Newtonian liquid. Smaller initial gap sizes yielded a better linear fit. Data sets obtained with Seymour (S, open symbols) and Trimaster (T, solid symbols) devices showed strong agreement. Data are decimated for clarity.

calculated for initial gap sizes of 0.380 mm, 0.514 mm and 0.612 mm respectively; it is notable that the accuracy of the fit of Equation 1 to these data sets decreased as a function of increasing gap size, with R^2 values of 0.997, 0.989 and 0.982 respectively. The mean value of the viscosity found by these extensional measurements was 2.75 Pa s, some 16% higher than the reported value of 2.37 Pa s. The measurements suggest that smaller initial gap sizes are preferable in order to obtain accurate measurements. All gap sizes used were smaller than the capillary length, $l_{cap} = \sqrt{\sigma/_{\rho g}}$, where ρ is the fluid density, as recommended by other researchers in this area [24].

The gradients and filament breakup time, t_F, increase with initial filament diameter, D_0, as predicted by Equations (1) and (4), respectively. Figure 4(a) shows excellent agreement between the t_F values measured on the two devices at 22°C. The relationship between t_F and g_0 also exhibits the trend expected for a Newtonian fluid (Equation (4)), as D_0 is expected to vary with g_0.

Figure 4(b) presents results obtained using the Seymour device at higher temperatures, spanning the range anticipated for field studies (up to 40°C). Filament evolution plots were linear, as in Figure 3, and t_F decreases at higher temperature. This is consistent with Equation (4), as η_0 decreases with temperature.

Figure 5(a) shows the results obtained for the CMC solution, which is known to be viscoelastic, measured on the Seymour in the laboratory in Cambridge and in the field in Borneo in summer 2014. The filament evolution is presented a function of dimensionless time. The use of dimensionless time accounts for the difference in temperature between the tests in Cambridge and those done in the field; additional insight into this superposition is given by Torres and co-workers [22]. Tests on the Trimaster gave similar results. There is a noticeably sharp transition to filament breakup at $t/t_F > 0.8$, which is not predicted by any of the simple constitutive models (Equations (1)-(3)). The effect of initial gap size on filament break up time is shown in Figure 5(b). The Seymour t_f values tended to be shorter than those obtained with the Trimaster. This difference may be related to the protocols: it took longer to load the Trimaster and for the final gap to stabilise, and water evaporation would increase the viscosity and thus filament breakup time. CMC solutions represent complex fluids [25] and these results confirm that the Seymour gives qualitatively similar results to the Trimaster.

The above results constitute proof-of-concept of the portable extensional rheometer (Seymour). The data from this device showed good agreement with those obtained with a precision unit, the Cambridge Trimaster Mk II. This was the case for both Newtonian (silicone oil) and complex biopolymer solutions (CMC). Moreover, the Seymour unit is small enough to fit into a climate-controlled chamber, allowing the effects of temperature and relative air humidity to be studied. The Seymour functions satisfactorily at temperatures up to 40°C, which is essential for field studies in the tropics as well as for medical studies under physiological temperatures. Humidity levels were not investigated as part of this study, but the limit in this regard in field tests is likely to be set by the camera and laptop computer. This broad operational range, together with the small size and weight of the system, renders the Seymour highly suitable for field studies.

Application to biological (pitcher plant) fluids

Pitcher plant fluids taken from individual *N. rafflesiana*, *N. eymae*, and *N. maxima* obtained from botanical gardens (*i.e.* greenhouse plants) were tested within one day and periodically thereafter over a period of two weeks. The fresh fluids formed filaments which remained intact for some time (an example for *N. maxima* in shown in Figure 1(c)), while the *N. rafflesiana* fluid was not very viscous and the filament often broke before the platens finished moving. We only report data obtained with Seymour here as the Trimaster yielded similar results. Some samples exhibited the formation of satellite droplets, known as 'beads on a string' (BOAS). The presence of viscoelasticity is a pre-requisite for the formation of BOAS within fluid samples [26] and an example of this behaviour is evident in the field test on *N. rafflesiana* in the Additional file 1: video.

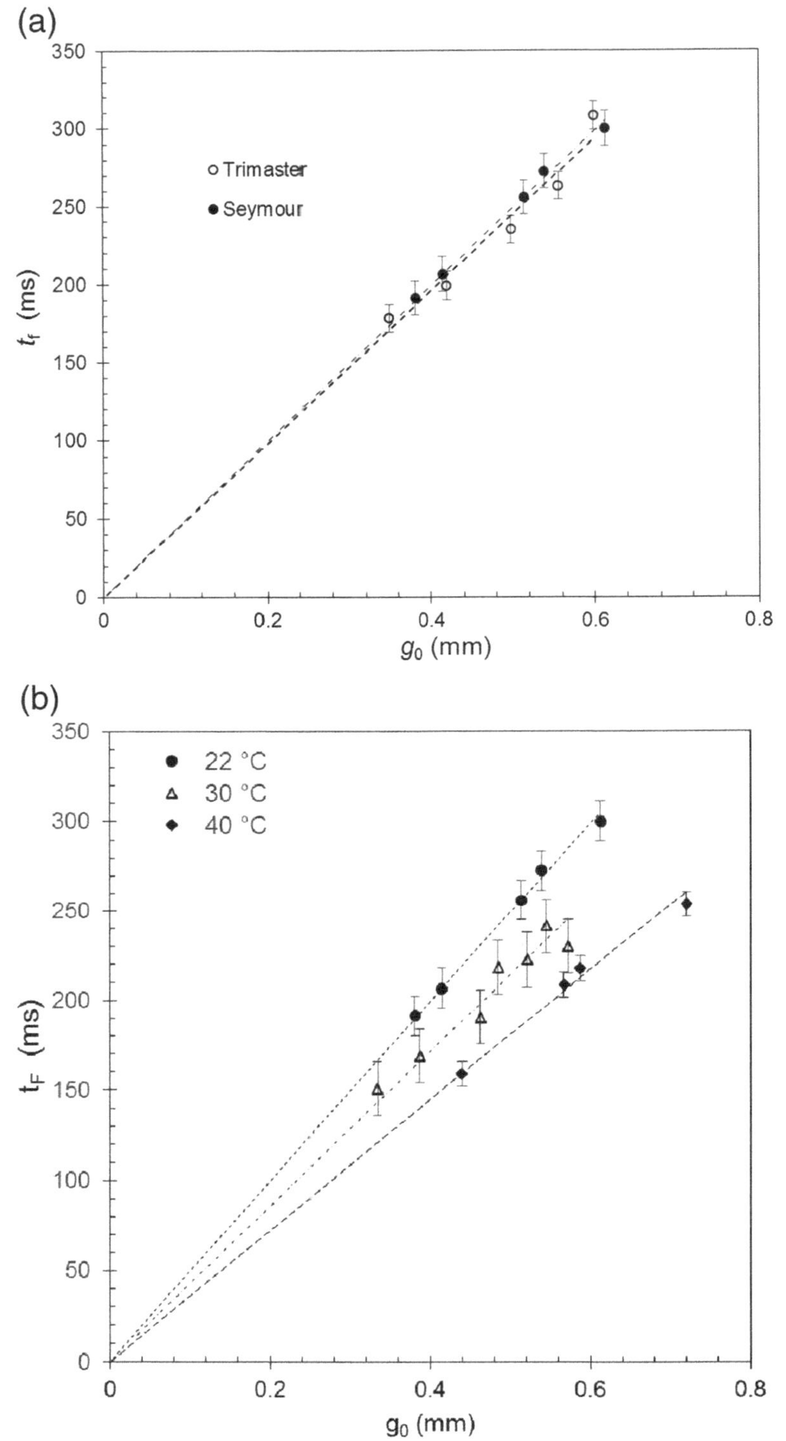

Figure 4 Filament break-up times for silicone oil (Newtonian fluid). (a) Measurements obtained with Trimaster Mk II and Seymour at 22°C show excellent agreement. **(b)** Higher temperatures lead to shcrter break-up times (data obtained with Seymour device).

The regression coefficients quantifying the fit of Equations (1), (2) and (3) to filament thinning profiles (plots of D/D_0 or $\ln(D/D_0)$ against t) are given in Table 1. The plots in Figure 6, of $\ln(D/D_0)$ against time, suggested by Equation (2), show an approximately linear trend for all three fluids. This indicates that the viscoelasticity is adequately described by the simple, single parameter UCM model. The values of the relaxation time,

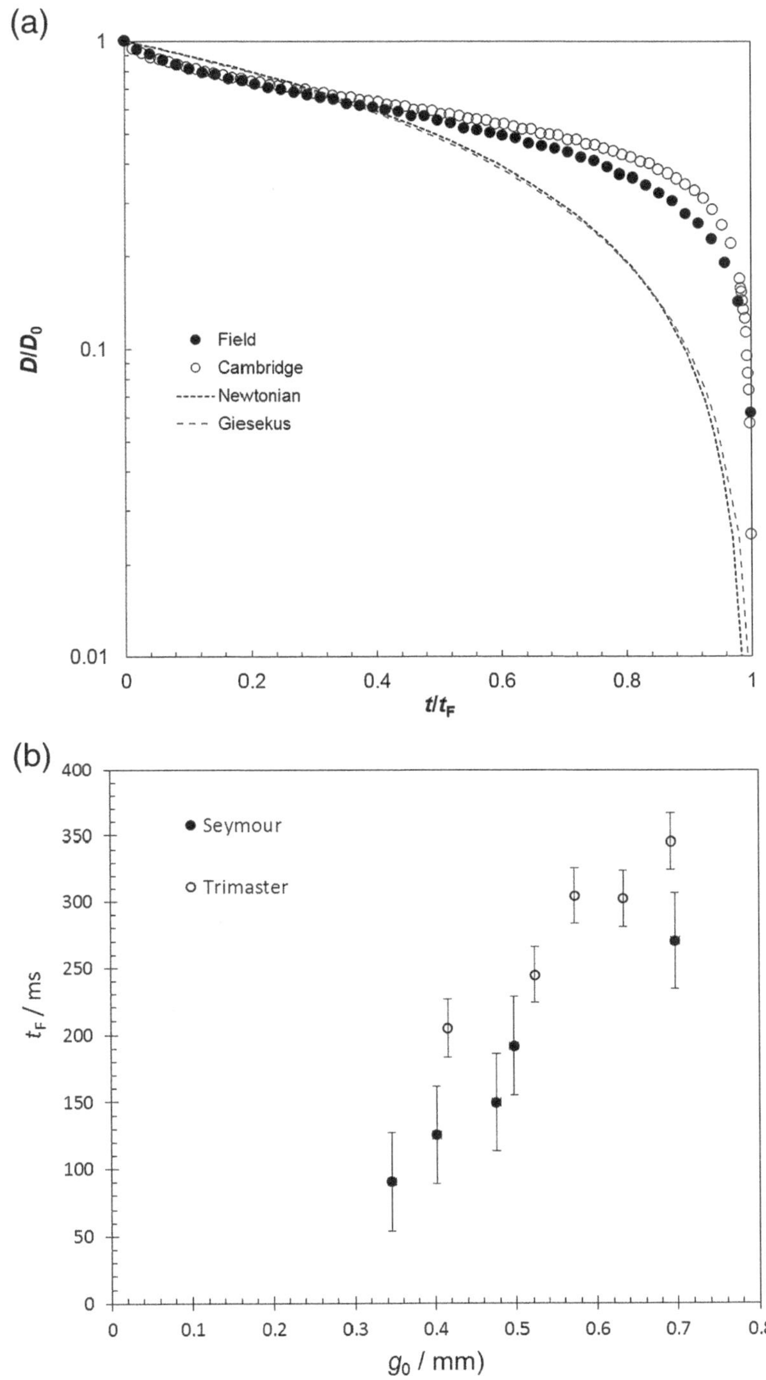

Figure 5 CMC solution filament thinning behaviour showing complex behaviour. (a) Seymour testing, alongside best fit lines for the Newtonian and Giesekus models. The data are plotted against dimensionless time, t/t_F. Laboratory and field measurements largely agree. **(b)** Filament break-up times increased linearly with initial gap size. Seymour yielded consistently shorter break-up times than the Trimaster.

λ_{UCM}, extracted from model fitting are reported in Table 1. The *N. rafflesiana* value, of 3 ms, is small and could not be measured reliably with either of the Trimaster or Seymour devices. The *N. eymae*, and *N. maxima* values are more than an order of magnitude smaller than relaxation times of ~ 1 s reported by Gaume and Forterre for *N. rafflesiana* [9], confirming the desirability to perform tests in the field if possible.

Effect of sample storage on pitcher plant viscoelasticity

It has been observed that pitcher plant fluids stored for over one month lose their stickiness (Bauer, unpublished),

Table 1 Parameter estimates and goodness of fit for different fluid models for pitcher fluids from three *Nepenthes* species (samples obtained from greenhouse plants, measurements performed with Seymour)

Species	Equation (1)		Equation (2)		Equation (3)			
	η_0 (Pa s)	R^2	λ_{UCM} (ms)	R^2	η_0 (Pa s)	a (–)	λ_G (ms)	R^2
N. rafflesiana	3.47	0.964	2.95	0.994	1.97	0	1.79	0.995
N. eymae	26.7	0.503	29.8	0.958	0.0644	0	32.4	0.979
N. maxima	20.6	0.599	20.6	0.991	0.0491	0	21.7	0.994

which is an indicator of a reduction in viscoelasticity. This was reproduced in laboratory studies on (greenhouse-sourced) fluid samples of *N. maxima* and *N. eymae.* Fluids were found to lose their viscoelastic properties when stored at ambient temperatures over 2–4 weeks.

The samples were stored in sealed containers at room temperature and small aliquots were extracted for testing on the Seymour device at different times over two weeks and a month. The testing period was longer for *N. eymae* as the initial t_f value was larger and quantitative data could be obtained over a month. Figure 7 shows noticeable differences in viscoelasticity with storage time. These data were fitted to all three expressions to quantify the change in viscoelastic behaviour. The R^2 values are reported in Additional file 2: Table S1. The

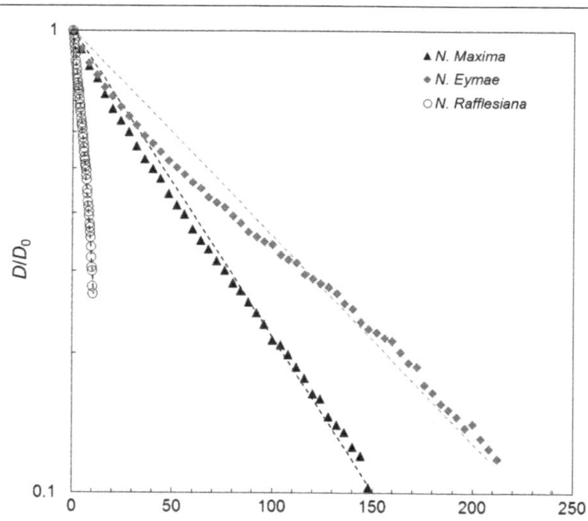

Figure 6 Filament evolution profiles for fresh greenhouse-sampled of pitcher fluids. *N. eymae* (diamonds) and *N. maxima* (triangles) both showed clearly viscoelastic behaviour while the *N. rafflesiana* filament broke before the pistons finished separating. The UCM model (dashed lines) provided the best fit for all three samples; λ_{UCM} values given in Table 1. Experimental data have been decimated for clarity.

plots show an approximately linear decrease in $\ln(D/D_0)$ with time, indicating that the viscoelasticity is adequately described by the simple, single parameter, UCM expression and this model provided a better description for most cases for both fluids across the sample sets. The Giesekus model gave comparable R^2 values but the low (sometimes zero) magnitude of η_0 cast doubt on the validity of the results. The quality of fit of the Newtonian model for *N. eymae* improved considerably with storage time. This trend indicates that the fluids are losing their viscoelastic properties with time when stored at ambient temperature.

Figure 7 also shows a noticeable reduction in t_f with storage time, which is again consistent with the fluids changing from viscoelastic to Newtonian behaviour. The relaxation times extracted from model fitting are reported in Table 2 and also decrease over the storage period, by over an order of magnitude. This will give rise to predominantly Newtonian behaviour, which can be illustrated using the result for the Giesekus fluid in Equation (3). If the product of the terms is small compared to the non-dimensional filament diameter, D/D_0, then it can be shown that

$$(4a-3)\lambda_G \ln(D/D_0) + \frac{2\eta_0}{\alpha}(D_0-D) = t \quad (5)$$

As λ_G decreases, the viscoelastic contribution (first term on the left hand side) becomes negligible and viscous behaviour dominates. Figure 8 shows how λ_{UCM} changes over the storage period. The almost linear trend for the *N. eymae* fluid on this log-linear plot suggests a first order decay. Elucidating this behaviour requires further work and analysis of the biopolymer components.

The above results confirmed and quantified the previously observed decay of viscoelasticity for pitcher fluids in storage. Both *N. maxima* and *N. eymae* fluids showed a tendency to become more Newtonian over the course of two to four weeks (Figure 7). The time–dependent reduction in the relaxation time of longer-stored fluid samples provided a second quantitative measure of the loss of fluid viscoelasticity (Figure 8). The two fluids considered here were sampled from newly opened pitchers so the decay is unlikely to be caused by environmental factors or by the interaction with captured prey or pitcher-colonising infauna organisms. Some insight into this behaviour was provided by storing a sample under chilled conditions, at 4°C. An aliquot was withdrawn and allowed to warm to room temperature for testing. Chilling effectively halted the change in viscoelasticity, as shown for *N. maxima* in Additional file 3: Figure S1, which may be related to inactivation of enzymes: this is the subject of ongoing work. The

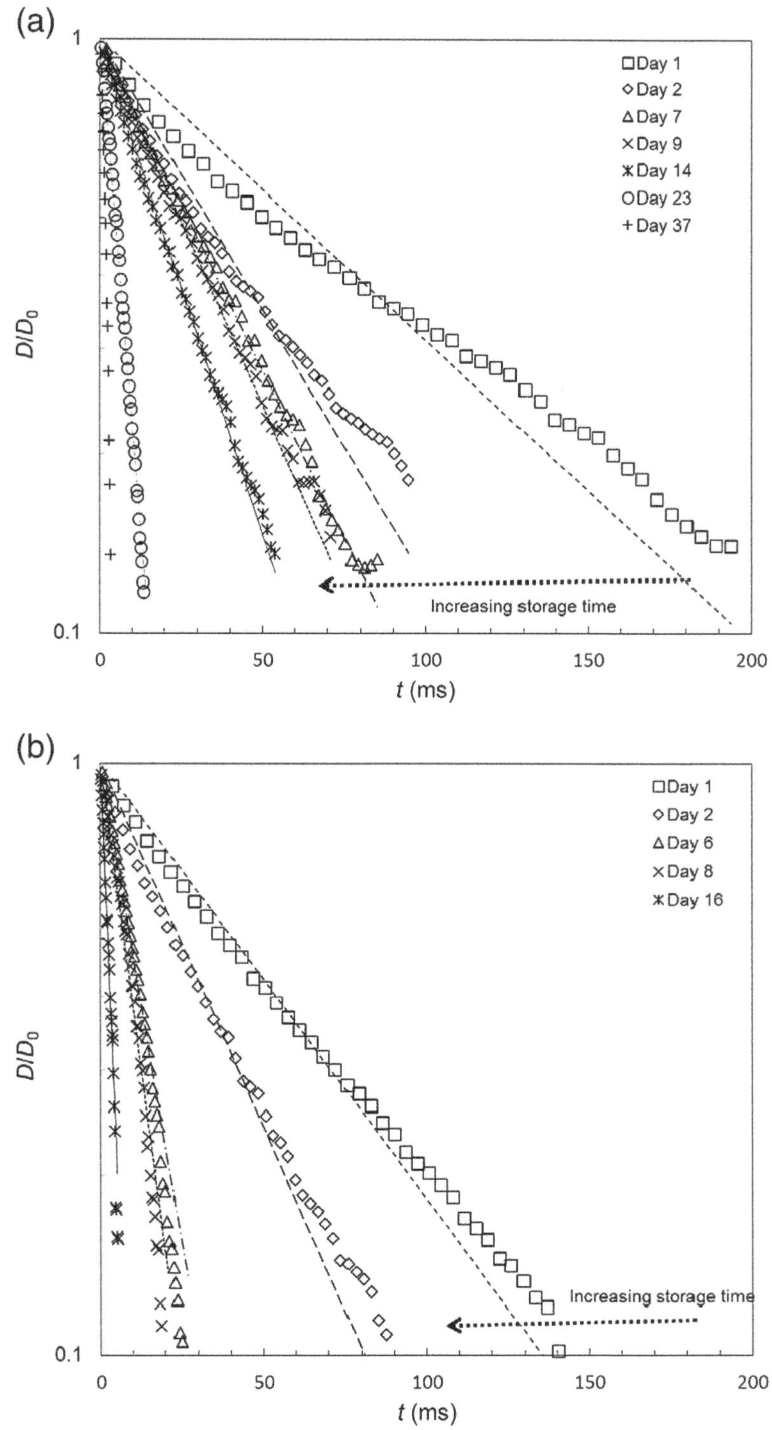

Figure 7 Effect of storage time on filament thinning behaviour. The viscoelasticity of **(a)** *N. eymae* and **(b)** *N. maxima* pitcher fluid decreased markedly with storage time. Symbols show experimental data, decimated for clarity. Lines show the fit to the UCM model, equation (2), with λ_{UCM} values in Table 2.

marked change in fluid viscoelasticity observed over a comparably short period of time further highlights the need for a portable device to measure fluid samples in the field, immediately after collecting them.

Application in the field (Borneo 2014)

The Seymour device was tested in Borneo in summer 2014 where we used it successfully to measure field-collected pitcher plant fluids as well as the above mentioned test

Table 2 Effect of storage at room temperature on the UCM relaxation time for two greenhouse-sourced pitcher plant fluids

Species	Storage time (day)	λ_{UCM} (ms)	R^2 (–)
N. eymae	1	28.5	0.957
	2	15.8	0.975
	7	12.9	0.997
	9	11.7	0.994
	14	8.64	0.996
	23	2.21	0.998
	37	0.73	0.950
N. maxima	1	19.6	0.991
	2	11.7	0.986
	6	4.25	0.991
	8	3.49	0.979
	16	1.08	0.989

Values of λ_{UCM} obtained by fitting Equation (2) to the *N. eymae* and *N. maxima* data in Figure 7.

liquids (silicon oil, CMC). A preliminary analysis of 11 newly-opened pitchers of *N. rafflesiana* indicates that extensional thinning of the fluid is best modelled by Equation (3), indicating a Giesekus response with R^2 values being essentially the same as, or higher than, those obtained using Equation (2). The model parameters and the associated R^2 values are reported in Additional file 4: Table S2.

The viscoelastic properties of the *N. rafflesiana* trap fluids showed pronounced individual variation with relaxation times spanning an order of magnitude from 85 ms to 1.2 s (mean ± S.E. = 551 ± 109 ms). The upper limit of this range is comparable to the value of 1 s reported by Gaume and Forterre [2]. The mean viscosity of 25.3 ± 4.8 Pa s was similarly variable.

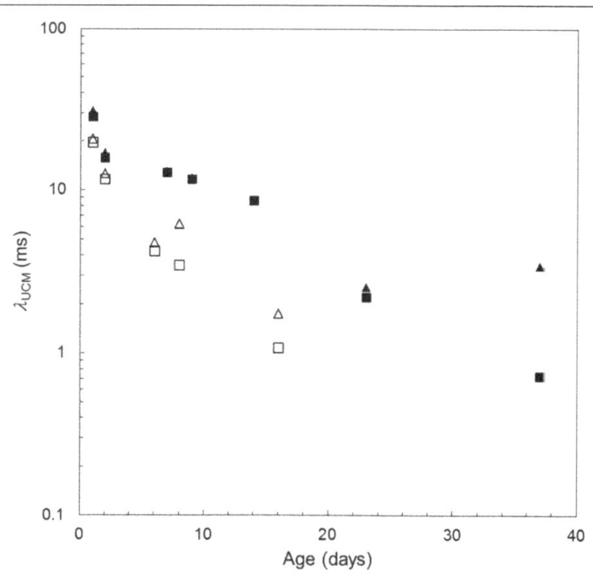

Figure 8 Effect of storage time on UCM and Giesekus model relaxation time. Relaxation time for *N. eymae* (solid symbols) and *N. maxima* (open symbols) decreased strongly with storage time. λ_G – triangles; λ_{UCM} – squares.

Fluid from 9 of the 11 pitchers was stored for a period of 20 days in sterile screw-top plastic vials at room temperature (25-30°C). At the end of this 20-day period, the fluid samples were re-measured using the Seymour device to determine whether the rheological parameters had changed. The model parameters obtained by fitting to the samples after storage, and associated R^2 values, are reported in Additional file 5: Table S3. The relaxation time of the samples decreased significantly over the 20-day storage period (Wilcoxon-matched-pairs test, $n = 9$, $R^\wedge = 4.00$, $P < 0.05$; Figure 9). Likewise, the mean viscosity was found to be significantly lower after storage (Paired-samples t test, d.f. = 8, $t = 2.34$, $P < 0.05$; Figure 9). These results confirm that there was a measurable loss of both elasticity and viscosity over 20 days of storage time which is consistent with previous observations of the loss of fluid stickiness with time [21]. They further underline the need for a reliable method for making on-site measurements as soon as possible after sample collection, especially in remote field locations where refrigeration is not available.

These preliminary field data also allow initial comparison to be made between typical extensional-thinning behaviour of fluid obtained from pitchers grown in the field and pitchers cultivated in greenhouses. Figure 10 shows the filament thinning profile data extracted from a newly-opened pitcher of *N. rafflesiana* having a relaxation time close to the cohort mean. Also plotted on this figure is the thinning profile extracted from the same pitcher 20 days later along with data from a newly-opened pitcher sourced from Cambridge Botanical Gardens. The fresh liquid from the plant in its native habitat was noticeably more elastic than the sample obtained from the greenhouse plant, as indicated in the Additional file 1 (video) by the strand which attached to the pipette when it was withdrawn from the pitcher. After 20 days the liquid extracted from the native pitcher, however, exhibited a noticeably reduced level of elasticity. The greenhouse sample, however, exhibited a response that shows a level of elasticity that is orders of magnitude smaller than either of the field samples. This observation illustrates the benefits of *in-situ* testing of plants in their native habitat.

Viscoelasticity models for pitcher fluids

Also plotted in Figure 10 are the loci obtained by fitting the UCM and the Giesekus models to the data from the fresh and stored samples. The R^2 values for the fresh samples are 0.984 and 0.989 for the UCM and Giesekus models, respectively. The three-parameter Giesekus model is able to provide a better fit to the experimental data as

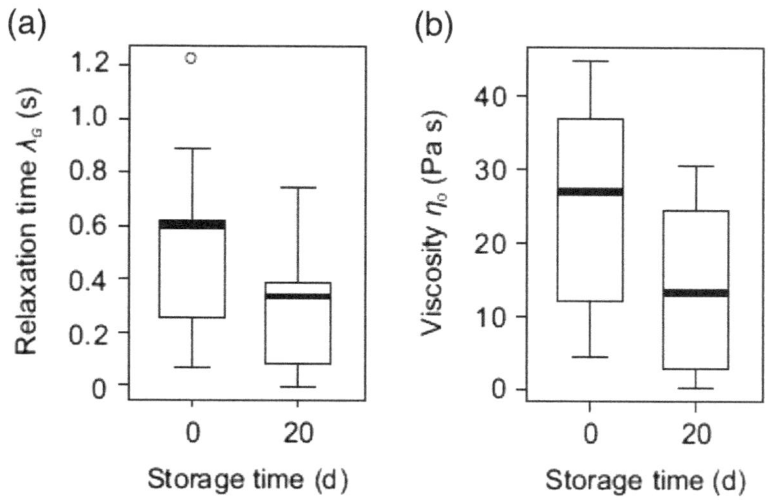

Figure 9 Effect of storage at room temperature (25-30°C) on *N. rafflesiana* pitcher fluid. Relaxation time **(a)** and viscosity **(b)** of 9 field-collected fluid samples from freshly opened *N. rafflesiana* pitchers decreased significantly ($P < 0.5$ for both; see text for statistics and Additional file 4: Tables S2 and Additional file 5: Table S3 for raw data) over a 20-day storage period at room temperature. Bars denote medians, boxes represent the inner two quartiles and whiskers include all values within 1.5 times interquartile range.

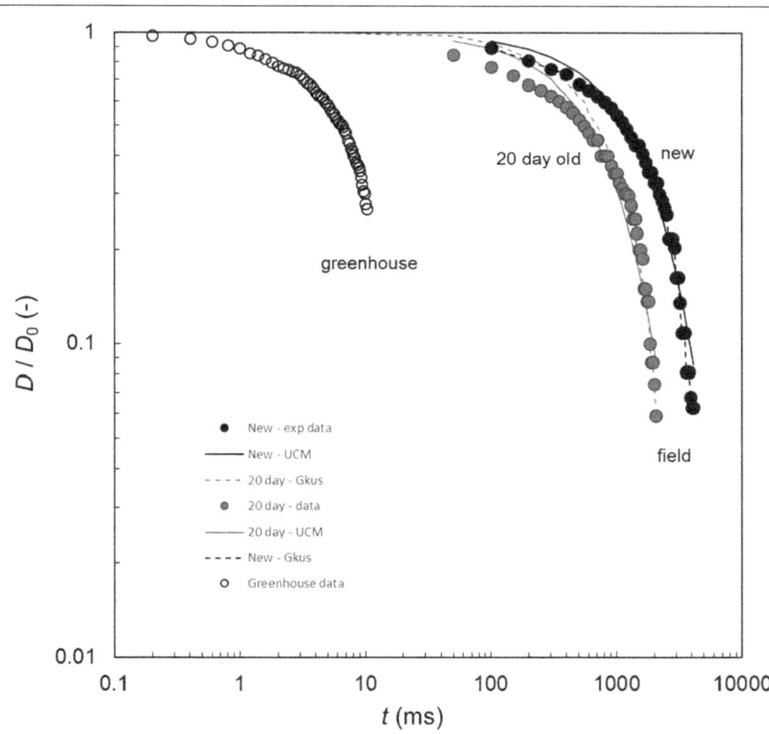

Figure 10 Normalised filament profile from field and greenhouse *N. rafflesiana* pitchers. Data show field measurements for fluid from newly-opened (black circles) and 20-day old (grey circles) *N. rafflesiana* pitchers alongside the profile for a newly-opened greenhouse plant (open circles) fluid reported in Figure 6. Data have been decimated for clarity. Data points are Seymour measurements. Solid line shows the fit to Equation (2) and dotted line shows the fit to the Giesekus fluid model, Equation (3); UCM and Giesekus parameters are those for pitcher 19 in Additional file 4: Tables S2 and Additional file 5: Table S3.

the filament rupture time is approached. The relaxation times for the freshly-opened pitcher, λ_G (620 ms) and λ_{UCM} (558 ms), are comparable to the the previously reported value of 1 s [9]; the relaxation times for the aged pitcher, λ_G (360 ms) and λ_{UCM} (284 ms), have fallen to between 50% and 60% of the newly-opened pitcher.

Figure 10 confirms the need to measure the extensional behaviour *in-situ*. There is a marked difference in the *N. rafflesiana* results for samples from greenhouse cultivated plants and those collected *in-situ* in Borneo, with the relaxation times for the latter being considerably longer, at 620 ms *vs.* 3 ms. The results for *N. rafflesiana*, *N. eymae*, and *N. maxima* from botanic gardens (Figure 6) indicate that the fluids from all three species initially possess viscoelasticity. Their viscoelastic response is adequately described by the single-parameter UCM model, which is the simplest viscoelastic fluid model available. Fluid filament rupture was observed in *N. rafflesiana* after about 10 ms, and after about 150 ms and 200 ms for *N. maxima* and *N. eymae*, respectively. The reduced viscoelasticity could be due to different growth conditions (temperature, air humidity, light levels) in the greenhouse environment, or to accidental dilution of the pitcher fluid when watering the plants.

Furthermore, whereas the simple UCM model could be used to describe the levels of viscoelasticity present in greenhouse-collected samples of all ages, the field-collected *N. rafflesiana* fluid is better described by a more complex constitutive model. The results in Figure 10 show good agreement with the single mode Giesekus fluid. Other models, such as the FENE dumbbell model, exist and may give a slightly better fit than those considered in this paper [5,27,28]; these, however, were not considered further due to the current lack of a simple analytical solution that can describe filament thinning. The benefit of this is that the factors affecting these constitutive model parameters, such as polymer nature, concentration and size, have been studied in depth for synthetic polymer solutions. Since the Seymour device now allows fluids such as pitcher plant liquids to be studied in the field (and thus free from any storage artefacts), variations between species and age may now be related to these polymer characteristics with greater confidence.

Conclusions

A robust and reliable portable extensional rheometer (Seymour) was developed and commissioned using silicone oil and CMC solutions as test fluids. Results obtained on the Seymour device showed good agreement with those obtained with a precision unit, the Cambridge Trimaster. The Seymour unit also fitted into a climate chamber, allowing the effect of temperature and humidity on the viscoelastic behaviour of fluids to be studied.

The extensional rheology of four pitcher fluid samples sourced from greenhouses, and a further 11 samples measured in the field, were investigated. Three of the four greenhouse samples exhibited significant viscoelasticity, which could be modelled using the Upper Convected Maxwell model. The field sample exhibited stronger viscoelasticity, with relaxation times two orders of magnitudes greater, and were best described by the Giesekus model. The viscoelasticity of greenhouse- and field-collected samples alike decreased significantly during storage.

Perspective: limitations and potential applications

The Seymour concept for a portable extensional rheometer has been demonstrated. The device was successfully employed in a field study investigating pitcher plant fluids in Borneo in 2014. The portable rheometer allows, for the first time, natural complex fluids such as pitcher plant trapping fluid to be studied in an ecological context. This will help to answer important questions such as how fluid viscoelasticity influences the trapping success of the plant or how it might depend on the presence of infauna or prey in the fluid. Although the device was developed with pitcher plants in mind, it can be used for studying a wide range of viscous and viscoelastic fluids, including sticky secretions, nectar, honeydew, resins, rubbers, saps, slimes, mucus and other biological fluids (of plant, animal or microbial origin).

We have confirmed Seymour's capacity to work at 40°C, which makes it suitable for the study of biomedical fluids at physiological temperatures. Virtually all the parts are sterilisable, so it can be used in sterile environments and decontaminated readily. We have not tested the performance at temperatures below 20°C: however, it is reasonable to assume that the lower limit of the operating range is around 0°C as the solenoid will be affected by frost. We expect the device to work well over a broad range of relative humidities but would advise to avoid extremes (*i.e.* working between 20-80%) as excessive evaporation or condensation will affect the sample volume, rendering the measurements invalid.

Methods

The labelled photograph in Figure 11 shows the key components of the device. Separation of the platens was provided by a solenoid (electromagnet; model 8.02.13.62, HE & BS Benson Ltd, Newmarket, UK [29]) drawing 4 W power at 6 V DC with a duty cycle rating of 100%. Power to the solenoid was provided by 4×1.5 V AA batteries *via* a switch. The solenoid rod (which provides the upper platen) was replaced with a lighter version with the top end capped to stop it falling through the coil when de-energised. The bottom end was machined from aluminium to give a flat, 1.2 mm diameter platen. The lower platen was similar, but constructed with a screw

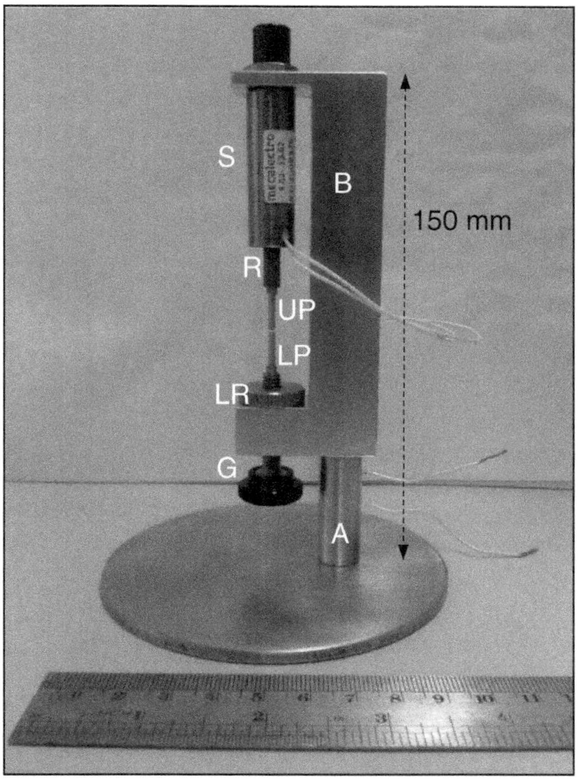

Figure 11 Photograph of Seymour device. Labelled parts: S – solenoid; R – end of solenoid rod; UP -upper (moving) piston, LP - lower (stationary) piston used to set the initial gap; LR – locking ring; G – gap adjustment screw; B – Seymour body; A – stand and base.

thread so that its location could be adjusted to set the initial gap and locked in place by a securing knob. The fluid volume was defined by that of a cylinder with initial gap, g_0, and the platen diameter, r:

$$V = \pi r^2 g_o \tag{6}$$

with r = 0.6 mm. The distance that the upper platen travelled, *i.e.* the final stretched distance, was adjusted by moving a rubber 'O'-ring along the shaft. This stopped the rod moving when it contacted the solenoid housing. The platen surfaces were readily wetted by all the liquids tested and filaments always broke by extending (high strain).

The range of operating conditions was established by preliminary tests on *N. maxima* and *Nepenthes* samples from the Cambridge botanical gardens using a Cambridge Trimaster Mk II device [16,30]. The results indicated that an initial gap, g_0, of 0.3-0.7 mm and final stretched distance, g_f, of 1–2 mm was desirable. The Seymour device was constructed with platens of similar diameter to the Trimaster, of 1.2 mm, and maximum final separation of 4 mm in order to accommodate more viscous liquids if needed. These dimensions were sufficiently small to avoid gravitational effects in the cylindrical sample. The influence of gravity is given by the Bond number, *Bo*, defined

$$Bo \equiv \frac{\rho g L^2}{\alpha} \tag{7}$$

where ρ is the fluid density, α the surface tension, g the gravitational acceleration and L the length scale, here the sample initial diameter, D_0. Bond numbers greater than 1 indicate that gravity effects will be significant. For the aqueous solutions studied here, *Bo* was always less than 0.5.

Key dimensions (e.g. initial and final stretched gap) in movie frames were calculated using the known platen and rod dimensions. The stainless steel body of Seymour was mounted on a (detachable) brass disc for stability. The mass of the assembly, of 350 g, was comparable to that of the camera and lens (720 g). The unit was small enough to fit inside a Stable Micro Systems Total Immersion Temperature Chamber for studies at temperatures and humidities likely to be encountered in tropical testing.

Image capture and processing

Images of the filaments were captured using a Leica Monozoom 7 lens, illuminated in laboratory experiments by a fibre optic light source. A small battery-powered torchlight mounted on a bracket, subsequently proved to give satisfactory contrast for the Seymour unit. Initial commissioning and comparison studies employed a Photron SA3-60 K-M1-LCA high speed camera operating at 5000 fps with a shutter speed of 10^{-5} s, which gave images with 512×256 pixels. Image acquisition was achieved through the Photron FastCam Viewer software and processed using the TriVision software tool [30] running LabView® on a standard PC. Field trials employed a Ximea MQ003MG-CM (maximum frame rate 500 fps giving images of 648×488 pixels) using the Streampix6® software tool on a laptop for image acquisition. Data processing was performed using (*i*) the TriVision software and (*ii*) a new application scripted for the open source image processing code Fiji, based on code provided by Damian Vadillo (currently at Akzo Nobel Research).

Image processing performed two functions. After the sample was loaded and before platen separation, dimensions in the image capture window were calibrated using the known piston diameter. During separation, extracting the minimum width of the filament requires locating the midpoint. On the Trimaster the gap is set by moving the platens an equal distance from the mid-plane and the operation is straightforward (Figure 1(a)). On the Seymour, the location varies with the gap and so the filament shape has to be identified and its minimum located (Figure 1(b)). The frame rate, illumination and threshold settings required adjustment to ensure that the filament could be identified readily.

The Fiji code automated image analysis for a sequence of frames: it allowed the individual frames to be rotated, if required, to be cropped to remove artefacts that would otherwise hinder the analysis, and it located and measured the filament midpoint in each frame. The output from this code is a text file that can be directly imported into Matlab® or Excel® for further analysis.

The Seymour design has not been patented as the authors consider it to be a logical extension of the existing art. A technical drawing and the Fiji image analysis code can be obtained on request from the authors. Additional file 6: Figure S2 demonstrates the three stages of the image acquisition and analysis, going from the raw images acquired with the high speed camera, to the image masks generated for mid-filament detection to the final location, and value, of the mid-filament diameter.

Measuring protocol

The platens were initially cleaned with isopropanol and dried with a lens cloth. The illumination and camera settings were checked and trials run to establish the initial and final gap settings. The initial and final gap heights were chosen such that a filament formed and broke after the platens stopped moving. An estimate of these heights could be obtained by slowly lowering the bottom piston using the screw (labelled G on Figure 11) until the filament just broke, and measuring the piston displacement. The piston was then reset to its starting position and the O ring that sets the stroke length for the upper (moving) piston set to the appropriate position using a feeler gauge. Feeler gauges were also used to set the initial gap height.

A micropipette was used to load the test sample so that it fully covered both platens and filled the gap in between. Any excess liquid was carefully removed with a lens tissue. Video recording was started before the solenoid was activated in order to collect frames for calculating dimensions. Video recording was stopped when the filament broke and the image files (saved in .tiff format) were then transferred to the analysis tools. The initial filament diameter, D_0, was taken as the value when the platens stopped moving and time t was counted from this event. The pistons were cleaned between experimental sessions using ethanol or isopropanol.

Note on regression analysis

Fitting data to the Giesekus model, Equation (3), was sensitive to the initial estimates provided to the regression algorithm. The value for the Newtonian viscosity obtained from fitting Equation (1) to the data set provided an upper bound for the viscosity parameter, η_0, so 10% of this value was taken as the initial estimate. Similarly, the relaxation time obtained from fitting the data to Equation (2) was used as the initial estimate of the relaxation time.

Benchmarking studies

A number of experiments were performed to benchmark the performance of the Seymour device against its archetype, the Trimaster Mk II. Initially, the gap separation was characterised for both devices using identical stretching settings (initial and final gap width). Key parameters such as separation speed, total separation time, piston overshoot and damping were measured for both devices. Subsequently, both devices were used in direct comparison to

Figure 12 Seymour device in field testing. Labels: C – camera; L – lens; S – Seymour extensional rheometer; F – feeler gauges; A – solenoid switch.

measure (*i*) a silicone oil (Newtonian liquid), and (*ii*) solutions of carboxymethylcellulose (CMC), a biopolymer exhibiting non-Newtonian behaviour in solution.

The silicone oil (Silicone fluid f191/1300, batch no. 805, Ambersil Ltd, UK) had a viscosity of 2.37 Pa s. Aqueous solutions of 2 wt% carboxymethylcellulose (molecular weight approximately 750 kDa, BDH Chemicals, UK) were prepared by gentle stirring for 4 h. CMC solutions are viscoelastic shear-thinning fluids and are relatively stable [25]. The shear viscosity of the silicone oil and CMC solutions was measured on a Bohlin CVO 120 HR controlled stress rheometer using 25 mm diameter, smooth parallel plates with a 0.5 mm gap at 22°C. The shear rates studied ranged from 0.1 to 3000 s^{-1}. The volume of fluid required for these tests (250 μL) was considerably larger than that needed for the Seymour tests (approximately 1 μL). The CMC solution exhibited shear thinning with a zero shear rate viscosity, η_0, of 2.95 Pa s, similar to the viscosity of the silicone oil.

Botanical garden pitcher plant studies

Samples of pitcher plant fluid for *N. maxima* and *N. alata* were obtained from individual plants at the Cambridge Botanic Gardens, while *N. rafflesiana* and *N. eymae* fluids were obtained from Kew Gardens, London. Greenhouse relative humidity levels were maintained between 34% and 92%. Freshly opened pitchers were chosen where possible. Ideally, fluid would be extracted from an unopened pitcher using a syringe, but in many cases the pitchers had already opened, and there was some contamination by captured flies and plant detritus in the fluid. Approximately 5 ml of liquid was removed and stored in a small sterilised Nalgene bottle. Samples were stored at room temperature (around 22°C), and tested repeatedly over a period of 37 days (*N. eymae*)/16 days (*N. maxima*).

Field trials

Field measurements were performed in Brunei Darussalam, NW Borneo, between July and September 2014. The test fluids used for the benchmarking studies (silicone oil, CMC) were used to calibrate the measurements in the field. Fluid samples (100 μL per sample) were taken from just opening *N. rafflesiana* pitchers using a micropipette and were either measured immediately on site (Figure 12, see also the Additional file 1: video), or transferred to 2 mL sterile screw-top vials and transported back to a nearby field station. Within a maximum of 4 h after sampling, these samples were measured in a room at temperatures between 24°C and 27°C. The pitcher plant fluid samples were then stored at room temperature (22-30°C) and re-measured after 20 days. Storage at room temperature was chosen to mimic conditions during expeditions or at remote field stations where reliable refrigeration may not be available.

Confirmation of model assumptions

Discussion of the characteristic timescales for CaBER™ experiments can be found in the literature [8,9,31]; these give indications of when filament stretching is the controlling mechanism. The inertial (Rayleigh) time scales for the tests presented here, given by $\sqrt{(\rho D_0^3/8\alpha)}$, ranged from 0.02-0.04 ms whereas the viscous times, (given by $\eta_0 D_0/2\alpha$), lay between 2 and 4 ms. The majority of tests were then considerably longer than both these times. The elastocapillary number, comparing the relaxation time in extension to the viscous time, $Ec= 2\lambda\alpha/\eta_0 D_0$, ranged from 9 for the greenhouse *N. maxima* tests to 2100 for the wild *N. rafflesiana* (Figure 10). These data confirm the presence of viscoelasticity.

Additional files

Additional file 1: Video Operation of Seymour in the field, measuring a sample of *N. rafflesiana* pitcher fluid in the natural habitat of the plant in Brunei Darussalam, Borneo.

Additional file 2: Table S1. Effect of storage at room temperature on the goodness of fit (quantified by the correlation coefficient, R^2), for *N. eymae* and *N. maxima* fluid obtained from greenhouses (one pitcher for each species).

Additional file 3: Figure S1. Effect of storage under chilled conditions. Filament thinning behaviour for greenhouse-sourced *N. maxima* pitcher fluid. Sample tested at Day 1 then stored at 4°C for up to 13 days, for comparison with Figure 7(b). Each aliquot was brought to room temperature before testing. Data decimated for clarity.

Additional file 4: Table S2. Fitted model parameters for 11 newly-opened pitchers of *N. rafflesiana* grown in the field.

Additional file 5: Table S3. Fitted model parameters for 9 of the 11 pitchers of *N. rafflesiana* in Additional file 4: Table S2 following storage for 20 days at 25-30°C. Two experimental measurements were made for each pitcher. Samples 13b and 23 were not tested.

Additional file 6: Figure S2: Frames from a typical image sequence. The figure shows the original and processed images at the start of a typical filament thinning experiment, two midpoint filament thinning times and at a time close to filament breakup.

Abbreviations

BOAS: Beads on a string; CMC: Carboxymethylcellulose; UCM: Upper Convected Maxwell (rheological model).

Competing interests

The authors declare that they have no competing interests.

Authors' contributions

CC, AA and EC all participated in the design and commissioning of the device, laboratory experimental studies, and reading the manuscript. GC and LP devised, designed and constructed the device. LT-P assisted with experimental studies and rheology work. UB performed the experimental work in Brunei and helped to draft the manuscript. BH and DIW conceived of the study, co-ordinated the work and drafted the manuscript. BH wrote the software tools for image analysis. All authors read and approved the final manuscript.

Authors' information

UB is an early career research fellow in the School of Biological Sciences at the University of Bristol, UK, working on the biomechanics and functional morphology of pitcher plants. All the other authors were based in the Department of Chemical Engineering and Biotechnology at Cambridge, UK, at the time of this study. CC, AA and EC performed this work as final year/diploma research projects under the direction of chemical engineering academics BH and DIW. GC and LP are assistant staff in the Department's workshops.

Acknowledgements
This work grew from a conversation over dinner at Jesus College, Cambridge between UB, DIW, Dr Walter Federle (Department of Zoology, Cambridge) and Prof. Francis Gadala-Maria (Department of Chemical Engineering University of South Carolina). Additional assistance and advice from the following is gratefully acknowledged: Dr Simon Butler, John Gannon and Kevin Swann (Department of Chemical Engineering & Biotechnology, Cambridge), Alex Summers (Cambridge University Botanic Garden), and Dr Ulmar Grafe (Faculty of Science, Universiti Brunei Darussalam). The image processing tool in Fiji was based on the code kindly provided by Dr Damian Vadillo. Mathias Scharmann (Institute of Integrative Biology, ETH Zürich) kindly took the video showing the experimental protocol.
The following financial support is gratefully acknowledged: a Henslow Research Fellowship from the Cambridge Philosophical Society and a Leverhulme Early Career Fellowship for UB; a visiting research fellowship (POS-A/2012/116) for MDT from Xunta de Galicia's Consellería de Cultura, Educación e OrdenaciónUniversitaria of Spain and the European Union's European Social Fund; and a summer project grant for CC from Sidney Sussex College, Cambridge. The fare at the dinner was also most agreeable.

Author details
[1]Department of Chemical Engineering and Biotechnology, New Museums Site, Pembroke Street, Cambridge CB2 3RA, UK. [2]School of Biological Sciences, Life Sciences Building, 24 Tyndall Avenue, Bristol BS8 1TQ, UK. [3]Faculty of Science, Universiti Brunei Darussalam, Tungku Link, Gadong 1410, Bandar Seri Begawan, Brunei Darussalam. [4]ESPCI ParisTech, 10 rue Vauquelin, 75005 Paris, France. [5]Department of Chemical Engineering, University of Santiago de Compostela, Lope Gómez de Marzoa St, Santiago de Compostela E-15782, Spain.

References

1. Haynes R, Burton A. Role of the non-Newtonian behavior of blood in hemodynamics. Am J Physiol. 1959;197:943–50.
2. Lopez-Vidriero M, Reid L. Bronchial mucus in health and disease. Br Med Bull. 1978;34:63–74.
3. Denny M. The role of gastropod pedal mucus in locomotion. Nature. 1980;285:160–1.
4. Dirks J, Federle W. Fluid-based adhesion in insects–principles and challenges. Soft Matter. 2011;7:11047–53.
5. Larson RG. The structure and rheology of complex fluids. New York: Oxford University Press; 1999.
6. Fauci L, Dillon R. Biofluidmechanics of reproduction. Annu Rev Fluid Mech. 2006;38:371–94.
7. Kojić N, Bico J, Clasen C, McKinley GH. Ex vivo rheology of spider silk. J Exp Biol. 2006;209(Pt 21):4355–62.
8. Erni P, Varagnat M, Clasen C, Crest J, McKinley GH. Microrheometry of sub-nanolitre biopolymer samples: non-Newtonian flow phenomena of carnivorous plant mucilage. Soft Matter. 2011;7:10889–98.
9. Gaume L, Forterre Y. A viscoelastic deadly fluid in carnivorous pitcher plants. PLoS One. 2007;2:e1185.
10. Wolf D, Blasco L, Khan M, Litt M. Human cervical mucus. II. Changes in viscoelasticity during the ovulatory menstrual cycle. Fertil Steril. 1977;28(1):47–52.
11. Bauer U, Willmes C, Federle W. Effect of pitcher age on trapping efficiency and natural prey capture in carnivorous Nepenthes rafflesiana plants. Ann Bot. 2009;103:1219–26.
12. Collyer A, Clegg DW. Rheological measurement. London: Elsevier; 1998.
13. Sachsenheimer D, Hochstein B, Willenbacher N. Experimental study on the capillary thinning of entangled polymer solutions. Rheol Acta. 2014;53:725–39.
14. Bauer U, Grafe TU, Federle W. Evidence for alternative trapping strategies in two forms of the pitcher plant, Nepenthes rafflesiana. J Exp Bot. 2011;62:3683–92.
15. McKinley GH, Tripathi A. How to extract the Newtonian viscosity from capillary breakup measurements in a filament rheometer. J Rheol. 2000;44:653.
16. Tuladhar TR, Mackley MR. Filament stretching rheometry and break-up behaviour of low viscosity polymer solutions and inkjet fluids. J Nonnewton Fluid Mech. 2008;148:97–108.
17. Galindo-Rosales FJ, Alves MA, Oliveira MSN. Microdevices for extensional rheometry of low viscosity elastic liquids: a review. Microfluid Nanofluidics. 2012;14:1–19.
18. Arnolds O, Buggisch H, Sachsenheimer D, Willenbacher N. Capillary breakup extensional rheometry (CaBER) on semi-dilute and concentrated polyethyleneoxide (PEO) solutions. Rheol Acta. 2010;49:1207–17.
19. Stelter M, Brenn G, Yarin AL, Singh RP, Durst F. Validation and application of a novel elongational device for polymer solutions. J Rheol. 2000;44:595.
20. Liang RF, Mackley MR. Rheological characterization of the time and strain dependence for polyisobutylene solutions. J Nonnewton Fluid Mech. 1994;52:387–405.
21. Giesekus H. A simple constitutive equation for polymer fluids based on the concept of deformation-dependent tensorial mobility. J Nonnewton Fluid Mech. 1982;11:69–109.
22. Torres MD, Hallmark B, Wilson DI, Hilliou L. Natural Giesekus fluids: Shear and extensional behavior of food gum solutions in the semidilute regime. AICHE J. 2014;60:3902–15.
23. Silcone Fluid Performance Test Results. [http://www.silicone.jp/e/catalog/pdf/kf96_e.pdf]
24. Rodd LE, Scott TP, Cooper-White JJ, McKinley GH. Capillary break-up rheometry of low-viscosity elastic fluids. Applied Rheol. 2005;15:12–27.
25. Benchabane A, Bekkour K. Rheological properties of carboxymethyl cellulose (CMC) solutions. Colloid Polym Sci. 2008;286:1173–80.
26. Bhat PP, Appathurai S, Harris MT, Pasquali M, McKinley GH, Basaran OA. Formation of beads-on-a-string structures during break-up of viscoelastic filaments. Nat Phys. 2010;6:625–31.
27. Entov VM, Hinch EJ. Effect of a spectrum of relaxation times on the capillary thinning of a filament of elastic liquid. J Nonnewton Fluid Mech 1997;72:31–53.
28. Clasen C, Verani M, Plog JP, McKinley GH, Kulicke W-M. Effects of polymer Concentration and molecular weight on the dynamics of visco-elasto-capillary breakup. In: Proc. XIVth Int. Cong. on Rheology. Seoul, Korea, publ. Korean Society of Rheology. 2004.
29. Industrial solenoid data sheet: 8.02.11.52 [http://www.hebsbenson.co.uk/wp-content/uploads/2013/08/8.02.11.52-datasheet.pdf]
30. Vadillo DC, Tuladhar TR, Mulji AC, Jung S, Hoath SD, Mackley MR. Evaluation of the inkjet fluid's performance using the "Cambridge Trimaster" filament stretch and break-up device. J Rheol. 2010;54:261.
31. Clasen C, Phillips PM, Palangetic L, Vermant J. Dispensing of rheologically complex fluids: the map of misery. AICHE J. 2012;58:3242–55.

Hyperspectral phenotyping on the microscopic scale: towards automated characterization of plant-pathogen interactions

Matheus Kuska[1], Mirwaes Wahabzada[1], Marlene Leucker[1], Heinz-Wilhelm Dehne[1], Kristian Kersting[2], Erich-Christian Oerke[1], Ulrike Steiner[1] and Anne-Katrin Mahlein[1*]

Abstract

Background: The detection and characterization of resistance reactions of crop plants against fungal pathogens are essential to select resistant genotypes. In breeding practice phenotyping of plant genotypes is realized by time consuming and expensive visual rating. In this context hyperspectral imaging (HSI) is a promising non-invasive sensor technique in order to accelerate and to automate classical phenotyping methods.
A hyperspectral microscope was established to determine spectral changes on the leaf and cellular level of barley (*Hordeum vulgare*) during resistance reactions against powdery mildew (*Blumeria graminis* f.sp. *hordei,* isolate K1). Experiments were conducted with near isogenic barley lines of cv. Ingrid, including the susceptible wild type (WT), mildew locus a 12 (*Mla12* based resistance) and the resistant mildew locus o 3 (*mlo3* based resistance), respectively. The reflection of inoculated and non-inoculated leaves was recorded daily with a hyperspectral linescanner in the visual (400 – 700 nm) and near infrared (700 – 1000 nm) range 3 to 14 days after inoculation.

Results: Data analysis showed no significant differences in spectral signatures between non-inoculated genotypes. Barley leaves of the near-isogenic genotypes, inoculated with *B. graminis* f.sp. *hordei* differed in the spectral reflectance over time, respectively. The susceptible genotypes (WT, *Mla12*) showed an increase in reflectance in the visible range according to symptom development. However, the spectral signature of the resistant *mlo*-genotype did not show significant changes over the experimental period. In addition, a recent data driven approach for automated discovery of disease specific signatures, which is based on a new representation of the data using Simplex Volume Maximization (SiVM) was applied. The automated approach - evaluated in only a fraction of time revealed results similar to the time and labor intensive manually assessed hyperspectral signatures. The new representation determined by SiVM was also used to generate intuitive and easy to interpretable summaries, e.g. fingerprints or traces of hyperspectral dynamics of the different genotypes.

Conclusion: With this HSI based and data driven phenotyping approach an evaluation of host-pathogen interactions over time and a discrimination of barley genotypes differing in susceptibility to powdery mildew is possible.

Keywords: Hyperspectral imaging, Hyperspectral microscope, Spectral reflectance, Plant disease, Disease resistance, Mildew locus o (*Mlo*), Near-isogenic lines, *Hordeum vulgare*, *Blumeria graminis* f.sp. *hordei*

* Correspondence: amahlein@uni-bonn.de
[1]Institute for Crop Science and Resource Conservation (INRES) - Phytomedicine, University of Bonn, Meckenheimer Allee 166a, 53115 Bonn, Germany
Full list of author information is available at the end of the article

Introduction

In agricultural production the demands on efficient crop plants are manifold. Improved quantitative and qualitative plant traits are desired, along with enhanced stress resistance, especially against plant pathogens. The development of resistant cultivars is a challenging task in plant breeding. Fungal plant pathogens affect almost all relevant crops in different stages of their development and impair the yield and product quality. In barley production powdery mildew is one of the main damaging diseases in Europe and other temperate regions [1]. The disease is caused by the biotroph ascomycete *Blumeria graminis* f.sp. *hordei* (*Bgh*) and is spread during the vegetation period by wind with conidiospores. *Bgh* is able to colonize barley plants within 24 hours after the first contact. A haustorium - the feeding organ of *Bgh* – develops within penetrated epidermal cells, which remain vital. The new epiphytic mycelium grows over the leaf surface to penetrate other epidermal cells and develop new haustoria. Finally, *Bgh* produces conidiophores bearing new conidia. This asexual life cycle is completed in approximately five days and is repeated by multiple generations per season. The application of fungicides and the cultivation of resistant barley varieties are the main methods for controlling barley powdery mildew [2]. Unfortunately, the farmers face fungicide resistances developed by *Bgh* [3]. This emphasises the importance of resistant barley genotypes generated in breeding programs.

A well-known resistance mechanism of barley against *Bgh* is the non-race specific mildew locus o (*mlo*) based resistance [4]. In all *mlo* mutants, *Bgh* cannot penetrate the epidermal cell, because a cell wall apposition (papilla) is developed under the penetration point [5] that possesses a high electron density [6]. The basic components of this cell wall apposition are a complex of lignin, cellulose, callose, peroxidases, phenols, proteins and further cell-wall materials [7]. The mildew locus a (*Mla*) gene based resistance is an another resistance reaction of barley against *Bgh* and is associated with a hypersensitive reaction of epidermal cells attacked by *Bgh* [7-9].

These resistance properties are used in plant breeding programs to improve the resistance of barley plants to powdery mildew. However, one main drawback of recent breeding programs is their time consuming and labor intensive nature. The traditional breeding procedure of common crop plants still takes 7 to 17 years and requires a high amount of plant material and human effort on the way to a desired cultivar [10]. In this complex breeding process, manifold steps in different environments - under controlled and under field conditions - at different plant levels - from single organs to the canopy - are executed. Hereby the selection process of predominant genotypes and relevant crop traits by genotyping and phenotyping methods is crucial and determines the time span and the success of the breeding process.

In recent years the genotyping of plants has been significantly accelerated by advances in molecular profiling and sequencing technologies [11]. Marker-assisted selection, bi-parental recombinant inbred lines, or the increasing number of completely sequenced species in genomic databases provides solutions to current breeding challenges [12]. To bridge the gap from genomic characterization to plant function and agricultural traits, the expression of the genome in a given environment has to be tested carefully. This step is defined as the phenotyping process. Several authors have addressed the labor-intensive and costly nature of conventional phenotyping processes as the limiting and time-consuming factor in plant breeding. This challenge has been identified as the phenotyping bottleneck [11-13].

Innovative technologies, e.g. optical and non-invasive sensors, have been characterised as new phenotyping methods with potential to overcome this bottleneck and to improve the breeding process. Various optical imaging methods using e.g. RGB [11], 3D [14], fluorescence [15], thermography [16] and HSI [17] sensors are able to characterize different plant parameters and could potentially be implemented in automated, high-throughput phenotyping pipelines.

Among these methods, hyperspectral imaging (HSI) is one of the most promising techniques to assess functional plant traits [17-19]. Using HSI, the spectral characteristics of plants can be visualised non-invasively over time and on different scales. The sensitivity of sensors enable a high spectral and spatial resolution and the reflectance per pixel can be analysed in narrow wavelengths. Characteristic spectral signatures provide information about the physiological status of plants and for plant breeding on the reaction of different genotypes to biotic or abiotic stress factors [20]. Leaf pigments, like chlorophylls and carotenoids are the main factors influencing the spectral information in the visual range (VIS, 400–700 nm) [21]. The near infrared range (NIR, 700–1000 nm) is mainly affected by scattering processes in the spongy mesophyll, and additionally by absorptions due to proteins, fatty acids, starch, water and cellulose with different specific absorption wavelengths. The leaf water content, including further chemical compounds, can be derived from the short wave infrared range (SWIR, 1000–2500 nm).

Various researchers have demonstrated that plant diseases and processes during pathogenesis can be detected by hyperspectral sensors [22-24]. Plants diseased by fungal pathogens could be distinguished from healthy plants at different stages of the pathogenesis and at different disease severities. Since first interaction sites and primary symptoms of fungal plant diseases are in the range of sub-millimeters, highly sensitive sensor systems and powerful subsequent data analysis routines are required for a reliable evaluation of plants under biotic stress. Few researchers postulate, that early modifications of the cellular leaf structure occur due to fungal toxins or plant resistance reactions and that these subtle changes are detectable via HSI [25,26]. They focused on small-scale studies with HSI or on the detection and identification of plant diseases before visible symptoms appear. Based on these insights the development of a HSI routine to estimate the susceptibility of plants and to characterise defence mechanisms against fungal pathogens is a desirable task. However, it requires methodological adaptions and technical advances to exploit the potential of hyperspectral sensors for the implementation in resistance screenings.

The hypothesis of this study is that subtle processes during compatible and incompatible plant-pathogen interactions have an effect on optical properties of plants. It is expected that specific regions of the electromagnetic spectrum are influenced depending on the type of interaction. The detection of these changes demands specific sensor setups with a high spatial and spectral resolution combined with sophisticated, data analysis methods. To proof this hypothesis a HSI microscope, a measuring protocol for detached barley leaves and an automated data analysis approach was established in this study. This approach can be used in resistance screening for the differentiation of barley genotypes and for a characterization of their interaction with *Bgh*, the causal agent of powdery mildew. As an application model susceptible and resistant genotypes with different, well known genetic backgrounds were monitored visually and by hyperspectral imaging in time-series experiments at small-scale level.

To uncover the full information from high-dimensional HSI data, characteristic spectra were extracted both manually and using data mining techniques. A qualitative and automated analysis of reflectance data was realized using a data driven approach based on the matrix factorisation technique Simplex Volume Maximisation (SiVM) [27]. With this methodological approach, (i) processes during pathogenesis on the different genotypes could be characterized, (ii) stages of pathogenesis were automatically visualized and (iii) spectral dynamics were evaluated over time.

Results and discussion

The pathogenesis of powdery mildew and early interactions on different barley genotypes was characterized on the leaf and tissue level using a hyperspectral microscope (Figure 1).

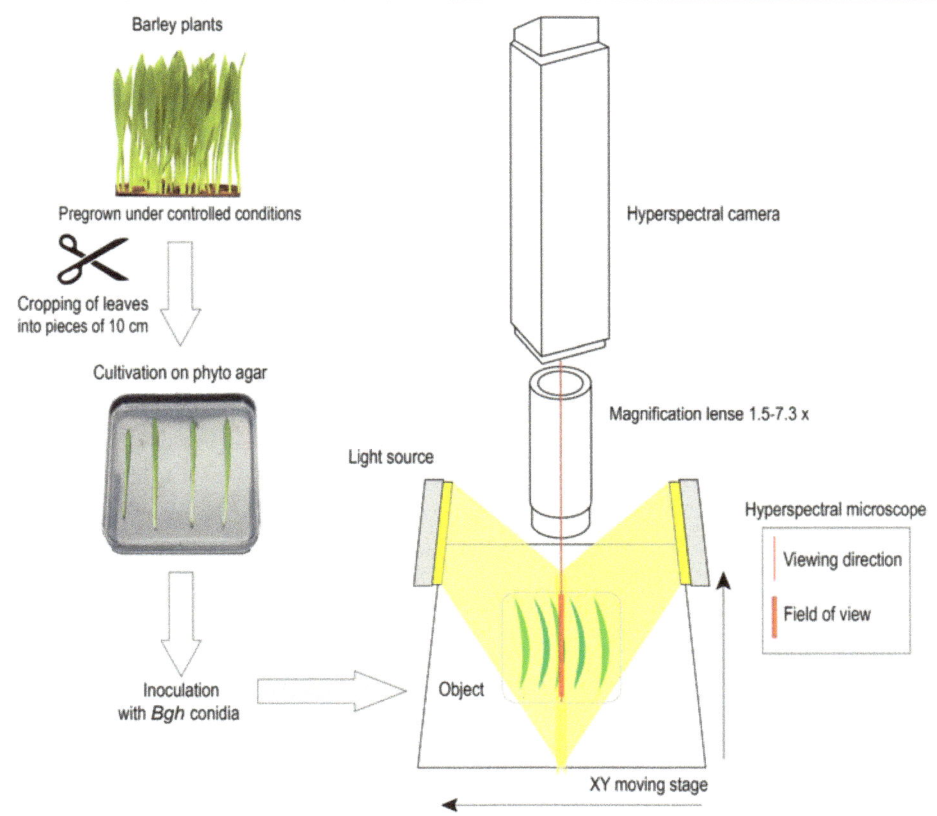

Figure 1 Hyperspectral imaging microscope setup for small-scale image analysis. The spectral reflectance of detached barley leaves on phyto agar was measured with a hyperspectral camera, which was mounted on a magnification lens to enable a magnification up to 7.3x. Two linear light emitters with a vertical orientation of 30° illuminated the samples in a distance of 20 cm. The samples were moveable due to a XY moving stage. A spectral resolution of up to 2.73 nm and a maximum spatial resolution of 7.5 μm per pixel was obtained. The field of view ranged from 4.1 to 0.95 cm, dependent on the magnification.

A high spatial resolution of a pixel size as low as 7.5 μm allowed the detection of subtle processes.

Hyperspectral imaging can improve disease detection through a better examination of host pathogen interactions [28]. This has been demonstrated by researchers, using a diversity of hyperspectral sensors with different crops and their relevant diseases on different scales ranging from remote to proximal sensing [24,29]. Since imaging sensors allow for a pixel-wise attribution of disease-specific symptoms, primary infection sites can be identified and analyzed spectrally [25]. In contrast to our new plant phenotyping approach, existing HSI microscopes are prohibited due to destructive nature [30,31].

Phenotypic development of healthy and inoculated barley leaves

The phenotypes of detached, healthy and *Bgh* inoculated leaves of barley genotypes WT, *Mla12* and *mlo3* were assessed visually on phyto agar (Figure 2). *Mla12* leaves were included as an additional susceptible genotype to analyze differences during the pathogenesis between near-isogenic lines, respectively. Non-inoculated leaves of the three genotypes did not show any visible symptoms during the first 6 days after inoculation (dai). Symptoms of senescence occurred 6 dai on healthy *mlo3* leaves, indicated by yellowing of the leaves. The WT and *Mla12* started to became chlorotic 10 dai. This senescence process developed further until 14 dai. The relative long life span and vitality of detached leaves in this investigations indicated consistent conditions for HSI of the plant system in a controlled environment, for a period of 14 days.

Bgh inoculated leaves showed no visible symptoms during the first 5 dai. Characteristic powdery mildew pustules occurred 6 dai on the susceptible WT and the near-isogenic *Mla12* line. On *Mla12* leaves the pustules were distributed homogeneously on the leaf surface compared to clustered pustules on WT leaves. Pustules expanded and covered nearly the complete leaf surface of the susceptible WT and *Mla12* leaves 10 dai. Furthermore, the leaves became light-green and chlorotic in areas without powdery mildew pustules 10 dai. Necrotic tissue occurred with the exception of powdery mildew dominated leaf areas, which showed light-green to yellow discoloration 14 dai. The resistant *mlo3* leaves did not show any powdery mildew symptoms during the experiment. The leaves were healthy and green with a delayed senescence. First signs of senescence of inoculated *mlo3* leaves appeared only 14 dai. Further studies are required for an explicit interpretation of the decelerated senescence of the inoculated *mlo3* leaves.

Spectral similarity of non-inoculated near-isogenic barley lines over time

Detached non-inoculated (healthy) and inoculated leaves of the near-isogenic lines cv. Ingrid WT, *Mla12* and *mlo3* were measured daily 3 to 14 dai in order to assess changes in the spectral signatures. Healthy leaves of the different near-isogenic lines exhibited a typical spectral pattern of healthy plants with low reflectance from 400 – 700 nm, a characteristic green peak at 500 – 570 nm, a steep reflectance increase at the red edge inflection point and a high reflectance plateau in the NIR 3 days after detachment (Figure 3a). This pattern slightly changed over time. The reflectance between 420 and 680 nm increased every day due to changes in the pigment composition [32-34]. Other ranges of the spectrum were not affected (Figure 3a). Spectral changes indicated senescence processes of non-inoculated, healthy leaves over time, which were in accordance to the leaf phenotype. However, differences in the development of individual leaves were reflected by the relative standard deviation (RSD) of 0.6 – 6.8% over time and measured wavelength range. The highest RSD was calculated in the NIR range for WT 10 dai, for *Mla12* 13 dai, and for *mlo3* 13 dai. The low RSD indicate the practicability and robustness of a detached leaf system for the assessment of phenotypic differences due to resistance reactions. The similarity between reflectance spectra of healthy *Mla12* and *mlo3* leaves to healthy WT leaves was due to the identical genomic background of the near-isogenic lines.

Consequently, healthy leaves of the near-isogenic lines cv. Ingrid *WT*, *Mla12* and *mlo3* showed a high spectral similarity and a similar performance on the phyto agar plates. The assessed barley spectra were characteristic reflectance patterns of healthy plant tissue [25,26,35,36]. Reflectance of the detached leaves between 420 – 740 nm increased with each day due to senescence. The absorption features in this range are related to chlorophyll and other pigments linked to photosynthesis [21,33,34]. Increased hyperspectral reflectance indicated a reduction to the chlorophyll activity and content. This effect is well described as one main process during plant senescence [32]. The course of reflectance changes due to senescence coincided with the phenotypic senescence processes observed (Figure 2).

Spectral signatures of near-isogenic barley lines during powdery mildew pathogenesis

The susceptible near-isogenic genotypes cv. Ingrid WT and *Mla12* and the *Bgh* isolate K1 were used to evaluate the progress of powdery mildew pathogenesis and to identify spectral fingerprints of the barley-*Bgh* system. Inoculated WT leaves showed minor differences to healthy WT leaves 3 dai (Figure 3b). The reflectance of inoculated WT leaves increased between 534 – 563 nm. An overall increase of reflectance in the entire range was observed already 4 dai and the shift to higher reflectance, continued the following days. Within this period of time, the increased reflectance was in accordance with

Figure 2 Phenotypes of detached barley leaves non-inoculated (healthy) and inoculated with *Blumeria graminis* f.sp. *hordei* (*Bgh*) of near-isogenic lines cv. Ingrid (WT, *Mla12* and *mlo3*) 0, 3, 6, 10 and 14 dai, on phyto agar. Non-inoculated leaves of the genotypes showed natural senescence over the experimental period. First characteristic powdery mildew pustules became visible 6 dai on inoculated leaves of the susceptible WT and the near-isogenic line *Mla12*. The resistant near-isogenic *mlo3* leaves did not show any powdery mildew symptoms. Senescence of inoculated near-isogenic *mlo3* leaves was delayed until 14 dai. Images were taken with a digital camera (EOS 6D, Canon, Tokio, Japan) and a 100 mm object lens (EF Lens Ultrasonic EF 100 mm 1:2.8 L Macro IS USM, Canon, Tokio, Japan).

the *Bgh* ontogenesis on barley leaves. The reflectance alterations in the VIS of inoculated WT leaves indicated changes in photochemical processes and pigment content, which are associated to the photosynthetic activity [32-34]. Reflectance in the NIR from 743 – 830 nm decreased 5 dai. This NIR response turned to an increased reflectance again 6 dai, when first powdery mildew pustules occurred on the WT leaf surface. Subsequently, the reflectance in the NIR from 743 – 830 nm decreased stepwise until 14 dai. Symptoms were accompanied by significant reflectance changes over the full range. This gradual increase of reflectance was prominent from 400 – 680 nm and from

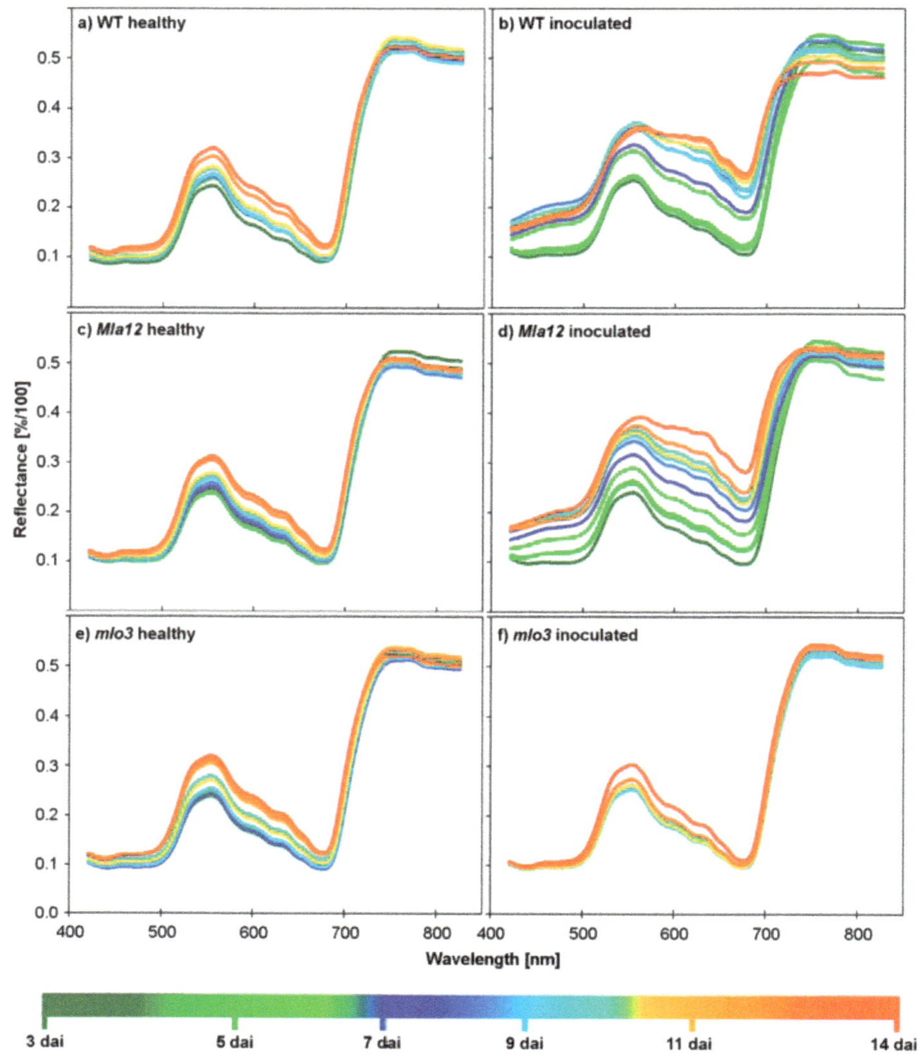

Figure 3 Spectral signatures of non-inoculated (healthy) barley leaves cv. Ingrid WT **(a)**, *Mla12* **(c)**, *mlo3* **(e)** and barley leaves inoculated with *Blumeria graminis* f.sp. *hordei* (*Bgh*) **(b, d, f)**, from 3 to 14 dai. Reflectance spectra of healthy leaves of the near-isogenic lines are similar. During the measuring period, the reflectance of healthy leaves increased in the visible range. Reflectance of inoculated, susceptible genotypes (WT and *Mla12*) increased in the VIS and WT had decreased reflectance in the NIR. The inoculated, resistant *mlo3*-genotype showed significant differences to healthy leaves in the reflection from 530 – 680 nm only 14 dai. (n = 3).

700 – 740 nm. The reflectance in the VIS increased daily according to the growth of powdery mildew mycelium until 9 dai. The course of the spectral pattern changed from 10 to 14 dai and the reflectance from 420 – 500 nm decreased again due to the occurrence of first necrosis and tissue collapse below powdery mildew pustules. Increased green reflectance was in accordance with senescence chloroses, associated to a reduction and breakdown of chlorophyll (Figure 2). The reflectance spectrum 14 dai represented a necrotic leaf tissue covered with powdery mildew. The RSD among the inoculated WT leaves was 1 – 14.7% over time and wavelengths. The highest RSD was calculated 14 dai over the full spectral range. The spatial distribution of the *Bgh* mycelium and the vitality of the individual leaves influenced the leaf phenotypes, which explained the higher RSD of the hyperspectral reflectance compared to non-inoculated leaves.

The hyperspectral reflectance pattern of *Bgh* pathogenesis described for WT was also monitored for the near-isogenic line *Mla12* (Figure 3d). The appearance of first tiny powdery mildew pustules 4 dai was associated to first increase of leaf reflectance between 420 to 680 nm. In contrast to inoculated WT leaves, reflectance did not decrease in the blue range 10 dai and later. Reflectance of *Mla12* leaves in the range 500 – 742 nm increased day by day in contrast to the WT leaves. This effect can be explained by a faster development of *Bgh* on *Mla12* leaves compared to WT leaves. In addition, the diseased area and the density of mycelium and conidia on the leaves was higher than on the WT leaves. Similar to inoculated WT leaves, reflectance increased stepwise until 14 dai, except from 743 to 830 nm the reflectance did not change significantly over time. Interestingly, the reflectance between 743 to 830 nm 5 dai was lower compared to the other days. This phenomenon was also observed for inoculated WT leaves 5 dai. The RSD of reflectance among inoculated *Mla12* leaves was 3 – 14% over time and wavelengths. The highest RSD was calculated between 500 – 680 nm 14 dai. Also the range 420 – 500 nm showed high RSD among the inoculated *Mla12* leaves from 6 dai until 14 dai.

In contrast, susceptible near-isogenic lines WT and *Mla12* showed slightly differences in the spectral reflectance during the pathogenesis. Nevertheless, reflectance patterns of *Bgh* pathogenesis on the susceptible genotypes were characterized by a reflectance increase between 400 – 700 nm over time. This increase is due to white powdery epiphytic mycelium and conidia. Similar patterns were observed for powdery mildew diseased leaves of sugar beet and winter wheat on different scales [25,37]. The results demonstrate a similarity of spectral patterns and dynamics during powdery mildew pathogenesis, independently of the scale of investigations, but with a higher sensitivity of the HSI microscope because of the higher spatial resolution.

Mlo3 inoculated leaves showed no powdery mildew infestation over time

The spectral reflectance of *Bgh* inoculated *mlo3* leaves (Figure 3f), differed from that of inoculated, susceptible WT and *Mla12* leaves. Resistant *mlo3* leaves showed a spectral pattern similar to non-inoculated leaves of all genotypes until 9 dai. No visible symptoms were assessed on the inoculated *mlo3* leaves. Interestingly, inoculated *mlo3* leaves showed no effects of natural senescence on reflectance until 13 dai. The reflectance was constant over time with low RSD of 5-7% in the full spectral range. A first increase in reflectance was observed from 540 – 680 nm 13 dai, and first symptoms of senescence occurred.

Swarbrick *et al.* [38] reported an induced cell-death and a reduction of the photosynthetic activity during the resistance reaction of *mlo5* leaves inoculated with *Bgh* isolate A6. In contrast, *Bgh* isolate K1 inoculated *mlo3* leaves in this study did not change the chlorophyll content until 13 dai. This is indicated by constant low reflectance from 420 – 680 nm [32-34]. Moreover, the constant reflectance over time, especially at 680 – 700 nm, allowed to distinguish between susceptible and resistant leaves already 4 dai. Specific resistance reactions of barley genotypes, such as lignification, controlled cell death, or formation of papilla may be assessed only by using a HSI system with higher spatial resolution [25]. The hyperspectral microscope is an important methodological innovation to elucidate subtle responses of plants to biotic stress. However, specific reflectance patterns of the barley *mlo3* resistance reaction were not assessed in this experiment. Further investigations of the first 48 hours after inoculation are required, since most defense mechanisms of plants against fungal pathogens take place within the first hours after plant-pathogen-contact, attempted penetration and early infection [39]. The *mlo5* based papilla formation against *Bgh*, for instance is completed within the first 16 – 30 h after the contact of conidia and plant surface [9]. To improve the results from the HSI microscope exact spatial referencing of images from subsequent days is necessary. A better spatial orientation within the image could be realized by placing localization plates beside the leaf area of interest.

An automated approach for hyperspectral image analysis for plant phenotyping

The effective analysis and interpretation of HSI data are limiting factors for an implementation into plant phenotyping [36]. The 3-dimensional complexity of HSI data requires high input of human intervention and labelling of disease specific image pixels. Automated analysis pipelines are required to optimize the use of HSI for phenotyping or precision agriculture. Within this study an automated analysis cascade using Simplex Volume Maximization was adopted (Figure 4). Reflectance spectra of inoculated WT leaves assessed by this automatic approach (Figure 5a) showed patterns similar to the manually assessed reflectance spectra. Differences between manually and automatically extracted reflectance were calculated to highlight the similarity of the results (Figure 5b,d). According to both methods of data analysis, no differences between healthy and diseased leaves were observed 3 dai. An increase of reflectance between 420 and 742 nm was detected in the automated approach already 4 dai indicating a higher sensitivity when spectral information is defined automatically from all pixels of an image. Further differences were detected in the NIR range 4 and 5 dai (Figure 5b). The next days showed a spectral trend similar to the manual analysis. The differences between automated and manually assessed reflectance values were low and reached only –0.04 [%/100] at 680 nm. The automatically assessed reflectance spectra of inoculated *Mla12* leaves were in accordance to those assessed manually (Figure 5c); differences varied from –0.02 to 0.03 [%/100] over time and wavelengths (Figure 5d). The present results highlight a standardized system with detached leaves for a HSI microscopy and automated data mining suitable for plant phenotyping. Interestingly, the automatically assessed reflectance spectra showed more details among days (Figure 5) as all image pixels were considered,

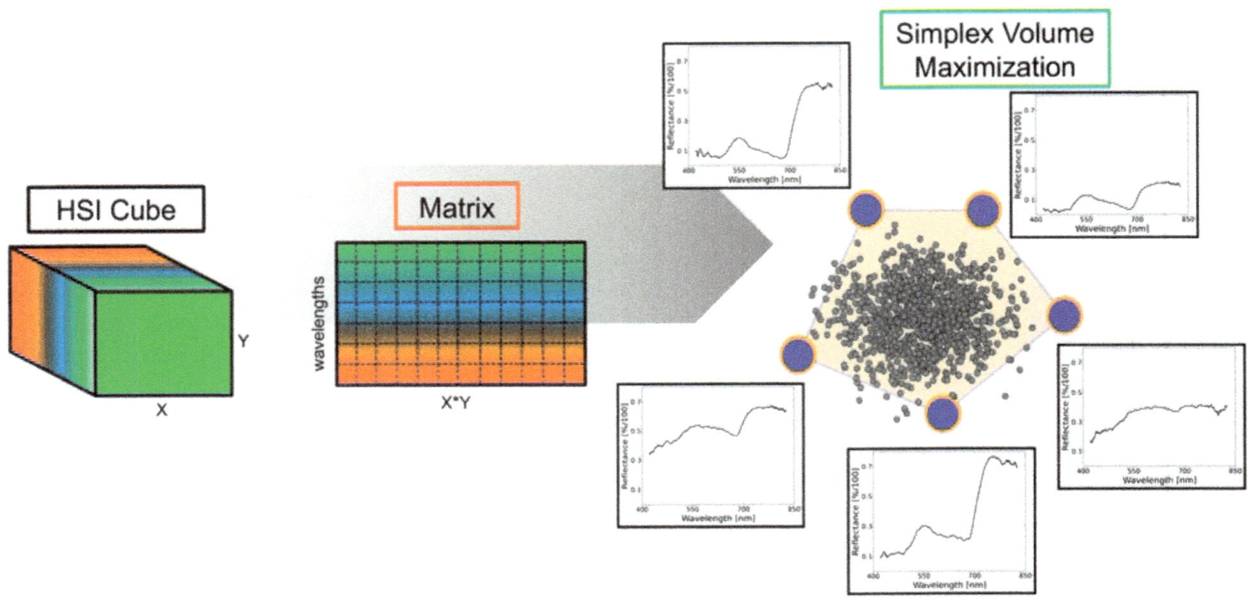

Figure 4 Interpretable matrix factorization for hyperspectral images. Each hyperspectral data cube is transformed into a dense matrix. Then, extreme components/signatures on all matrices are computed, using Simplex Volume Maximization. The final step includes the computation of the new representation of all signatures in a space, spanned by the extremes.

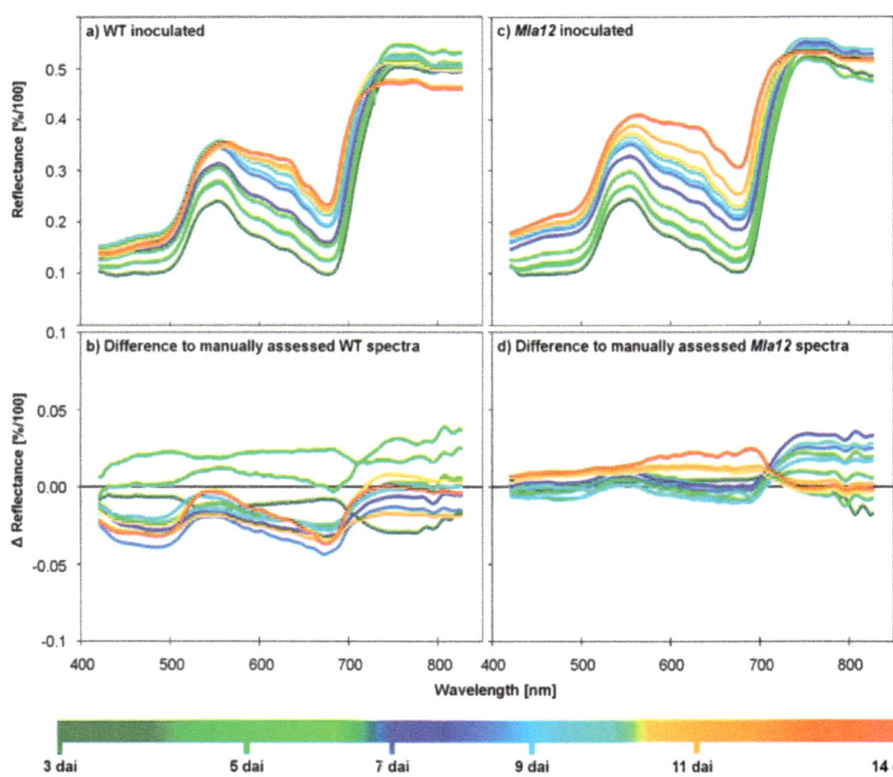

Figure 5 Automatically determined mean signatures of barley leaves cv. Ingrid WT **(a)** and *Mla12* **(c)** inoculated with *Blumeria graminis* f.sp. *hordei* from 3 to 14 dai. The automatically assessed spectra were similar to signatures assessed manually. The differences between automatically and manually analyzed data for WT were −0.04 – 0.04 [%/100] **(b)**, −0.02 – 0.03 [%/100] for *Mla12* **(d)**, respectively, over the wavelengths and time.

whereas the manually assessed spectra are only from a few selected pixels.

Based on the automatically assessed reflectance spectra a binary map system of *Bgh* inoculated susceptible leaves was established (Figure 6). The binary maps visualize *Bgh* diseased leaf tissue over time. Black pixels indicate healthy leaf tissue, while white pixels indicate sites with powdery mildew. This allows the observation of disease development on susceptible plant genotypes with rapid visual identification of relevant pixels. Powdery mildew symptoms were absent on RGB images 3 dai. The corresponding binary map was almost completely black, however some white pixels appeared before visible symptoms occurred. First tiny powdery mildew pustules became visible on RGB images 4 dai and were accurately detected in the binary map. Senescent leaf tissue was not included in the binary maps due to the consideration of natural senescence of detached, healthy leaves. Characteristic spectral patterns could be identified without human intervention. The binary maps illustrates disease specific pixels and allows the operator to control the automated results by comparing the binary maps with the corresponding RGB images. In complex biological systems and for resistance screenings, it will be an advantage to take spatial properties of spectral dynamics into account [25,40]. This unsupervised and data driven approach requires no a-priori knowledge such as pre-defined endmembers from a spectral libraries used in existing classification or machine learning approaches [20].

Extraction of spectral dynamics of healthy and *Bgh* inoculated barley leaves

The spectral dynamics of healthy and inoculated near-isogenic lines over time were used to elucidate differences among the near-isogenic lines illustrated as traces (Figure 7) according to Kersting *et al.* [41]. The spectral traces are an example of an interpretable summary of high dimensional hyperspectral imaging data, highlighting the phenotypic evolution and processes during the interaction of *Bgh* with susceptible and resistant barley genotypes. With this interactive approach, an adaption

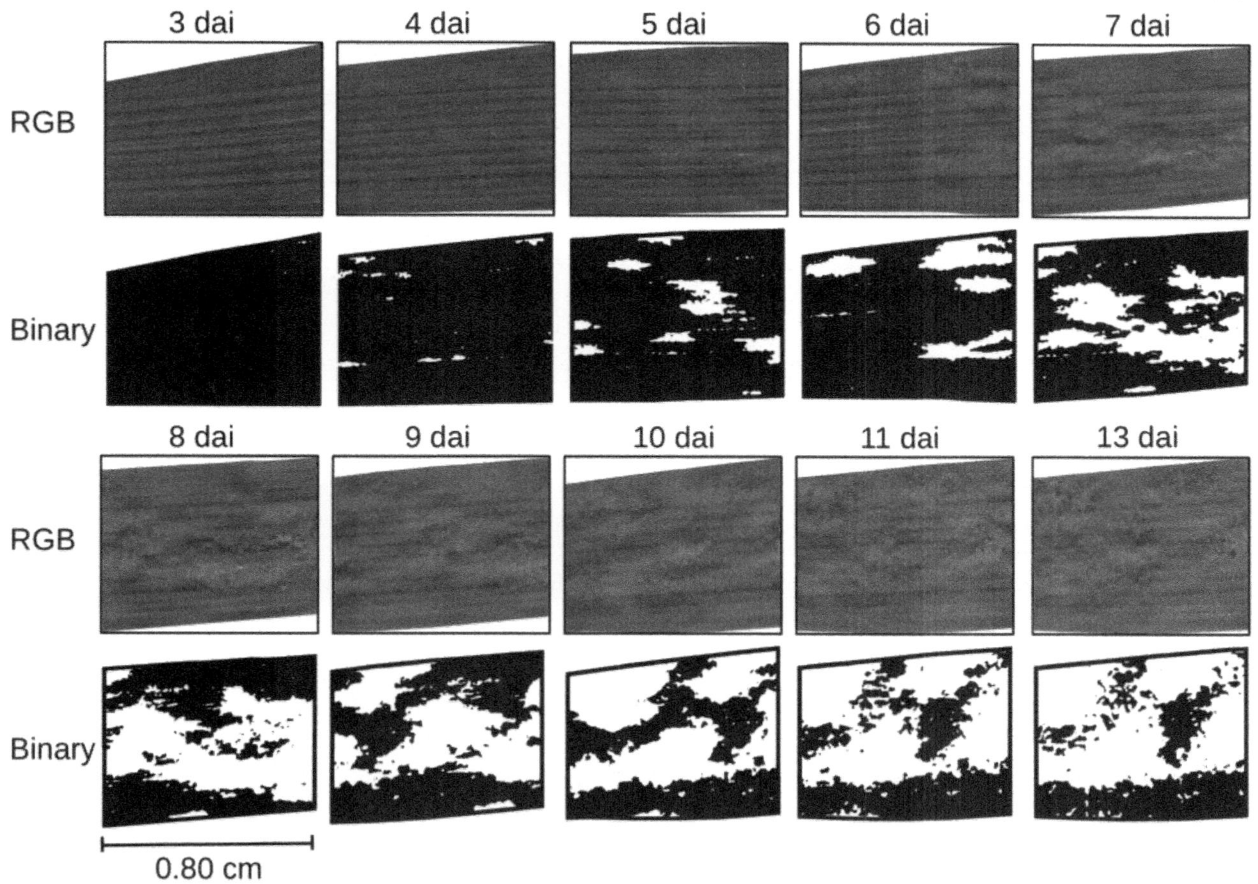

Figure 6 RGB images and binary infestation maps for automatic localization of barley tissue diseased by *Blumeria graminis* f.sp. *hordei* 3 to 13 dai. Black color indicates *Bgh* free tissue, white color highlights *Bgh* diseased barley tissue. Image section varies from day to day. No powdery mildew symptoms were visible 3 dai on RGB images. The binary map was almost completely black with small exceptions. First tiny powdery mildew pustules occurred in the RGB image 4 dai, which were detected and illustrated as white areas on the binary map.

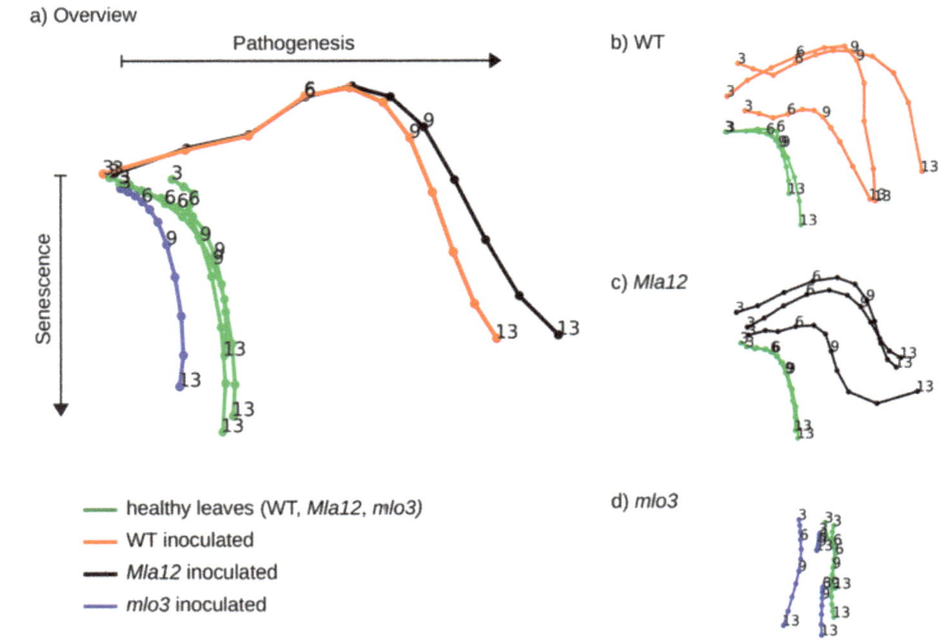

Figure 7 Leaf traces to uncover hyperspectral dynamics of healthy and *Blumeria graminis* f.sp. *hordei* inoculated near-isogenic lines of cv. Ingrid leaves (WT, *Mla12*, *mlo3*) over time. Healthy leaves had a similar spectral pattern and trend, indicated by a minor distance among their traces **(a)**. This is shown also for inoculated WT and *Mla12* leaves **(a)**. Between the individual leaves, hyperspectral dynamics are illustrated by their spectral traces **(b, c)**. The traces of inoculated, resistant *mlo3* leaves differed from healthy and inoculated, susceptible (WT and *Mla12*) leaves over the measuring period **(a)**. Inoculated *mlo3* leaves, showed differences **(d)**.

of known data mining methods to plant phenotyping tasks is demonstrated.

Figure 7a illustrates the mean traces of healthy and inoculated WT, *Mla12* and *mlo3* genotypes. Each line describes the spectral trace of three leaves 3 to 13 dai. The similarity among healthy leaves is indicated by the close trends of the traces. They developed in the same direction and had similar dynamics in time, indicated by short traces. Differences in the spectral traces of diseased WT and *Mla12* leaves were apparent. A variation in symptom development and time can be concluded from slightly different traces of the inoculated genotypes. During the experimental time, the mean trace of resistant *mlo3* overlapped with the traces for all healthy leaves and finally differed from them. The traces in Figure 7a allowed for an overall overview of the disease development over time. In order to reach this global view, we used the averaged mean signatures over all images of the leaves of the same type and treatment. To visualize specific details (Figure 7b-d), traces for each particular leave are provided for each genotype (WT, *Mla12* and *mlo3*) and treatment (healthy and inoculated) separately.

Differences in the spectral traces in direction and length, between the healthy and diseased genotypes are indicated (Figure 7b,c). For inoculated, susceptible genotypes WT and *Mla12*, differences within genotype and treatment resulted from different disease severities and development stages of powdery mildew over time. Likewise, inoculated *mlo3* leaves gave different spectral traces (Figure 7d).

Conclusions

Spectral information and phenotypes, assessed with a detailed microscopic HSI approach correspond to reflectance data from single plants or crop stands, whereby the HSI microscope enables a higher spatial resolution and richness of details. In microscopic HSI, the small pixel size (7.5 μm) eliminates the problem of pixels with mixed information from initial sites of pathogen infection. The proposed phenotyping set up is a promising new approach for the hyperspectral assessment and characterization of plant diseases and early processes during pathogenesis. In incompatible host-pathogen systems, specific resistance reactions may be identified from spectral reflectance data. The data analysis cascade based on data driven, automated machine learning methods, reduces the required human input in disease resistance screening systems (Figure 8) and in the evaluation of the performance of a set of plant genotypes under different environmental conditions.

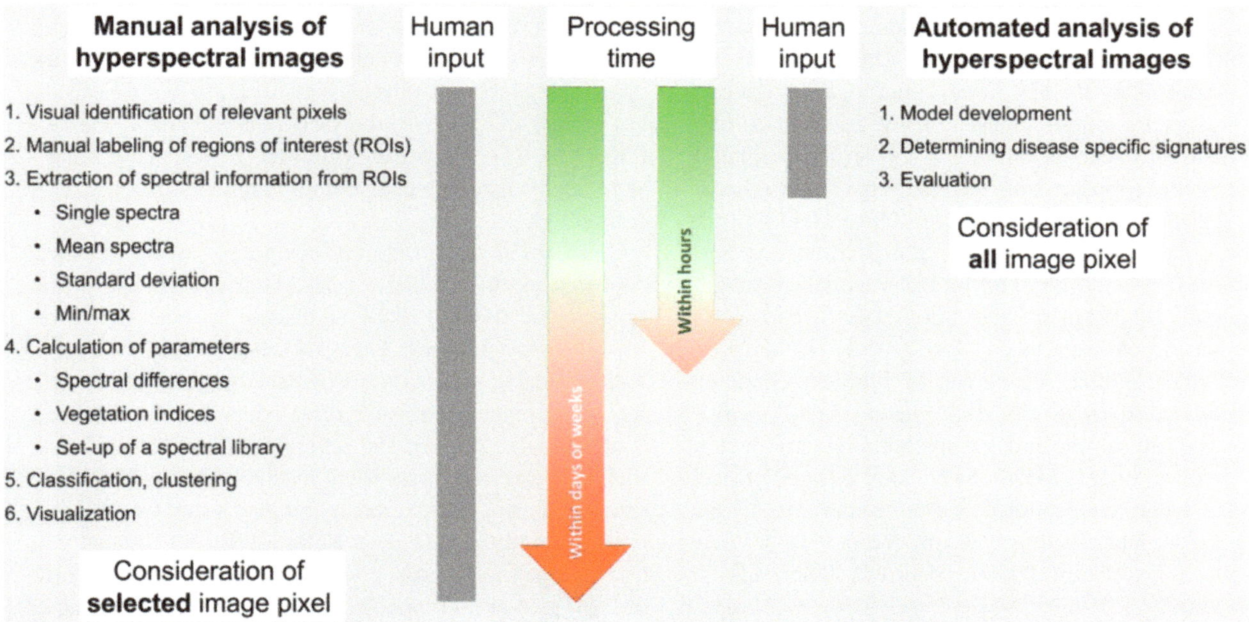

Figure 8 Workflow of the manual and automated hyperspectral image analysis, starting after preprocessing of hyperspectral images. Manual analysis requires high input of human experts and hence is time and cost intensive while still subjective. The automated analysis cascade improves the analysis of hyperspectral images due to the reduction of human input, the economization of time and the consideration of all image pixels.

Materials and methods

Plant cultivation and inoculation of *Blumeria graminis* f.sp. *hordei*

Near-isogenic barley (*Hordeum vulgare*) lines cv. Ingrid-wild type (WT), Ingrid -I10 containing resistant mildew locus a 12 (*Mla12*) [42] and Ingrid -M. C. 20 containing resistant mildew locus o 3 (*mlo3*) [43] were grown in a commercial substrate (Klasmann-Deilmann GmbH, Germany) in plastic pots (10×10×20 cm) in a greenhouse at 23/20°C (day/night), 60% relative humidity (RH) and a photoperiod of 16 h. One week after sowing, the primary leaves (with an approx. length of 10 cm) were detached and transferred on to aseptic phyto agar (Duchefa Biochemie, Haarlem, Netherlands) containing 0.034 mM benzimidazole.

For each genotype, three leaves were inoculated with fresh spores of *Blumeria graminis* f.sp. *hordei* isolate K1 (*Bgh*-K1), and four leaves were kept untreated as healthy control. *Bgh-K1* is virulent to cv. Ingrid WT and Ingrid I10 [42] and avirulent to Ingrid M. C. 20 [43]. Fresh conidia were obtained from heavily infected barley (cv. Leibniz). Twenty-four hours before plant inoculation, the conidia of *Bgh*-K1 infested plants were shaken off and discarded in order to assure homogenous and vital conidia for the inoculation. Conidia of a recently formed powdery mildew pustule (7 dai) are transferred to the prepared leaves on phyto agar using an aseptic inoculation loop. The agar plates were sealed with Parafilm M® (Bemis, Oshkosh, USA) and incubated in a controlled environment at 19°C, 1100 $m^{-2} \cdot cd$ illuminance and a photoperiod of 16 h per day.

Hyperspectral time series imaging and data preprocessing

Spectral reflectance was measured with an hyperspectral imaging line scanner (spectral camera PFD V10E, Specim, Oulu, Finland) mounted on a stereo microscope foreoptic (Z6 APO, Leica, Wetzlar, Germany) with a magnification up to 7.3x (Figure 1). The line scanning spectrograph has a spectral range from 400 to 1000 nm and a spectral resolution of up to 2.73 nm. The maximum image size of the 30 μm sensor slot results in 1300 pixels per line, with a sensor pixel size of 0.0074 mm. Depending on this measurement setup and the magnification, a maximum spatial resolution of 7.5 μm per pixel was obtained. For image recording the leaf samples were placed nadir on a XY-moving stage (H105/2/0 ProScan Upright Stage, Prior Scientific, Jena, Germany), controlled with a joystick and Oasis software (Oasis Controller, Objective Imaging Ltd., Cambridge, England). The samples were illuminated by two linear light emitters (Dual line Lightlines, Schott, Mainz, Germany) with a vertical orientation of 30° and a distance of 20 cm to the sample besides the foreoptic. As a light source a 150 watt halogen tungsten lamp connected to the line lights via a non-absorbing fiber was used (DCR® Light Source EKE, Polytec, Waldbronn, Germany). Hyperspectral measurements were performed in a dark room after 60 minutes pre-heating of the light source in order to realize constant and reproducible illumination conditions. The software SpectralCube (Spectral Imaging Ltd., Oulu, Finland) was used for controlling the HSI line scanner and for acquiring the hyperspectral images. Images on the leaf surface level were taken with spectral binning 1 and spatial binning 1. Frame rate and exposure time were adjusted to the object.

The reflection in the range from 400 to 1000 nm was measured daily 3 to 14 days after inoculation (dai) with a magnification of 7.3x. For image normalization and subsequent calculation of reflectance, four hyperspectral images per sample were taken. First, a white reference bar (SphereOptics GmbH., Uhldingen-Mühlhofen, Germany) was recorded, followed the dark current image. Subsequently, the leaf sample and a corresponding dark current image were recorded. Additionally, RGB images of each leaf were taken daily with a digital camera (EOS 6D, Canon, Tokio, Japan) and a 100 mm object lens (EF Lens Ultrasonic EF 100 mm 1:2.8 L Macro IS USM, Canon, Tokio, Japan).

Pre-processing of hyperspectral images

Because reflection data was noisy at the extremes, only data values between 420 to 830 nm were analysed. The reflectance of samples, was calculated by normalizing the images relative to the white reference reflection and the dark current measurements using the software ENVI 5.1 + IDL 8.3 (ITT Visual Information Solutions, Boulder, USA). Furthermore, the spectral signals are smoothed by employs the Savitzky-Golay filter [44]. Parameters for the smoothing process were 25 supporting points to the left and right, respectively, and a third degree polynomial. The pre-processed images were used for further analysis using ENVI 5.1 + IDL 8.3 and data mining methods.

Spectral signature extraction, analysis and characterization

In a first step, spectral signatures of pixels from healthy and diseased regions were extracted manually. Therefore >300 pixel were extracted daily from the same area by an ellipsoid region of interest from each non-inoculated leaf. When powdery mildew pustules became visible the symptomatic area was extracted as region of interest, thus the amount of pixels extracted increased depending on symptom development.

Data driven approach for fast analysis of hyperspectral dynamics

Following the method of Wahabzada *et al.* [36] a data driven approach was applied, allowing an automated analysis of hyperspectral data. Simplex Volume Maximization (SiVM) applied for fast and interpretable factorization [27], using an implementation based on the Python™ Matrix Factorization Module (PyMF) library (https://code.google.com/p/pymf/). SiVM represents the hyperspectral data in terms of only few extreme components determined across all images considered. Since the components are real extreme signatures, they are easily interpretable and uncover the variations existing in the data (Figure 4). The signatures within all hyperspectral images were then represented as combination of these extreme components.

Given the new representation opens door to statistical data mining on a massive scale. That is, the representation can be used to discover disease specific signatures within diseased leaves. This procedure avoids the risk of losing valuable information when selecting disease specific signatures manually at some diseased spots only. Following Wahabzada *et al.* [36] the differences for each particular signature was computed using likelihood ratios LLR(s) together with the distributions computed using simplex representation of the data. That is, the LLR(s) for a signature s of a diseased leaf at day d were computed in terms of the distributions of non-inoculated healthy leaf at day d and of a subsequent day r (we used $r = d + 2$) of the diseased leaf as the reference. For the latter days ($d \geq 8$ dai) we used the distribution determined from the image 10 dai for the diseased leaf as reference, as we assumed to observe the most diseased specific hyperspectral characteristics at this day.

For the binary maps of the location of disease spots a Gaussian filter was placed on the computed differences LLR(s) and set all positive values to 1 and 0 otherwise.

The leaf traces were computed by applying the Dirichlet aggregation regression on the representation determined by SiVM [45]. For interpolated mean signatures a 2-dimensional map was computed by the simplex traces approach [41]. This uncovers hyperspectral dynamics of diseased and non-inoculated (healthy) leaves of the different genotypes over time.

Abbreviations

Bgh: *Blumeria graminis* f.sp. *hordei*; DAI: Days after inoculation; HSI: Hyperspectral imaging; LLR: Likelihood ratio; ML: Mildew locus; NIR: Near infrared; RH: Relative humidity; ROI: Region of interest; RSD: Relative standard deviation; SiVM: Simplex volume maximization; SWIR: Short wave infrared; VIS: Visible range; WT: Wild type.

Competing interests

The authors declare that they have no competing interests.

Authors' contributions

MK, MW, HWD, KK, ECO, US, and AKM designed the study, interpreted the data and drafted the manuscript. ML, US, ECO and AKM constructed the hyperspectral microscope. MK and AKM developed the measuring procedure and the assessment of hyperspectral images of detached barley leaves on phyto agar. MK carried out hyperspectral measurements, manual and statistical analysis. MW and KK developed and applied data mining methods for the automated spectral data analysis. All authors read and approved the final manuscript.

Acknowledgements

This work could be carried out due to the financial support of the German Federal Ministry of Education and Research (BMBF) within the scope of the competitive grants program "Networks of excellence in agricultural and

nutrition research - CROP.SENSe.net" (Funding code: 0315529), junior research group "Hyperspectral phenotyping of resistance reactions of barley" and due to financial support of the Daimler and Benz foundation.
The authors are thankful to Ralph Panstruga, RWTH Aachen and to Frank Ordon, JKI Quedlinburg for providing seed material and the *Blumeria graminis* f.sp. *hordei* isolates, and to Uwe Rascher, Forschungszentrum Jülich and Stefan Thomas for helpful discussions.

Author details

[1]Institute for Crop Science and Resource Conservation (INRES) - Phytomedicine, University of Bonn, Meckenheimer Allee 166a, 53115 Bonn, Germany. [2]Department of Computer Science, TU Dortmund, Otto-Hahn-Str. 14, 44227 Dortmund, Germany.

References

1. Lyngkjaer M, Newton A, Atzema J, Baker S. The barley *mlo*-gene: an important powdery mildew resistance source. Agronomie. 2000;20(7):745–56.
2. Walters DR, Avrova A, Bingham IJ, Burnett FJ, Fountaine J, Havis ND, et al. Control of foliar disease in barley: towards an integrated approach. Eur J Plant Pathol. 2012;133:33–73.
3. Ma Z, Michailides TJ. Advances in understanding molecular mechanisms of fungicide resistance and molecular detection of resistant genotypes in phytopathogenic fungi. Crop Prot. 2005;24:853–63.
4. Jørgensen JH. Discovery, characterization and exploitation of *Mlo* powdery mildew resistance in barley. Euphytica. 1992;63:141–52.
5. Jørgensen JH, Mortensen K. Primary infection by *Erysiphe graminis* f.sp. *hordei* of barley mutants with resistance genes in the *mlo* locus. Phytopathology. 1977;67:678–85.
6. Hippe-Sanwald S, Hermanns M, Somerville SC. Ultrastructural comparison of incompatible and compatible interactions in the barley powdery mildew disease. Protoplasma. 1992;168:27–40.
7. Schulze-Lefert P, Vogel J. Closing the ranks to attack by powdery mildew. Trends Plant Sci. 2000;5(8):343–8.
8. Jørgensen JH. Genetics of powdery mildew resistance in barley. Crit Rev Plant Sci. 1994;13:97–119.
9. Hückelhoven R, Fodor J, Preis C, Kogel KH. Hypersensitive cell death and papilla formation in barley attacked by the powdery mildew fungus are associated with hydrogen peroxide but not with salicylic acid accumulation. Plant Physiol. 1999;119(4):1251–60.
10. Shimelis H, Laing M. Timelines in conventional crop improvement: pre-breeding and breeding procedures. Aust J Crop Sci. 2012;6(11):1542–9.
11. Fiorani F, Schurr U. Future scenarios for plant phenotyping. Annu Rev Plant Biol. 2013;64:267–91.
12. Furbank RT, Tester M. Phenomics – technologies to relieve the phenotyping bottleneck. Trends Plant Sci. 2011;16(12):635–44.
13. Cobb JN, DeClerck G, Greenberg A, Clark R, McCouch S. Next-generation phenotyping: requirements and strategies for enhancing our understanding of genotype-phenotype relationships and its relevance to crop improvement. Theor Appl Genet. 2013;126:867–87.
14. Paulus S, Behmann J, Mahlein AK, Plümer L, Kuhlmann H. Low-Cost 3D systems: suitable tools for plant phenotyping. AH S Sens. 2014;14(2):3001–18.
15. Chaerle L, Leinonen I, Jones HG, Van Der Straeten D. Monitoring and screening plant populations with combined thermal and chlorophyll fluorescence imaging. J Exp Bot. 2007;58(4):773–84.
16. Oerke EC, Steiner U, Dehne HW, Lindenthal M. Thermal imaging of cucumber leaves affected by downy mildew and environmental conditions. J Exp Biol. 2006;57(9):2121–32.
17. Ustin SL, Gamon JA. Remote sensing of plant functional types. New Phytol. 2010;186:795–816.
18. Rascher U, Blossfeld S, Fiorani F, Jahnke S, Jansen M, Kuhn AJ, et al. Non-invasive approaches for phenotyping of enhanced performance traits in bean. Funct Plant Biol. 2011;38:968–83.
19. Mahlein AK, Oerke EC, Steiner U, Dehne HW. Recent advances in sensing plant diseases for precision crop protection. Eur J Plant Pathol. 2012;133:197–209.
20. Behmann J, Mahlein AK, Rumpf T, Römer C, Plümer L. A review of advanced machine learning methods for the detection of biotic stress in precision crop protection. Precision Agric. 2014, doi:10.1007/s11119-014-9372-7.
21. Curran PJ. Remote sensing of foliar chemistry. Remote Sens Environ. 1989;30:271–8.
22. Delalieux S, Aardt J, Keulemans W, Schrevens E, Coppin P. Detection of biotic stress (*Venturia inaequalis*) in apple trees using hyperspectral data: Non-parametric statistical approaches and physiological implications. Eur J Agron. 2007;27:130–43.
23. Mahlein AK, Steiner U, Dehne HW, Oerke EC. Spectral signatures of sugar beet leaves for the detection and differentiation of diseases. Precision Agric. 2010;11:413–31.
24. West JS, Bravo C, Oberti R, Lemaire D, Moshou D, McCartney HA. The potential of optical canopy measurement for targeted control of field crop disease. Annu Rev Phytopathol. 2003;41:593–614.
25. Mahlein AK, Steiner U, Hillnhütter C, Dehne HW, Oerke EC. Hyperspectral imaging for small-scale analysis of symptoms caused by different sugar beet disease. Plant Methods. 2012;8(1):3.
26. Rumpf T, Mahlein AK, Steiner U, Oerke EC, Dehne HW, Plümer L. Early detection and classification of plant disease with Support Vector Machines based on hyperspectral reflectance. Comput Electron Agr. 2010;74:91–9.
27. Thurau C, Kersting K, Wahabzada M, Bauckhage C. Descriptive matrix factorization for sustainability: Adopting the principle of opposites. Data Min Knowl Disc. 2012;24:325–54.
28. Bock CH, Poole GH, Parker PE, Gottwald TR. Plant disease severity estimated visually, by digital photography and image analysis, and by hyperspectral imaging. Crit Rev Plant Sci. 2010;29:59–107.
29. Oerke EC, Mahlein AK, Steiner U. Proximal sensing of plant diseases. In: Gullino ML, Bonants PJM, editors. Detection and Diagnostic of Plant Pathogens. Plant Pathology in the 21st Century 2014. Dordrecht: Springer Science and Business Media; 2014. p. 55–68.
30. Leavesley SJ, Annamdevula N, Boni J, Stocker S, Grant K, Troyanovsky B, et al. Hyperspectral imaging microscopy for identification and quantitative analysis of fluorescently-labeled cells in highly autofluorescent tissue. J Biophotonics. 2012;5:67–84.
31. Schultz RA, Nielsen T, Zavaleta JR, Runch R, Wyatt R, Garner HR. Hyperspectral imaging: A novel approach for microscopic analysis. Cytometry. 2001;43:239–47.
32. Carter GA, Knapp AK. Leaf optical properties in higher plants: Linking spectral characteristics to stress and chlorophyll concentration. Am J Bot. 2001;88:677–84.
33. Gamon JA, Surfus JS. Assessing leaf pigment content and activity with a reflectometer. New Phytol. 1999;143:105–17.
34. Gamon JA, Serrano L, Surfus JS. The photochemical reflectance index: An optical indicator of photosynthetic radiation use efficiency across species, functional types, and nutrient levels. Oecologia. 1997;112:492–501.
35. Jacquemoud S, Ustin SL. Leaf optical properties: A state of the art. In: Proceedings 8th International Symposium Physical Measurements & Signatures in Remote Sensing (CNES). France: Aussois; 2001. p. 223–32.
36. Wahabzada M, Mahlein AK, Bauckhage C, Steiner U, Oerke EC, Kersting K. Metro Maps of plant disease dynamics - automated mining of differences using hyperspectral images. PLOS One. 2015;10(1):1–20.
37. Zhang JC, Pu RL, Wang JH, Huang WJ, Yuan L, Luo JH. Detecting powdery mildew of winter wheat using leaf level hyperspectral measurements. Comput Electron Agr. 2012;85:13–23.
38. Swarbrick PJ, Schulze-Lefert P, Scholes JD. Metabolic consequences of susceptibility and resistance (race-specific and broad-spectrum) in barley leaves challenged with powdery mildew. Plant Cell Environ. 2006;29:1061–76.
39. Bell AA. Biochemical mechanisms of disease resistance. Ann Rev Plant Physiol. 1981;32:21–81.
40. Fiorani F, Rascher U, Jahnke S, Schurr U. Imaging plants dynamics in heterogenic environments. Curr Opin Biotechnol. 2012;23:227–35.
41. Kersting K, Wahabzada M, Roemer C, Thurau C, Ballvora A, Rascher U, Leon J, Bauckhage C, Pluemer L: Simplex distributions for embedding data matrices over time. Proceedings of the 12th SIAM International Conference on Data Mining 2012, 295–306.
42. Shen QH, Zhou FS, Bieri S, Haizel T, Shirasu K, Schulze-Lefert P. Recognition specifity and RAR1/SGT1 dependence in barley *Mla* disease resistance genes to the powdery mildew fungus. Plant Cell. 2003;15:732–44.
43. Hinze K, Thompson RD, Ritter E, Salamini F, Schulze-Lefert P. Restriction fragment length polymorphism-mediated targeting of the *ml-o* resistance locus barley (*Hordeum vulgare*). Proc Natl Acad Sci. 1991;88:3691–5.
44. Savitzky A, Golay JME. Smoothing and differentiation of data by simplified least squares procedures. Anal Chem. 1964;36:1627–39.

45. Kersting K, Xu Z, Wahabzada M, Bauckhage C, Thurau C, Roemer C, Ballvora A, Rascher U, Leon J, Pluemer L: Pre-symptomatic prediction of plant drought stress using Dirichlet-aggregation regression on hyperspectral images. Proceedings of the Twenty-Sixth AAAI Conference on Artificial Intelligence (AAAI). Special Track on Computational Sustainability and AI 2012, 302–308.

A software tool for the input and management of phenotypic data using personal digital assistants and other mobile devices

Karin Köhl* and Jürgen Gremmels

Abstract

Background: Plant breeding and genetics demand fast, exact and reproducible phenotyping. Efficient statistical evaluation of phenotyping data requires standardised data storage ensuring long-term data availability while maintaining intellectual property rights. This is state of the art at phenomics centres, which, however, are unavailable for most scientists. For them we developed a simple and cost-efficient system, the Phenotyper, which employs mobile devices or personal digital assistants (PDA) for on-site data entry and open-source software for data management.

Results: A graphical user interface (GUI) on a PDA replaces paper-based form sheet and data entry on a desktop. The user can define his phenotyping schemes in a web tool without in-depth knowledge of the system and thus adjust it more easily to new research aspects than in a classical laboratory information management system (LIMS). In the Phenotyper, schemes are built from controlled vocabulary gained from published ontologies. Vocabulary and schemes are stored in a database that also manages the user access. From the web page, schemes are downloaded as extended markup language (XML) files for the transfer to the PDA and the exchange between users. On the PDA, the GUI displays the schemes and stores data in comma separated value format and XML format. After manual quality control, data are uploaded via a web page to an independently hosted results database, in which data are stored in an entity-attribute-value structure to provide maximum flexibility. Datasets are linked to the original and curated data files stored on a file server. The ownership stamp, project affiliation and date stamp of a dataset are used to regulate data access, which is restricted to data belonging to the user or to his projects and data, for which the embargo period has ended. By export of standardised ASCII reports to long-term data storage facility, long-term accessibility allows searching, citing and use of raw data beyond the lifetime of the database. The Phenotyper is available to the scientific community for use and further development.

Conclusions: The Phenotyper provides a well-structured, but flexible data acquisition and management structure for mobile on-site measurements for efficient evaluation and shared use of data.

Keywords: Phenotyping, Field research, Relational database, Data management, Controlled vocabulary

Background

Evaluation of plant genotypes in breeding programs or genomics projects require precise and reproducible phenotyping [1]. The coining of the term 'phenomics' indicates that morphological and physiological data are as important as biochemical and molecular information [2]. The lack of phenotype information rather than the lack of genetic information causes a bottleneck in generating the genotype-phenotype map that links genes to relevant features like fitness or survival [3,4]. With respect to the amount of publicly available data, phenomics lags behind genomics [4]. Large genomics projects have been pioneers in making data rapidly available to scientific communities on web pages. Large research centres ([5-10]) phenotype plants highly reproducible under standardised conditions and store information in databases. These systems are, however, expensive, as infrastructure for data collection and storage require high financial and personnel investment [11]. These phenomics facilities are thus unavailable for most of the scientific community, which

* Correspondence: koehl@mpimp-golm.mpg.de
Max-Planck-Institute of Molecular Plant Physiology, 14424 Potsdam-Golm, Germany

nevertheless produces an increasing amount of valuable phenotypic data. The limiting factors for the optimal use of these data within models that link genetic, environmental and phenotypic data are data management and analysis [12,13]. Most scientists combine phenotyping with conventional methods of data recording and management. Especially in field-based research, data are frequently recorded in individual formats on paper, often without a safety copy. During manual entry into the computer for further evaluation errors are introduced into the dataset. In plant sciences, data are generally entered, stored and evaluated in spreadsheet programs like MS Excel (Microsoft, Redmond, US). Data storage in Excel is considered risky as data are easily compromised by unintentional use of automatic formatting function [14]. Furthermore, raw data are often mixed up with evaluation procedures and results, including figures. Column headings that indicate the content of a column are generally unstandardised leading to different names for the same parameter or the same name for parameter that differ e.g. in the unit. Furthermore, data are often spread to several files, the connection between datasets is ambiguous and the file storage structure determined by the personal preferences of the individual researcher. Access is often limited to the individual researcher. Frequently, data are deleted after the researcher left the institution and even if the data are archived, insufficient documentation of individual file names and column heads make the information ambiguous and impede searches. All these facts render the compilation of a complete set of raw data as basis of a comprehensive evaluation or to provide original data when publishing a manuscript very time-consuming. In phenotyping, this is especially deplorable. Many phenotypic features are semi-quantitative (colour, shape, developmental stage) [15,16] or highly variable (growth rate) und thus require large datasets to gain sufficient statistical power. Therefore, the combination of phenotypic data from many experiments and several users is a prerequisite for a successful project. Data from various sources should therefore be available rapidly for automatic evaluation. This can be achieved by using a central data repository with a highly standardised data storage structure based on publicly available controlled vocabulary. Several initiatives offer central data storage and sharing structures (overview see [11]). The first phenotype databases concentrated on data generated by standardised methodology that results in well-defined data structures like data from transcriptomics or ionomics [17-19]. In the American cyberinfrastructure project [20], the PODD Data Repository Project in Australia [21,22] and the Phenoscape project [23], information of highly variable structure are stored as free text descriptions enhanced by a semantic search algorithms that link texts with ontological terms, thus providing the basis for efficient search functions. In addition to raw data, data analysis procedures need to be documented and made available to convert data efficiently into results [4] and to provide a reproducible connection between raw data and published results [24]. Within plant phenotyping, image analysis is furthest advanced with respect to automated data storage and analysis [8,25,26].

To facilitate the generation of large, centrally stored phenotyping datasets from manually or semi-automatically collected data, we generated a simple and inexpensive tool, the Phenotyper that uses mobile devices for on-site data entry. The system allows tailoring a graphical user interface (GUI) to the individual project in a web-based building-tool without programming skills or database knowledge. Phenotyping results are transferred as XML files via an internet page into a central database, from where they can be downloaded in user-defined, access-controlled mode by web tools.

Results and discussion

Phenotyping schemes based on controlled vocabulary from ontologies

The core feature of the phenotyping workflow is the combination of controlled vocabulary for entities, e.g. plant organs, and attributes, e.g. features that are to be scored, in user-defined phenotyping schemes. These are exported to mobile devices and displayed in a GUI (see Figure 1). By using controlled vocabulary, it achieves a high degree of standardisation during data recording while maintaining flexibility. Controlled vocabulary is used directly or indirectly by several data management systems for phenotypic data (e.g. the Triticeae Toolbox [27], Phenoscape [23], DbNP [28]) as it makes search and automatic analysis by computational algorithms more efficient [11]. The phenotyping schemes are composed on a web page (see Figure 2A), that retrieves the controlled vocabulary from the scheme database (S-DB) (Entity relationship (ER) diagram see Additional file 1: Figure S1). The controlled vocabulary was derived from the Plant Ontology (PO) Consortium [29] for entities. For parameters, it was partially user-defined, partially gained from the plant trait ontology (TO). By storing PO or TO identifiers, the connection towards the original source was maintained. The definition of project-specific vocabulary subsets improves the handling of the Phenotyper. The recording of schemes in a user-specific mode avoids accidental modification of the scheme by other users. Schemes are exported in XML format to be displayed in the GUI of mobile devices (see below). Cloning of schemes on the web page supports efficient scheme modification. Exchange of schemes between users is performed by importing scheme XML files into the S-DB. A language switch function, which changes the language on the web page, for the controlled vocabulary

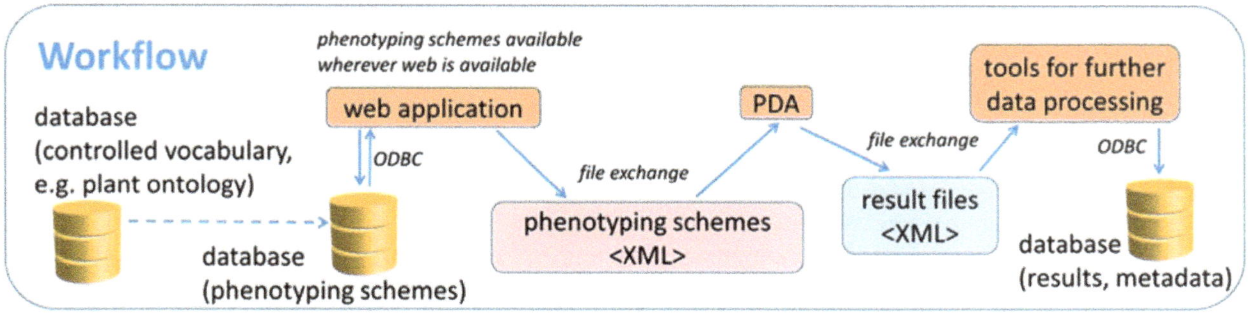

Figure 1 Workflow of the phenotyping system. Phenotyping schemes are built in web applications from controlled vocabulary, which is stored in the S-DB (left), and submitted to the mobile device (personal digital assistant PDA). After phenotyping, results are transferred by web tools to the R-DB (right).

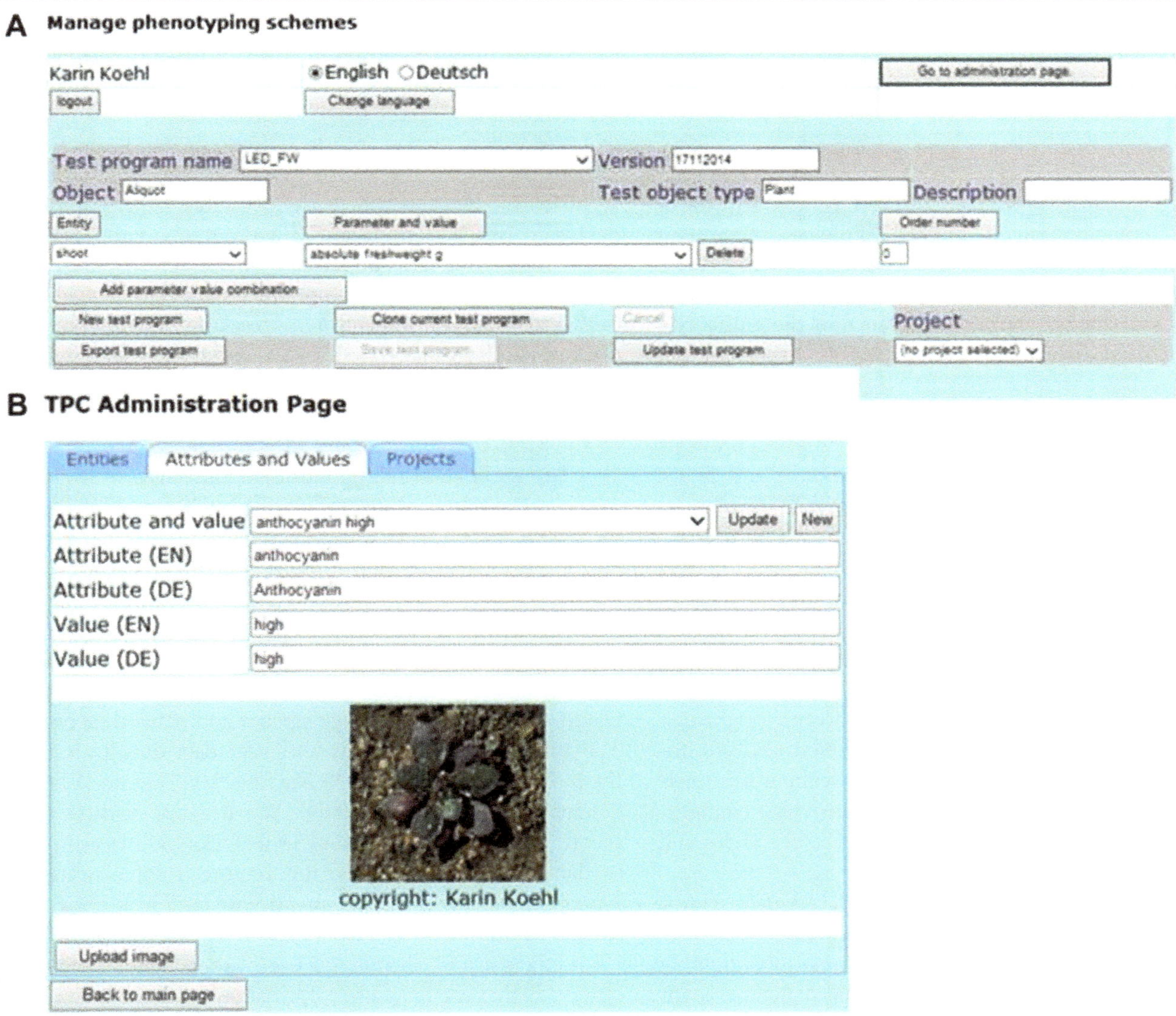

Figure 2 Web tool phenotyping scheme composer and administration page. A. The web tool combines controlled vocabulary of entity and attribute-value information in user-defined schemes that are stored in the S-DB and exported to the mobile device. **B.** Controlled vocabulary is managed on the Administration pages, which also provide an upload function for illustrating image files.

and on the mobile device, facilitates the use by technical personnel in multinational projects.

The user management of the S-DB also handles the access management to the web page. Users with administrator status can access additional pages, on which they can create new user accounts and projects, affiliate users to projects, define project-specific vocabulary subsets, introduce additional or modify existing vocabulary and upload picture files to illustrate categorical parameters (see Figure 2B). These tasks can be thus performed by project leaders without in-depth database knowledge independent of the database administrator. The ease with which the Phenotyper can be adjusted to the needs of an individual project and the open-source software model are advantages over commercial LIMS solutions. In LIMS, modification of workflows requires in-depth knowledge of the data structure and the workflow tools and is thus restricted to trained personnel. Spontaneous changes of the scheme, to accommodate an observation in the experiment in the data collection, is thus more easily and rapidly done with the Phenotyper.

Mobile devices

As any standardised data acquisition system requires a higher 'activation energy' during implementation than the classical paper-based method, the attractiveness of the mobile device to the potential user is *a* if not *the* critical issue for the overall acceptance of the entire system. At the start of the project in 2011, we therefore recruited a beta-user group that was sociologically (age, computer-affinity, intellectual background) representative for the envisaged user population and had previous experience with phenotyping in greenhouse and fields. Brainstorming supplied the following criteria for high attractiveness of the device: large screens with good readability in sunlight, robust barcode-scanner, large, glove-compatible keyboard, and long battery lifetimes (≥8 h). Additionally, low weight, easy handling of the device and a sun-light compatible barcode-scanner were asked for. Some of the criteria are obviously hard to combine, like low weight and handiness with large screens, bright displays and low weight with long battery lifetimes. We implemented the Phenotyper for barcode-scanner terminals (details see Material and methods) and provided configuration tools to adjust the GUI to different screen width and keyboard types of tablets and smartphones.

Graphical user interface on the mobile device

On-site, a GUI displays the user-defined phenotyping schemes for data entry into the personal digital assistant (PDA). A check function alerts if no scheme file is found or if a scheme file is invalid. The phenotyping scheme is selected on the start page of the GUI (Figure 3A). On this page, the identifier (id) of the object, on which the measurements are performed (e.g. a plant, a plot, a sample), is entered. The entities defined in the phenotyping scheme are displayed as subsequent pages, on which the parameters that are to be measured for each entity are listed (Figure 3B). Data can be entered (numeric variables, Figure 3B) or chosen (categorical variables, Figure 3B) from the listed attributes. Categorical variables can be illustrated by images (Figure 3C). A multiplication function is provided for scoring a single feature on a number of objects sharing one id, e.g. several plants in a plot. The entry of a multiplication factor on the first page defines the number of repeated measurements per id and permits efficient data entry for multiple measurements.

The Phenotyper generates unique names for the results files based on the device name, date and time information. With each measurement, date and time are recorded. To ensure the validity of these time stamps, the user is prompted to check the date and time settings and re-adjust it with a configuration tool. With the tool, the user can also define the storage location for phenotyping schemes and result data, assign functions (e.g. decimal separator, entry) to specific keys, and enable/disable the use of the end signals of barcode-scanners as entry signal. The tool thus facilitates adjusting the device to personal preferences without manual editing of the registry. When the factory default settings are restored by a hard reset, the tool automatically reactivates these personal settings, thus increasing the Phenotyper's reliability under harsh field conditions. Field tests furthermore revealed the need to check and edit data in the GUI. Individual entries can be selected and recalled to the GUI by choosing the id that acts as a primary key. The entry can then be modified and saved.

The optimization of the system for small screens makes the Phenotyper superior to LIMS systems, when data are to be recorded while moving through a greenhouse, field or natural eco-system. LIMS in general and the Phenopsis system [9,10], are optimized for automatic data import from machines (e.g. analytical devices, sensors) or for manual data entry on the LIMS user interface or on a web page, displayed on machines with large screens and network access. The Phenotyper can, however, transfer data directly from the mobile device to the LIMS Nautilus 9.0 (Thermo Fisher Scientific, Dreieich, Germany). We use this feature to record locations for plants and seed stocks [30]. However, as data storage *per se* does not require a sophisticated LIMS, we decided to store phenotypic data in an open-source database.

An efficient and meaningful use of the Phenotyper requires ids that are linked to information on the object, on which the phenotyping is performed. The optimal solution for the object identification is a unique id from a plant description table used by all users of the database, in which the phenotyping results are stored (R-DB) (see below). For

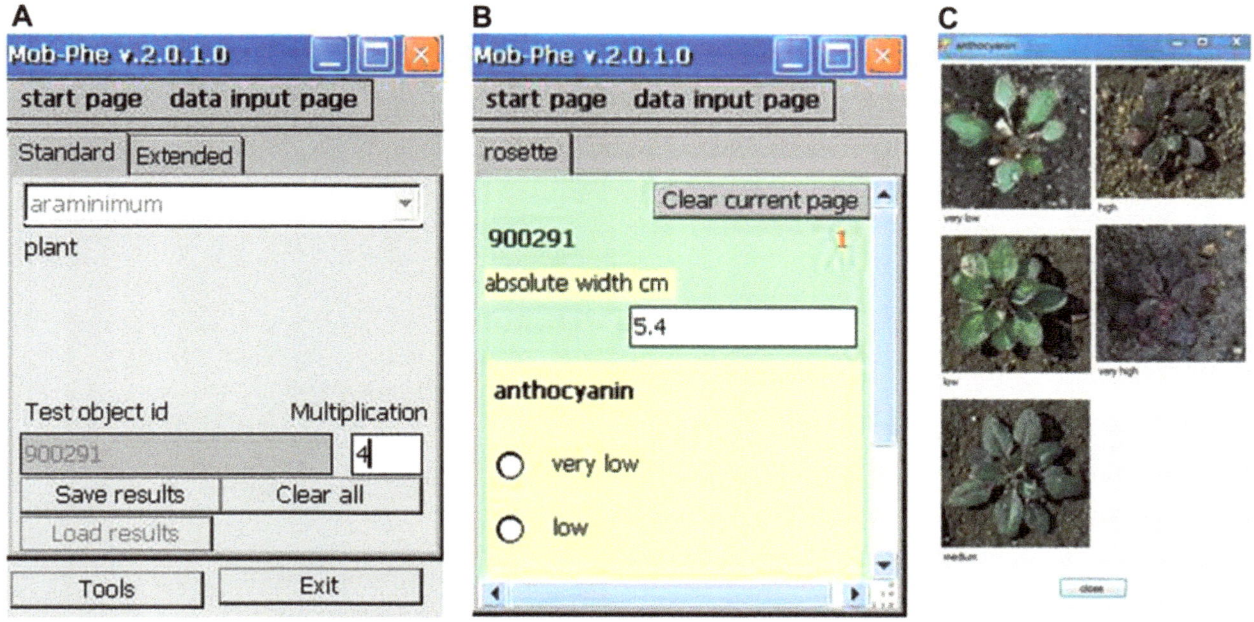

Figure 3 Graphical user interface of the mobile device. Graphical user interface on the mobile device displays the selected phenotyping scheme (**A**, above), which had been predefined by the user and stored on the PDA. After scanning or entering the id of the measured object (plant, plot) values can be entered or chosen **(B)** on the input page. When a number > 1 is entered into the multiplication box **(A)**, the input page is shown repeatedly, thus permitting to enter several measurements for each id. The count number is added to the identifier string to generate a unique identifier (e.g. id-1, id-2). **(C)** Images can be displayed for categorical values.

the TROST project (BMEL/FNR 22011208), that served as a test case, these ids were imported from the LIMS-based plant cultivation database of the MPI-MP [30] into the R-DB [31]. Alternative solutions for the documentation of plant cultivation experiments have been suggested [32-34]. When all plants of an institution or project are identified by a unique id, these soon have more than four digits making their entry tedious and error-prone, unless they are converted to barcodes. An alternative solution is the combination of experiment ids and unique ids within an experiment.

The Phenotyper can also be used stand-alone without the link to a plant database containing meta-information. In this case, information about e.g. the genotype of the plant or the seed lot, the treatment, the cultivation location and the planting date could be recorded in a start scheme at the beginning of each experiment. The use of standard start schemes ensures that all relevant information is documented. Treatments and genotypes can be defined as attribute-value combination (e.g. genotype – Col-0, mutant – Wt, mutant – cbb). The efficiency of recording can be enhanced by grouping plants in experimental units that inherit information from the handling unit to its members [30]. In addition to enforcing metadata entry in standardised start schemes by institutional operation procedures, users can be motivated to enter metadata completely and correctly by the increased efficiency of data evaluation. Phenotyper allows downloading all data required for efficient data analysis in one dataset of highly standardised structure (see section data access).

Data transfer and quality control

On site, the GUI on the mobile device operates without access to WLAN or mobile phone networks, making the Phenotyper usable in areas devoid of both (Faraday cage, remote areas). Data are stored immediately in the permanent memory of the mobile device to secure against data loss due to terminal crash, hard reset or power loss. The result data are stored in comma separated values format (csv) and XML format. In the csv-file (see Figure 4C), entries recorded for one identifier are recorded in one row. The cells' contents are identified by the headers in the first rows. Both formats can be read easily by spreadsheet programs like MS Excel. The csv format is suitable for import into statistics packages like SAS (SAS Institute Inc., Cary, NC, USA) or R [35]. When csv files from schemes containing different entity-attribute-value combinations are combined, the resulting file will contain empty cells for those variables that are not represented in all schemes.

If data from the csv-files are to be uploaded into the entity-attribute-value structure of a database, the file needs to be reformatted. Mapping-parsing scripts need to take into account that the order in which the variables are presented varies and that the array may contain empty cells. In XML files, however, every entry is tagged by the name of the variable. The upload procedure is

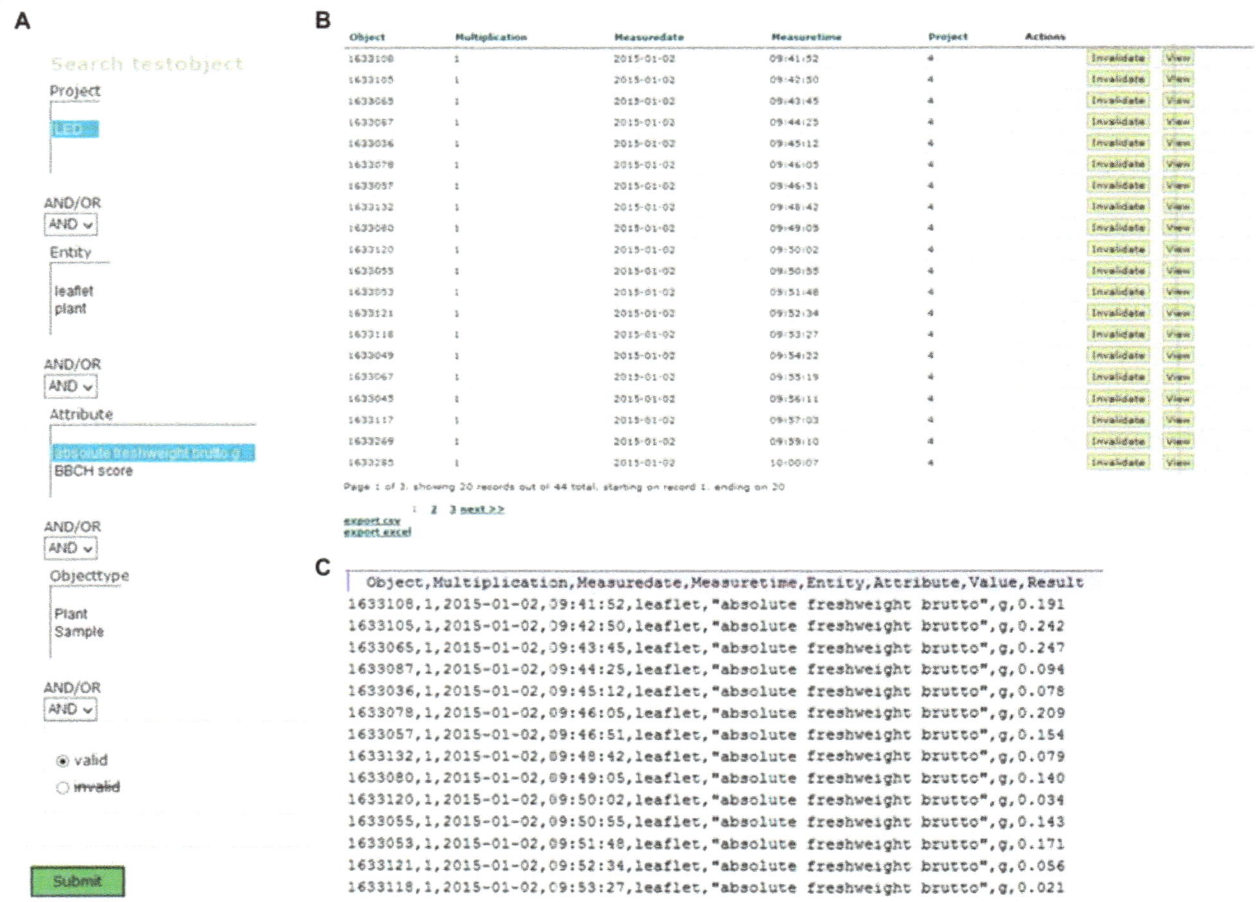

Figure 4 Web page for data access. A The screenshot of the web page displays query parameters as a basis for a user-defined export from the database. **B** The response page displays the query results, which can be exported in csv **(C)** or xls format.

thus completely independent of the order, in which the variables are presented in the file, as long as the structure hierarchy is maintained. The number and order of variables per identifier can also vary within one file without causing empty cells (see Additional file 2). Furthermore, there are powerful validation tools to check XML files. Files in XML format can thus be easily used by automatic data processing tools, e.g. for the transfer of data into the database of a LIMS. The LIMS used at the MPI-MP, Nautilus 9.0 provides procedures for data import from csv and from xml files. We found the xml-import engine easier and safer to handle when data from one import file were to be written to several different LIMS tables.

For secure data storage, we developed a results database (see Data storage) combined with a file server for the storage of original (raw) and quality controlled results files. The quality control (QC) of the data for plausibility and typing errors is done after the retrieval of data from the mobile device and before upload. Immediate QC is a most important step to ensure high data quality, as it allows re-evaluation of out of range results. However, modification of the original dataset may lead to accusation of fraudulent data modification. The original file from the scanner should therefore be stored together with the file resulting from QC to safeguard the user and establish good data management practise. To our knowledge, the Phenotyper is the first system that provides this feature.

For data transfer from the mobile device or local computer disk to the database server, an upload function is provided on a web page (Figure 5). The web page performs a quality control on the file structure, copies the files to the file storage system, uploads the data from the QC file to the results database and links the imported datasets to the files. The name of the user who logged onto the transfer web page is added to the uploaded datasets as an ownership-stamp together with the upload time stamp. This information forms the basis for the maintenance of intellectual property rights and the access policy (see section Data access). To facilitate data sharing in projects, the user can choose from the list of projects, to which he is affiliated, which project the uploaded dataset belongs to. The default setting is 'no project' for data that are not meant to be shared.

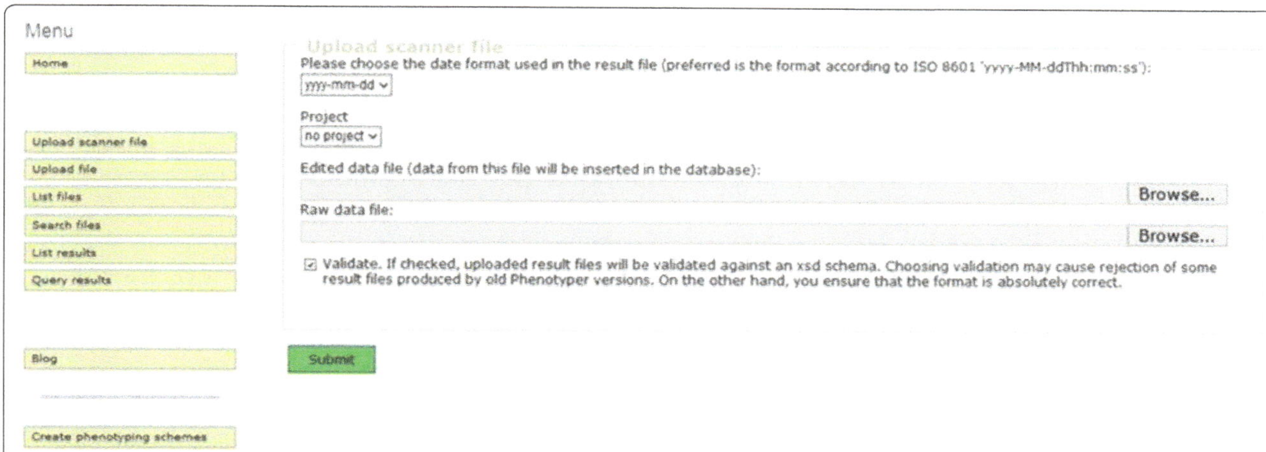

Figure 5 Screenshot of the data upload and validation function. The data upload function transfers the data from the quality-controlled, edited XML file to the database. During upload the data can be allocated to a project and validated against existing XML schema definition (XSD). Raw data file and edited (QC) data file are transferred to the file server. The buttons on the left side of the image link to additional functions of the web page.

Data storage

Result data are stored in a results database (R-DB) modified from a prototype developed for the TROST project [31]. In the relational R-DB, data are stored in an entity-attribute-value (EAV) data model (ER diagram see Additional file 1: Figure S2). This data storage structure is especially suitable for so-called sparse and volatile data, which means that the number of parameters measured on an object is small compared to the number of parameters that could be measured and it tends to change during a project [31,36]. As an example, at the beginning of a project, 20 parameters may be assessed on each plant. After the first evaluations, measurements are continued only on the five most predictive parameters plus a newly discovered parameter. In classical data storage structure, this addition would require a change in the database structure and result in many empty cells. In the EAV system, the identifier will be combined with attribute-value combinations for 20 parameters at the beginning of the project and change to six identifier-parameter combinations in a later phase. The query, with which data for a parameter of interest are extracted, remains the same as the automatic data evaluation procedures throughout the project.

Several other projects [24,37,38] and the commercial system Lemnatec (Aachen, Germany) use relational (SQL) databases for raw data storage, whereas the image storage and processing system IAP uses the no-SQL database MongoDB [8,25].

The user data in the R-DB are the basis for the access regulation to the upload web page, the ownership stamp, the project information and the data access during download. The link between projects and user names is modelled in a m - n relationship, thus each user can belong to several projects and each project can be shared by several users.

To permit data interpretation independent of an access to the S-DB, controlled vocabulary needs to be imported into the R-DB. In our implementation, controlled vocabulary was mirrored from the S-DB to the R-DB. An alternative solution is to import the controlled vocabulary together with its primary key from the results files to the R-DB. In our implementation, the R-DB should only be connected to one S-DB to avoid ambiguity. One S-DB can, however, serve many R-DBs. The R-DB could either be hosted locally by an individual researcher or research group or centrally by an institute or project coordination site.

To facilitate the process of setting up web server and R-DB locally, we develop a setup routine for both (see Additional files 3, 4 and 5). The person or institution hosting the Phenotyper would, however, need to have some experience with web server and database administration; thus, some degree of centralization in an institution or larger project is desirable. We do not, however, envisage a central R-DB maintained for many research projects for a decade or longer. As many phenotyping projects will be financed by time-limited grants, interest in and maintenance off the central R-DB may wean at some stage. When checking the availability of central data storage solutions, we found that some of them were no longer accessible or maintained less than ten years after publication [22,39]. Institutions that concentrate on scientific discovery usually have no resources to maintain databases for decades when these are no longer actively used. Thus, data ought to be transferred to a central long-term storage facility that is dedicated to provide quality-controlled data storage for many decades. Examples for centrally maintained long-term storage facilities are the EWIG project, the RADAR and the DRYAD facility [40-42]. These archives provide a document identifier and meta-information for each dataset and ensure controlled access to the data, but not necessarily a fast data transfer structure. The DataCite metadata store provides metadata and document identifiers for dataset that are stored in different archives and thus allows efficient search for datasets and their citation [43]. The best data storage format is an ASCIII or XML format as these formats can be read independently of specific software products. Our R-DB provides a function to generate standardised ASCII report, in which each dataset contains the complete information in a structured format. These data can later on be easily imported and evaluated by other scientists, then increasing the awareness of the work and of the person, who deposited the data [24].

A trigger for the transfer of raw phenotyping data to the long-term storage facility could be the publication of results based on these data. This procedure is already state of the art in research areas like transcriptomics [44] or in large ecological studies [42]. Universities could make the deposition of all raw data, on which a PhD thesis is based, an obligatory part of the defence. Furthermore, data should be committed to long-term storage at the end of a project to maintain long-term data access as demanded by an increasing number of funding agencies (e.g. Deutsche Forschungsgemeinschaft [45], EU, BBSRC [46]).

Data access

For data evaluation or submission to long-term storage facilities, data need to be downloaded from the database. To provide easy access to a potentially large number of datasets, we generated a web page with search and export functions (Figure 4A). The search function provides filters for entries e.g. for entities and attributes to limit the export to data required for an evaluation. The data are displayed on the web page (Figure 4B) and exported in csv or xls format (Figure 4C). Data access is regulated based on the ownership stamp, project information, the upload time stamp and the embargo period defined for the project or institution. By default, data are only accessible to the user defined in the owner field. Data assigned to a project are accessible to all users affiliated to the project. The timestamp opens data access to all users after the end of predefined embargo period.

Whether the ownership stamp is sufficient to guard intellectual property rights is debatable. The ownership stamp links to a table containing first and family name of the user, which may be sufficient to identify a person within an institution when combined with information recorded in the Human Resources Department of the respective institution. A potential alternative could be the inclusion of the scientist's Scopus identifier into the user management.

Data access to centrally stored result data is a sensitive issue. In contrast to the situation in material sciences [24], many scientists in life sciences are very reluctant to transfer their raw data to a central depository for fear of uncontrolled data access and use by other scientists. This lack of enthusiasm is enhanced by the knowledge that raw data are not covered per se by the regulations on intellectual property rights. Thus, in addition to providing a system that facilitates structured recording and storage of data and to provide long-term storage facilities, some issues linked to open data access remain to be solved by discussion in the scientific community. One option is to agree on making it obligatory to provide raw data on publication, when submitting a thesis and at the end of a project. Furthermore, raw datasets with metadata ought to be made searchable and citable as suggested by the DataCite initiative, thus conferring value to the dataset *per se* and rewarding data sharing.

Conclusion

Altogether, our phenotyping tool provides a well-structured, but flexible data acquisition and management structure for on-site measurements with mobile devices like smartphones or barcode-scanner terminals. As the user chooses the type of controlled vocabulary, the tool can be used for any type of manual data entry by any discipline, phenotyping as well as documentation of field sampling to inventories in a (museum) collection. We have used the Phenotyper for all these purposes successfully. Likewise, measurement data can be stored alongside genotype or treatment data including spatial or experimental design information. The predefined structure and the use of controlled vocabulary facilitate efficient automatic data evaluation and shared data use. Mechanisms for the protection of intellectual property rights and making datasets citable per se increase the scientist's trust when submitting his data to a central depository. Upon publication, well-structured files with controlled vocabulary can be automatically generated from the database, which saves much time and avoids errors compared to manually compiling a data file from many sources (e.g. spreadsheet files). In contrast to self-defined parameter, no definition needs to be written for controlled vocabulary. The submission of raw data to long-term storage facilities is thus made very simple and will guarantee data accessibility beyond the live time of the results database of an individual project. Thus, the tool may help to improve data management within phenotyping projects and, by automatic generation of structured csv export files, long-term data accessibility.

Material and methods

Software and database

The web pages for combining controlled vocabulary to phenotyping schemes and for the administration of the S-DB was implemented with Framework Struts 2 (Apache Software Foundation) within the application development system MyEclipse (Genuitec, Flower Mound, TX 75028, US). The servlet container was Apache Tomcat 6.0

(Apache Software Foundation). The web pages for the upload and download of result data were implemented in Framework Cake-PHP 1.3 (Cake Software Foundation). The support web pages and the fileserver system were based on the system developed for the TROST project [31]. The databases S-DB and R-DB were implemented in MySQL 5.5 (Oracle, Redwood City, US); MySQL-Workbench 5.2 (Oracle) was used for administration. The mirror function that imports user data and controlled vocabulary from the S-DB to the R-DB was implemented in SQL.

The graphical user interface for the mobile devices was implemented in .NET-Compact-Framework 3.5 (Microsoft Corporation) in C#. The application development system was Microsoft Visual Studio 2008 (Microsoft Corporation).

The internationalization was realised in frameworks Struts 2, Cake-PHP and the .NET library.

Hardware

The Phenotyper can be operated on mobile devices, e.g. tablets, barcode-scanner terminals or mobile phones with the operating systems Windows 7 Pro, Windows Mobile 6.0, Windows CE 5.0. The use on other Windows based systems is possible but was not tested extensively. During implementation and beta-testing, we used the barcode-scanner terminals Datalogic elf 00ALOLS-1 N1-Meno with laser-scanner (Datalogic, Diez, Germany) and M3T-1D (M3-Mobile, Seoul, Korea) and the semi-robust tablet computer Advantech-DLoG PWS 770 (ADVANTECH, München, Germany).

Availability

Users who like to see whether the functionality of the systems fits their requirements without having to run a complete setup can obtain test accounts for the scheme composer web pages and a test version of the results database. Login information is available on demand from the authors for a limited time period.

The complete software package is available as Additional files 3, 4 and 5 submitted with the manuscript. These files contain the installation package for the mobile device GUI, the files for the web pages and dumps of the databases. The software (see Material and methods) required to run e.g. the web pages is open-source. Links to repositories, from which the open-source software can be downloaded are provided in the help files of each package.

The software package and the source files required for further development of the software by the community is available from the software project repository of the Bioinformatics organization (http://www.bioinformatics.org/groups/?group_id=1210 and http://www.bioinformatics.org/cgi-bin/viewvc.cgi/phenotyper/ for the source code).

Additional files

Additional file 1: ER diagrams. Entity-relationship diagrams for the S-DB **(Figure S1)** and the R-DB **(Figure S2)**.

Additional file 2: Example file. Example of data structures of information collected on one object on several different dates. **Table A1.** shows the data the csv format, in which they are exported from the database, **Table A2.** shows the data after reformatting in the 'broad' format, in which data are generally presented in spreadsheet programs, **Table A1.** shows the data in xml format.

Additional file 3: PDA-Software. This archive contains the software to be installed on the personal digital assistant (PDA) including configuration files and installation instructions.

Additional file 4: Phenotyping Scheme Composer. This archive contains a dump file for the S-DB, a WAR file for installing the web application for generating phenotyping schemes, an encrypting tool, and installation instructions.

Additional file 5: Phenotyper web site and results database. This archive contains a dump file for the R-DB, the root directory of the result web application, an encrypting tool, and installation instructions.

Competing interests

The authors declare that they have no competing interests.

Authors' contributions

JG performed the implementation, selected the hardware and contributed to the manuscript, KK designed the system, organised the field tests and wrote the manuscript. Both authors tested the system. Both authors read and approved the final manuscript.

Acknowledgements

The German Research Foundation funded the work by grant DFG KO1731/3-1 to Karin Köhl. The following people were part of the beta-test group and contributed, by their suggestions and feedbacks, to the optimization of the system: Jessica Alpers, Ulrike Seider, Heike Sprenger and Ellen Zuther (MPI-MP), Katharina Rudack, Sylvia Seddig and Regina Schmidt (JKI Groß Lüsewitz), Vanessa Prigge (SAKA, Windeby), B. Truberg (Norika, Groß Lüsewitz). We thank Axel Fischer (MPI-MP) for valuable comments on the manuscript and Kenny Billiau for assistance with the implementation of the web pages.

References

1. Tester M, Langridge P. Breeding technologies to increase crop production in a changing world. Science. 2010;327(5967):818–22.
2. Finkel E. With 'Phenomics', plant scientists hope to shift breeding into overdrive. Science. 2009;325(5939):380–1.
3. Furbank RT, Tester M. Phenomics - technologies to relieve the phenotyping bottleneck. Trend Plant Sci. 2011;16(12):635–44.
4. Houle D, Govindaraju DR, Omholt S. Phenomics: the next challenge. Nat Rev Genet. 2010;11(12):855–66.
5. Traitmill -Platform and Process. [http://www.cropdesign.com/tech_traitmill.php]
6. The Plant Accelerator. [http://www.plantaccelerator.org.au/]
7. JPPC - the Jülich Plant Phenotyping Centre. [http://www.fz-juelich.de/ibg/ibg-2/DE/Organisation/JPPC/JPPC_node.html]
8. Klukas C, Chen D, Pape J-M. Integrated analysis platform: an open-source information system for high-throughput plant phenotyping. Plant Phys. 2014;165(2):506–18.
9. Granier C, Aguirrezabal L, Chenu K, Cookson S, Dauzat M, Hamard P, et al. PHENOPSIS, an automated platform for reproducible phenotyping of plant responses to soil water deficit in *Arabidopsis thaliana* permitted the identification of an accession with low sensitivity to soil water deficit. New Phytol. 2006;16:623–35.
10. Fabre J, Dauzat M, Negre V, Wuyts N, Tireau A, Gennari E, et al. PHENOPSIS DB: an Information System for *Arabidopsis thaliana* phenotypic data in an environmental context. BMC Plant Biol. 2011;11(1):77.
11. Cobb JN, DeClerck G, Greenberg A, Clark R, McCouch S. Next-generation phenotyping: requirements and strategies for enhancing our understanding

of genotype-phenotype relationships and its relevance to crop improvement. Theor Appl Genet. 2013;126(4):867–87.
12. Araus JL, Cairns JE. Field high-throughput phenotyping: the new crop breeding frontier. Trend Plant Sci. 2014;19(1):52–61.
13. Dhondt S, Wuyts N, Inze D. Cell to whole-plant phenotyping: the best is yet to come. Trend Plant Sci. 2013;18(8):433–44.
14. Zeeberg B, Riss J, Kane D, Bussey K, Uchio E, Linehan WM, et al. Mistaken identifiers: gene name errors can be introduced inadvertently when using Excel in bioinformatics. BMC Bioinformatics. 2004;5(1):80.
15. Lancashire PD, Bleiholder H, Van Den Boom T, Landgeluddeke P, Strauss R, Weber E, et al. A uniform decimal code for growth stages of crops and weeds. Ann Appl Biol. 1991;119(3):561–601.
16. Institute IRR. Standard evaluation system for rice. Los Banos, Laguna Philippines: International Rice Research Institute; 1996.
17. Baxter I, Ouzzani M, Orcun S, Kennedy B, Jandhyala S, Salt D. Purdue ionomics information management system. An integrated functional genomics platform. Plant Physiol. 2007;143:600–11.
18. van Ommen B, Bouwman J, Dragsted L, Drevon C, Elliott R, de Groot P, et al. Challenges of molecular nutrition research 6: the nutritional phenotype database to store, share and evaluate nutritional systems biology studies. Genes Nutr. 2010;5(3):189–203.
19. Barrett T, Troup DB, Wilhite SE, Ledoux P, Rudnev D, Evangelista C, et al. NCBI GEO: mining tens of millions of expression profiles—database and tools update. Nucleic Acids Res. 2007;35 suppl 1:D760–5.
20. Goff SA, Vaughn M, McKay S, Lyons E, Stapleton AE, Gessler D, et al. The iPlant collaborative: cyberinfrastructure for plant biology. Front Plant Sci. 2011;2:34.
21. Li Y-F, Kennedy G, Ngoran F, Wu P, Hunter J. An ontology-centric architecture for extensible scientific data management systems. Int J Grid Comput Escience. 2013;29(2):641–53.
22. PODD Data Repository Project to aid Phenomics. [http://www.plantphenomics.org.au/projects/podd/]
23. Phenoscape. [http://kb.phenoscape.org]
24. Yang E, Matthews B, Wilson M. Enhancing the core scientific metadata model to incorporate derived data. Future Generat Comput Syst. 2013;29(2):612–23.
25. Klukas C, Pape J-M, Entzian A. Analysis of high-throughput plant image data with the information system IAP. J Integr Bioinform. 2012;9(2):191.
26. Slovak R, Goeschl C, Su X, Shimotani K, Shiina T, Busch W. A scalable open-source pipeline for large-scale root phenotyping of Arabidopsis. Plant Cell. 2014;26(6):2390–403.
27. Triticeae Toolbox. [http://triticeaetoolbox.org]
28. DbNP. [http://www.dbnp.org]
29. Jaiswal P, Avraham S, Ilic K, Kellogg EA, McCouch S, Pujar A, et al. Plant Ontology (PO): a controlled vocabulary of plant structures and growth stages. Comp Funct Genom. 2005;6(7–8):388–97.
30. Köhl KI, Basler G, Luedemann A, Selbig J, Walther D. A plant resource and experiment management system based on the Golm Plant Database as a basic tool for omics research. Plant Methods. 2008;4:11.
31. Billiau K, Sprenger H, Schudoma C, Walther D, Köhl KI. Data management pipeline for plant phenotyping in a multisite project. Func Plant Biol. 2012;39(11):948–57.
32. Hannemann J, Poorter H, Usadel B, Blasing OE, Finck A, Tardieu F, et al. Xeml Lab: a tool that supports the design of experiments at a graphical interface and generates computer-readable metadata files, which capture information about genotypes, growth conditions, environmental perturbations and sampling strategy. Plant Cell Environ. 2009;32(9):1185–200.
33. Nieuwland J, Sornay E, Marchbank A, de Graaf B, Murray J. Phytotracker, an information management system for easy recording and tracking of plants, seeds and plasmids. Plant Methods. 2012;8(1):43.
34. Exner V, Hirsch-Hoffmann M, Gruissem W, Hennig L. PlantDB - a versatile database for managing plant research, vol. 4. 2008. p. 1.
35. R-Forge. [http://r-forge.r-project.org]
36. Dinu V, Nadkarni P. Guidelines for the effective use of entity-attribute-value modeling for biomedical databases. Int J Med Inform. 2007;76(11–12):769–79.
37. Mabee P, Balhoff J, Dahdul W, Lapp H, Midford P, Vision T, et al. 500,000 fish phenotypes: the new informatics landscape for evolutionary and developmental biology of the vertebrate skeleton. J Appl Ichthyol. 2012;28:300–5.
38. Klukas C, Schreiber F. Integration of -omics data and networks for biomedical research with VANTED. J Integr Bioinform. 2010;7(2):112.
39. Chern C-G, Fan M-J, Yu S-M, Hour A-L, Lu P-C, Lin Y-C, et al. A rice phenomics study-phenotype scoring and seed propagation of a T-DNA insertion-induced rice mutant population. Plant Mol Biol. 2007;65(4):427–38.
40. EWIG. [http://ewig.gfz-potsdam.de].
41. RADAR. [http://www.radar-projekt.org/].
42. DRYAD. [http://datadryad.org/].
43. DataCite. [https://www.datacite.org].
44. GEO. [http://www.ncbi.nlm.nih.gov/geo/].
45. Empfehlungen 200901. [http://www.dfg.de/download/pdf/foerderung/programme/lis/ua_inf_empfehlungen_200901.pdf].
46. Data sharing policy. [http://www.bbsrc.ac.uk/web/FILES/Policies/data-sharing-policy.pdf].

Automated integrative high-throughput phenotyping of plant shoots: a case study of the cold-tolerance of pea (*Pisum sativum* L.)

Jan F Humplík[1†], Dušan Lazár[2†], Tomáš Fürst[3], Alexandra Husičková[2], Miroslav Hýbl[4] and Lukáš Spíchal[1*]

Abstract

Background: Recently emerging approaches to high-throughput plant phenotyping have discovered their importance as tools in unravelling the complex questions of plant growth, development and response to the environment, both in basic and applied science. High-throughput methods have been also used to study plant responses to various types of biotic and abiotic stresses (drought, heat, salinity, nutrient-starving, UV light) but only rarely to cold tolerance.

Results: We present here an experimental procedure of integrative high-throughput in-house phenotyping of plant shoots employing automated simultaneous analyses of shoot biomass and photosystem II efficiency to study the cold tolerance of pea (*Pisum sativum* L.). For this purpose, we developed new software for automatic RGB image analysis, evaluated various parameters of chlorophyll fluorescence obtained from kinetic chlorophyll fluorescence imaging, and performed an experiment in which the growth and photosynthetic activity of two different pea cultivars were followed during cold acclimation. The data obtained from the automated RGB imaging were validated through correlation of pixel based shoot area with measurement of the shoot fresh weight. Further, data obtained from automated chlorophyll fluorescence imaging analysis were compared with chlorophyll fluorescence parameters measured by a non-imaging chlorophyll fluorometer. In both cases, high correlation was obtained, confirming the reliability of the procedure described.

Conclusions: This study of the response of two pea cultivars to cold stress confirmed that our procedure may have important application, not only for selection of cold-sensitive/tolerant varieties of pea, but also for studies of plant cold-response strategies in general. The approach, provides a very broad tool for the morphological and physiological selection of parameters which correspond to shoot growth and the efficiency of photosystem II, and is thus applicable in studies of various plant species and crops.

Keywords: Plant phenotyping, RGB digital imaging, Chlorophyll fluorescence imaging, Shoot growth, Biomass production, Cold adaptation, Pea (*Pisum)*

Introduction

In plants, acclimation to cold, causes reduced growth, increase in antioxidant content, reduced water content, and changes in gene regulation, hormone balance, membrane composition, osmotic regulation, and photosynthetic function [1]. The adaptability and productivity of legumes (chickpea, faba bean, lentil, and pea) are limited by abiotic stresses in general [2], and their high sensitivity to chilling and freezing temperatures is well described [3].

Since cold tolerance is an important agronomical problem in Central and Northern Europe and geographically similar regions, we aimed to develop a routine measuring procedure for automated integrative high-throughput screening for selection of potentially cold tolerant cultivars. Pea (*Pisum sativum* L.) was chosen as a model crop because its tolerance to cold stress is one of the limiting factors in autumn sowings which allows for the enhanced productivity of pea plants. Overwintering plants have developed adaptive responses to seasonal weather changes. For example, overwintering evergreens

* Correspondence: lukas.spichal@upol.cz
†Equal contributors
[1]Department of Chemical Biology and Genetics, Centre of the Region Haná for Biotechnological and Agricultural Research, Faculty of Science, Palacký University, Šlechtitelů 11, Olomouc CZ-78371, Czech Republic
Full list of author information is available at the end of the article

have developed so-called sustained non-photochemical quenching (reviewed, e.g., by Verhoeven [4]) as a protection mechanism against absorbed light which is in excess with respect to the capacity of the carbon photosynthetic reactions and which is decreased during winter. These plants sense the upcoming cold period through the perception of environmental impulses, mainly temperature and day length. However, the sustained non-photochemical quenching does not work in modern pea cultivars. For this reason, we chose two modern cultivars and investigated their reaction to cold stress. We employed digital RGB imaging to study shoot growth, and chlorophyll (Chl) fluorescence imaging (CFIM) to analyze various parameters of plant photosystem II (PSII) efficiency. The cultivars used in this study were morphologically similar which facilitated the validation of sensitivity and resolution of our visible imaging analysis.

There is a paucity of information on the acclimation of pea plants to cold. An extensive study was published by Markarian et al. [5]. These authors evaluated 26 pea lines based on their winter survival. Further physiological parameters (total dry matter and photosynthetic area) of autumn- and spring-sown pea plants were evaluated by Silim et al. [6]. Autumn-sown plants produced similar seed yields to spring sowings when the winter survival was adequate, and autumn sowings matured 2–4 weeks before the spring-sown crops, depending on the variety and season [6]. The effects of short term acclimation (four days) of pea plants to cold temperatures (5°C) were explored by Yordanov et al. [7] who measured the rate of oxygen production and CO_2 assimilation, and Chl fluorescence parameters in order to evaluate photochemical activity and functional heterogeneity of PSII. They found that cold-acclimated plants showed higher photosynthetic rates and better Chl fluorescence parameters than non-acclimated plants [7]. The effects of short term cold acclimation (three days, 4°C) and subsequent recovery (2 days) of standard pea plants were studied by Chl fluorescence measurements in more detail by Georgieva and Lichtenthaler [8]. The Chl fluorescence parameters reflecting photosynthetic function decreased during cold acclimation but were reversible in the subsequent recovery [8]. A similar study was later carried out with three different pea cultivars by Georgieva and Lichtenthaler [9].

These studies revealed the importance of two potential traits that could be used to distinguish between pea cultivars with different cold-sensitivity: rate of shoot growth and values of Chl fluorescence parameters. Both traits can now be studied by non-invasive high-throughput platforms to provide integrative insight into plant physiology during cold acclimation. The spatio-temporal changes in shoot biomass or leaf area can be assessed using automated RGB imaging and image-analysis software, as has been shown for many species such as cereals, tomatoes, soybean and beans [10-13]. The Chl fluorescence parameters are routinely analyzed by non-imaging fluorometres (NICF) or the imaging system (CFIM). For physiological studies, kinetic types of CFIM that allow computation of various Chl fluorescence parameters on the whole leaf or shoot are the most valuable. However, the kinetic type CFIM has not been commonly integrated into high-throughput systems [14] and in recent reports only systems measuring a single Chl fluorescence level have been employed [11,15]. The intensity of Chl fluorescence depends on the amount of chlorophylls; thus, a single Chl fluorescence level can be used, e.g., to distinguish between non-stressed and senescent leaves (when amount of Chls is decreased) at late stages of stress. However, this does not provide any information about earlier processes in PSII that are not necessarily linked to later senescence events.

In this report, we describe a procedure employing an automated integrative high-throughput platform suitable for studies of the physiological basis of cold-stress adaptation and selection of pea cultivars with cold sensitivity/tolerance. The platform measures shoot area and Chl fluorescence to provide a complex analysis of plants during cold-acclimation. For this purpose, we developed new software for automatic RGB image analysis and we evaluated various parameters of Chl fluorescence obtained from CFIM. The data from the automated phenotyping platform were validated through estimation of shoot biomass by manual weighing of the shoots and by measurement of Chl fluorescence by a NICF hand operated fluorometer. Despite the complexity of pea shoots, very good correlation between pixel based shoot area and fresh biomass were obtained. Similarly, the Chl fluorescence parameters measured by NICF fully confirmed the reliability of the automated CFIM analysis.

Results and discussion

Visible imaging used for shoot growth

To compare the influence of cold acclimation on biomass production, two putative cold-resistant cultivars of pea Terno and Enduro were selected (labeled as TER and END, respectively). After germination, the seedlings were grown in a growth chamber at 22/20°C (see Materials and methods) and after the development of the first true leaf, the cold stress conditions were established. The seedlings continued growing in 5°C for 21 days and were screened twice per week in the automated platform. The green area of each individual seedling was extracted from particular projections (Figure 1) and combined to account for the overall shoot biomass. As shown in Figure 2, the total green area of the plants was calculated at 7 time-points. The cultivar TER showed a significantly higher (for p values see Table 1) increase in the total green area compared to the cultivar END

Figure 1 The example images of three optical projections of single END seedling used for calculation of total green area on 8th day of cold acclimation. The green area that was digitally extracted from the images is marked by white border line.

(Figure 3A). Because the green area of the cultivars was different at the beginning of the experiment, the normalized green area (NGA) was calculated, where the green area on the nth (5, 8, ... 21) day of measurement was divided by the green area obtained on the 1st measuring day. The TER cultivar showed higher shoot growth which on the 21st day was almost a 3.5 fold increase in the green area, whereas END multiplied its projected area by only about 2.5-times (Figure 3B). To analyze how the cultivars differed in their growth rates, the relative growth rate (RGR) was used according to Hoffmann and Poorter [16]. We used the following formula:

$$RGR = \frac{\overline{lnW_2} - \overline{lnW_1}}{t_2 - t_1}$$

where $\overline{lnW_1}$ and $\overline{lnW_2}$ are the means of the natural logarithms of the plant's green areas and t_1 and t_2 are the times at which the green areas were measured. The TER cultivar relative growth rate was significantly higher (for p values see Table 1) during the whole period of cold acclimation. Moreover, at the beginning of the cold stress, the TER cultivar tended to speed-up its growth, then reached a steady state and finally decreased its RGR by the end of the experiment. The second cultivar, END, was very stable, slightly decreasing its growth rate during the experiment (Figure 3C). To examine the statistical significance of the differences between obtained TER and END growth-related parameters, the non-parametric Mann–Whitney *U* test was performed for each measuring day. The p values obtained for each measuring day are shown in Table 1.

It has been reported that cold-treatment affects total shoot biomass production and growth-rate in spring-sown and overwintering pea cultivars [6,17]. Besides shoot growth cold-treatment affects also growth of the root as showed in work by Bourion et al. [17]. However, the effect on the root is less severe compared to the above ground parts of the plants [17]. Due to this fact and due to the technical set up of our automated platform in this study we focused only on the analyses of cold-treatment effects on shoot growth. We describe here the development of the measuring setup for automated screening of pea cultivars with different cold-sensitivity through analysis of the shoot growth by RGB imaging followed by precise image-analysis. A similar approach has been shown for different species and different types of stresses. Considering crop species alone, most of the protocols for automated phenotyping using RGB imaging were designed for cereals, most often to

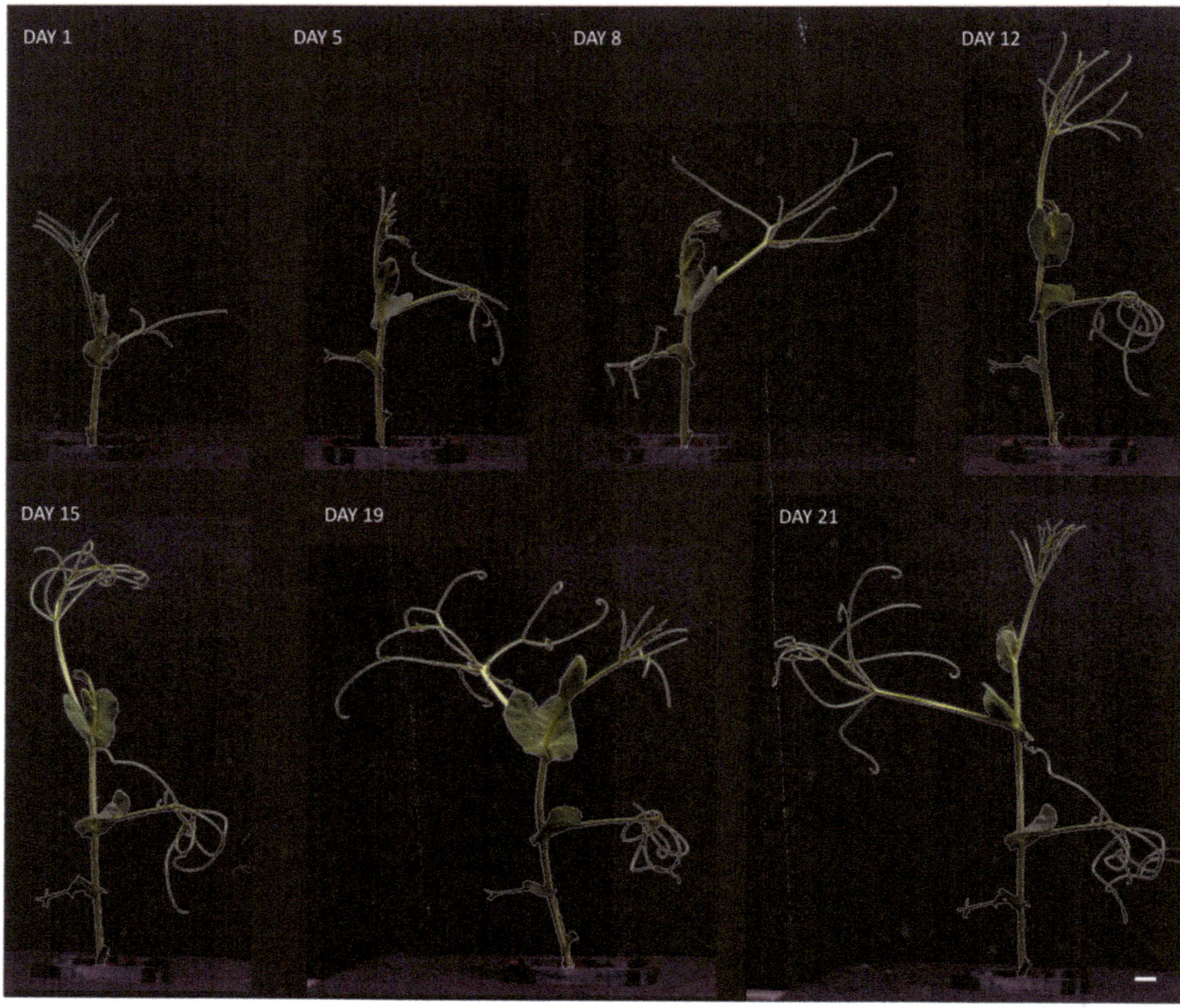

Figure 2 Example images taken from front view camera showing the growth progress of the TER seedling during cold acclimation for 21 days. The green area that was digitally extracted from the images is marked by white border line. The white bar in the right bottom site represents length of 1 cm.

Table 1 The p values of the Mann–Whitney test of statistical significant difference of growth parameters based on RGB imaging

Measuring day	Green area p value	NGA p value	Growth rate p value
1	<0.001	n. a.	n. a.
5	<0.001	<0.003	<0.003
8	<0.001	<0.001	<0.001
12	<0.001	<0.001	<0.001
15	<0.001	<0.001	<0.001
19	<0.001	<0.001	<0.001
21	<0.001	< 0.001	< 0.001

For each day the comparisons between TER and END datasets were tested in all three parameters respectively.

screen for drought, or salt tolerant plants [10,15,18-23]. Surprisingly, use of such a method has not been presented so far for any crops studied for cold-acclimation. Although there was no presumed effect of cold-treatment on the reliability of RGB imaging, the complicated morphology of field pea cultivars could potentially affect the accuracy of the automated measurements. For this reason, we tested our method of the green area (or projected area) estimation from automated RGB imaging by its comparison with a method of manual weighing of the shoots. The shoots of both cultivars were harvested on the last measuring day and FW of individual plant shoots was measured. Subsequently, correlations between the green area and FW were calculated using the non-parametric Spearman correlation coefficient. A similar approach has been reported recently by Hairmansis et al. [15] for rice. These authors found a

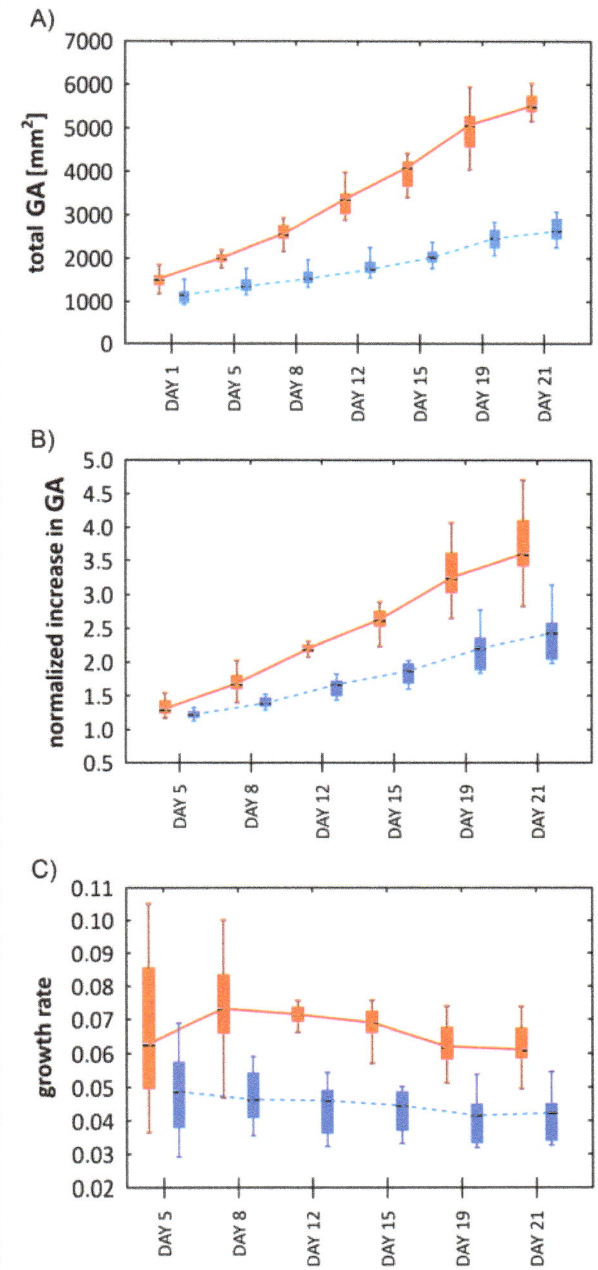

Figure 3 Analyses of the growth progress of shoots of TER (red boxes – full line) and END (blue boxes – dashed line) pea cultivars. The values derived from the green area on n^{th} days (1, 5, 8,..., 21) are presented as medians (black bars) and quartiles (boxes). For better readability, the boxes are shifted in x-axes to not to overlap, but still represent the values measured on the same days. **A)** A total green area. **B)** A normalized green area. **C)** A relative growth rate. The error bars show minimal and maximal values.

correlation of projected area and FW ranging from 0.96 to 0.97. A more sophisticated calculation was developed by Golzarian et al. [22] who used estimated shoot area as a function of plant area and plant age. This method was applied by Pereyra-Irujo et al. [12] in experiments with soybean, providing a correlation of 0.97 in dry mass. Shoots of cereals and soybean have relatively low spatial-complexity. In contrast, shoots of field pea cultivars TER and END are formed mainly by stem and tiny tendrils (Figures 1, 2) requiring very precise identification by image-analysis software. Despite the challenging pea shoot morphology, Spearman correlation coefficients of 0.91 and 0.96 for TER and END cultivars, respectively, were found in our analysis ($p < 0.05$; Figure 4). This is fully comparable with the phenotyping protocols designed for other crop species and provides an efficient and reliable tool for the evaluation of pea growth.

Chlorophyll fluorescence imaging used for determination of photosynthetic function

Further variables used for phenotyping of the two pea cultivars were those obtained from measurements of Chl fluorescence induction (CFIN), which reflects photosynthetic function, mainly of PSII. Based on our knowledge of the parameters that can be determined from CFIN (reviewed in Lazár [24]), we selected the following parameters: i) the maximal quantum yield of PSII photochemistry for a dark-adapted state, $\Phi_{Po} = (F_M - F_0)/F_M = F_V/F_M$, where F_0, F_M, and F_V are the minimal, maximal, and variable fluorescence levels, respectively, for a dark-adapted state; ii) the actual quantum yield of PSII photochemistry for a light-adapted state, $\Phi_P = (F_M' - F(t))/F_M'$, where F_M' and $F(t)$ are the maximal and actual (at time t; usually in the steady state) fluorescence levels for a light-adapted state; iii) the quantum yield of constitutive non-light induced (basal or dark) dissipation processes consisting of Chl fluorescence emission and heat dissipation, $\Phi_{f,D} = F(t)/F_M$; and iv) the quantum yield of regulatory light-induced heat dissipation, $\Phi_{NPQ} = F(t)/F_M' - F(t)/F_M$. It is worth mentioning here that $\Phi_P + \Phi_{f,D} + \Phi_{NPQ} = 1$; further that $\Phi_P = q_P\Phi_{PSII}$, where q_P $(= (F_M' - F(t))/(F_M' - F_0'))$ is the coefficient of photochemical quenching which estimates a fraction of the so-called open PSII reaction centers; and that Φ_{PSII} $(= (F_M' - F_0')/F_M')$ is the maximal quantum yield of the PSII photochemistry for a light-adapted state. The F_0' in the last two equations is the minimal fluorescence level for a light-adapted state which was estimated from: $F_0' = F_0/(((F_M - F_0)/F_M) + (F_0/F_M'))$ (for details see [24]).

The changes in these Chl fluorescence parameters measured during acclimation of TER and END cultivars to 5°C for 21 days are shown in Figure 5. Φ_{Po} is affected very little by the cold acclimation of TER but there is a continual decrease in Φ_{Po} of END (Figure 5A). Φ_P initially decreases more in TER than in END but after 6 days it maintains its value in TER but continues to decrease in END (Figure 5B). The continual decrease in Φ_P in END is mostly caused by a continual decrease in Φ_{PSII}; q_P slightly increasing in the last two measurements in END (Figure 5B). On the other hand, the initial decrease in Φ_P in TER is mostly caused by decrease in

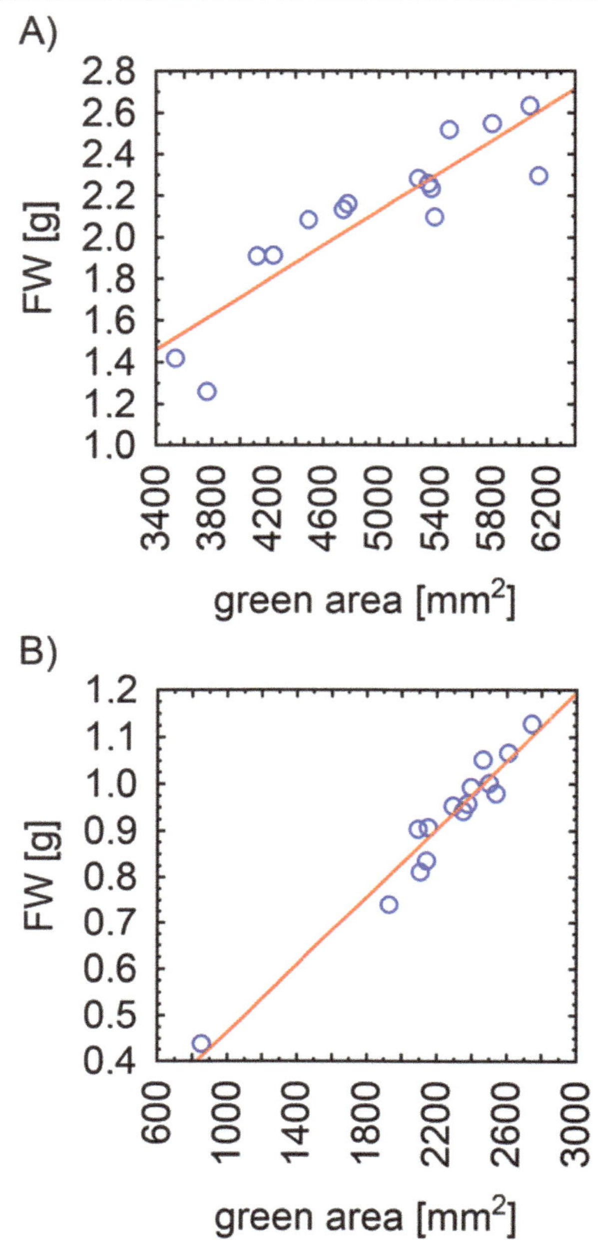

Figure 4 The correlation of the green area and biomass. The Spearmann correlation coefficients of FW and green area of TER cultivar **A)** and END cultivar **B)** were 0.91 and 0.96, respectively (p value < 0.05).

q_P but the almost unchanged value of Φ_P in TER after 6 days is caused by the counter action of q_P, which increases, and of Φ_{PSII}, which decreases (Figure 5B). Therefore, it can be concluded that photosynthesis of the two pea cultivars uses different strategies for cold acclimation. Whereas in END, the number of open reaction centers as well as their maximal photosynthetic quantum yield in light generally decrease with prolonged cold acclimation, in TER, a decrease of the maximal quantum yield of PSII photochemistry in light (Φ_{PSII}) is compensated by an increase of number of the open PSII reaction centers (q_P) (Figure 5B). Furthermore, END shows an increased quantum yield of constitutive non-light induced dissipation processes ($\Phi_{f,\ D}$) at the end of the cold acclimation compared to TER (Figure 5C), whereas the rise of the quantum yield of regulatory light-induced heat dissipation (Φ_{NPQ}) during the acclimation is faster in TER than in END (Figure 5D).

It is interesting to note that cold-induced changes of the Chl fluorescence parameters for given cultivar and differences (or about the same values) of the parameters between the cultivars (Figure 5) are not accompanied by expected changes and differences of green areas and growth rates (Figure 3). Even when the photosynthetic function was decreased by cold treatment (decrease of the Φ_{Po}, Φ_P, q_P, and Φ_{PSII} parameters; Figure 5A and 5B), the total and normalized green area of both cultivars was still increased (Figure 3A and 3B). It might show that the grow rate changed (for TER; Figure 3C) or decreased (for END; Figure 3C) with increasing duration of the cold treatment, however, these changes were not statistically significant (data not shown). The uncorrelated behavior of photosynthetic and growth parameters reflects different temperature dependences of photosynthesis and processes hidden behind the plant growth. While photosynthetic function was decreased by treatment of the cultivars at 5°C, probably much lower temperatures would be needed to stop plant growth. Therefore, FCIM data and RGB imaging data carry different and complementary information about acclimation of plants to lower temperatures. To take advantage of the high-throughput capacity of our phenotyping platform, we used a relatively short protocol to measure CFIN. This set up, however, did not allow for determination of photoinactivated centers which might be formed during a joint action of light and cold [25-28]. Depending on the theory used, the formation of the photoinactivated PSII centers can influence all quantum yields of the light-adapted state (for a review see [24]) used in this work. Therefore, in the next study we aim to modify the CFIN measuring protocol in order to determine the quantum yield of photoinactivated PSII centers as well.

Furthermore, we tested the reliability and accuracy of the Chl fluorescence parameters measured by the automated CFIM in a high-throughput set up by comparing the selected parameter (Φ_{Po}) with the same parameter measured by a hand-operated non-imaging Chl fluorometer. For this purpose the overall Chl fluorescence images were separated into images of the second and third leaves and their Φ_{Po} were evaluated. On the other hand, Φ_{Po} was evaluated from the fast Chl fluorescence rise as measured by the non-imaging Chl fluorometer with a different set of leaves (see Materials and methods). The results of these comparisons are presented in Figure 6A for the second leaves and in Figure 6B for the third leaves, respectively.

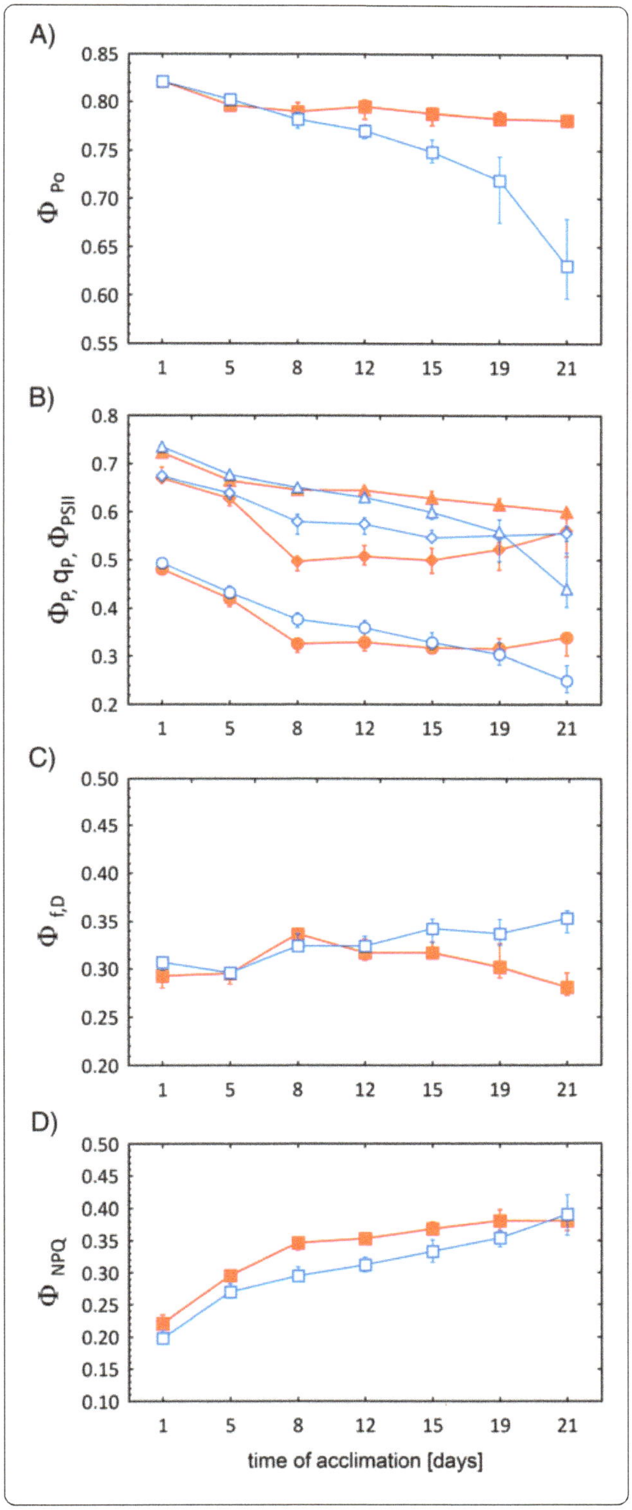

Figure 5 Changes of CFIN parameters of TER (full symbols) and END (open symbols) pea cultivars measured during the 21 days of cold acclimation. Changes in **A)** the maximal quantum yield of PSII photochemistry for a dark-adapted state (Φ_{Po}); **B)** the maximal and the actual quantum yield of photosystem II photochemistry for a light-adapted state (Φ_{PSII}, Φ_P respectively), the coefficient of photochemical quenching (q_P); **C)** the quantum yield of constitutive non-light induced dissipation processes ($\Phi_{f,\,D}$); **D)** the quantum yield of regulatory light-induced heat dissipation (Φ_{NPQ}); are shown. The values represent medians from 15 measurements. The error bars represent quartiles. The medians of all the TER and END parameters at the end of measurements were statistically significant (p value < 0.05), except of q_p and Φ_{NPQ}.

A representative image of the spatial distribution of Chl fluorescence is presented in Figure 6C. Not surprisingly, the data show that there is no statistically significant difference (at $p < 0.05$) between Φ_{Po} measured for given leaves by the two different approaches. Moreover, Figure 6C documents another advantage of using the CFIM in automated high-throughput platforms. Although the software is primarily adjusted to calculate the mean value of fluorescence from the total surface of every plant, if needed, the CFIN images can be later separated for subsequent calculation of the Chl fluorescence parameters taken from the individual selected areas which represent individual plant parts (Figure 6C).

To the best of our knowledge, only one study was published reporting on use of CFIM integration into a high-throughput phenotyping platform to analyze cold- or chilling-stress. Using an automated phenotyping platform Jansen et al. [14] evaluated only the F_V/F_M parameter (Φ_{Po}) for two different *Arabidopsis* plants (wild-type and a mutant), and wild-type tobacco plants. Φ_{Po} decreased in the wild-type tobacco plants during the cold treatment, and the same decreasing trends were found with *Arabidopsis* plants, however, the differences between the wild-type and a mutant were not convincing. Using a CFIM system, Lootens et al. and Devacht et al. [25,29] studied the effect of different cold temperatures on industrial chicory plants. In agreement with our results, the authors found again only a small decrease of Φ_{Po} after 10-day incubation at 4°C and the values of the Φ_P and Φ_{PSII} parameters caused by the incubation were similar to those obtained in our study. Mishra et al. [30,31] used CFIM to study the effect of a two-week incubation at 4°C on nine *Arabidopsis thaliana* accessions differing in cold tolerance. In addition to evaluation of standard Chl fluorescence parameters, like Φ_{Po}, Φ_P, and q_P, the authors also showed that combinatorial imaging of Chl fluorescence transients combined with classifier and feature selection methods could discriminate between detached leaves from cold sensitive and cold tolerant accessions.

Materials and methods

Plant material

Two morphologically similar field pea (*P. sativum* subsp. *sativum* var. *sativum*) cultivars Terno (TER) and Enduro

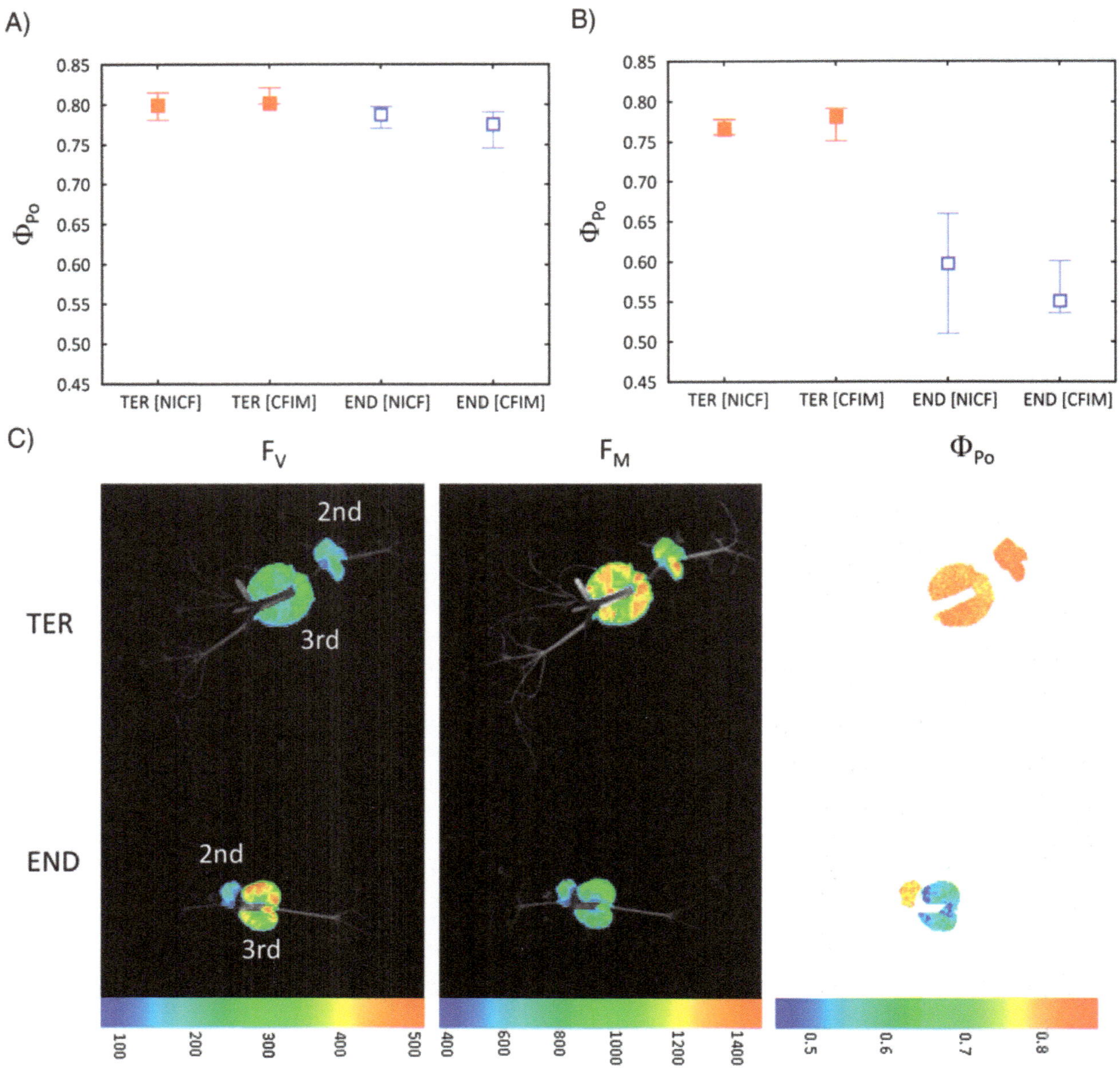

Figure 6 Comparison of F_V/F_M (Φ_{Po}) values measured in A) 2nd and B) 3rd leaves by hand-operated non-imaging chlorophyll fluorometer [NICF] or chlorophyll fluorescence imaging [CFIM] part of the platform with manually extracted single leave areas. The data obtained from each leaf in TER and END genotypes were tested by the Mann–Whitney *U* test showing no significant differences between Φ_{Po} determined by NICF and CFIM ($p > 0.05$). The values represent medians and the error bars quartiles, respectively. **C)** Imaging of chlorophyll fluorescence in separated leaves of both genotypes. The variable minimal fluorescence (F_V), maximal fluorescence (F_M) and maximal quantum yield of PSII photochemistry (Φ_{Po}) are shown in false colour scales with relative units.

(END) were used in the experiment. TER is pea cultivar, used for spring sowing term with a certain capacity to cold-acclimation, whereas END is a cold-tolerant over-wintering cultivar. The END cultivar was obtained from the Selgen a.s. company (Prague, Czech Republic). The TER cultivar was taken from the Czech collection of pea genetic resources kept in Agritec Ltd., Šumperk, Czech Republic. The collection is run according to the general rules of the National Programme for Plant Genetic Resources of the Czech Republic and the passport data are available on http://genbank.vurv.cz/genetic/resources/.

Cultivation conditions and experimental setup

The TER and END pea cultivars were sown into standardized pots (65 x 65 x 95 mm, Plant-It-Rite, Australia) filled with 100 g of soil (Substrate 2, Klasmann-Deilmann

GmbH, Germany) and watered to full water capacity. The seeds were germinated in mini-greenhouses (50 x 32 x 6 cm with clear plastic lid) in a growth chamber with white LED lighting (150 μmol photons of PAR m^{-2} s^{-1}). The conditions were set-up to simulate a long day (16 h day, 8 h night) with temperatures of 22°C during the light period and 20°C in the night. The relative humidity was set to 60%. After the development of the first true leaves, the temperature was decreased to 5°C for the entire experiment, the other parameters remained unchanged. The plants were regularly watered with the same amount of water. Fifteen seedlings from each cultivar were used for the automated phenotyping, and another fifteen plants were used for control measurements of maximal quantum yield of PSII photochemistry through the use of a hand-operated non-imaging Chl fluorometer. For measurements in PlantScreen^{TM} phenotyping platform (Photon Systems Instruments, Brno, Czech Republic), the pots with the seedlings were placed in standardized trays; two pots per tray and automatically loaded and measured by the platform. The movement of the trays was performed by a robotic-driven conveyor belt that routinely transferred experimental plants between the growing and measuring areas according to a user-defined protocol. A single measuring round of 8 trays consisted of 20 minutes of dark-adaptation, followed by the measurement of Chl fluorescence and digital RGB imaging from three optical projections. Approximately 16 plants per hour were analyzed, due to the length of the measuring round that is dependent on the length of the dark adaptation and CFIM measurement. In the case of RGB imaging the platform throughput increases to about 60 experimental trays (120 plants) per hour. The data from Chl fluorescence and RGB imaging were stored in a database server, and analyzed either by the software provided by the manufacturer or by the software developed by the authors of this study as described below.

RGB software image analysis

The plants were automatically loaded into the measuring cabinets of the PlantScreen^{TM} platform where the three RGB images – the top, front, and side views – (Figure 1) of each experimental tray containing two plants were taken. To assess the total green area, the green mask of the individual plants has to be found in the image. To this end, we used a combination of automatic thresholding procedures and automatic edge detection techniques. First, the image was converted from the RGB colour space into the HSV colour space. It is much easier to find the green mask in the H channel of the HSV colour space because the S and V channels only contain information on the saturation and brightness of the colour but not the hue itself. The region in the three dimensional RGB space which defines the 'plant green' colour may have a rather complicated shape, however, it is reduced to a line-segment in the one-dimensional H space as the S and V coordinates can be ignored. For thresholding in the H channel, several standard automatic algorithms can be used, e.g., the most popular Otsu method [32] that calculates the optimum threshold separating the foreground and background pixels so that their combined intra-class variance is minimal. In our case, we used an even simpler technique - foreground (i.e., the plant) was predefined as a particular line segment in the H channel. This was possible due to the standardized image acquisition setting.

The thresholding step usually provides very good discrimination between the plant and its background and no further processing is necessary. However, the pea plants possess very thin offshoots (only one or two pixels thick) that may be difficult to find by thresholding alone. If the thresholding routine makes a single-pixel mistake, which often happens due to noise in the image, the entire offshoot is lost, which is undesirable. We solved this problem by exploiting the Canny automatic edge detection algorithm which tracks the contours of the plant image [33]. The thin offshoots were tracked particularly well because the edge detection algorithm focused on such thin structures. The results of the thresholding step were then combined with the edge detection step and the final green mask of the object was found. Finally, a couple of post processing steps were performed (e.g. median filtering and image opening and/or closing) to enhance the quality of the mask.

It only took several seconds on a standard PC to find the green mask of a single pea plant. The mask provided information about the projection of the plant surface area onto the three image planes. The projections can be expressed in square millimeters because the RGB camera had been calibrated beforehand. The calibration proceeded as follows. Two bars covered by millimeter paper were placed in the pots instead of the pea plants. The bars were approximately the same height as the plants. Three images (top, front, side) of the two bars were acquired with the same camera setting used for the entire experiment. These images served as the standard for converting the leaf area from pixels to square millimeters. The total green area of the plant is then estimated as $A = \sqrt{(A_x^2 + A_y^2 + A_z^2)}$, where A_x, A_y, and A_z are the respective projections onto the three image planes. This procedure is naturally not precise but it gives an estimate which is in good correlation (Figure 4) with the fresh biomass of the above ground plant parts.

CFIM and non-imaging Chl fluorescence measurements

A standard protocol was used for the measurement of Chl fluorescence quenching using the CFIM part of the PlantScreen^{TM} platform. The plants underwent 20 – 40 minutes of dark adaptation before CFIM measurements. During all signal recordings, short (33.3 μs) red (650 nm) "measuring" flashes were applied and a Chl fluorescence signal was detected a few microseconds before the measuring flash and during the flash, and then the two signals were subtracted. This is a pulse amplitude modulation (PAM) type of measurement. To measure the minimal fluorescence for a dark-adapted state, F_0, only the measuring flashes were applied for an initial 5 seconds. Then, a saturation pulse of 800 ms duration (white light, intensity of 1000 μmol photons of PAR m^{-2} s^{-1}) was applied and the maximal fluorescence for a dark-adapted state, F_M, was measured. After the F_M measurement, fluorescence was kept relaxed in darkness for 17 seconds. Red actinic light (650 nm, intensity of 100 μmol photons m^{-2} s^{-1}) was then switched on for 70 seconds to drive photosynthesis. It was visually checked so that a steady state fluorescence signal was attained at 70 s of illumination. During the actinic illumination, saturation pulses were applied at 8, 18, 28, 48, and 68 seconds from the beginning of the actinic illumination. The value of the maximal fluorescence measured during the last saturation pulse was taken as the maximal fluorescence signal for the light-adapted state, F_M'. The fluorescence signal caused by the actinic illumination measured just before the last saturation pulse was taken as the steady state fluorescence for a light-adapted state, F (t). The four fluorescence levels (F_0, F_M, F (t), F_M') were used for calculation of the minimal fluorescence level for a light-adapted state, F_0,' the quantum yields, and the other fluorescence parameters as defined and described in the Results section.

A hand-operated FluorPen fluorometer (Photon Systems Instruments, Brno, Czech Republic) was used for control measurements in order to compare the results obtained using automatized CFIM with hand-operated non-imaging Chl fluorescence measurements. Blue light (455 nm) of intensity 1000 μmol photons m^{-2} s^{-1} and a duration of 1 second was used by FluorPen for illumination of the sample and a whole fast fluorescence rise (the O-J-I-P curve) was recorded. However, only the minimal and maximal fluorescence levels, F_0 and F_M, respectively, for the dark adapted state, were evaluated from the curve using built-in routines. The two fluorescence levels were used for calculation of the maximal quantum yield of PSII photochemistry (see Results). The data for Chl fluorescence measurements are presented as medians and lower and upper quartiles [34].

Conclusion

In this proof-of-concept study, the high-throughput method for automated screening of cold-tolerant pea (*Pisum sativum* L.) cultivars was designed. TER and END cultivars were screened simultaneously in an automated way with throughput of 16 plants per hour for i) growth of the aerial parts by RGB imaging and ii) for the efficiency of photosynthesis by chlorophyll fluorescence imaging. We demonstrated that the presented integrative approach based on analyses of differences in relative growth rate and selected CFIM parameters can provide deeper insight into the physiological base of cold-acclimation. Data from both analytical tools pointed to significant differences in the growth and photosynthesis of TER and END cultivars, and indicated that the two pea cultivars use different strategies for cold acclimation differing in number of open PSII reaction centers, their maximal photosynthetic quantum yield in light and quantum yield of constitutive non-light induced dissipation processes. The reliability of the screening was verified by independent measuring of the fresh weight of the shoots and by Chl fluorescence measurement by hand fluorometer. Since the CFIM analysis is not limited to plant morphology and our image analysis was sensitive enough to detect tiny tendrils of pea, we believe that the described procedure can be easily employed for shoot analyses of other different plant species.

Abbreviations

Chl: Chlorophyll; CFIM: Chlorophyll fluorescence imaging; CFIN: Chlorophyll fluorescence induction; END: Enduro; F_0 and F_0': Minimal chlorophyll fluorescence levels for dark- and light-adapted states, respectively; F_M and F_M': Maximal chlorophyll fluorescence levels for dark- and light-adapted states, respectively; FW: Fresh weight; F (t): Actual (at time t; usually in the steady state) fluorescence level for light-adapted state; F_V: Variable chlorophyll fluorescence level for a dark-adapted state; Φ_{Po}: The maximal quantum yield of photosystem II photochemistry for a dark-adapted state; Φ_P: The actual quantum yield of photosystem II photochemistry for a light-adapted state; Φ_{PSII}: The maximal quantum yield of photosystem II photochemistry for a light-adapted state; $\Phi_{f, D}$: The quantum yield of constitutive non-light induced (basal or dark) dissipation processes consisting of fluorescence emission and heat dissipation, Φ_{NPQ}, The quantum yield of regulatory light-induced heat dissipation; GA: Green area; NICF: Non-imaging chlorophyll fluorescence fluorometer; NGA: Normalized green area; RGB: Red-green-blue; RGR: Relative growth rate; PAM: Pulse amplitude modulation; PAR: Photosynthetic active radiation; PSII: Photosystem II; q_P: The coefficient of photochemical quenching; TER: Terno.

Competing interests

The authors declare that they have no competing interests.

Authors' contributions

JFH and DL carried out the visible and fluorescence imaging analyses, data processing and interpretation, participated in the design of the study and drafted the manuscript. TF developed and carried out the software image analysis, data processing, performed the statistical analysis, and drafted the manuscript. AH carried out the visible and fluorescence imaging analyses, data processing and interpretation, participated in the design of the study and drafted the manuscript. MH contributed to the design of the study, and helped to draft the manuscript. LS conceived the study, participated in its design and coordination and drafted the manuscript. All authors read and approved the final manuscript.

Acknowledgments

This work was supported by the grant No LO1204 (Sustainable development of research in the Centre of the Region Haná) from the National Program of Sustainability I, Ministry of Education, Youth and Sports, Czech Republic. A.H. was supported by the Operational Program Education for Competitiveness - European Social Fund (project CZ.1.07/2.3.00/20.0165).

Author details

[1]Department of Chemical Biology and Genetics, Centre of the Region Haná for Biotechnological and Agricultural Research, Faculty of Science, Palacký University, Šlechtitelů 11, Olomouc CZ-78371, Czech Republic. [2]Department of Biophysics, Centre of the Region Haná for Biotechnological and Agricultural Research, Faculty of Science, Palacký University, Šlechtitelů 11, Olomouc CZ-78371, Czech Republic. [3]Department of Mathematical Analysis and Applications of Mathematics, Faculty of Science, Palacký University, 17. listopadu 12, Olomouc CZ-77146, Czech Republic. [4]Department of Genetic

Resources for Vegetables, Medicinal and Special Plants, Centre of the Region Haná for Biotechnological and Agricultural Research, Crop Research Institute, Šlechtitelů 11, Olomouc CZ-78371, Czech Republic.

References

1. Xin Z, Browse J. Cold comfort farm: the acclimation of plants to freezing temperatures. Plant Cell Environ. 2000;23:893–902.
2. Stoddard FL, Balko C, Erskine W, Khan HR, Link W, Sarker A. Screening techniques and sources of resistance to abiotic stresses in cool-season food legumes. Euphytica. 2006;147:167–86.
3. Maqbool A, Shafiq S, Lake L. Radiant frost tolerance in pulse crops – a review. Euphytica. 2010;172:1–12.
4. Verhoeven A. Sustained energy dissipation in winter evergreens. New Phytol. 2014;201:57–65.
5. Markarian D, Harwood RR, Rowe PR. The inheritance of winter hardiness of Pisum. II. Description and release of advance generation breeding lines. Euphytica. 1968;17:110–3.
6. Silim SN, Hebblethwaite PD, Heath MC. Comparison of the effects of autumn and spring sowing date on growth and yield of combining peas (Pisum sativum L.). J Agri Sci. 1985;104:35–46.
7. Yordanov I, Georgieva K, Tsonev T, Vilikova V. Effect of cold hardening on some photosynthetic characteristics of pea (Pisum sativum L., CV. RAN 1) plant. Bulg J Plant Physiol. 1996;22:13–21.
8. Georgieva K, Lichtenthaler HK. Photosynthetic activity and acclimation ability of pea plants to low and high temperature treatment as studied by means of chlorophyll fluorescence. J Plant Physiol. 1999;155:416–23.
9. Georgieva K, Lichtenthaler HK. Photosynthetic response of different pea cultivars to low and high temperature treatments. Photosynthetica. 2006;44:569–78.
10. Berger B, de Regt B, Tester M. High-throughput phenotyping of plant shoots. In: Normanly J, editor. High-throughput phenotyping in plants. New York City: Humana Press; 2012. p. 9–20.
11. Petrozza A, Santaniello A, Summerer S, Di Tommaso G, Di Tommaso D, Paparelli E, et al. Physiological responses to Megafol® treatments in tomato plants under drought stress: a phenomic and molecular approach. Sci Hortic (Amsterdam). 2014;174:185–92.
12. Pereyra-Irujo GA, Gasco ED, Peirone LS, Aguirrezábal LA. GlyPh: a low-cost platform for phenotyping plant growth and water use. Funct Plant Biol. 2012;39:905–13.
13. Chaerle L, Hagenbeek D, Vanrobaeys X, Van Der Straeten D. Early detection of nutrient and biotic stress in Phaseolus vulgaris. Int J Remont Sens. 2007;28:3479–92.
14. Jansen M, Gilmer F, Biskup B, Nagel KA, Rascher U, Fischbach A, et al. Simultaneous phenotyping of leaf growth and chlorophyll fluorescence via GROWSCREEN FLUORO allows detection of stress tolerance in Arabidopsis thaliana and other rosette plants. Funct Plant Biol. 2009;36:902–14.
15. Hairmansis A, Berger B, Tester M, Roy SJ. Image-based phenotyping for non-destructive screening of different salinity tolerance traits in rice. Rice. 2014;7:16.
16. Hoffmann WA, Poorter H. Avoiding bias in calculations of relative growth rate. Ann Bot London. 2002;90:37–42.
17. Bourion V, Lejeune-Hénaut I, Munier-Jolain N, Salon C. Cold acclimation of winter and spring peas: carbon partitioning as affected by light intensity. Eur J Agron. 2003;19:535–48.
18. Honsdorf N, March TJ, Berger B, Tester M, Pillen K. High-throughput phenotyping to detect drought tolerance qtl in wild barley introgression lines. PLoS One. 2014;9:e97047.
19. Fehér-Juhász E, Majer P, Sass L, Lantos C, Csiszár J, Turóczy Z, et al. Phenotyping shows improved physiological traits and seed yield of transgenic wheat plants expressing the alfalfa aldose reductase under permanent drought stress. Acta Physiol Plant. 2014;36:663–73.
20. Rajendran K, Tester M, Roy SJ. Quantifying the three main components of salinity tolerance in cereals. Plant Cell Environ. 2009;32:237–49.
21. Harris BN, Sadras VO, Tester M. A water-centred framework to assess the effects of salinity on the growth and yield of wheat and barley. Plant Soil. 2010;336:377–89.
22. Golzarian MR, Frick RA, Rajendran K, Berger B, Roy S, Tester M, et al. Accurate inference of shoot biomass from high-throughput images of cereal plants. Plant Methods. 2011;7:1–11.
23. Schilling RK, Marschner P, Shavrukov Y, Berger B, Tester M, Roy SJ, et al. Expression of the Arabidopsis vacuolar H^{+}-pyrophosphatase gene (AVP1) improves the shoot biomass of transgenic barley and increases grain yield in a saline field. Plant Biotechnol J. 2014;12:378–86.
24. Lazár D. Parameters of photosynthetic energy partitioning. J Plant Physiol. 2015;175:131–47.
25. Lootens P, Devacht S, Baert J, Van Waes J, Van Bockstaele E, Roldán-Ruiz I. Evaluation of cold stress of young industrial chicory (Cichorium intybus L.) by chlorophyll a fluorescence imaging. II. Dark relaxation kinetics. Photosynthetica. 2011;49:185–94.
26. Somersallo S, Krause GH. Reversible photoinhibition of unhardened and cold-acclimated spinach leaves at chilling temperatures. Planta. 1990;2:181–7.
27. Liu P, Meng QW, Zou Q, Zhao SJ, Liu QZ. Effects of cold hardening on chilling induced photoinhibition of photosynthesis and on xantophyll cycle pigments in sweet pepper. Photosynthetica. 2001;39:467–72.
28. Hogewoning SW, Harbinson J. Insights on the development, kinetics, and variation of photoinhibition using chlorophyll fluorescence imaging of a chilled, variegated leaf. J Exp Bot. 2007;58:453–63.
29. Devacht S, Lootens P, Baert J, Van Waes J, Van Bockstaele E, Roldán-Ruiz I. Evaluation of cold stress of young industrial chicory (Cichorium intybus L.) plants by chlorophyll a fluorescence imaging. I. Light induction curve. Photosynthetica. 2011;49:161–71.
30. Mishra A, Mishra KB, Höermiller II, Heyer AG, Nedbal L. Chlorophyll fluorescence emission as a reporter on cold tolerance in Arabidopsis thaliana accessions. Plant Signal Behav. 2011;6:301.
31. Mishra A, Heyer AG, Mishra KB. Chlorophyll fluorescence emission can screen cold tolerance of cold acclimated *Arabidopsis thaliana* accessions. Plant Methods. 2014;10:38.
32. Otsu N. A threshold selection method from gray-level histograms. IEEE T Syst Man Cyb. 1979;9:62–6.
33. Canny JA. Computational approach to edge detection. IEEE T Pattern Anal. 1986;8:679–98.
34. Lazár D, Nauš J. Statistical properties of chlorophyll fluorescence parameters. Photosynthetica. 1998;35:121–7.

Mathematical modeling of positron emission tomography (PET) data to assess radiofluoride transport in living plants following petiolar administration

Alexander K Converse[1*], Elizabeth O Ahlers[1], Tom W Bryan[2], Jackson D Hetue[2], Katherine A Lake[3], Paul A Ellison[3], Jonathan W Engle[3], Todd E Barnhart[3], Robert J Nickles[3], Paul H Williams[2] and Onofre T DeJesus[3]

Abstract

Background: Ion transport is a fundamental physiological process that can be studied non-invasively in living plants with radiotracer imaging methods. Fluoride is a known phytotoxic pollutant and understanding its transport in plants after leaf absorption is of interest to those in agricultural areas near industrial sources of airborne fluoride. Here we report the novel use of a commercial, high-resolution, animal positron emission tomography (PET) scanner to trace a bolus of [^{18}F]fluoride administered via bisected petioles of *Brassica oleracea*, an established model species, to simulate whole plant uptake of atmospheric fluoride. This methodology allows for the first time mathematical compartmental modeling of fluoride transport in the living plant. Radiotracer kinetics in the stem were described with a single-parameter free- and trapped-compartment model and mean arrival times at different stem positions were calculated from the free-compartment time-activity curves.

Results: After initiation of administration at the bisected leaf stalk, [^{18}F] radioactivity climbed for approximately 10 minutes followed by rapid washout from the stem and equilibration within leaves. Kinetic modeling of transport in the stem yielded a trapping rate of 1.5 +/− 0.3%/min (mean +/− s.d., n = 3), velocity of 2.2 +/− 1.1 cm/min, and trapping fraction of 0.8 +/− 0.5%/cm.

Conclusion: Quantitative assessment of physiologically meaningful transport parameters of fluoride in living plants is possible using standard positron emission tomography in combination with petiolar radiotracer administration. Movement of free fluoride was observed to be consistent with bulk flow in xylem, namely a rapid and linear change in position with respect to time. Trapping, likely in the apoplast, was observed. Future applications of the methods described here include studies of transport of other ions and molecules of interest in plant physiology.

Keywords: Ion transport, Positron emission tomography, Radiotracer, Kinetic modeling, Brassica oleracea

Background

Ion transport is a fundamental process in plant physiology [1-3]. Radiotracer methods have been useful for studying transport phenomena because of their inherent sensitivity and quantitative accuracy [4]. Positron emission tomography (PET) is a non-invasive radiotracer imaging technique widely used in the clinic to diagnose diseases and monitor their progression [5]. PET imaging provides tracer kinetic data on the uptake, distribution, retention and clearance of positron-emitting compounds administered to living organisms. These tracer kinetic data can then be modeled to obtain insights into trapping mechanisms from which information on metabolism [5], receptor density [6], and other *in vivo* molecular interactions can be obtained.

In this study PET imaging was utilized to model fluoride ion transport in plants following absorption through the leaves by administering radiofluoride via a cut

* Correspondence: akconverse@wisc.edu
[1]T123 Waisman Center, University of Wisconsin-Madison, Madison, WI 53705, USA
Full list of author information is available at the end of the article

petiole. Fluoride uptake and distribution in the whole living plant was then non-invasively monitored in real-time by PET. Time-activity curves (TACs) in selected regions of interest (ROIs) in the plant were obtained and mathematically modeled to assess fluoride transport rates and trapping probabilties.

Fluoride is a trace element ubiquitous in the environment. Fluoride has beneficial effects in human health such as in preservation of teeth, but excessive fluoride leads to toxic effects such as fluorosis of bones and teeth in humans and animals as well as browning of leaves in plants [7] (p.2). Environmental fluoride comes from both natural sources, e.g. volcanic emissions, and industrial sources, e.g. aluminum smelting [7] (pp. 6–18). Inorganic fluoride can be taken up by plants from soil through the root system and from air through the stomata of leaves [8]. Fluoride uptake via roots is mitigated by its low solubility in soil. While the highest concentration of fluoride in plants is found in the roots and diminishes with distance from the roots, 95% of the fluoride present in the roots can be readily washed away with water [7] (p. 33). Plant uptake of fluoride from air is more significant than uptake from soil [9]. Fluoride concentration in the leaves can range from < 1 mg/kg to several thousand mg/kg [7] (p. 32). Reviews of mechanisms of fluoride uptake through the leaves from air and their effects on plant metabolism and growth can be found in [7] (pp. 36–46), [10], and [8].

The kinetics of uptake and transport of [^{18}F]fluoride in plants determined by positron emission imaging was first reported by McKay et al. [11] in a study utilizing a custom-made imaging system comprised of two planar detectors designed to detect coincident positron annihilation events. Later work by other investigators employed a more advanced positron-emitting tracer imaging system (PETIS) to follow plant uptake and transport of [^{18}F]fluoride to study radiation effects [12] and drought effects [13]. In the present study, we used a commercially available, small-animal, high-resolution 32-ring PET scanner, which we had found to be well-suited for plant imaging [14,15].

In contrast to previous studies that utilized cut stems that were immersed in aqueous [^{18}F]fluoride solution [11-13,16], the present study involved application of microliter amounts of aqueous [^{18}F]fluoride solution into the freshly-cut petiole of our plant model, rapid cycling *Brassica oleracea* [17]. This technique allowed for the behavior of a bolus of fluoride traveling up the stem of the living plant to be characterized using kinetic modeling, which permitted estimation of trapping probability and mean position as a function of time. Based on previous studies, we hypothesized *a priori* that the tracer would appear to move at a uniform velocity along the stem [11,18].

Results

Three plants (*Brassica oleracea,* age 22.0 +/− 3.5 days), were imaged by PET for one hour starting with petiolar administration of [^{18}F]fluoride (7.2 +/− 3.6 MBq) in 14 +/− 4 microliters of reverse osmosis water. PET images revealed uptake throughout the plant within ten minutes (Figure 1). Concentration over time is shown in Additional file 1: Movie 1, and the final distribution is shown in Additional file 2: Movie 2. As seen in Figure 2, time-activity curves for stems and leaves were consistent with bolus transport, and leaves exhibited greater relative trapping of radioactivity. A tracer kinetic model was implemented that describes the time courses of free and trapped [^{18}F]fluoride. This model was used to determine a single parameter, the trapping probability per unit time, sv (1/s), which is the product of the trapping probability per unit length, s (1/mm), and the local velocity, v (mm/s), at a given position along the stem. Using this model, the time-activity curve for the free fluoride was inferred at various positions along the stem of the plants, and this was used to calculate the mean arrival time of the free fluoride at each of these positions (Figure 3). The velocity of transport up the stem, V, was determined based on the mean arrival time of the free fluoride at various positions along the stem (Figure 4). Averaging over results for the three plants, the observed trapping probability per unit time was sv = 0.000253 +/− 0.000053 1/s, and the mean velocity was 0.368 +/− 0.180 mm/s. Trapping probabilities and velocities for each of the three plants are detailed in Table 1.

Discussion

The temporal resolution and quantitative precision afforded by PET, combined with the petiolar administration of a [^{18}F]fluoride bolus into the living plant, permitted kinetic modeling that yielded estimates of ion velocity and trapping fraction in the stem. We observed fluoride transport consistent with bulk flow of water in xylem [18], namely a linear change in position with respect to time. Qualitatively, the time-activity curves that we observed in leaves, reaching a plateau at approximately 10 minutes, agree with a report of [^{18}F]fluoride fed to cut stem of soybean [12]. Compared to another report of fluoride transport in the cut stem of soybean [11], the speed we observed in live *Brassica oleracea*, 2.2 cm/min, was 16x slower, and the trapping probability we observed, 0.8%/cm, 1.7x higher. These differences may be attributable to the species studied, environmental conditions in the PET scanner room, physiologic states of the samples, and the presumably lower hydraulic conductance of our plant preparation with intact roots, compared with an open stem.

This study demonstrates the feasibility of imaging basic physiological processes in living plants using a

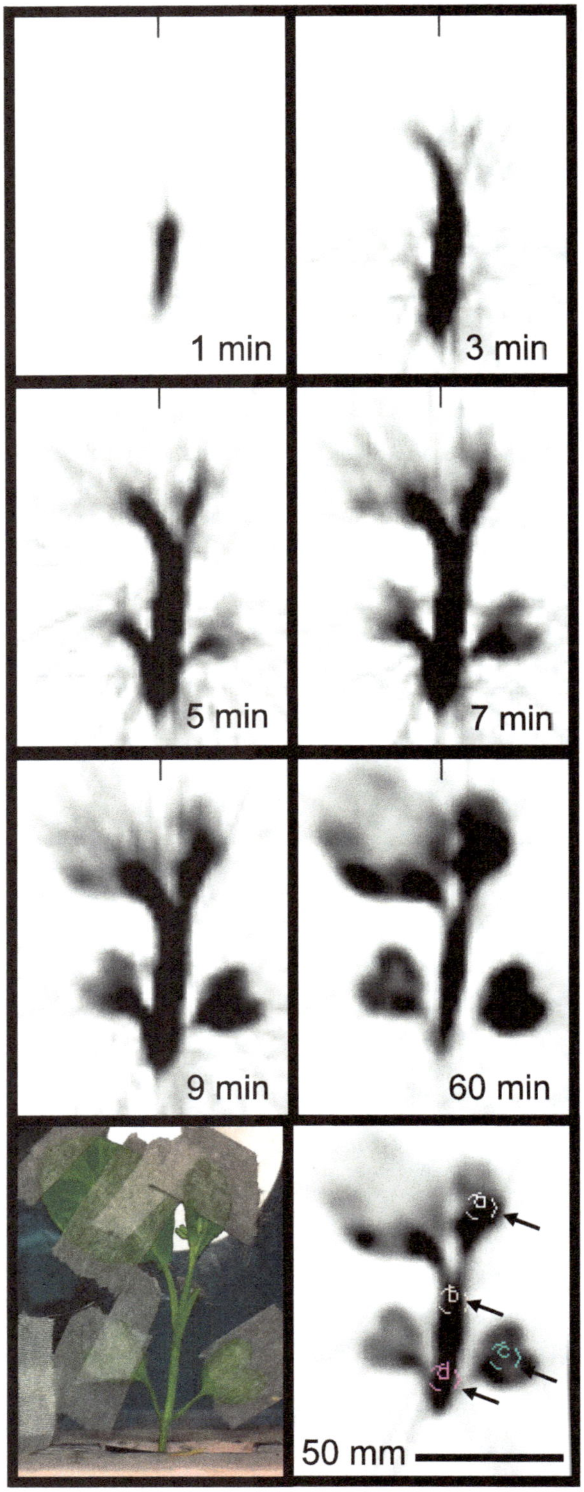

Figure 1 [^{18}F]fluoride uptake. PET images shown in a 1.2 mm thick transverse scanner plane. Times shown relative to start of petiolar administration. bottom row: photograph of plant and four regions of interest (arrows) delineated on 0 to 60 minute image used for time-activity curves shown in Figure 2 (a = leaf, b = stem - high, c = cotyledon, d = stem - low).

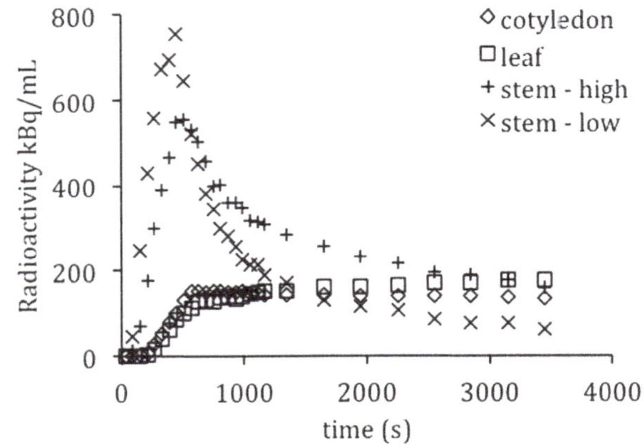

Figure 2 Radioactivity in stem and leaves. Time-activity curves corresponding to regions of interest shown in Figure 1. Radioactivity is seen moving as a bolus through the stem and accumulating in the leaves.

commercially available and widely installed high resolution PET scanner primarily intended for small animal studies. The microPET P4 used in this study easily fits *Brassica oleracea* of up to 4 weeks of age in its 22 cm diameter field of view. The spatial resolution of 2 mm full width at half maximum (FWHM) facilitates delineation of regions of interest to determine time courses in different parts of the plant. Transmission scan capability also permits quantification of the radiotracer temporally and spatially.

In order to model the kinetics of the tracer, we desired a finite duration pulse of fluoride moving through the plant. To that end, we developed a petiolar method of tracer administration. As we reported previously, continuous administration at the roots resulted in constantly increasing uptake in the stem and leaves [14]. With limited administration of a few drops of tracer to the cut petiole, we could observe a finite bolus of activity traveling through the stem permitting application of the kinetic model described here. This petiolar bolus administration was adapted from the petiole-feeding method described by Lin et al. [19]. Because petiole feeding results in the distribution of the fed constituents throughout the entire plant, Lin et al. inferred that translocation was via the phloem and/or apoplast [19]. The images in Figure 1 show whole plant distribution similar to that seen by Lin et al. in their studies using blue dye and radioactive mannitol [19].

The use of relatively widely available scanner technology and a well-established plant model species supports the potential of this method to study plant physiology *in vivo*. Although this study was limited by incomplete quantification because positrons were only partially absorbed [20], the kinetic model that we used was unaffected because it depended only on the shape of the

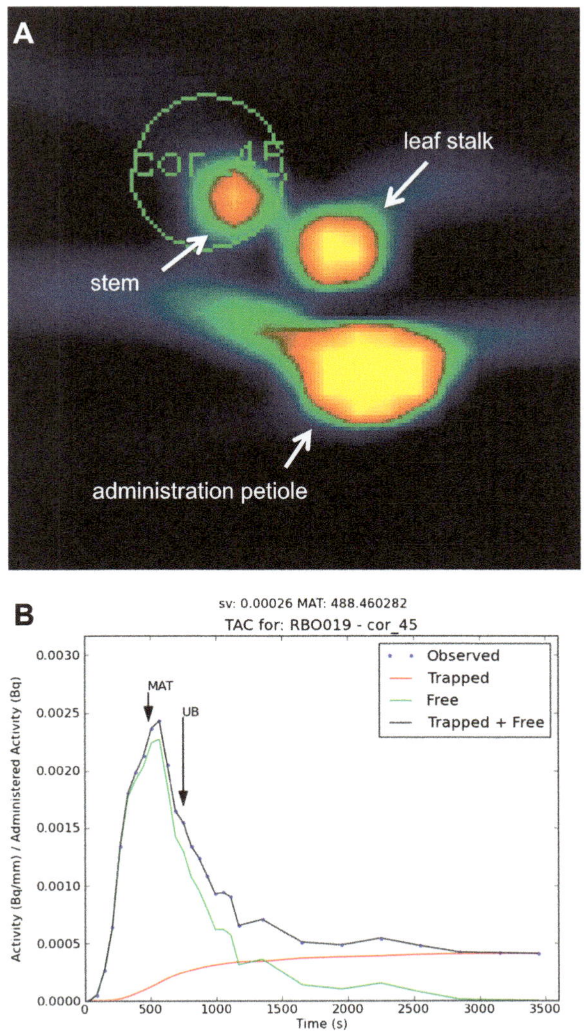

Figure 3 Modeling of observed radioactivity in stem. (A) 1.9 mm thick horizontal cross section of plant shown in Figure 1 at upper stem. **(B)** Time-activity curves showing observed activity in stem (blue), modeled free activity (green), modeled trapped activity (red), total modeled activity (black). MAT: mean arrival time, UB: upper bound of integral to calculate MAT.

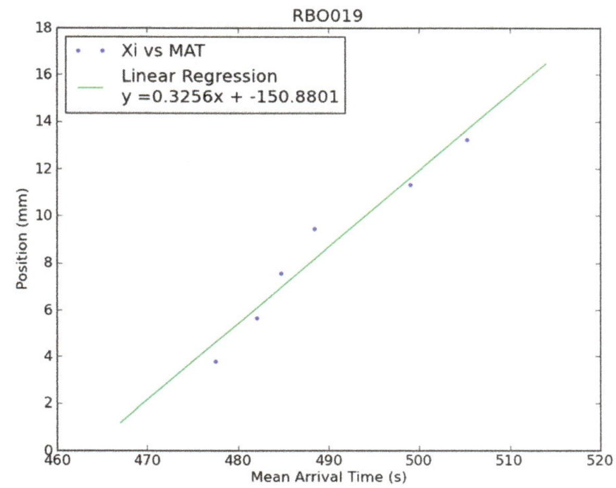

Figure 4 Speed of bolus traveling up stem. Plot of mean arrival time (s) at various positions along the stem (mm) with respect to the node of the bisected petiole used for administration. The slope gives the speed (mm/s) at which the bolus travels.

time-activity curves and not the overall scale. As seen in the dynamic image (Figure 1) and the TACs (Figure 2), the tracer appears first below the administration petiole. In an effort to elucidate this behavior, this simple kinetic model of [^{18}F]fluoride translocation was also applied downward along the stem below the cut petiole. However, this analysis was unsuccessful (not shown), likely due to the complexity of transport in this region of the stem, which may require multi-compartmental analysis [21].

The accuracy and precision of the measures presented here may be subject to a number of influences. The observed velocity, 2.2 +/− 1.1 cm/min (s.d. n = 3), has a standard error of the mean of +/− 0.6 cm/min, i.e. 95% C.L. 1.0 - 3.4 cm/min, and coefficient of variation of 49%. Some of the variance could be due to differences between the plants and environmental conditions during scanning, while some variance may involve the analysis methods. For instance, all of the free fluoride was assumed to have cleared the regions of interest by the end of the 60 minute scans leaving only an unchanging level of irreversibly trapped fluoride. While this appears to be a plausible assumption in the case of plant RBo019, it may lead to an overestimation of the trapped concentration for the other two plants (Additional file 3: Figure S1). We therefore performed a sensitivity analysis by determining values for sv such that the trapped activity reached only 80% of the final observed activity. This led to relatively small changes in the observed velocities of −2.1% for RBo021 and +3.7% for RBo026, which suggests that 60 minutes is a sufficiently long scan duration. As a further check of the methods, we calculated fluoride arrival times corresponding to the half-height of the free fluoride time activity curves, a simpler technique that uses only the values from the leading edge of the bolus [22]. This yielded comparable velocities (1.6 +/− 1.0 cm/min), albeit with a somewhat higher coefficient of variation, and lends additional confidence to the mean arrival time analysis presented in the Results.

Applications of these techniques currently underway in our laboratory include studies of transport of other substances in living plants. These substances include the ions copper, a crucial micronutrient, chloride, which is important both as a nutrient and a toxin, as well as arsenic, which is an environmental pollutant with significant human health impacts. Understanding the transport behavior of arsenic in plants may aid in advancing methods for bioremediation of arsenic contaminated sites. We are also assessing the mechanisms of uptake

Table 1 Tracer kinetic modeling results

Plant	Age (days)	Activity (MBq)	Admin. duration (s)	ROIs	sv (1/s) mean(s.d.)	V (mm/s)	s = sv/V (1/mm)
1 RBo019	20	3.18	240	6	0.000287(220)	0.326	0.00088
2 RBo021	20	8.53	300	8	0.000191(058)	0.566	0.00034
3 RBo026	26	9.96	180	5	0.000280(029)	0.214	0.00131
Mean (s.d., n = 3)	22.0(3.5)	7.2(3.6)	240(60)		0.000253(53)	0.368(180)	0.00084(49)

ROIs = number of regions of interest at different positions on stem, sv = trapping probability per unit time; V = velocity up stem, slope of linear regression of position vs mean arrival time; s = Trapping probability per unit length of stem. Corresponding values in units of cm and min, mean(s.d., n = 3): sv = 0.0152(32) 1/min, V = 2.21(1.08) cm/min, s = 0.0084(49) 1/cm.

and transport of auxins, plant hormones that control growth and other critical processes such as phototropism and gravitropism, using radiotracer imaging by PET. Additionally, physical stress associated with the experimental techniques such as taping to the acrylic plate and bisecting the petiole for radiotracer administration should be further explored and perhaps mitigated [23]. Finally, these techniques may be used to study the influence of environmental conditions such as light, temperature, wind, and humidity on transport processes of various compounds with physiologic relevance to plants.

Conclusions

The protocol described here utilized a novel combination of techniques to achieve quantitative assessment of fluoride transport in living plants. High resolution commercial small animal PET scanners such as the one used in this work are available for similar studies at many institutions. Petiolar administration permits bolus injection of microliter quantities of radiotracer, and leaves the plant largely intact. This, along with the spatial and temporal resolution of PET allow compartmental kinetic modeling to obtain measures of fluoride velocity and trapping fraction along the stem.

Methods

Plants

Experiments were conducted with greenhouse-grown *Brassica oleracea* plants. Seeds of rapid cycling *Brassica oleracea* (RCBC stock #3-012, https://rcbc.wisc.edu) were grown under controlled conditions in 30 cc cell packs filled with a 1:1 mixture of Redi-earth™ and coarse vermiculite (Sun Gro Horticulture Ltd, Canada). Reverse osmosis water was provided through capillary wicking from a reservoir. Each cell/plant received approximately 0.5 g Osmocote™ pellet slow-release fertilizer (14-14-14 NPK) at planting. Additionally, each cell/plant received approximately 25 mL of 2 g/L Peter's Professional fertilizer (20-20-20 NPK, with micronutrients) at planting and weekly thereafter.

Plants were grown in Madison, WI (43°04'34.1"N 89° 25'20.6"W) under a 16 hr photoperiod from May to July in natural light with a daily maximum of 1500 μmol photons m^{-2} s^{-1} photosynthetically active radiation (PAR) supplemented with 500 μmol photons m^{-2} s^{-1} PAR from high pressure sodium lamps. Greenhouse temperature was 21 +/− 3°C. Plants were grown for 20–26 days until they were a suitable size for injection of tracer compounds and scanning.

[^{18}F]Fluoride

[^{18}F]Fluoride was produced via the $^{18}O(p,n)^{18}F$ nuclear reaction at 16 MeV using the UW-Madison Department of Medical Physics PETtrace cyclotron (GE Healthcare, Uppsala, Sweden) and pure ^{18}O-enriched water [24]. This irradiated target water containing [^{18}F]fluoride was used directly for plant petiolar administration.

Scanning procedure and petiolar administration

Plants were imaged in the UW-Madison microPET P4 scanner (Siemens, Knoxville) with resolution of 2 mm × 2 mm × 2 mm FWHM and 8 cm axial × 19 cm diameter field of view. A plant was placed in a 4.5 cm × 3.0 cm × 4.7 cm hole in a 19.0 cm × 11.5 cm × 5.0 cm block of styrofoam. Directly behind the plant, a slit was cut into the styrofoam such that a 15.2 cm × 11.3 cm × 0.3 cm sheet of acrylic could be mounted behind it. The edges of the plant's leaves were then taped to the acrylic using paper surgical tape (Micropore™). This resulted in the topside of the leaves being flush against the acrylic with the stomata-rich leaf underside exposed to air. The acrylic was intended to stop positrons and thus increase coincidence counts based on the expectation that over half of [^{18}F] positrons escape the parenchyma of a typical leaf [20]. The first true leaf was left untaped to point approximately parallel to the scanner bore, and would be used for the petiolar injection. The plant was then centered in the field of view of the scanner. A ^{57}Co transmission scan was acquired (120–125 keV energy window) and then a 60 minute emission scan was begun (350–650 keV, 6 ns coincidence window). Sixty seconds after scan start the petiole was cut using a wet scalpel to ensure that the vasculature remain hydrated and open, and [^{18}F] in ^{18}O-enriched pure water was applied a drop at a time as each bead was absorbed using a 25 μL liquid chromatography injection syringe (Hamilton) for up to 5 minutes (Figure 5) [19]. No further measures were taken after [^{18}F] "injection". All data were acquired in

Figure 5 The petiolar administration method is illustrated here with dye in place of radiotracer and outside the PET scanner for clarity. The petiole of the first true leaf is transected with a wet scalpel **(A)**, tracer is administered with a liquid chromatography syringe **(B)**, and additional drops are applied before the preceding drop is completely absorbed **(C)**.

3D list mode. Studies were performed at 21 +/− 1°C, typical humidity of 25-45%, and typical light intensity of 200–250 μmol photons m^{-2} s^{-1} PAR.

Image reconstruction

Image reconstruction was carried out using the scanner vendor's software (Siemens, microPET Manager 2.4.1.1) For each scan, emission list data were binned into 3D sinograms with time frames of 20×1 + 8×5 minutes (4 byte integers, 168 projection angles x192 bins, span 3 ring difference 31, hist.exe 2.338). For use in attenuation and scatter correction, ^{57}Co transmission list data were binned into 3D sinograms, which were then reconstructed by filtered back projection (FBP, full transaxial field of view). Emission images were reconstructed with FBP (fourier 2D rebinning, pixel size 1.90 mm × 1.90 mm in-plane × 1.21 mm slice thickness, ramp filter, recon.exe 2.330). The resulting images included corrections for detector sensitivity, deadtime, decay, attenuation, and scatter.

Image post processing

The images were visualized using the scanner vendor's software (Siemens ASIPro VM 6.7.1.2). 2 mm transverse × 2 mm axial Gaussian smoothing was performed on each PET image. The plant petiole, stem, and the junction of the petiole and stem were identified. Circular ROIs of a fixed diameter were drawn using the ASIPro ROI tool. ROI drawing was performed in horizontal planes to cover the cross sectional area of the stem. Depending on the anatomy of the plant, 5 to 8 ROIs were drawn on successive planes beginning at the junction between the petiole and stem and ending when the stem branched into leaves. The ROI closest to the petiole-stem junction was drawn on the first plane where the petiole and stem appeared as two separate circular regions of activity (Additional file 3: Figure S1). Time-activity curves were calculated from the ROIs using the ASIPro software and saved as text files. Before modeling, each time-activity curve was converted to concentration per vertical length of stem per total administered activity.

Tracer kinetic modeling

A model was implemented based largely on the one described by McKay et al. [11]. Radiotracer is assumed to move unidirectionally up the stem and to be irreversibly trapped at a given position with probability s (1/mm) per unit length. For radiotracer moving at velocity v (mm/s), the trapping probability per unit time is then sv (1/s). Observed radiotracer per unit length at a given position of the stem as a function of time, M(t) (Bq/mm), is assumed to consist of free radiotracer G(t) (Bq/mm) and trapped radiotracer B(t) (Bq/mm), such that

$$M(t) = G(t) + B(t). \quad (1)$$

The rate at which radiotracer is trapped is then given by

$$dB(t)/dt = sv\, G(t) = sv\; (M(t)\text{-}B(t)), \quad (2)$$

and integrating over time yields

$$B(t) = sv \int (M(t')\text{-}B(t'))\; dt'. \quad (3)$$

Given the initial condition $B(t = 0) = 0$, the trapped concentration B(t) can then be calculated iteratively for a given set of observed M(t) and the trapping per unit time parameter sv. A value of sv is determined to meet the final condition that at a late time t_f all of the observed activity is trapped, i.e. $M(t_f) = B(t_f)$. The mid-time of the final frame of PET data, 55 minutes, is taken as t_f. Once the trapped time-activity curve B(t) has been calculated to meet the final condition, the free time-activity curve is calculated, $G(t) = M(t) - B(t)$ (Additional file 3: Figure S1). Then the mean arrival time of the free radiotracer is given by

$$T = \int tG(t)dt / \int G(t)dt. \quad (4)$$

This mean arrival time was calculated for several positions up the stem and plotted as position vs time

(Figure 4). For each plant, the slope of a linear fit to the time-position plot gave the average velocity of the radiotracer traveling up the stem (Additional file 4: Figure S2).

While the integral for Equation 4 would ideally be taken from zero to infinity, the resulting time-position plots were seen to deviate significantly from linearity. This deviation is likely a result of simplifying assumptions in the kinetic model. By limiting the upper bound of the equation 4 integrals, more linear time-position plots were generated, with the upper bound at approximately the half-height of the free time-activity curve for optimal fit. Physically, this corresponds to calculating the mean arrival time of the first 50 to 70% of the free [^{18}F]fluoride, likely minimizing the confounding effects of bidirectional fluoride movement and reversible trapping.

The analysis code is available at https://go.wisc.edu/RBoKinetics.

Additional files

Additional file 1: Movie 1. Uptake over time, 60 1--min frames, 3 mm Gaussian smooth.

Additional file 2: Movie 2. Distribution at 40--60 min, 3 mm Gaussian smooth.

Additional file 3: Figure S1. Kinetic modeling. left: Regions of interest delineated on PET image of horizontal section of stem. right: Corresponding time-activity curves. blue = observed, red = trapped (model), green = free (model), black = trapped (model) + free (model).

Additional file 4: Figure S2. Position vs time plots. Slope of best fit line gives velocity of radiotracer traveling up stem.

Competing interests

The authors declare that they have no competing interests.

Authors' contributions

AKC, RJN, PHW, and OTD designed the study. AKC, EOA, TWB, PHW, JWE, and TEB performed the experiments. AKC, EOA, KAL, PAE, PHW, and OTD analyzed and interpreted the data. AKC, EOA, JDH, KAL, PAE, and OTD wrote the manuscript. All authors read and approved the final manuscript.

Acknowledgments

The authors thank Simon Gilroy for helpful discussions. The reviewers also provided many useful suggestions. Funding was provided by US Department of Energy grant DE-SC0005281 (OTD). Additional support was provided by US National Institutes of Health grants P30HD03352 (AKC) and T32CA009206 (PAE).

Author details

[1]T123 Waisman Center, University of Wisconsin-Madison, Madison, WI 53705, USA. [2]Department of Plant Pathology, University of Wisconsin-Madison, Madison, WI 53705, USA. [3]Department of Medical Physics, University of Wisconsin-Madison, Madison, WI 53705, USA.

References

1. Assmann SM. From proton pump to proteome. Twenty-five years of research on ion transport in higher plants. Plant Physiol. 2001;125:139–41.
2. Brouwer R. Ion absorption and transport in plants. Annu Rev Plant Physiol. 1965;16:241–66.
3. Nardini A, Salleo S, Jansen S. More than just a vulnerable pipeline: xylem physiology in the light of ion-mediated regulation of plant water transport. J Exp Bot. 2011;62:4701–18.
4. Minchin PEH, Thorpe MR. Using the short-lived isotope C-11 in mechanistic studies of photosynthate transport. Funct Plant Biol. 2003;30:831–41.
5. Farwell MD, Pryma DA, Mankoff DA. PET/CT imaging in cancer: current applications and future directions. Cancer. 2014;120(22):3433–45.
6. Bartlett RM, Dejesus OT, Barnhart TE, Nickles RJ, Christian BT, Graner JL, et al. Fetal dopamine receptor characteristics assessed in utero. J Cereb Blood Flow Metab. 2010;30:1437–40.
7. Weinstein LH, Davison AW. Fluorides in the Environment: Effects on Plants and Animals. Wallingford, UK; Cambridge, MA: CABI Publishing; 2004.
8. Mesquita GL, Ossamu Tanaka FA, Cantarella H, Mattos Jr D. Atmospheric absorption of fluoride by cultivated species. Leaf structural changes and plant growth. Water Air Soil Pollut. 2011;219:143–56.
9. Groth E. Fluoride pollution. Environment. 1975;17:29–38.
10. Baunthiyal M, Bhatt A, Ranghar S. Fluorides and its effects on plant metabolism. Int J Agric Technol. 2014;10:1–27.
11. McKay RML, Palmer GR, Ma XP, Layzell DB, McKee BTA. The use of positron emission tomography for studies of long-distance transport in plants - uptake and transport of F-18. Plant Cell Environ. 1988;11:851–61.
12. Kume T, Matsuhashi S, Shimazu M, Ito H, Fujimura T, Adachi K, et al. Uptake and transport of positron-emitting tracer (F-18) in plants. Appl Radiat Isot. 1997;48:1035–43.
13. Nakanishi TM, Tanoi K, Yokota H, Kang DJ, Ishii R, Ishioka NS, et al. F-18 used as tracer to study water uptake and transport imaging of a cowpea plant. J Radioanal Nucl Chem. 2001;249:503–7.
14. Converse AK, Ahlers EO, Williams PH, Engle JW, Barnhart TE, Nickles RJ, et al. Fluoride transport in brassica: a positron emission tomography botanical study [poster MIC9.S-1]. IEEE Nucl Sci Symp Med Imaging Conf Conf Rec. 2011.
15. Converse AK, Ahlers EO, Bryan TW, Williams PH, Barnhart TE, Engle JW, et al. Positron emission tomography (PET) of radiotracer uptake and distribution in living plants: methodological aspects. J Radioanal Nucl Chem. 2013;297:241–6.
16. Keutgen N, Matsuhashi S, Mizuniwa C, Ito T, Fujimura T, Ishioka NS, et al. Transfer function analysis of positron-emitting tracer imaging system (PETIS) data. Appl Radiat Isot. 2002;57:225–33.
17. Williams PH, Hill CB. Rapid-cycling populations of Brassica. Science. 1986;232:1385–9.
18. Ohya T, Tanoi K, Hamada Y, Okabe H, Rai H, Hojo J, et al. An analysis of long-distance water transport in the soybean stem using H215O. Plant Cell Physiol. 2008;49:718–29.
19. Lin Y-H, Lin M-H, Gresshoff PM, Ferguson BJ. An efficient petiole-feeding bioassay for introducing aqueous solutions into dicotyledonous plants. Nat Protoc. 2011;6:36–45.
20. Alexoff DL, Dewey SL, Vaska P, Krishnamoorthy S, Ferrieri R, Schueller M, et al. PET imaging of thin objects: measuring the effects of positron range and partial-volume averaging in the leaf of Nicotiana tabacum Nucl Med Biol. 2011;38:191–200.
21. Bühler J, von Lieres E, Huber G. A class of compartmental models for long-distance tracer transport in plants. J Theor Biol. 2014;341:131–42.
22. Ferrieri RA, Gray DW, Babst BA, Schueller MJ, Schlyer DJ, Thorpe MR, et al. Use of carbon-11 in Populus shows that exogenous jasmonic acid increases biosynthesis of isoprene from recently fixed carbon. Plant Cell Environ. 2005;28:591–602.
23. Gilroy S, Suzuki N, Miller G, Choi W-G, Toyota M, Devireddy AR, et al. A tidal wave of signals: calcium and ROS at the forefront of rapid systemic signaling. Trends Plant Sci. 2014;19:623–30.
24. Barnhart TE, Converse AK, Dabbs KA, Nickles RJ, Buckley K, Jivan S, et al. Water-cooled grid support for high-power irradiation with thin target windows. Appl Radiat Isot. 2003;58:21–6.

A comparison of high-throughput techniques for assaying circadian rhythms in plants

Andrew J Tindall, Jade Waller, Mark Greenwood, Peter D Gould, James Hartwell and Anthony Hall*

Abstract

Over the last two decades, the development of high-throughput techniques has enabled us to probe the plant circadian clock, a key coordinator of vital biological processes, in ways previously impossible. With the circadian clock increasingly implicated in key fitness and signalling pathways, this has opened up new avenues for understanding plant development and signalling. Our tool-kit has been constantly improving through continual development and novel techniques that increase throughput, reduce costs and allow higher resolution on the cellular and subcellular levels. With circadian assays becoming more accessible and relevant than ever to researchers, in this paper we offer a review of the techniques currently available before considering the horizons in circadian investigation at ever higher throughputs and resolutions.

Keywords: Luciferase, Leaf movement, Delayed fluorescence, Infra-red gas exchange, Circadian clock, High-throughput assay

Introduction to the plant circadian clock

Circadian clocks are endogenous, persistent, temperature-compensating timekeepers which provide temporal organization of biological processes from cyanobacteria to man [1]. In plants, circadian rhythmicity is widespread; a transcriptional circadian cycling has been reported in a range of diverse species [2], including the model plant *Arabidopsis thaliana* in which the clock has been well characterized [3]. Approximately one-third of the *Arabidopsis* transcriptome shows circadian oscillations in abundance when in free-run conditions [4], indicating direct or indirect circadian control.

It is becoming increasingly clear that robust circadian rhythms are integral to overall fitness [3], are key players in the control of flowering time [5], are regulators of the susceptibility and response to pathogen attack [6-8], and are linked to important agronomic traits in multiple crop species including potato, rice, wheat, barley and soybean [9-15].

In the last twenty years, we have begun to elucidate essential features of the clock in the model plant Arabidopsis thaliana. The clock appears thus far to be considerably conserved with the rest of the plant kingdom [3,16,17]. Thanks to molecular studies, and the use of circadian reporter systems, we know that the central oscillator consists of three interlocking auto-regulatory transcriptional feedback loops. LATE ELONGATED HYPOCOTYL 1 (LHY) and CIRCADIAN CLOCK ASSOCIATED 1 (CCA1) repress transcription of their inducer *TIMING OF CAB EXPRESSION 1 (TOC1)* at dawn, forming the "morning loop" [18]. A second "evening loop", comprising of GIGANTEA (GI) as an inducer of *TOC1* transcription which is in turn repressed by CCA1/LHY1, was predicted in silico to fit empirical observations [19] and has had it's existence confirmed experimentally [20]. The final transcriptional loop represses transcription of *CCA1/LHY* through the activity of PSEUDO-RESPONSE REGULATOR 7 (PRR7) and PRR9 and is known as the "Night Inhibitor" loop (NI) [21]. This transcriptional-level model is further augmented by post-transcriptional and post-translational modifications that affect clock function. GI protein regulates the evening loop at the post- translational level through the stabilisation of the F-box protein ZEITLUPE (ZTL) in the presence of light [22] which, in turn, is necessary for the targeting of TOC1 protein for degradation by the proteasome. The connections between LHY and the night inhibitor loop [23], as well as between

*Correspondence: anthony.hall@liverpool.ac.uk
Institute of Integrative Biology, University of Liverpool, Crown Street, Liverpool, UK

TOC1 and CCA1/LHY [24], are also post-translational in nature.

Our understanding of the clock, however, is far from complete. Environmental inputs and outputs of the clock are not well characterised. Whilst the transcription of a significant number of genes appears to be circadian regulated, relatively few possess a predicted circadian motif in their promoter, suggesting control through as-yet-to-be implicated factors [25,26]. Additionally, whilst it is known that the central transcriptional clock is tuned or "entrained" by diurnally oscillating external stimuli such as red and blue light [27,28], temperature [29], and phytohormones [30], it remains unclear how these pathways are regulated.

In the last two decades, our knowledge of the plant circadian clock has been vastly improved through the development of new assay techniques. The high-throughput nature of these assays has led us towards a more integrated understanding of how the circadian system works and, more over, has allowed us to perform work that would previously be impossible. In this review, we outline the major techniques that have been developed for investigating the circadian clock in plants before considering the possible horizons that will open up following future assay development.

The need for high-throughput assays in circadian investigation

The circadian system can be probed in considerable depth through the use of molecular techniques such as quantitative-PCR and Northern blotting to assay clock gene expression directly over a sampling time-course. Indeed, these techniques are used extensively in circadian studies alongside high-throughput techniques, having been used to characterise numerous clock components, most notably *CCA1* [31] and *LHY* [32]. More recently, the proliferation of micro-arrays and RNA-seq technology has expanded the molecular toolkit, allowing us to view the abundance and splicing patterns of multiple mRNAs simultaneously [33,34]. They are still vital tools for in-depth characterisation and investigation of the mechanism of action for individual circadian genes, and for teasing apart the roles of separate clock loops.

However, molecular techniques are considerably time and resource consuming, requiring the researcher to perform multiple, regular samplings over several days. They also require destructive sampling – the harvest of leaf tissue, whole plants or groups of plants– which not only requires lots of plant growth space, but involves large amounts of resource-consuming wet-work at the bench following the time-course. This necessarily reduces the resolution and throughput of the assay, limiting not only the sampling frequency but also the number of parallel and consecutive experiments that the lab's finances (and level of fatigue) can support. Additionally, destructive sampling raises issues with averaging biological variation between tissues as well as between plants, making observation of an individual plant or tissue throughout a time-course impossible. Together, this makes molecular techniques unsuitable for the wide-scale screening and initial identification of clock genes.

By means of contrast, high throughput assays are far better suited to screening. As non-invasive, non-destructive techniques they allow concurrent sampling on the same plant without perturbing the clock through stress or killing it. They also allow a high level of automation. As opposed to physical harvest, the assays outlined below can be set up and left to run with minimal human intervention, compared with methods where harvest have to be performed regularly over several days. This has resulted in a paradigm shift in recent years, whereby the limiting factor in the throughput of circadian investigations has become the data handling and preparation of plant materials, and the scale of these experiments is limited only by the capacity and number of assay machines. This allows researchers to carry out broader, further-reaching experiments investigating the influence of several factors on the clock to identify candidates for future in-depth study.

Lumino-fluorescent circadian reporters

Transgenic luciferase as a circadian reporter

Increasing numbers of investigations of the transcriptional circadian oscillator are using transgenic luciferase reporters to probe clock period and robustness [35,36]. Luciferase is an enzyme that catalyses the light-emitting ATP-dependent monooxidation of luciferin, a compound with which test plants are dosed. Typically, the brighter, modified *LUC+* gene sourced from the firefly *Photinus pyralis* is used [37]. Luminescence intensity can be recorded through the use of photo-multiplier tubes attached to detectors, as in the TopCount system, or, increasingly, high sensitivity charge-coupled cameras enclosed within light-tight growth chambers with automated lights [36].

In a typical cell, following saturation with luciferin, oxygen and ATP are in excess; as such the amount of luminescence observed is directly dependent on how much luciferase is present [35]. As luciferase is an unstable enzyme that rapidly loses function, the amount of functional luciferase is directly determined by the rate of luciferase expression. When under the control of a circadian-regulated promoter, the observed luminescence provides a quantitative measure of promoter-driven gene expression. This makes luciferase, when placed under the control of a circadian-regulated promoter such as *CAB2* or *CCR2* [38], an excellent reporter of clock transcriptional output [35,39,40]. This technique has been used to identify several core clock genes, most notably *TOC1*

[41], and to investigate the roles external entraining stimuli such as temperature [42-44] and sucrose [45] play in clock function.

Interestingly, in recent years the sensitivity of luminescence imaging cameras has increased to the point where more specific resolution is now possible. In the duckweed *Lemna gibba*, circadian rhythms in *AtCCA1:LUC+* have been detected within individual cells following transient transformation by particle bombardment [46]. In *Arabidopsis*, tissue-specific clocks have been identified in the leaf mesophyll and vasculature through the use of split-luciferase assays, wherein half of the luciferase protein is driven by a clock promoter and another half by a tissue-specific promoter, which are only luminous when expressed together in the same cell [47]. However, whilst luciferase has been used successfully to assay clock function in various species, most notably tobacco, arabidopsis [35] and rice [48], the need to introduce transgenic luciferase into the plants is a time-consuming step that greatly reduces throughput and renders the system unsuitable for species without a transformation protocol. Whilst currently transgene expression is dependent on promoter activity and the genomic context of the random transgenic insertion, in the future and through the use of precision genome engineering techniques [49] we are likely to see luciferase being fused directly to endogenous genes within the native context.

Delayed fluorescence as an endogenous, universal assay

First described in 1951 [50], delayed fluorescence is the emission of light from plants, algae and cyanobacteria following their transfer from light to dark conditions resulting from charge recombination in the photosynthetic machinery [51], primarily within the P680 light harvesting complex of photosystem II [52-54].

In normal photosynthesis, incoming photons excite electron pairs in the light harvesting complexes of the photosystems, raising them to higher energy states. These electrons are passed to receiver molecules further down the photorespiratory chain which use their stored energy to generate a membrane potential, pump protons across the thylakoid membrane, and do biological work [55]. Charge-recombination between the receiver molecule plastoquinone QA re-excites the P680 complex, which produces fluorescence as it returns to the ground state through the release of stored energy in the form of a photon [53]. Approximately 0.03% of absorbed solar energy is re-emitted in this manner [53]. The intensity of the delayed fluorescence emission decays rapidly to undetectable near-background levels over the course of a minute [56,57]; it is therefore critical that images are taken over exactly the same period following lights off in order to observe rhythmicity. This persistence allows the long exposure windows required for detailed imaging. However, because the intensity decays exponentially and rapidly, accurate and precise control of the light source and length of delay is required to acquire accurate data.

The intensity of delayed fluorescence has been shown to be under the control of the circadian clock. The periods and robustness of known clock mutants *lhy-21*, *cca1-11*, *gi-11* and *toc1-2* as detected by delayed fluorescence agree with those previously determined by other techniques [58]. Whilst the exact nature of the relationship between the clock and delayed fluorescence is still unclear, many of the key genes that make up the light harvesting complexes within PSI and II are under circadian control at the transcriptional level [59], providing a possible link between the nuclear transcriptional clock and this particular output.

As a platform for circadian phenotyping, delayed fluorescence provides a non-transgenic system that can be used to assay the clock in a vast number of species. The CCD cameras used for luciferase imaging are sufficiently sensitive for delayed fluorescence assays, provided that the growth chamber and lights are under accurate precision control [58]. To date, it has been used to investigate plant species including *Arabidopsis*, the C3 monocots barley [58] and einkorn wheat [60], the C4 monocot maize [58], the model CAM species *K. fedtschenkoi* [58], and the coniferous gymnosperm Norway spruce (*Picea abies*) [61]. The system provides a quick and simple, non-invasive, universal platform for phenotyping plant clocks across species and taxa without the need for transformation.

Leaf movement as a circadian reporter

Along with diel rhythms in stem and root elongation, leaf movement has long been well-characterised as a growth-dependent circadian clock output [62]. Circadian-regulated oscillations in leaf position are a result of differential patterns of growth in cells on opposing adaxial and abaxial sides of the petiole (the structure that connects the leaf blade to the stem, which causes the positions of young leaves to rise and fall throughout the 24 hour period [63,64]. This movement is differentially phased from petiole elongation, and requires a functional *ELF3* and evening complex for proper phasing [64].

Although the biological importance of leaf movement remains unproven, it has long been used as a circadian reporter. The plant clock was first described through the study of rhythmic pulvinus-driven opening and closing of leaves in *Mimosa pudica* [65]. Low-throughput leaf movement assays have also provided insight into the clock in *Arabidopsis*, with the absence of leaf movement rhythms in *lhy* mutants helping to implicate that gene in the circadian system [32]. Measuring directly the clock controlled physiological outputs such as leaf movement provides a non-invasive assay that, unlike luciferase reporters, does not require transformation. Leaf movement is favoured over other physiological outputs due to the increased rhythm robustness and ease of assay [66]. Computer-automated image capture by CCD within the growth chamber can be performed throughout the time-course [67]. Over the past decade several straight-forward systems have been developed based on computer-automated capture of leaf position [68], whilst advancements in digital camera technology has reached a point where consumer-level cameras are suitable for circadian imaging, dramatically driving down the hardware costs of such systems [69]. Compared to lumino-fluorescent techniques, with associated equipment costs in the tens of thousands of pounds, relatively cheap digital cameras have removed a major barrier to entry for circadian phenotyping. The more recent development of sophisticated machine vision algorithms and reliable computer automation have helped automate the leaf tracking analysis, previously a time-consuming bottleneck in the pipeline for the researcher. Together this has gone a long way to enabling the use of leaf movement as a high throughput clock assay.

Whilst this system can be used in non-transformable plants, it cannot be used for species with sessile leaves (i.e. those that lack petioles). This makes the vast majority of monocots, including all major cereal crops, unsuitable for this technique. Furthermore, in *Arabidopsis*, leaf movements halt once the leaves are mature which equates to a window for assay of approximately one week [67]. Regardless, leaf movement is a robust and accurate assay that, with further development and ever-reducing hardware costs, is an attractive and valuable tool for investigating circadian function. With the advent of sophisticated image analysis algorithms, in the future other circadian regulated growth processes are likely to be developed into high throughput systems, for example hypocotyl growth [66].

Infra-red gas exchange

Observations from various studies throughout the past century have alluded to the fact that carbon dioxide fixation displays a 24-hour rhythm in photosynthetic organisms [70] and this has long since been shown to be a reliable measure of circadian output [71,72]. Carbon fixation can be assayed quickly through use of an Infra-Red Gas Exchange Analyser (IRGA), whereby whole or partial plants are grown in chambers containing a known, controlled atmosphere, and the resulting output air analysed for it's composition, allowing photosynthetic rate to be defined rapidly and with high resolution, with sample readings being taken multiple times per hour.

The IRGA has become an effective tool to assay the circadian clock, especially within the crassulacean acid metabolism community [73] where it has proven useful at identifying peripheral and partially-redundant components that when perturbed feedback to affect the clock which have otherwise proved elusive [74]. The technique has also been used to analyse circadian rhythms in stomatal conductance in the *toc1-1* mutant of *Arabidopsis thaliana* [75]. As carbon fixation is universally clock-regulated in plants, this technique is applicable to many plant species.

Whilst the IRGA system is becoming more practical, with a range of models in production including large multichannel machines that allow parallel analysis and portable field versions, and flexible, with the sealed chambers allowing manipulation and comparison of multiple environmental conditions, it is not without drawbacks. As with any indirect assay, it allows a gauge of overall circadian health, rather than specific investigation of particular clock loops. The level of clock control on net carbon fixation is complex, with contributions at the level of photosynthetic gene expression all the way up to regulation of stomatal conductance and growth. None-the-less it allows investigation of the plant clock on a level that is otherwise unaccounted for by other techniques.

Data analysis solutions

With the development of high-throughput assays, developing large data-sets in ever decreasing time frames, data analysis and period identification has increasingly become a time-consuming and rate-reducing step in circadian investigation. Fortunately, new methods for estimating the underlying period have been developed that can reduce and simplify this process. With a variety of different techniques and algorithms of varying complexity, choosing which is most applicable can prove challenging.

The related curve-fitting algorithms mFourfit [76] and Fast Fourier Transform Non-Linear Least Squares (FFT-NLLS) [77,78] have long been central period estimators in circadian investigations. FFT-NLLS has the additional advantage of providing the confidence intervals for the predicted periods, phases and amplitudes, allowing us to identify arrhythmic samples in a manner that mFourfit cannot. It does, however, tend to "over-fit" the series and, as such, is poor at identifying noise in the data.

Curve-fitting is not the only technique for analysis - stochastic modelling approaches can be used to estimate periods, as in Maximum Entropy Spectral Analysis (MESA) [79]. This performs better when identifying periods for datasets with large baseline trends. An additional technique, Spectrum Resampling (SR), is based around iterative boot-strapping of a smoothed power series to provide a steady-state model which reflects the data. Whilst MESA and SR especially are designed to be more robust when confronted with observational noise [80], they are considerably more computationally taxing than other methods.

Table 1 Summary comparison of circadian screening techniques

Assay	Type	Species suitability	Requires transformation?	Max resolution	Throughput	Ref.
Western Blot/qPCR	Direct assay of gene expression.	All plants	No	Single Cell/Tissue	Low	[89]
Luciferase	Direct assay of gene transcription.	Transformable plants	Yes	Single Cell/Tissue	High	[36]
Delayed fluorescence	Indirect assay of clock phenotype.	Most plants	No	Whole plant	High	[58]
Leaf movement	Indirect assay of clock phenotype.	Most Dicots	No	Whole plant	High	[67,69]
IRGA	Indirect assay of clock phenotype	All plants	No	Whole organ (Individual leaves)	Medium	[73]

We recommend, if possible, performing simultaneous analysis with techniques from both curve-fitting and stochastic modelling schools for analysing data. As all of these techniques are available through the BioDARE service, hosted by the University of Edinburgh [81], it is relatively painless to perform multiple analyses to gather accurate period estimations and simply share data in a standard format following publication.

Emerging assays

So far, the vast majority of high-throughput circadian assays have looked at the plant group in groups of seedlings, whole-plant or whole leaf scale (Table 1). Recently, novel circadian assays have been developed that allow the clock to be surveyed within individual tissues, individual cells and beyond to the sub-cellular level. In addition to the single-cell and tissue-specific luciferase based systems outlined above, fluorescence based techniques have potential as circadian assays [82]. This has been done successfully using fluorescent protein tagged CCA1 as a marker to probe the intracellular dynamics of CCA1 [83] identifying the independence of the guard cell clock from that of the surrounding leaf [84]. Although this paper has made significant strides in measuring the clock at the single cell level, major bottlenecks with this type of technique still exist. To make single cell circadian research plausible, new techniques need to be identified that allow oscillations to be measured for several days in different tissue types across hundreds of cells. Further development and adoption of these and related techniques will bring a new fine-scale understanding of the circadian clock.

Contrary to previous belief that the clock was cell autonomous [85], increasing amounts of evidence has emerged to suggest inter cellular communication between circadian clocks. Weak communication between individual circadian clocks has been observed in between shoots and roots [86] and within leaves [87] . Comparative assay of clock-driven luciferase expression in roots and leaves has provided evidence of phased "waves" of coordination through the tissue from root to shoot [88]. Using tissue-specific luciferase assays, it has demonstrated that there is significant communication between the vascular and mesophyll clocks, with the vascular clock serving as the dominant coordinator [47]. The required resolution of these techniques is currently on the threshold of the detection limit. It seems likely, in light of these discoveries, that further inter-cellular coordination will be uncovered and prove instrumental in understanding the clock system in the wider context of the organism in the environment.

Conclusion

For the first time, we are equipped with the tools to rapidly, accurately and extensively investigate the clock architecture in ways that go beyond it's temporal arrangement. We are also capable of assaying various clock outputs and reporters in different tissues. With the development of higher-resolution techniques and more powerful imaging methods, our understanding of the clock and it's roles will expand to incorporate different spacio-developmental contexts, and approach a fuller picture of underlying circadian coordination in plants.

Competing interests

The authors declare that they have no competing interests.

Authors' contributions

AJT wrote the manuscript. JW provided research and helped with the section on the IRGA. MG provided research and helped with the section on leaf movement. PDG provided research and input on the section on Emerging Assays. AH provided guidance, research for the data analysis section, and helped with the drafting and proofreading of the manuscript. All authors read and approved the final manuscript.

Acknowledgements

AJT, JW and MG were funded through BBSRC DTP grants. PDG was funded through BBSRC grant BBK017152.

References

1. Wijnen H, Young MW. Interplay of circadian clocks and metabolic rhythms. Annu Rev Genet. 2006;40:409–8.
2. McClung CR. Beyond Arabidopsis: the circadian clock in non-model plant species. Semin Cell Dev Biol. 2013;24(5):430–6.

3. Harmer SL. The circadian system in higher plants. Annu Rev Plant Biol. 2009;60:357–77.
4. Covington MF, Maloof JN, Straume M, Kay Sa, Harmer SL. Global transcriptome analysis reveals circadian regulation of key pathways in plant growth and development. Genome Biol. 2008;9(8):130.
5. Suárez-López P, Wheatley K, Robson F, Onouchi H, Valverde F, Coupland G. CONSTANS mediates between the circadian clock and the control of flowering in Arabidopsis. Nature. 2001;410(6832):1116–20.
6. Roden LC, Ingle RA. Lights, rhythms, infection: the role of light and the circadian clock in determining the outcome of plant-pathogen interactions. Plant Cell. 2009;21(9):2546–52.
7. Wang W, Barnaby JY, Tada Y, Li H, Tör M, Caldelari D, et al. Timing of plant immune responses by a central circadian regulator. Nature. 2011;470(7332):110–4.
8. Zhang C, Xie Q, Anderson RG, Ng G, Seitz NC, Peterson T, et al. Crosstalk between the circadian clock and innate immunity in Arabidopsis. PLoS Pathog. 2013;9(6):e1003370.
9. Dunford RP, Yano M, Kurata N, Sasaki T, Huestis G, Rocheford T, et al. Comparative mapping of the barley Ppd-H1 photoperiod response gene region, which lies close to a junction between two rice linkage segments. Genetics. 2002;161(2):825–34.
10. Turner A, Beales J, Faure S, Dunford RP, Laurie DA. The pseudo-response regulator Ppd-H1 provides adaptation to photoperiod in barley. Science. 2005;310(5750):1031–34.
11. Mockler TC, Michael TP, Priest HD, Shen R, Sullivan CM, Givan SA, et al. The diurnal project: Diurnal and circadian expression profiling, model-based pattern matching, and promoter analysis. Cold Spring Harbor Symposia Quantitative Biol. 2007;72:353–63.
12. Shaw LM, Turner AS, Laurie DA. The impact of photoperiod insensitive Ppd-1a mutations on the photoperiod pathway across the three genomes of hexaploid wheat (Triticum aestivum). Plant J. 2007;71:71–84.
13. Kloosterman B, Abelenda JA, Gomez MDMC, Oortwijn M, de Boer JM, Kowitwanich K, et al. Naturally occurring allele diversity allows potato cultivation in northern latitudes. Nature. 2013;495:246–50.
14. Filichkin SA, Breton G, Priest HD, Dharmawardhana P, Jaiswal P, Fox SE, et al. Global profiling of rice and poplar transcriptomes highlights key conserved Circadian-controlled pathways and cis-regulatory modules. PLoS One. 2011;6(6):16907.
15. Preuss SB, Meister R, Xu Q, Urwin CP, Tripodi Fa, Screen SE, et al. Expression of the Arabidopsis thaliana BBX32 gene in soybean increases grain yield. PLoS One. 2012;7(2):30717.
16. McClung CR. The genetics of plant clocks. Adv Genet. 2011;74:105–39.
17. Pruneda-Paz JL, Kay Sa. An expanding universe of circadian networks in higher plants. Trends Plant Sci. 2010;15(5):259–65.
18. Alabadí D, Oyama T, Yanovsky MJ, Harmon FG, Más P, Kay SA. Reciprocal regulation between TOC1 and LHY/CCA1 within the Arabidopsis circadian clock. Science. 2001;293(5531):880–3.
19. Locke JCW, Millar AJ, Turner MS. Modelling genetic networks with noisy and varied experimental data: the circadian clock in Arabidopsis thaliana. J Theor Bio. 2005;234(3):383–93.
20. Locke JCW, Kozma-Bognár L, Gould PD, Fehér B, Kevei E, Nagy F, et al. Experimental validation of a predicted feedback loop in the multi-oscillator clock of Arabidopsis thaliana. Mol Syst Biol. 2006;2:59.
21. Salomé PA, McClung CR. PSEUDO-RESPONSE REGULATOR 7 and 9 are partially redundant genes essential for the temperature responsiveness of the Arabidopsis circadian clock. Plant Cell. 2005;17(3):791–803.
22. Kim WY, Fujiwara S, Suh SS, Kim J, Kim Y, Han L, et al. ZEITLUPE is a circadian photoreceptor stabilized by GIGANTEA in blue light. Nature. 2007;449(7160):356–60.
23. Pokhilko A, Hodge SK, Stratford K, Knox K, Edwards KD, Thomson AW, et al. Data assimilation constrains new connections and components in a complex, eukaryotic circadian clock model. Mol Syst Biol. 2010;6:416.
24. Pruneda-Paz JL, Breton G, Para A, Kay SA. A functional genomics approach reveals CHE as a component of the Arabidopsis circadian clock. Science. 2009;323(5920):1481–5.
25. Michael TP, McClung CR. Enhancer trapping reveals widespread circadian clock transcriptional control in Arabidopsis. Plant Physiol. 2003;132:629–39.
26. Yakir E, Hilman D, Harir Y, Green RM. Regulation of output from the plant circadian clock. FEBS J. 2007;274(2):335–45.
27. Somers DE, Devlin PF, Kay SA. Phytochromes and cryptochromes in the entrainment of the Arabidopsis circadian clock. Science. 1998;282(5393):1488–90.
28. Devlin PF, Kay SA. Cryptochromes are required for phytochrome signaling to the circadian clock but not for rhythmicity. Plant Cell. 2000;12(12):2499–510.
29. Gould PD, Locke JCW, Larue C, Southern MM, Davis SJ, Hanano S, et al. The molecular basis of temperature compensation in the Arabidopsis circadian clock. Plant Cell. 2006;18(5):1177–87.
30. Hanano S, Domagalska MA, Nagy F, Davis SJ. Multiple phytohormones influence distinct parameters of the plant circadian clock. Genes Cells. 2006;11(12):1381–92.
31. Wang ZY, Tobin EM. Constitutive expression of the CIRCADIAN CLOCK ASSOCIATED 1 (CCA1) gene disrupts circadian rhythms and suppresses its own expression. Cell. 1998;93(7):1207–17.
32. Schaffer R, Ramsay N, Samach A, Corden S, Putterill J, Carré IA, et al. The late elongated hypocotyl mutation of Arabidopsis disrupts circadian rhythms and the photoperiodic control of flowering. Cell. 1998;93(7):1219–29.
33. Schaffer R, Landgraf J, Accerbi M, Simon V, Larson M, Wisman E. Microarray analysis of diurnal and circadian-regulated genes in Arabidopsis. Plant Cell. 2001;13(1):113–23.
34. Filichkin SA, Priest HD, Givan SA, Shen R, Bryant DW, Fox SE, et al. Genome-wide mapping of alternative splicing in Arabidopsis thaliana. Genome Res. 2010;20(1):45–58.
35. Millar AJ, Short SR, Hiratsuka K, Chua NH, Kay Sa. Firefly luciferase as a reporter of regulated gene expression in higher plants. Genome Res. 1992;10(14):324–37.
36. Southern MM, Brown PE, Hall A. Luciferases as Reporter Genes. In: Arabidopsis Methods, vol. 323; 2006. p. 293–305.
37. Sherf BA, Wood KV. Firefly Luciferase Engineered for Improved Genetic Reporting. Promega Notes Mag. 1994;49:14–21.
38. Millar A, Kay S. Circadian control of cab gene transcription and mRNA accumulation in Arabidopsis. Plant Cell. 1991;3(5):541–50.
39. Millar AJ, Straume M, Chory J, Chua NH, Kay SA. The regulation of circadian period by phototransduction pathways in Arabidopsis. Science. 1995;267(5201):1163–6.
40. Hall A, Brown P. Monitoring circadian rhythms in Arabidopsis thaliana using luciferase reporter genes. Methods Mol Biol. 2007;362:143–52.
41. Millar AJ, Carré IA, Strayer CA, Chua NH, Kay SA. Circadian clock mutants in Arabidopsis identified by luciferase imaging. Science. 1995;267(5201):1161–3.
42. Thines B, Harmon FG. Ambient temperature response establishes ELF3 as a required component of the core Arabidopsis circadian clock. Proc Natl Acad Sci U S A. 2010;107(7):3257–62.
43. Kusakina J, Gould PD, Hall A. A fast circadian clock at high temperatures is a conserved feature across Arabidopsis accessions and likely to be important for vegetative yield. Plant Cell Environ. 2014;37:327–40.
44. Nagel DH, Pruneda-Paz JL, Kay Sa. FBH1 affects warm temperature responses in the Arabidopsis circadian clock. Proc Natl Acad Sci U S A. 2014;111:14595–600.
45. Dalchau N, Baek SJ, Briggs HM, Robertson FC, Dodd AN, Gardner MJ, et al. The circadian oscillator gene GIGANTEA mediates a long-term response of the Arabidopsis thaliana circadian clock to sucrose. Proc Natl Acad Sci U S A. 2011;108(12):5104–9.
46. Muranaka T, Kubota S, Oyama T. A single-cell bioluminescence imaging system for monitoring cellular gene expression in a plant body. Plant Cell Physiol. 2013;54(12):2085–93.
47. Endo M, Shimizu H, Nohales MA, Araki T, Kay SA. Tissue-specific clocks in Arabidopsis show asymmetric coupling. Nature. 2014;515(7527):419–22.
48. Sugiyama N, Izawa T, Oikawa T. Light regulation of circadian clock-controlled gene expression in rice. Plant J. 2001;26(6):607–15.
49. Belhaj K, Chaparro-Garcia A, Kamoun S, Nekrasov V. Plant genome editing made easy: targeted mutagenesis in model and crop plants using the CRISPR/Cas system. Plant Med. 2013;9(1):39.
50. Strehler BL, Arnold W. Light Production By Green Plants. J Gen Physiol. 1951;34(6):809–20.
51. Rutherford AW, Govindjee, Inoue Y. Charge accumulation and photochemistry in leaves studied by thermoluminescence and delayed light emission. Proc Natl Acad Sci U S A. 1984;81(4):1107–11.

52. Arnold W, Davidson JB. The identity of the fluorescent and delayed light emission spectra in Chlorella. J Gen Physiol. 1954;37(5):677–84.
53. Jursinic PA. Delayed fluorescence: current concepts and status In: Govindjee Amesz J, Fork DC, editors. Light Emission by Plants and Bacteria. New York: Academic Press; 1986. p. 291–328.
54. Van Wijk R, Scordino A, Triglia A, Musumeci F. 'Simultaneous' measurements of delayed luminescence and chloroplast organization in Acetabularia acetabulum. J Photochem Photobiol B. 1999;49(2–3):142–9.
55. Blankenship RE. Molecular Mechanisms of Photosynthesis. Oxford: Blackwell Science Ltd; 2002.
56. Goltsev V, Zaharieva I, Lambrev P, Yordanov I, Strasser R. Simultaneous analysis of prompt and delayed chlorophyll a fluorescence in leaves during the induction period of dark to light adaptation. J Theor Biol. 2003;225(2):171–83.
57. Berden-zrimec M, Drinovec L, Zrimec A. Delayed Fluoresence In: Suggett DJ, Prášil O, Borowitzka MA, editors. Chlorophyll a fluorescence in aquatic sciences: methods and applications. Dordrecht: Springer; 2011. p. 293–309.
58. Gould PD, Diaz P, Hogben C, Kusakina J, Salem R, Hartwell J, et al. Delayed fluorescence as a universal tool for the measurement of circadian rhythms in higher plants. Plant J. 2009;58(5):893–901.
59. Harmer SL, Hogenesch JB, Straume M, Chang HS, Han B, Zhu T, et al. Orchestrated transcription of key pathways in Arabidopsis by the circadian clock. Science. 2000;290(5499):2110–3.
60. Gawronski P, Ariyadasa R, Himmelbach A, Poursarebani N, Kilian B, Stein N, et al. Genetics. 2014;196(4):1253–61.
61. Gyllenstrand N, Karlgren A, Clapham D, Holm K, Hall A, Gould PD, et al. No time for spruce: Rapid dampening of circadian rhythms in picea abies (L. Karst). Plant Cell Physiol. 2014;55(3):535–40.
62. Farré EM. The regulation of plant growth by the circadian clock. Plant Biol. 2012;14(3):401–10.
63. Engelmann W, Johnsson A. Rhythms in Organ Movement In: Lumsden PJ, Millar AJ, editors. Biological rhythms and photoperiodism in plants. 1st ed. Oxford: BIOS Scientific Publishers; 1998. p. 35–48.
64. Dornbusch T, Michaud O, Xenarios I, Fankhauser C. Differentially phased leaf growth and movements in Arabidopsis depend on coordinated circadian and light regulation. Plant Cell. 2014;26(10):3911–21.
65. d'Ortous De Mairan J-J. Hist Acad R Sci. 179235–6.
66. Dowson-Day MJ, Millar AJ. Circadian dysfunction causes aberrant hypocotyl elongation patterns in Arabidopsis. Plant J. 1999;17(1):63–71.
67. Edwards KD, Millar AJ. Analysis of circadian leaf movement rhythms in Arabidopsis thaliana. Methods Mol Biol. 2007;362:103–13.
68. Kim JS, Nam HG. Instrumentation and software for analysis of Arabidopsis Circadian leaf movement. 2009. doi:10.4051/ibc.2009.1.0005.
69. Bours R, Muthuraman M, Bouwmeester H, van der Krol A. OSCILLATOR: A system for analysis of diurnal leaf growth using infrared photography combined with wavelet transformation. Plant Methods. 2012;8(1):29.
70. Spoehr HA, McGee JM. Studies in Plant Respiration and Photosynthesis. Washington, DC: The Carnegie Institution of Washington; 1923.
71. Neeb CX. Hydrodictyon als Objekt einer vergleichenden Untersuchung physiologischer Grossen. Flora. 1952;139:39–95.
72. Schon J. Periodische Schwankungen der Photosynthese und Atmung bei Hydrodictyon. Flora. 1955;142:347–80.
73. Bohn A, Geist A, Rascher U, Lüttge U. Responses to different external light rhythms by the circadian rhythm of Crassulacean acid metabolism in Kalanchoe daigremontiana. Plant Cell Environ. 2001;24:811–20.
74. Dever LV, Boxall SF, Kneřová J, Hartwell J. Transgenic perturbation of the decarboxylation phase of Crassulacean acid metabolism alters physiology and metabolism but has only a small effect on growth. Plant Physiol. 2014;167:44–59.
75. Somers DE, Webb AA, Pearson M, Kay SA. The short-period mutant, toc1-1, alters circadian clock regulation of multiple outputs throughout development in Arabidopsis thaliana. Development. 1998;125(3):485–94.
76. Edwards KD, Akman OE, Knox K, Lumsden PJ, Thomson AW, Brown PE, et al. Quantitative analysis of regulatory flexibility under changing environmental conditions. Mol Syst Biol. 2010;6(424):4.
77. Johnson BML, Frasier SG. Nonlinear Least Squares Analysis. Methods Enzymol. 1985;117:301–42.
78. Martin S, Fraiser-Cadoret SG, Johnson ML, Vol. 2. Least-squares analysis of fluorescence data; 2002, pp. 177–240.
79. Burg JP. The relationship between maximum entropy and maximum likelihood spectra. Geophysics. 1972;37(2):375–6.
80. Costa MJ, Finkenstädt B, Roche V, Lévi F, Gould PD, Foreman J, et al. Inference on periodicity of circadian time series. Biostatistics. 2013;14(4): 792–806.
81. Zielinski T, Moore AM, Troup E, Halliday KJ, Millar AJ. Strengths and limitations of period estimation methods for circadian data. PLoS One. 2014;9(5):85754.
82. Shor E, Hassidim M, Green RM. The Use of Fluorescent Proteins to Analyze Circadian Rhythms In: Staiger D, editor. Plant circadian networks: methods & protocols. Methods in Molecular Biology. New York: Springer; 2014. p. 209–13. Chap. 14.
83. Yakir E, Hilman D, Kron I, Hassidim M, Melamed-Book N, Green RM. Posttranslational regulation of CIRCADIAN CLOCK ASSOCIATED1 in the circadian oscillator of Arabidopsis. Plant Physiol. 2009;150(2):844–57.
84. Yakir E, Hassidim M, Melamed-Book N, Hilman D, Kron I, Green RM. Cell autonomous and cell-type specific circadian rhythms in Arabidopsis. Plant J. 2011;68:520–31.
85. Thain SC, Hall A, Millar AJ. Functional independence of circadian clocks that regulate plant gene expression. Curr Biol. 2000;10(16):951–6.
86. James AB, Monreal JA, Nimmo GA, Kelly CL, Herzyk P, Jenkins GI, et al. The circadian clock in Arabidopsis roots is a simplified slave version of the clock in shoots. Science. 2008;322(December):1832–5.
87. Wenden B, Kozma-Bognár Lo, Edwards KD, Hall AJW, Locke JCW, Millar AJ. Light inputs shape the Arabidopsis circadian system. Plant J. 2011;66(3):480–91.
88. Fukuda H, Ukai K, Oyama T. Self-arrangement of cellular circadian rhythms through phase-resetting in plant roots. Phys Rev E Stat Nonlin Soft Matter Phys. 2012;86(4):041917.
89. Rosato E. Circadian rhythms: methods and protocols vol. 362. New York: Springer; 2007.

"Rhizoponics": a novel hydroponic rhizotron for root system analyses on mature *Arabidopsis thaliana* plants

Laura Mathieu†, Guillaume Lobet†, Pierre Tocquin and Claire Périlleux*

Abstract

Background: Well-developed and functional roots are critical to support plant life and reach high crop yields. Their study however, is hampered by their underground growth and characterizing complex root system architecture (RSA) therefore remains a challenge. In the last few years, several phenotyping methods, including rhizotrons and x-ray computed tomography, have been developed for relatively thick roots. But in the model plant *Arabidopsis thaliana, in vitro* culture remains the easiest and preferred method to study root development, which technically limits the analyses to young seedlings.

Results: We present here an innovative design of hydroponic rhizotrons (rhizoponics) adapted to *Arabidopsis thaliana*. The setup allows to simultaneously characterize the RSA and shoot development from seedling to adult stages, i.e. from seed to seed. This system offers the advantages of hydroponics such as control of root environment and easy access to the roots for measurements or sampling. Being completely movable and low cost, it can be used in controlled cabinets. We chose the case of cadmium treatment to illustrate potential applications, from cell to organ levels.

Conclusions: Rhizoponics makes possible, on the same plants of Arabidopsis, RSA measurements, root sampling and characterization of aerial development up to adult size. It therefore provides a valuable tool for addressing fundamental questions in whole plant physiology.

Keywords: Roots, Phenotyping, Rhizotron, Hydroponic

Background

Plant roots are responsible for nutrient and water uptake and are thus critical components of the overall plant productivity [1]. The root system architecture (RSA) is determined by both endogenous factors (reviewed in [2]) and environmental constraints such as nutrient availability [3,4]. Therefore, the understanding of the mechanisms regulating RSA is important for future crop improvement [5,6].

Arabidopsis (*Arabidopsis thaliana*) remains the most widely used plant model for studying fundamental processes in plant biology, while there is an increasing interest for transposing and acquiring further knowledge on economically relevant plants such as maize (*Zea Mays*) or rice (*Oryza sativa*) (Additional file 1). Regarding RSA studies, the relative simplicity of Arabidopsis root system makes it an ideal case to study the impact of endogenous traits and/or exogenous factors on root development, such as the genetic variation in plant response to nitrate supply [7] or the toxicity of cadmium on root and shoot development [8].

As a largely mastered plant lab technique, *in vitro* culture is the most common method used to study RSA in Arabidopsis. Seedlings are grown in vertically placed Petri plates and hence the roots develop along the surface of the medium, in a 2-D space suitable for image-based analyses, and can be easily harvested. However, several disadvantages are intrinsically linked to *in vitro* culture. Firstly, roots are exposed to light, which has been shown to influence their development [9]. Secondly, the confined atmosphere limits gas exchanges and metabolism, so that exogenous sucrose is frequently

* Correspondence: cperilleux@ulg.ac.be
†Equal contributors
Department of Life Sciences, Laboratory of Plant Physiology, PhytoSYSTEMS, University of Liège, Liège, Belgium

Table 1 Feature comparison of existing culture setups

Methods	Root tissues collection	RSA analysis	Long time-course analysis	Nutrients control	Suitable for Arabidopsis	Heterogenous root environment
Rhizotron	NO	YES	YES	YES	NO	YES
Pouch	YES	YES	NO	YES	NO	YES
Transparent soil	NO	YES	NO	YES	YES	YES
Hydroponics	YES	NO	YES	YES	YES	NO
Rhizoponics	YES	YES	YES	YES	YES	NO

supplied in the medium. Finally, the size of the Petri plates limits the studies to young plants (2 to 3 weeks).

In the last few years, alternative methods have emerged to phenotype RSA in the laboratory (Table 1). Germination on paper ('pouches') [10,11] has become popular for the study of RSA of crop plants. However it is not suitable for fine root system analyses (such as the one of Arabidopsis) because of the difficulty to distinguish the roots from the fibrous background. Soil-based 2D- or 3D-methods, such as rhizotrons [12,13] or x-ray tomography [14], have been adapted to Arabidopsis. A transparent solid medium was developed to facilitate the observations [15]. Major drawbacks of the existing soil-like methods are their reduced flexibility and the fact that the roots are not accessible for sampling.

Hydroponics is frequently used for studies requiring control of nutrients and accessibility to the root system. Hydroponics is appropriate for cultivation of the plants throughout their entire life cycle, does not limit gas exchanges, and allows to perform independent experiments in reproducible root-environment conditions. However quantification of root system architecture is not possible in such systems due to root tangling in the liquid medium. This is an even more acute issue with small plants, such as Arabidopsis, having a fine root system [16,17].

An efficient hydroponic device was previously developed in our lab for synchronous growth and flowering of Arabidopsis [16]. Here we present an innovative system (named "rhizoponics") that combines the advantages of hydroponics and rhizotrons to study RSA of Arabidopsis adult plants. Roots are grown in 2D, without any physical constraint, on a framed nylon support immersed into the nutrient solution. Rhizoponics is readily amenable to image analysis and allows sampling of selected roots in controlled conditions.

Results and discussion

Setup description

The rhizoponics setup is made of a nylon fabric (mesh of 0.5 mm^2) that is maintained in an aluminium frame of 335 × 250 mm (Height x Width) (Figure 1). The frame hangs in a tank filled with nutritive solution and is therefore slid into a slot of the tank cover. The tank and its cover are dark to prevent light exposure of the roots and the nutritive solution. The top side of the U-frame is larger (30 mm) that the cover slot in order to block the frame. It bears a hole in its middle, designed to insert an Araponics seed holder (http://www.araponics.com/) and supports the aerial rosette. Rhizoponics blueprints and 3D files are provided in the Additional files 2 and 3. Piece profiles could be easily adapted to other homemade seed starting systems, e.g. pierced

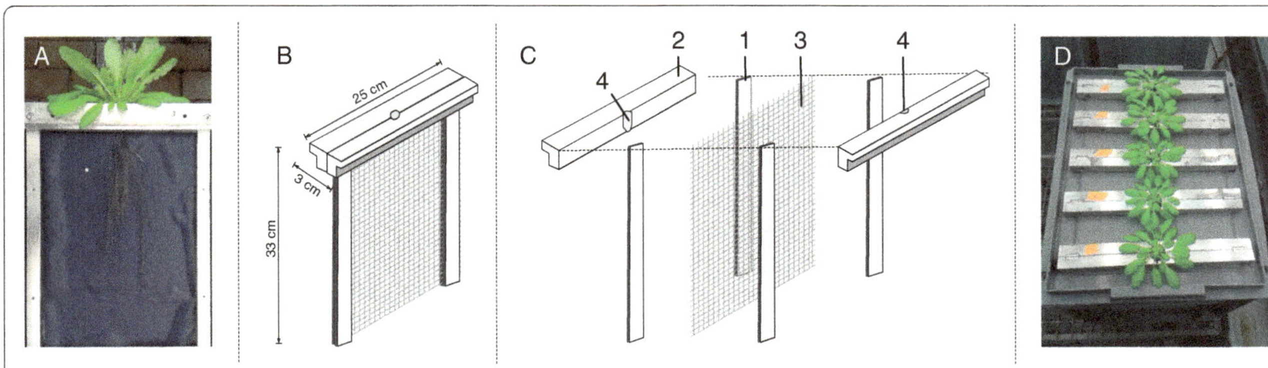

Figure 1 Schematic representation of the rhizoponics setup. A. Picture of the rhizoponics setup. A blue paper was placed behind the setup to increase the contrast. **B**. Assembled setup. **C**. Exploded view of the setup. 1. Lateral vertical element. 2. Top horizontal element. 3. Nylon mesh. 4. Hole for the Araponics seed holder. **D**. Full system picture.

microcentrifuge tubes [16,17], or to other hydroponics tanks.

Rhizoponics were designed to be light in weight and easily assembled. Aluminium was chosen as a building material for (i) its rigidity, (ii) its anti-corrosive properties and (iii) the possibility to sterilize the frames. However, we believe that any rigid plastic material could be used, opening the door to the use of 3D printing technologies.

Selection of the fabric

Various fabrics were tested. The size of the mesh appeared to be the critical point: a too narrow mesh allows the thin roots of Arabidopsis to penetrate into the fabric and thus makes the image analysis impossible. Conversely, a wide mesh does not provide an efficient support to untangle the roots. The selected fabric (black polyamide tulle netting, or Crinoline, 112 g m^{-2}, http://www.whaleys-bradford.ltd.uk/nylon-crinalin-black) has a mesh of 0.5 mm^2 and is perfectly adapted to Arabidopsis RSA. The fabric can be cleaned, autoclaved and recycled between experiments.

Pre-culture

The seed holders are filled with agar as described in [16] and the seeds can be sown directly on top of it in the rhizoponics setup. However, we observed that a pre-germination step in a standard Araponics tank allowed more synchronous germination and subsequent selection of homogeneous seedlings before transfer into the rhizoponics.

Image acquisition

Different setups were tested to acquire pictures of the root system. The main challenge was to find the right light source/background combination in order to distinguish the roots from the mesh. Direct lightning (light source on the same side as the camera) was not satisfactory as it did not discriminate the roots from the mesh. Moving the light source to the sides of the rhizoponics (raking lightning) greatly improved the contrast by highlighting the texture differences due to the presence of roots. However, this method allows the observation of the roots at one side of the mesh only. Direct backlighting was tested using an illuminated box, but the roots lacked contrast against the bright background. Finally, the best setup was based on indirect backlighting. The roots were placed above a black surface towards which the light source was directed (Figure 2). Raw pictures are provided in the Additional file 4.

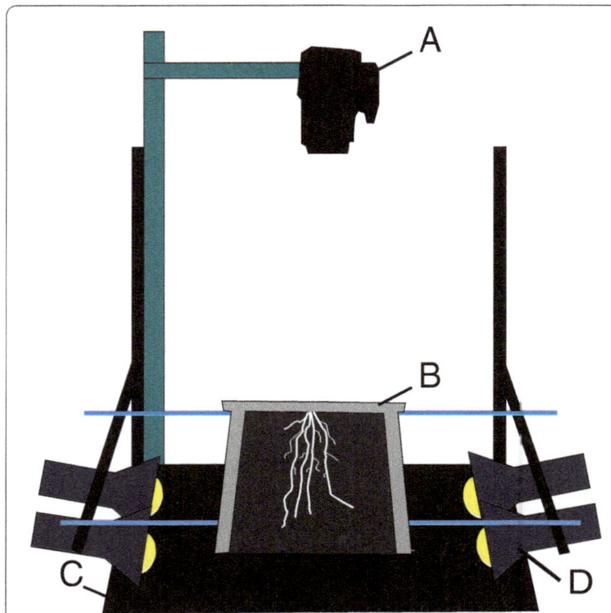

Figure 2 Schematic representation of the imaging setup adapted to rhizoponics setup. A. Camera. **B**. Rhizoponics. **C**. Dark background. **D**. Indirect light.

Rhizoponics setup enables a precise quantification of root and shoot development

The rhizoponics setup enables the simultaneous and non destructive observation of root and shoot development during the whole plant cycle (Table 2). On the shoot side, global growth estimators such as the projected rosette area or diameter can be easily followed (Figure 3A). On the root side, similar global estimators can be obtained (Figure 3B) as well as local features (lateral root density, length of unbranched root apical zone, lateral root length, (Figure 3D). In addition to these image-based analyses, the setup enables easy and local root sampling, without

Table 2 Non-exhaustive list of measurements enabled by the rhizoponics setup

Organ	Measurements	Units	Destructive
Shoot	Projected rosette area	cm^2	No
	Rosette diameter	cm	No
	Rosette convex hull area	cm^2	No
	Leaf count	-	No
	Rosette fresh/dry weight	g	Yes
Root	Projected root surface	cm^2	No
	Root system depth	cm	No
	Root system width	cm	No
	Root system convex hull area	cm^2	No
	Lateral root density	root/cm	No
	Length of unbranched apical root zone	cm	No
	Lateral root length	cm	No
	Microscopy analyses	-	Partial
	Reporter gene observation	-	Partial
	Root system fresh/dry weight	g	Yes

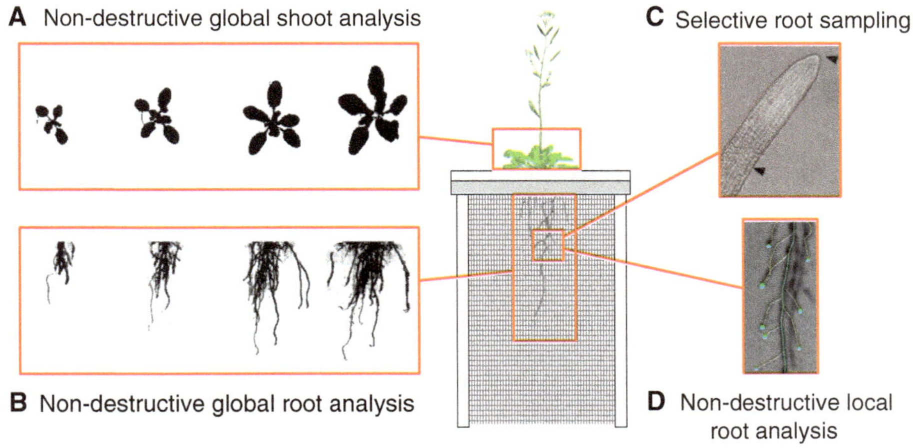

Figure 3 Illustration of global (A, B) and detailed (C, D) analyses that can be performed with the rhizoponics setup.

damaging the rest of the plant. Time course experiments are thus feasible, e.g. for microscopic observations or gene-expression analyses (Figure 3C).

Case study: effect of cadmium toxicity

As a proof of concept, we analysed the effect of cadmium (Cd) on Arabidopsis development. Cd is a heavy metal which is known to be toxic and limits shoot and root growth. However it is not lethal for the plant and the reproductive phase occurs normally [8,18].

Taking advantage of the rhizoponics setup, root and shoot growth kinetics were followed simultaneously during the whole life cycle of plants treated with 10 μM Cd. We observed a highly significant inhibition of root and shoot growth following Cd-treatment (Figure 4A & B). The effect of the treatment was first observed on the root system (6 days after start of treatment, DAT) and later on the shoot (9 DAT). We also observed, in line with previous results [18], that although the size of the rosette was strongly reduced in Cd-treated plants, the leaf apparition rate was not significantly affected (Figure 4C) indicating that individual leaf expansion - and not leaf number - was limited.

One of the major advantages of our rhizoponics setup is to enable the analysis of both global traits (such as those presented in Figure 3) and local root system parameters. In this example, we analysed the effect of Cd on the lateral root density, the length of the unbranched apical zone (LAUZ) and the lateral root growth. Our measurements revealed a significant negative effect of Cd on all parameters (Figure 5), indicating a strong inhibition of lateral root development.

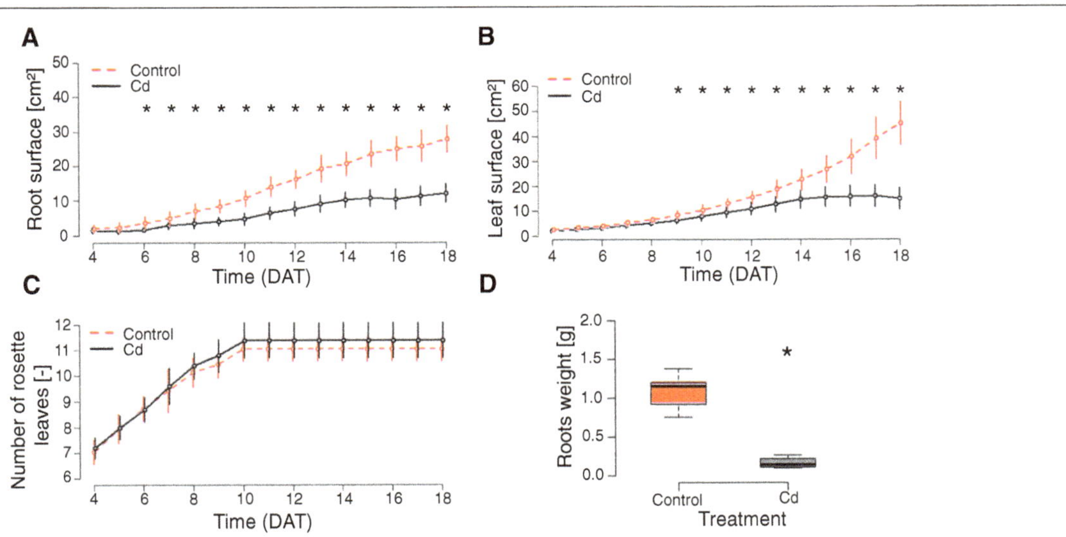

Figure 4 Effect of cadmium on global root and shoot measurements. A. Projected root surface. **B**. Projected rosette surface. **C**. Number of visible rosette leaves. **D**. Root weight measured 18 days after start of treatment. Stars indicate a significant difference between treatments (t-test, p-value < 0.05, n = 10).

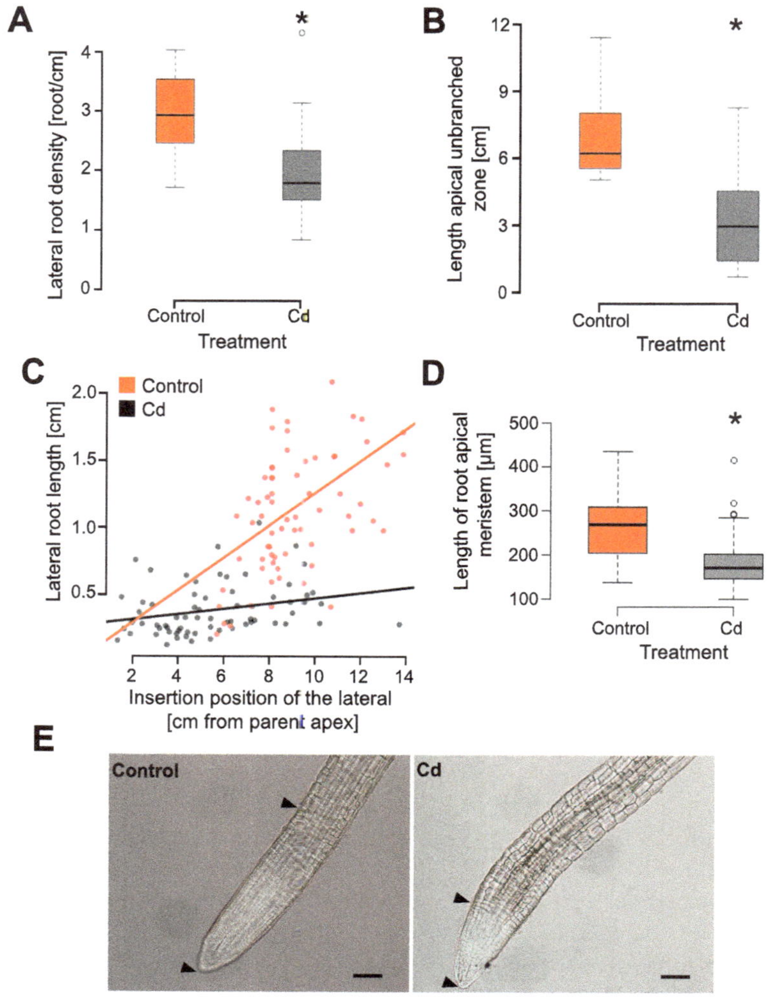

Figure 5 Effect of cadmium on lateral root growth measured 18 days after start of the treatment. A. Lateral root density (n = 10) **B**. Length of the apical unbranched zone (n = 10). **C**. Lateral root growth, illustrated by the relationship between the insertion position (distance from the parent root apex) and the length of the lateral roots. **D**. Length of the lateral root apical meristem (nmock = 102, nCd = 100). Significant differences (t-test, p-value < 0.05) are indicated by stars. **E**. Pictures of root apical meristem of control (left) and cadmium treated plants (right). Black lines represent 100 μm. Arrowheads indicate the limits of the root apical meristem.

We harvested lateral root tips to determine the size of their meristem and observed that they have a smaller length in Cd-treated plants than in controls (Figure 5D & E). This difference possibly explains the overall reduced growth of the lateral roots.

At the end of the experiment, Cd-treated plants had developed inflorescences and produced siliques. Root systems were harvested for final biomass quantification. The fresh root weight of treated plants was reduced by four in comparison with control plants (Figure 4D), which is consistent with the literature [8]. Thanks to RSA measurements, we determined that root biomass was best correlated with the total surface of the root system (Pearson correlation = 0.94). This result further shows the efficiency of rhizoponics to untangle the root system and provide relevant pictures of it.

Conclusion

We presented here a novel "rhizoponics" system that allows time-lapse studies of root system architecture and aerial part in Arabidopsis, up to mature stages. This system combines the advantages of hydroponics and rhizotrons. On one hand, roots grow in a controlled environment, ensuring high reproducibility of the results. On the other hand, a neutral support untangles the roots and so allows both global (e.g. total size, rooting depth) and local (e.g. lateral density, lateral length) analyses. Root sampling is easy, typically for molecular and microscopic analyses. As such, the rhizoponics setup opens new avenues for root/shoot researches.

Material and methods

Plant material and growing conditions

Arabidopsis (*Arabidopsis thaliana*, Col-0) seeds were stratified during 3 days on wet filter paper at 4°C in darkness. Thereafter, they were sown on Araponics (http://www.araponics.com/) seed-holders filled with 0.66% agar (Kalys HP697). Hydroponic solution was prepared with Flora Series fertilizers (FloraBloom, FloraMicro and FloraGro; GHE, France; 0.5 mL l^{-1} each). Seedlings were grown in 16-hour long days, at a (fluorescent) light fluence rate of 60 µmol m^{-2} s^{-1} (PAR), day/night temperature of 20°C and air relative humidity of 70%.

Two weeks after sowing, seedlings with 1-cm primary root were selected for transfer - together with their supporting seed-holder - into rhizoponics (1 plant/rhizoponic).

Rhizoponics were suspended in a plastic box (EURONORM 400X300X320 mm - STANDARD) filled with the same nutrient solution as described above. For Cd treatment, Cd- sulfate (3CdO4S.8H2O, Fluka 20920) was directly added into the box to obtain a final concentration of 10 µM.

Measurements

Leaves were counted every day starting 4 days after seedling transfer into rhizoponics. Pictures of both the root system and the shoot were also taken daily. Fresh root biomass was weighed at the end of the experiment. Prior to weighting, the roots were blotted on a tissue to remove the excess of water. The length of the root apical meristem was measured from the root tip, up to the beginning of the elongation zone with the imaging software *NIS*-Elements 3.20 [19].

Image collection

Shoot and root parts were photographed with a consumer CCD camera (Canon EOS 1100D, lens Canon EF 50 mm f/2.5). Aerial parts were imaged from above (top views), without removing the plants from the growing setup. For root imaging, rhizoponics were taken out of the tanks and placed horizontally, in a dark room, 15 cm above a black surface lightened with 4 incandescent bulbs of 40 W, placed in two rows (Figure 2). The camera was placed above the rhizoponic and aperture size and speed were set such as the acquired image was slightly under-exposed. ISO value was set to 200.

Image analysis

Root and shoot images were analysed using custom build ImageJ plugins and macros [20].

For shoot images, color images were split, based on their Hue-Saturation-Value (HSB) channels. The rosette segmentation was performed on the Hue channel since it provided an optimal contrast between the plant (green) and the growing setup colour (both the frame and the box are grey) [21]. The diameter, projected surface and convex hull surface were extracted from the mask using native ImageJ functions.

Two different treatments were performed on the root images, depending on the type of analysis required (local or global). For the local analysis, color images (Figure 6A) were split based on their HSV value. In this case, unlike the shoot, the Hue channel did not discriminate the roots from the mesh. The Value channel was used instead. The Value channel images (Figure 6B) were used for the local analysis, while the segmented images (Figure 6C) were used for the global analysis. SmartRoot [22] was used to analyse selected roots in the images. From these roots, we computed the length of the unbranched apical zone (LAUZ) and the lateral root density. For the global analysis, the width, height,

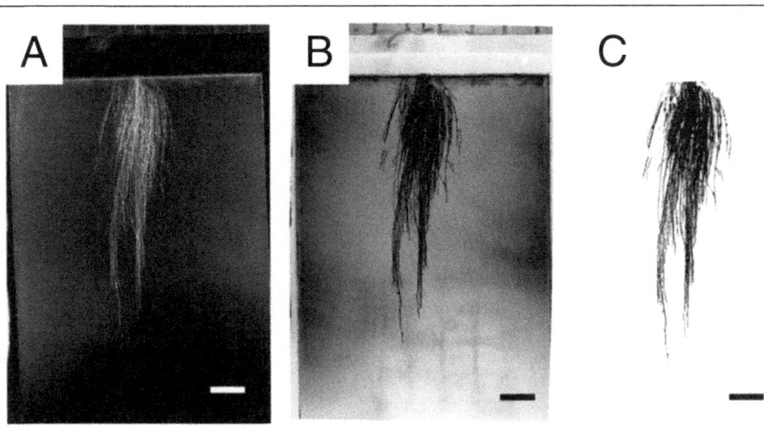

Figure 6 Root image treatment. A. Original color image. **B**. Grayscale image (Value channel from the HSV stack) obtained from **A**, with an increased contrast. **C**. Root system mask obtained from **B**, for quantitative analyses Bars = 3 cm.

projected surface and convex hull surface were computed from the mask.

The image analysis scripts (ImageJ macros) are freely available on GitHub (https://github.com/guillaumelobet/rhizoponics).

Additional files

Additional file 1: Number of publications on different plant species in 2013. Vernacular names were used for search in abstracts, titles and keywords of papers referenced in the Biochemistry, Genetics and Molecular Biology section of the Scopus© database.

Additional file 2: Rhizoponics blueprints.

Additional file 3: Rhizoponics blueprints (readable with FreeCAD 3D: http://www.freecadweb.org/).

Additional file 4: High definition picture of RSA 13 days after the transfer (DAT). A and B,control plants. C and D,Cd-treated plants.

Competing interests

The authors declare that they have no competing interests.

Authors' contributions

CP, LM, GL and PT designed the U-shape rhizoponics LM performed the experiments. GL developed the image analysis of roots and shoot. All authors participated to the writing of the manuscript. All the authors agree with the version to be published. All authors read and approved the final manuscript.

Acknowledgements

This work was supported by the Belgian Science Policy Interuniversity Attraction Poles Program (P7/29), by the Fonds National belge de la Recherche Scientifique (postdoctoral researcher grant of G.L.) and by the Fonds pour la formation à la Recherche dans l'Industrie et dans l'Agriculture (PhD grant of L.M.). The authors would like to thank the members of Plant Physiology lab for their constructive comments on the manuscript.

References

1. Lynch J. Root architecture and plant productivity. Plant Physiol. 1995;109:7–13.
2. Nibau C, Gibbs DJ, Coates JC. Branching out in new directions: the control of root architecture by lateral root formation. New Phytol. 2008;179:595–614.
3. Kellermeier F, Armengaud P, Seditas TJ, Danku J, Salt DE, Amtmann A. Analysis of the root system architecture of Arabidopsis provides a quantitative readout of crosstalk between nutritional signals. Plant Cell. 2014;26:1480–96.
4. Gruber BD, Giehl RFH, Friedel S, von Wirén N. Plasticity of the Arabidopsis root system under nutrient deficiencies. Plant Physiol. 2013;163:161–79.
5. Lynch JP. Roots of the second green revolution. Aust J Bot. 2007;55:493–512.
6. Den HG, Van Isterdael G, Beeckman T, De Smet I. The roots of a new green revolution. Trends Plant Sci. 2010;15:600–7.
7. De Pessemier J, Chardon F, Juraniec M, Delaplace P, Hermans C. Natural variation of the root morphological response to nitrate supply in Arabidopsis thaliana. Mech Dev. 2013;130:45–53.
8. Jozefczak M, Keunen E, Schat H, Bliek M, Hernandez LE, Carleer R, et al. Differential response of Arabidopsis leaves and roots to cadmium: glutathione-related chelating capacity vs antioxidant capacity. Plant Physiol Biochem. 2014;83:1–9.
9. Yokawa K, Kagenishi T, Baluška F. Root photomorphogenesis in laboratory-maintained Arabidopsis seedlings. Trends Plant Sci. 2013;18:117–9.
10. Hund A, Trachsel S, Stamp P. Growth of axile and lateral roots of maize: I development of a phenotying platform. Plant Soil. 2009;325:335–49.
11. Le Marié C, Kirchgessner N, Marschall D, Walter A, Hund A. Rhizoslides: paper-based growth system for non-destructive, high throughput phenotyping of root development by means of image analysis. Plant Methods. 2014;10:13.
12. Devienne-Barret F, Richard-Molard C, Chelle M, Maury O, Ney B. Ara-rhizotron: an effective culture system to study simultaneously root and shoot development of Arabidopsis. Plant Soil. 2006;280:253–66.
13. Nagel KA, Putz A, Gilmer F, Heinz K, Fischbach A, Pfeifer J, et al. GROWSCREEN-Rhizo is a novel phenotyping robot enabling simultaneous measurements of root and shoot growth for plants grown in soil-filled rhizotrons. Funct Plant Biol. 2012;39:891–904.
14. Mooney SJ, Pridmore TP, Helliwell J, Bennett MJ. Developing X-ray computed tomography to non-invasively image 3-D root systems architecture in soil. Plant Soil. 2012;352:1–22.
15. Downie H, Holden N, Otten W, Spiers AJ, Valentine TA, Dupuy LX. Transparent Soil for Imaging the Rhizosphere. PLoS One. 2012;7:6.
16. Tocquin P, Corbesier L, Havelange A, Pieltain A, Kurtem E, Bernier G, et al. A novel high efficiency, low maintenance, hydroponic system for synchronous growth and flowering of Arabidopsis thaliana. BMC Plant Biol. 2003;3:2.
17. Conn SJ, Hocking B, Dayod M, Xu B, Athman A, Henderson S, et al. Protocol: optimising hydroponic growth systems for nutritional and physiological analysis of Arabidopsis thaliana and other plants. Plant Methods. 2013;9:4.
18. Keunen E, Truyens S, Bruckers L, Remans T, Vangronsveld J, Cuypers A. Survival of Cd-exposed Arabidopsis thaliana: are these plants reproductively challenged? Plant Physiol Biochem. 2011;49:1084–91.
19. French AP, Wilson MH, Kenobi K, Dietrich D, Voss U, Ubeda-Tomas S, et al. Identifying biological landmarks using a novel cell measuring image analysis tool: Cell-o-Tape. Plant Methods. 2012;8:7.
20. Abràmoff MD, Magalhães PJ, Ram SJ. Image processing with image. J Biophotonics Int. 2004;11:36–41.
21. De Vylder J, Vandenbussche F, Hu Y, Philips W, Van Der Straeten D. Rosette tracker: an open source image analysis tool for automatic quantification of genotype effects. PLANT Physiol. 2012;160:1149–59.
22. Lobet G, Pages L, Draye X. A novel image analysis toolbox enabling quantitative analysis of root system architecture. Plant Physiol. 2011;157:29–39.

Permissions

All chapters in this book were first published in Plant Methods, by BioMed Central; hereby published with permission under the Creative Commons Attribution License or equivalent. Every chapter published in this book has been scrutinized by our experts. Their significance has been extensively debated. The topics covered herein carry significant findings which will fuel the growth of the discipline. They may even be implemented as practical applications or may be referred to as a beginning point for another development.

The contributors of this book come from diverse backgrounds, making this book a truly international effort. This book will bring forth new frontiers with its revolutionizing research information and detailed analysis of the nascent developments around the world.

We would like to thank all the contributing authors for lending their expertise to make the book truly unique. They have played a crucial role in the development of this book. Without their invaluable contributions this book wouldn't have been possible. They have made vital efforts to compile up to date information on the varied aspects of this subject to make this book a valuable addition to the collection of many professionals and students.

This book was conceptualized with the vision of imparting up-to-date information and advanced data in this field. To ensure the same, a matchless editorial board was set up. Every individual on the board went through rigorous rounds of assessment to prove their worth. After which they invested a large part of their time researching and compiling the most relevant data for our readers.

The editorial board has been involved in producing this book since its inception. They have spent rigorous hours researching and exploring the diverse topics which have resulted in the successful publishing of this book. They have passed on their knowledge of decades through this book. To expedite this challenging task, the publisher supported the team at every step. A small team of assistant editors was also appointed to further simplify the editing procedure and attain best results for the readers.

Apart from the editorial board, the designing team has also invested a significant amount of their time in understanding the subject and creating the most relevant covers. They scrutinized every image to scout for the most suitable representation of the subject and create an appropriate cover for the book.

The publishing team has been an ardent support to the editorial, designing and production team. Their endless efforts to recruit the best for this project, has resulted in the accomplishment of this book. They are a veteran in the field of academics and their pool of knowledge is as vast as their experience in printing. Their expertise and guidance has proved useful at every step. Their uncompromising quality standards have made this book an exceptional effort. Their encouragement from time to time has been an inspiration for everyone.

The publisher and the editorial board hope that this book will prove to be a valuable piece of knowledge for researchers, students, practitioners and scholars across the globe.

List of Contributors

Patrick Cosson
INRA, UMR 1332 de Biologie du fruit et Pathologie, Villenave d'Ornon F-33140, France
Univ Bordeaux, UMR 1332 de Biologie du fruit et Pathologie, Villenave d'Ornon F-33140, France

Véronique Decroocq
INRA, UMR 1332 de Biologie du fruit et Pathologie, Villenave d'Ornon F-33140, France
Univ Bordeaux, UMR 1332 de Biologie du fruit et Pathologie, Villenave d'Ornon F-33140, France

Frédéric Revers
INRA, UMR 1332 de Biologie du fruit et Pathologie, Villenave d'Ornon F-33140, France
Univ Bordeaux, UMR 1332 de Biologie du fruit et Pathologie, Villenave d'Ornon F-33140, France

Abhiram Das
School of Biology, Georgia Institute of Technology, Atlanta, Georgia, USA

Alexander Bucksch
School of Biology, Georgia Institute of Technology, Atlanta, Georgia, USA
School of Interactive Computing, Georgia Institute of Technology, Atlanta, Georgia, USA

Charles A Price
School of Plant Biology, University of Western Australia, Perth, Australia

Joshua S Weitz
School of Biology, Georgia Institute of Technology, Atlanta, Georgia, USA
School of Physics, Georgia Institute of Technology, Atlanta, Georgia, USA

Lucas AF Hanke
Department of Computer Science, Universidade Federal de Minas Gerais, Antônio Carlos Avenue, 6627, Pampulha, 30123-970 Belo Horizonte MG, Brazil

Cristiano S Botelho
Department of Computer Science, Universidade Federal de Minas Gerais, Antônio Carlos Avenue, 6627, Pampulha, 30123-970 Belo Horizonte MG, Brazil

Fernando AF Braz
Department of Computer Science, Universidade Federal de Minas Gerais, Antônio Carlos Avenue, 6627, Pampulha, 30123-970 Belo Horizonte MG, Brazil

Paulo HS Batista
Department of Computer Science, Universidade Federal de Minas Gerais, Antônio Carlos Avenue, 6627, Pampulha, 30123-970 Belo Horizonte MG, Brazil

Aurea V Folgueras-Flatschart
INMETRO, Nsa.Sra.das Graças Avenue, 50, 25250-020 Xérem-Duque de Caxias RJ, Brazil

Roberto W Noda
Embrapa Maize and Sorghum, MG 424 Highway KM 45, 35701-970 Sete Lagoas MG, Brazil

Andrea A Carneiro
Embrapa Maize and Sorghum, MG 424 Highway KM 45, 35701-970 Sete Lagoas MG, Brazil

Alessandra C Faria-Campos
Department of Computer Science, Universidade Federal de Minas Gerais, Antônio Carlos Avenue, 6627, Pampulha, 30123-970 Belo Horizonte MG, Brazil
INMETRO, Nsa.Sra.das Graças Avenue, 50, 25250-020 Xérem-Duque de Caxias RJ, Brazil

Sérgio VA Campos
Department of Computer Science, Universidade Federal de Minas Gerais, Antônio Carlos Avenue, 6627, Pampulha, 30123-970 Belo Horizonte MG, Brazil

John M Smith
Division of Biochemistry, University of Missouri-Columbia, 117 Schweitzer Hall, Columbia, MO 65211, USA
Division of Plant Sciences, University of Missouri-Columbia, Columbia, MO 65211, USA
Interdisciplinary Plant Group (IPG), University of Missouri-Columbia, Columbia, MO 65211, USA

Antje Heese
Division of Biochemistry, University of Missouri-Columbia, 117 Schweitzer Hall, Columbia, MO 65211, USA
Interdisciplinary Plant Group (IPG), University of Missouri-Columbia, Columbia, MO 65211, USA

Annette Nassuth
Department of Molecular and Cellular Biology, University of Guelph, Guelph, Ontario N1G 2 W1, Canada

Mahbuba Siddiqua
Department of Molecular and Cellular Biology, University of Guelph, Guelph, Ontario N1G 2 W1, Canada

Huogen Xiao
Department of Molecular and Cellular Biology, University of Guelph, Guelph, Ontario N1G 2 W1, Canada

Michelle A Moody
Department of Molecular and Cellular Biology, University of Guelph, Guelph, Ontario N1G 2 W1, Canada

Chevonne E Carlow
Department of Molecular and Cellular Biology, University of Guelph, Guelph, Ontario N1G 2 W1, Canada

Friederike Saxe
Department of Biomaterials, Max Planck Institute of Colloids and Interfaces, 14476 Potsdam-Golm, Germany

Michaela Eder
Department of Biomaterials, Max Planck Institute of Colloids and Interfaces, 14476 Potsdam-Golm, Germany

Gunthard Benecke
Department of Biomaterials, Max Planck Institute of Colloids and Interfaces, 14476 Potsdam-Golm, Germany

Barbara Aichmayer
Department of Biomaterials, Max Planck Institute of Colloids and Interfaces, 14476 Potsdam-Golm, Germany

Peter Fratzl
Department of Biomaterials, Max Planck Institute of Colloids and Interfaces, 14476 Potsdam-Golm, Germany

Ingo Burgert
Department of Biomaterials, Max Planck Institute of Colloids and Interfaces, 14476 Potsdam-Golm, Germany
Swiss Federal Institute of Technology Zurich, Institute for Building Materials, 8093 Zurich, Switzerland
Swiss Federal Laboratories for Materials Science and Technology, Applied Wood Materials Laboratory, 8600 Duebendorf, Switzerland

Markus Rüggeberg
Department of Biomaterials, Max Planck Institute of Colloids and Interfaces, 14476 Potsdam-Golm, Germany
Swiss Federal Institute of Technology Zurich, Institute for Building Materials, 8093 Zurich, Switzerland
Swiss Federal Laboratories for Materials Science and Technology, Applied Wood Materials Laboratory, 8600 Duebendorf, Switzerland

Maria Stolarz
Department of Biophysics, Institute of Biology and Biochemistry, Maria Curie-Skłodowska University, Akademicka 19, Lublin 20-033, Poland

Maciej Żuk
Department of Biophysics, Institute of Biology and Biochemistry, Maria Curie-Skłodowska University, Akademicka 19, Lublin 20-033, Poland

Elżbieta Król
Department of Biophysics, Institute of Biology and Biochemistry, Maria Curie-Skłodowska University, Akademicka 19, Lublin 20-033, Poland

Halina Dziubińska
Department of Biophysics, Institute of Biology and Biochemistry, Maria Curie-Skłodowska University, Akademicka 19, Lublin 20-033, Poland

Adam Healey
Queensland Alliance for Agriculture and Food Innovation, University of Queensland, Brisbane 4072, Australia

Agnelo Furtado
Queensland Alliance for Agriculture and Food Innovation, University of Queensland, Brisbane 4072, Australia

Tal Cooper
Queensland Alliance for Agriculture and Food Innovation, University of Queensland, Brisbane 4072, Australia

Robert J Henry
Queensland Alliance for Agriculture and Food Innovation, University of Queensland, Brisbane 4072, Australia

Achala S Jayasena
The University of Western Australia, School of Chemistry and Biochemistry & ARC Centre of Excellence in Plant Energy Biology, 35 Stirling Highway, Crawley, Perth 6009, Australia

David Secco
The University of Western Australia, School of Chemistry and Biochemistry & ARC Centre of Excellence in Plant Energy Biology, 35 Stirling Highway, Crawley, Perth 6009, Australia

Kalia Bernath-Levin
The University of Western Australia, School of Chemistry and Biochemistry & ARC Centre of Excellence in Plant Energy Biology, 35 Stirling Highway, Crawley, Perth 6009, Australia

Oliver Berkowitz
The University of Western Australia, School of Chemistry and Biochemistry & ARC Centre of Excellence in Plant Energy Biology, 35 Stirling Highway, Crawley, Perth 6009, Australia
The University of Western Australia, School of Plant Biology, 35 Stirling Highway, Crawley, Perth 6009, Australia

James Whelan
La Trobe University, Department of Botany, School of Life Sciences & ARC Centre of Excellence in Plant Energy Biology, AgriBio, the Centre for AgriBioscience, 5 Ring Road, Melbourne, Bundoora Victoria 3086, Australia

Joshua S Mylne
The University of Western Australia, School of Chemistry and Biochemistry & ARC Centre of Excellence in Plant Energy Biology, 35 Stirling Highway, Crawley, Perth 6009, Australia

Emily R Larson
Cellular, Molecular, and Biomedical Science Program, University of Vermont, Burlington, VT, USA
Laboratory of Plant Physiology and Biophysics, Institute of Molecular Cell and Systems Biology, Bower Building, University of Glasgow, Glasgow G12 8QQ, UK

Mary L Tierney
Cellular, Molecular, and Biomedical Science Program, University of Vermont, Burlington, VT, USA
Department of Plant Biology, University of Vermont, Burlington, VT, USA

Berke Tinaz
Department of Biology, Skidmore College, Saratoga Springs, NY, USA

David S Domozych
Department of Biology, Skidmore College, Saratoga Springs, NY, USA

Remy Kronbak
Science and Technology, Department of Molecular Biology and Genetics, Aarhus University, Forsøgsvej 1, DK-4200 Slagelse, Denmark

Christina Rønn Ingvardsen
Science and Technology, Department of Molecular Biology and Genetics, Aarhus University, Forsøgsvej 1, DK-4200 Slagelse, Denmark

Claus Krogh Madsen
Science and Technology, Department of Molecular Biology and Genetics, Aarhus University, Forsøgsvej 1, DK-4200 Slagelse, Denmark

Per Langkjær Gregersen
Science and Technology, Department of Molecular Biology and Genetics, Aarhus University, Forsøgsvej 1, DK-4200 Slagelse, Denmark

Karen J Kloth
Laboratory of Entomology, Wageningen University, P.O. Box 16, 6700, AA Wageningen, The Netherlands
Laboratory of Plant Physiology, Wageningen University, P.O. Box 16, 6700, AA Wageningen, The Netherlands
Plant Research International, Wageningen University and Research Center, P.O. Box 16, 6700, AA Wageningen, The Netherlands

Cindy JM ten Broeke
Laboratory of Entomology, Wageningen University, P.O. Box 16, 6700, AA Wageningen, The Netherlands

Manus PM Thoen
Laboratory of Entomology, Wageningen University, P.O. Box 16, 6700, AA Wageningen, The Netherlands
Laboratory of Plant Physiology, Wageningen University, P.O. Box 16, 6700, AA Wageningen, The Netherlands
Plant Research International, Wageningen University and Research Center, P.O. Box 16, 6700, AA Wageningen, The Netherlands

Marianne Hanhart-van den Brink
Laboratory of Entomology, Wageningen University, P.O. Box 16, 6700, AA Wageningen, The Netherlands

Gerrie L Wiegers
Laboratory of Entomology, Wageningen University, P.O. Box 16, 6700, AA Wageningen, The Netherlands
Plant Research International, Wageningen University and Research Center, P.O. Box 16, 6700, AA Wageningen, The Netherlands

Olga E Krips
Noldus Information Technology bv, P.O. Box 268, 6700, AG Wageningen, The Netherlands

Lucas PJJ Noldus
Noldus Information Technology bv, P.O. Box 268, 6700, AG Wageningen, The Netherlands

Marcel Dicke
Laboratory of Entomology, Wageningen University, P.O. Box 16, 6700, AA Wageningen, The Netherlands

Maarten A Jongsma
Plant Research International, Wageningen University and Research Center, P.O. Box 16, 6700, AA Wageningen, The Netherlands

Yuan Wang
Institute of Soil Science, Chinese Academy of Sciences, 71 East Beijing Road, Nanjing 210008, PR China

Dejian Wang
Institute of Soil Science, Chinese Academy of Sciences, 71 East Beijing Road, Nanjing 210008, PR China

Peihua Shi
National Engineering and Technology Center for Information Agriculture, Nanjing Agricultural University, 1 Weigang Road, Nanjing 210095, PR China

Kenji Omasa
Graduate School of Agricultural and Life Sciences, The University of Tokyo, Yayoi 1-1-1, Bunkyo-ku, Tokyo 113-8657, Japan

Ryan J Dean
Nanoscale Organisation and Dynamics Group, University of Western Sydney, Penrith NSW 2751, Australia

Timothy Stait-Gardner
Nanoscale Organisation and Dynamics Group, University of Western Sydney, Penrith NSW 2751, Australia

Simon J Clarke
National Wine & Grape Industry Centre, Charles Sturt University, Locked Bag 588, Wagga Wagga, New South Wales 2678, Australia

Suzy Y Rogiers
National Wine & Grape Industry Centre, Charles Sturt University, Locked Bag 588, Wagga Wagga, New South Wales 2678, Australia
New South Wales Department of Primary Industries, Locked Bag 588, Wagga Wagga, New South Wales 2678, Australia

Gabriele Bobek
School of Medicine, University of Western Sydney, Penrith, NSW 2751, Australia

William S Price
Nanoscale Organisation and Dynamics Group, University of Western Sydney, Penrith NSW 2751, Australia
National Wine & Grape Industry Centre, Charles Sturt University, Locked Bag 588, Wagga Wagga, New South Wales 2678, Australia
School of Medicine, University of Western Sydney, Penrith, NSW 2751, Australia

Sergej Bergsträsser
Institute for Bio- and Geosciences, IBG-2: Plant Sciences, Forschungszentrum Jülich GmbH, 52425 Jülich, Germany

Dimitrios Fanourakis
Institute for Bio- and Geosciences, IBG-2: Plant Sciences, Forschungszentrum Jülich GmbH, 52425 Jülich, Germany
Present address: Department of Crop Science, Technological Educational Institute of Crete, GR 71004 Heraklio, Greece
Present address: Institute of Viticulture, Floriculture and Vegetable Crops, Hellenic Agricultural Organization 'Demeter' (NAGREF), P.O. Box 2228, GR 71003 Heraklio, Greece

Simone Schmittgen
Institute for Bio- and Geosciences, IBG-2: Plant Sciences, Forschungszentrum Jülich GmbH, 52425 Jülich, Germany

Maria Pilar Cendrero-Mateo
Institute for Bio- and Geosciences, IBG-2: Plant Sciences, Forschungszentrum Jülich GmbH, 52425 Jülich, Germany

Marcus Jansen
Institute for Bio- and Geosciences, IBG-2: Plant Sciences, Forschungszentrum Jülich GmbH, 52425 Jülich, Germany
Present address: LemnaTec GmbH, Pascalstraße 59, 52076 Aachen, Germany

Hanno Scharr
Institute for Bio- and Geosciences, IBG-2: Plant Sciences, Forschungszentrum Jülich GmbH, 52425 Jülich, Germany

Uwe Rascher
Institute for Bio- and Geosciences, IBG-2: Plant Sciences, Forschungszentrum Jülich GmbH, 52425 Jülich, Germany

Xavier Serrat
Oryzon Genomics, S.A., Cornellà de Llobregat, Spain
Departament de Biologia Vegetal, Universitat de Barcelona, Barcelona, Spain

Roger Esteban
Oryzon Genomics, S.A., Cornellà de Llobregat, Spain

Nathalie Guibourt
Oryzon Genomics, S.A., Cornellà de Llobregat, Spain

Luisa Moysset
Departament de Biologia Vegetal, Universitat de Barcelona, Barcelona, Spain

Salvador Nogués
Departament de Biologia Vegetal, Universitat de Barcelona, Barcelona, Spain

Eric Lalanne
Oryzon Genomics, S.A., Cornellà de Llobregat, Spain

Hélène San Clemente
Université de Toulouse; UPS; UMR 5546, Laboratoire de Recherche en Sciences Végétales, BP 42617, F-31326 Castanet-Tolosan, France
CNRS; UMR 5546, BP 42617, F-31326 Castanet-Tolosan, France

Elisabeth Jamet
Université de Toulouse; UPS; UMR 5546, Laboratoire de Recherche en Sciences Végétales, BP 42617, F-31326 Castanet-Tolosan, France
CNRS; UMR 5546, BP 42617, F-31326 Castanet-Tolosan, France

Ralf Metzner
Institute of Bio- and Geosciences, IBG-2: Plant Sciences, Forschungszentrum Jülich GmbH, Wilhelm-Jonen- Str., 52425 Jülich, Germany

Anja Eggert
Development Center X-Ray Technology EZRT, Fraunhofer Institute for Integrated Systems IIS, Flugplatzstraße 75, 90768 Fürth, Germany

Dagmar van Dusschoten
Institute of Bio- and Geosciences, IBG-2: Plant Sciences, Forschungszentrum Jülich GmbH, Wilhelm-Jonen- Str., 52425 Jülich, Germany

Daniel Pflugfelder
Institute of Bio- and Geosciences, IBG-2: Plant Sciences, Forschungszentrum Jülich GmbH, Wilhelm-Jonen- Str., 52425 Jülich, Germany

Stefan Gerth
Development Center X-Ray Technology EZRT, Fraunhofer Institute for Integrated Systems IIS, Flugplatzstraße 75, 90768 Fürth, Germany

Ulrich Schurr
Institute of Bio- and Geosciences, IBG-2: Plant Sciences, Forschungszentrum Jülich GmbH, Wilhelm-Jonen- Str., 52425 Jülich, Germany

Norman Uhlmann
Development Center X-Ray Technology EZRT, Fraunhofer Institute for Integrated Systems IIS, Flugplatzstraße 75, 90768 Fürth, Germany

Siegfried Jahnke
Institute of Bio- and Geosciences, IBG-2: Plant Sciences, Forschungszentrum Jülich GmbH, Wilhelm-Jonen- Str., 52425 Jülich, Germany

Grant Calder
Department of Cell & Developmental Biology, John Innes Centre, Norwich Research Park, Norwich NR4 7UH, UK

Chris Hindle
Department of Cell & Developmental Biology, John Innes Centre, Norwich Research Park, Norwich NR4 7UH, UK

Jordi Chan
Department of Cell & Developmental Biology, John Innes Centre, Norwich Research Park, Norwich NR4 7UH, UK

Peter Shaw
Department of Cell & Developmental Biology, John Innes Centre, Norwich Research Park, Norwich NR4 7UH, UK

Catherine Collett
Department of Chemical Engineering and Biotechnology, New Museums Site, Pembroke Street, Cambridge CB2 3RA, UK

Alia Ardron
Department of Chemical Engineering and Biotechnology, New Museums Site, Pembroke Street, Cambridge CB2 3RA, UK

Ulrike Bauer
School of Biological Sciences, Life Sciences Building, 24 Tyndall Avenue, Bristol BS8 1TQ, UK
Faculty of Science, Universiti Brunei Darussalam, Tungku Link, Gadong 1410, Bandar Seri Begawan, Brunei Darussalam

Gary Chapman
Department of Chemical Engineering and Biotechnology, New Museums Site, Pembroke Street, Cambridge CB2 3RA, UK

Elodie Chaudan
Department of Chemical Engineering and Biotechnology, New Museums Site, Pembroke Street, Cambridge CB2 3RA, UK
ESPCI ParisTech, 10 rue Vauquelin, 75005 Paris, France

Bart Hallmark
Department of Chemical Engineering and Biotechnology, New Museums Site, Pembroke Street, Cambridge CB2 3RA, UK

Lee Pratt
Department of Chemical Engineering and Biotechnology, New Museums Site, Pembroke Street, Cambridge CB2 3RA, UK

Maria Dolores Torres-Perez
Department of Chemical Engineering and Biotechnology, New Museums Site, Pembroke Street, Cambridge CB2 3RA, UK
Department of Chemical Engineering, University of Santiago de Compostela, Lope Gómez de Marzoa St, Santiago de Compostela E-15782, Spain

D Ian Wilson
Department of Chemical Engineering and Biotechnology, New Museums Site, Pembroke Street, Cambridge CB2 3RA, UK

Matheus Kuska
Institute for Crop Science and Resource Conservation (INRES) - Phytomedicine, University of Bonn, Meckenheimer Allee 166a, 53115 Bonn, Germany

Mirwaes Wahabzada
Institute for Crop Science and Resource Conservation (INRES) - Phytomedicine, University of Bonn, Meckenheimer Allee 166a, 53115 Bonn, Germany

Marlene Leucker
Institute for Crop Science and Resource Conservation (INRES) - Phytomedicine, University of Bonn, Meckenheimer Allee 166a, 53115 Bonn, Germany

Heinz-Wilhelm Dehne
Institute for Crop Science and Resource Conservation (INRES) - Phytomedicine, University of Bonn, Meckenheimer Allee 166a, 53115 Bonn, Germany

Kristian Kersting
Department of Computer Science, TU Dortmund, Otto-Hahn-Str. 14, 44227 Dortmund, Germany

Erich-Christian Oerke
Institute for Crop Science and Resource Conservation (INRES) - Phytomedicine, University of Bonn, Meckenheimer Allee 166a, 53115 Bonn, Germany

Ulrike Steiner
Institute for Crop Science and Resource Conservation (INRES) - Phytomedicine, University of Bonn, Meckenheimer Allee 166a, 53115 Bonn, Germany

Anne-Katrin Mahlein
Institute for Crop Science and Resource Conservation (INRES) - Phytomedicine, University of Bonn, Meckenheimer Allee 166a, 53115 Bonn, Germany

Karin Köhl
Max-Planck-Institute of Molecular Plant Physiology, 14424 Potsdam-Golm, Germany

Jürgen Gremmels
Max-Planck-Institute of Molecular Plant Physiology, 14424 Potsdam-Golm, Germany

Jan F Humplík
Department of Chemical Biology and Genetics, Centre of the Region Haná for Biotechnological and Agricultural Research, Faculty of Science, Palacký University, Šlechtitelů 11, Olomouc CZ-78371, Czech Republic

Dušan Lazár
Department of Biophysics, Centre of the Region Haná for Biotechnological and Agricultural Research, Faculty of Science, Palacký University, Šlechtitelů 11, Olomouc CZ-78371, Czech Republic

Tomáš Fürst
Department of Mathematical Analysis and Applications of Mathematics, Faculty of Science, Palacký University, 17. listopadu 12, Olomouc CZ-77146, Czech Republic

Alexandra Husičková
Department of Biophysics, Centre of the Region Haná for Biotechnological and Agricultural Research, Faculty of Science, Palacký University, Šlechtitelů 11, Olomouc CZ-78371, Czech Republic

Miroslav Hýbl
Department of Genetic Resources for Vegetables, Medicinal and Special Plants, Centre of the Region Haná for Biotechnological and Agricultural Research, Crop Research Institute, Šlechtitelů 11, Olomouc CZ-78371, Czech Republic

Lukáš Spíchal
Department of Chemical Biology and Genetics, Centre of the Region Haná for Biotechnological and Agricultural Research, Faculty of Science, Palacký University, Šlechtitelů 11, Olomouc CZ-78371, Czech Republic

Alexander K Converse
T123 Waisman Center, University of Wisconsin-Madison, Madison, WI 53705, USA

Elizabeth O Ahlers
T123 Waisman Center, University of Wisconsin-Madison, Madison, WI 53705, USA

Tom W Bryan
Department of Plant Pathology, University of Wisconsin-Madison, Madison, WI 53705, USA

Jackson D Hetue
Department of Plant Pathology, University of Wisconsin-Madison, Madison, WI 53705, USA

Katherine A Lake
Department of Medical Physics, University of Wisconsin-Madison, Madison, WI 53705, USA

Paul A Ellison
Department of Medical Physics, University of Wisconsin-Madison, Madison, WI 53705, USA

Jonathan W Engle
Department of Medical Physics, University of Wisconsin-Madison, Madison, WI 53705, USA

Todd E Barnhart
Department of Medical Physics, University of Wisconsin-Madison, Madison, WI 53705, USA

Robert J Nickles
Department of Medical Physics, University of Wisconsin-Madison, Madison, WI 53705, USA

Paul H Williams
Department of Plant Pathology, University of Wisconsin-Madison, Madison, WI 53705, USA

Onofre T DeJesus
Department of Medical Physics, University of Wisconsin-Madison, Madison, WI 53705, USA

Andrew J Tindall
Institute of Integrative Biology, University of Liverpool, Crown Street, Liverpool, UK

Jade Waller
Institute of Integrative Biology, University of Liverpool, Crown Street, Liverpool, UK

Mark Greenwood
Institute of Integrative Biology, University of Liverpool, Crown Street, Liverpool, UK

Peter D Gould
Institute of Integrative Biology, University of Liverpool, Crown Street, Liverpool, UK

James Hartwell
Institute of Integrative Biology, University of Liverpool, Crown Street, Liverpool, UK

Anthony Hall
Institute of Integrative Biology, University of Liverpool, Crown Street, Liverpool, UK

Laura Mathieu
Department of Life Sciences, Laboratory of Plant Physiology, PhytoSYSTEMS, University of Liège, Liège, Belgium

Guillaume Lobet
Department of Life Sciences, Laboratory of Plant Physiology, PhytoSYSTEMS, University of Liège, Liège, Belgium

Pierre Tocquin
Department of Life Sciences, Laboratory of Plant Physiology, PhytoSYSTEMS, University of Liège, Liège, Belgium

Claire Périlleux
Department of Life Sciences, Laboratory of Plant Physiology, PhytoSYSTEMS, University of Liège, Liège, Belgium